EDISON AND THE BUSINESS OF INNOVATION

Johns Hopkins Studies in the History of Technology

Edison

Edison and the Business of Innovation

ANDRE MILLARD

The Johns Hopkins University Press
Baltimore and London

This book was brought to publication with the generous assistance of the New Jersey Historical Commission.

The Johns Hopkins University Press
701 West 40th Street
Baltimore, Maryland 21211
The Johns Hopkins Press Ltd., London

The paper used in this publication meets the minimum requirements of American National Standard for Information Sciences—Permanence of Paper for Printed Library Materials, ANSI Z39.48–1984.

Library of Congress Cataloging-in-Publication Data

Millard, A. J.
Edison and the business of innovation / Andre Millard. p. cm.—(Johns Hopkins studies in the history of technology; new ser., no. 10)
Bibliography: p.
Includes index.
ISBN 0-8018-3306-X (alk. paper)
1. Edison, Thomas A. (Thomas Alva), 1847-1931. 2. Inventors—United States—Biography. 3. Businessmen—United States—Biography. I. Title. II. Series.
TK140.E3M647 1990
338.4'76213'092—dc20
[B] 89-15419 CIP

Contents

Tables and Figures

Tables

Figures

Preface

This book does not claim to be the definitive history of Edison the inventor or Edison the businessman, a task that would require several more years, several more books, and the completion of the Edison Papers' project of editing the millions of documents at the Edison National Historic Site. Until that time I aim to provide an interpretation of Edison's work that covers the entirety of his business career, dealing with both business and technology and the relationship between the two. My book is an advance party crossing the vast stretches of Edison archives. It will doubtless be followed by many others as the Thomas A. Edison Papers Project gains momentum. My hope is that *Edison and the Business of Innovation* will provide some guide posts to the new territory.

I first came to the Edison National Historic Site in West Orange, New Jersey, as a graduate student. In the process of writing a dissertation about electrical history under the guidance of my mentor Rondo Cameron, I was drawn to the massive archive of papers relating to the life and work of Thomas Edison. As a student intern, and later as an assistant editor of the Edison Papers Project, I became better acquainted with this unique collection of documents. The associations formed at the West Orange laboratory have become the keystone in my understanding of Edison and have played an influential role in my research. Bernie Carlson and Paul Israel have generously shared their knowledge with me and have consistently refined and redirected my ideas. My other colleagues at the Edison Papers have also played a part in my schooling in Edison scholarship; a hundred years after the first young men came to Edison's laboratory to seek knowledge of the master, the West Orange facility still serves as an educational institution.

Much of the research for this book was carried out under a contract to write a history of the West Orange laboratory for the National Park Service, which now administers the site. Regional Historian Dwight Pitcaithley managed the research contract with aplomb. As curator of the Edison site, Ed Pershey set the project in motion and brought it to a successful conclusion. After the history was finished (in time for the centenary celebrations of the lab), Ed continued to play the major role in shaping the development of a new book. As scholar and friend, Ed Pershey has guided the research and writing of this book. My thanks for his great contribution. For the staff of the Edison National Historic Site I have nothing but gratitude and praise. Mimi Bowling, the archivist, and Eric Olsen, the archives technician, managed my access to the archive, taking years of abuse and impatience with good humor.

The New Jersey Historical Commission did more than generously support my research, for under the guidance of Howard Green I was able to place my study firmly in the economic and social history of New Jersey. I am indebted to the staffs of the libraries and archives who helped in the research sponsored by the commission. I especially remember the helpfulness and cordiality of Morton Snowhite of the New Jersey Institute of Technology Library, Newark, New Jersey; Robert Blackwell of the Newark Public Library; Bob Lupp of the New Jersey State Library, Trenton; and Emmett Chisum of the American Heritage Center at the University of Wyoming, Laramie.

Many friends and colleagues gave me the benefit of their advice and encouragement, including Steve Klein, Bill Pretzer, Dave Sicilia, Robbie Smith, and Darwin Stapleton. Jim Brittain, Howard Green, and Charlie Musser did me the great service of reading various parts of the manuscript and helping me remedy its faults.

The staff of the reading room, Boston Public Library, provided me with courteous service in the most magnificent surroundings. The reference staff of Solomon Baker Library, Bentley College, especially Lindsay Carpenter, Sheila Ekman, and Amy Lewontin, supported my research with professional expertise and unfailing resourcefulness. My thanks also to Henry Tom of the Johns Hopkins University Press and to my editor, Elsa Williams. A final word of appreciation to Cheryl Nuehring, whose services as editor over the years have been invaluable.

Without the help of the above-named, there would have been many more errors and weaknesses in this book. Those that remain are mine and mine alone.

EDISON AND THE BUSINESS OF INNOVATION

CHAPTER ONE

The Largest Laboratory Extant

Introduction

he Edison National Historic Site attracts to West Orange thousands of people who would normally avoid the harsh, deindustrialized landscape of north New Jersey. Situated about forty-five minutes from New York City, the Edison laboratory and museum is one of the most popular national parks on the East Coast. The Park Service estimated that over fifty thousand people visited the site in 1987, its centenary year. These visitors came from all parts of the nation and a large portion came from abroad. It is hardly surprising that Edison's laboratory is most popular with Japanese tourists, who share his work ethic and commitment to innovation.

The architects of today's microelectronics "revolution" find relevance in the history of a Second Industrial Revolution, which began in the 1870s and lasted until the Great War of 1914–18. This period of rapid economic growth came after the first wave of industrialization had begun to transform the economy and society of the United States. Edison's invention of the incandescent light bulb marked the beginning of another burst of innovation that created several major new industries. The second wave of industrialization does not have the same powerful images as the first; the steam engine and textile mill are universally recognized as the symbols for that great movement, which began in Great Britain and spread to the United States in the early nineteenth century. The new industries of the 1870s and 1880s do not have the same familiar symbols. Edison's Pearl Street station—the first architectural relic of the electrical industry—is no longer standing in downtown Manhattan.

The Bessemer steel and synthetic chemical industries have left no structure that the general public immediately associates with a period of rapid economic growth in which the United States led the world. But Edison's laboratory in West Orange is both an accessible and appropriate symbol of this movement.

The complex of buildings in West Orange was erected in the late 1880s, when the Second Industrial Revolution was just beginning. As the greatest industrial research facility in the United States, the laboratory was the breeding ground for a new generation of technology and the starting point of some important new industries of the twentieth century. Here Edison worked at spreading his electrical lighting systems all over the industrialized West and lowering the price of electricity until it was available to everyone. The motion-picture camera was invented at the laboratory, along with a host of other important products, such as the Edison storage battery and dictating machine. Edison perfected the phonograph at this facility and manufactured thousands of them in his nearby factories. Two of the twentieth century's most influential service industries—motion pictures and musical entertainment—had their humble beginnings in this cluster of brick buildings.

The experimental rooms and machine shops in the lab are a reminder of the complex technologies introduced by Edison. The impressive library is evidence of the growing importance of science in the technologies of the Second Industrial Revolution—a revolution that was based on the so-called scientific industries of electricity, steel, chemicals, and communications. The rows of technical journals (many of them from outside the United States), scholarly books, and bound patents show that the information age had begun well before the twentieth century, and that Edison saw the importance of keeping up with scientific and technical progress wherever it occurred. He often began his day by reading newspapers and technical journals at his desk in the library.

Now only tourists walk around the library and cross the courtyard to the experimental buildings that serve as museums of Edison's achievements. The lathes and drills in the grimy machine shops are quiet; the lights no longer burn all night in the experimental rooms; and Edison's bunk in the library is never used. The surviving buildings of this once vast industrial complex serve not only as a museum but also as a reflection of the life and work of one of America's greatest businessmen. The West Orange laboratory bore witness to a critical period in the economic history of the industrialized West and it provides an invaluable resource to understand our immediate past.

The Wizard of Menlo Park

When Edison reached the age of forty in February 1887, he had achieved more than many men do in their lifetimes (fig. 1.1). The development of commercial electric lighting had brought him worldwide fame and a considerable fortune. The famous invention of his incandescent lamp had taken place in 1879 at his laboratory in Menlo Park, New Jersey, the "invention factory" where groups of experimenters had developed a stream of new products that included electric lighting and the phonograph.[1] In the years that followed the invention of the lamp, Edison and his men built the first complete supply system based on a central power station. New York's Pearl Street station was completed in 1882. It distributed electricity to a few blocks of the business district in lower Manhattan. This was not the first electric light in New York City, for Charles Brush and Edward Weston had already installed arc lights in public places, but it was the prototype of the commercial distribution of electricity. The plain, shop-front facade of the Pearl Street station did not do justice to this historic installation, which proved that large-scale electricity supply was technologically feasible. It was a triumph for Edison and the small beginning of a great new industry. Copies of the station soon appeared on both sides of the Atlantic, as affluent city dwellers clamored for the new light and entrepreneurs rushed to form local Edison lighting companies.[2] To his contemporaries Edison, now known as the "Wizard of Menlo Park," stood astride a mighty business empire.

The engineering feat at Pearl Street was the opening act of the great drama of electrifying the industrial West. Edison had little time to bathe in the congratulations, for there was work to be done in diffusing the technology. His attention turned to creating the business organization to bring electric light and power to the world. He said, in 1883: "I'm going to be a business man, I'm a regular contractor for electric lighting plants and I'm going to take a long vacation in the matter of invention."[3] After creating the Edison Construction Department to build central stations, he traveled the United States erecting lighting systems. Small installations for one user were built by the Edison Company for Isolated Lighting, while the larger power stations were erected by local lighting companies

Figure 1.1. Edison examining a piece of ore in the metallurgical lab in Building 4, around the time the West Orange lab was built.

that were franchised to use the Pearl Street technology. Edison put up several factories to produce equipment: the Edison Machine Works was established in New York City to make dynamos and other equipment for power stations; a large factory in Harrison, New Jersey, mass produced the incandescent bulbs; the Electric Tube Company made underground conductors; and the Bergmann Company provided the switches, fixtures, and other accessories for the lighting system. The important electrical patents were in the hands of the Edison Electric Light Company, which sold the franchises to local lighting companies and provided overall direction to the enterprise.

The pressure of business forced Edison to leave Menlo Park and set up house in lower Manhattan, close to the headquarters of the Edison Electric Light Company at 65 Fifth Avenue. As a businessman, Edison had to be close to the decision making, yet the influence of the old laboratory remained in the important jobs he gave to men who had worked with him at Menlo Park, the faithful "boys" who formed the initial management of the electrical industry. The Electric Light Company was presided over by Edward H. Johnson, a friend and colleague of Edison's from their days as telegraphers. Samuel Insull, Edison's secretary at Menlo Park and New York, was made manager of the Machine Works. As private secretary, Insull had administered Edison's business and personal finances and now he ran the financial affairs of the lighting industry. The superintendent of the Machine Works was John Kruesi, a skilled mechanic who had worked with the inventor for many years. Two other laboratory men, Francis Upton and Charles Batchelor, were put in charge of factories.[4] Another crony from the old days, Sigmund Bergmann, was set up in his own manufacturing company.

By 1886 the work of "pushing the system" was well under way. Over fifty power stations had been built in the United States, and many were in use in large European cities. The Edison electric lighting companies were capitalized at around $10 million in the United States.[5] Edison's factories were full of orders and running at top speed to fill the backlog. These manufacturing operations were providing the profits in the electric light business. The money was pouring in. Edison was the hero of his generation, a rich and powerful industrialist, and an inventor of worldwide fame. He was also in love.

His second marriage was to Mina Miller in February 1886. (His first had ended in 1884 when his wife, Mary, died.) Mina came from a well-to-do family in Akron, Ohio, had attended finishing school in Boston, and was a more than acceptable partner for the famous inventor. Marriage inevitably brought changes to Edison's life. He must have been tempted to remain in New York, the center of his business affairs and the home of close associates (such as his chief assistant, Charles Batchelor) and important financial connections (including J. P. Morgan and Henry Villard). But instead of joining the other millionaires putting up mansions south of Central Park,

Edison deferred to his wife and moved to West Orange, New Jersey, in the scenic Orange valley, a rural area about twelve miles west of New York City.

They chose to live in Llewellyn Park, an enclave of magnificent houses that was the first planned residential suburb in the United States.[6] Edison purchased Glenmont: a sprawling stone and wooden castle that cost him nearly a quarter of a million dollars in 1887 and was appraised at a much higher figure. Glenmont came fully furnished and gave Edison instant social substance; he had achieved a solid Victorian respectability at a stroke. His biographers noted that he was now enjoying a life-style far removed from his Bohemian existence as an itinerant telegrapher.[7]

An ordinary mortal might have taken time off to savor the pleasures of married life and business success, but Edison was no ordinary man. He was still at the peak of his inventive powers and the flow of ideas had not ceased during the years of perfecting his lighting system. Also, his honeymoon in Fort Myers, Florida, in 1886 and a convalescence from sickness in early 1887 gave him two long breaks to contemplate future inventions. There were still plenty of opportunities in electricity, and other schemes were forming in his imagination: magnetic ore separation could be the technology to revolutionize the iron industry, and the tinfoil phonograph might be the greatest consumer product of the age. He had a wild plan to provide moving pictures to accompany the feeble squeak of the phonograph, but there was no place to develop these ideas. What he needed was another large laboratory, another invention factory, to replace the one he had left at Menlo Park.

Some historians have argued that Edison's idea of the invention factory—the organized application of scientific research to commercial ends—was his greatest invention.[8] He had promised "a minor invention every ten days and a big thing every six months or so" and for once he was as good as his word. The phonograph and the electric light had proved the soundness of this idea. Now he wanted to create "the best equipped and largest laboratory extant and the facilities superior to any other for rapid and cheap development of an invention."[9] The Menlo Park complex had been the largest private laboratory in the United States in the 1870s, and now Edison planned an even bigger facility, one that would enable him to carry out industrial research on an unprecedented scale and reflect his new status as the preeminent American inventor.

Yet Edison was by no means the only inventor in the electrical industry. Charles Brush, Edward Weston, and Hiram Maxim were all pioneers in electric lighting who remained active in the field in the years after Pearl Street. Weston's operations were close to home. He was the first to establish a lighting scheme in Newark, New Jersey, where he built a central station on Mechanic Street, and his arc lamps graced the newly constructed Brooklyn Bridge.[10] Weston was an immigrant from England, and although he had no formal training, he gained the reputation of a scientific inventor—in contrast to Edison who was seen as an inspired tinkerer.[11] Yet Weston's career had some similarities to Edison's: Weston began in the United States as a laborer, worked as a chemist and electrician, and then formed a series of successively larger electrical companies to become a leading figure in the new industry. His patents in the electrical field were almost as valuable as those of Edison.

In 1886 Weston decided to build a private laboratory behind his home on Main Street, Newark. It was finished in 1887, and Edison probably read the laudatory descriptions in the press as he contemplated his own laboratory in West Orange. Weston's lab was described as the largest and best equipped in America.[12] It was depicted as solely devoted to electricity and to "original scientific research and not for the purpose of gain of any money making concern."[13] Perhaps the last comment was aimed at Edison, whose reputation was somewhat tarnished by the unsavory activities of some of the electric lighting companies bearing his name. The two inventors did not enjoy cordial relations, and Weston could be depended upon to snipe at the Wizard of Menlo Park in the press whenever he got the opportunity.[14] These cutting remarks or the publicity surrounding Weston's new lab in Newark could have stung Edison into going one step farther than his rival. If anyone was to have the largest private laboratory in the United States, it would be Edison.

Grand Plan for a Laboratory

His written plans indicate the grandeur of his concept and reveal something about his method of inventing. The sketch from a lab notebook (fig. 1.2) shows that he was thinking about a structure that would reflect his prestige as a famous inventor and man of substance. This laboratory is an impressive three-story building with a mansard roof. The tower and courtyard are its two distinctive features (often found in the large textile mills of Newark), and these give the invention factory the dignity of a public building. The east wing contains the experimental tables and Edison's own personal

Figure 1.2. One of Edison's sketches for the new laboratory, drawn in a notebook in 1886.

room. The north side of this planned lab contains the machine shop and power plant. The quad in the center of the structure enables easy communication between the experimenters in the east and west wings and the machinists in the north wing. The inventor had discovered the vital importance of good communications in industrial research.

The chemical laboratory takes up the whole west wing of this plan, a measure of Edison's love of chemistry and its importance in the application of science to industry. The development of industrial research was closely associated with the progress of chemistry. The first industrial laboratories were established by German chemical companies to carry out testing and basic research, and similarly the first American corporate laboratories were devoted to industrial chemistry. The Pennsylvania Railroad, for example, created a laboratory as early as 1875 to analyze materials. Even Andrew Carnegie, who maintained that "pioneering don't pay," saw the advantages of hiring a chemist to analyze the operation of his blast furnaces. His experience that "9/10ths of the uncertainties of pig iron making were dispelled under the burning sun of chemical knowledge" was repeated by other industrialists.[15] Now that Edison had made electricity more plentiful, he planned to exploit it in ventures such as the electroplating of metals or the production of aluminum. He saw electrochemistry as a profitable new business for the future.

A library was a fixture of all Edison's plans for his new laboratory; it played an important part in every experimental project, which always began with a thorough search of the scientific literature. The first two steps of Edison's method are described thus: "1st. Study present construction. 2nd. Ask for all past experience . . . study and read everything you can on the subject."[16] Not one to explore new fields without considerable preparation, Edison often built on others' work in his inventions, and nowhere was this more evident than in the development of the electric light. Electrical technology was international, and its rapid advance was the result of the transatlantic movement of men, machines, and ideas. Edison had gained immeasurably from the foreign equipment and literature imported into his laboratory (such as the Gramme dynamo from France); they had sometimes been the inspiration for Edison's electrical machines. The experience of the electrical industry had shown that a key piece of information, the key to a breakthrough, could easily be hidden in an obscure scientific journal—such as Pacinotti's discovery of the ring armature—and Edison planned to take advantage of any new piece of scientific or technical knowledge that came his way.[17] The first function of the invention factory was to act as a net to capture ideas from the many streams of technical information; these turned into a flood as the nineteenth century progressed. The work on electricity had opened up new scientific horizons for Edison and provided the money to acquire the knowledge to explore them. The impressive library he built at West Orange, estimated at about

100,000 volumes, was an important resource for all members of the laboratory.[18]

By early 1887 Edison's ideas for the lab were taking shape. In addition to a machine shop, his plans show supply and apparatus rooms, which were positioned close to the experimental rooms. The supply room served both experimenters and machine shop. It contained metal parts, screws, wire, joints, gears, tubes, and sheets. It would also house the batteries, magnets, and electrical subassemblies that were needed to make up experimental apparatus. The instrument room contained the measuring equipment required in electrical and chemical experiments. The separate galvanometer room for electrical work is evidence of Edison's plans to perfect electric lighting in the future.

Although many electrical pioneers (Edison among them) had claimed that electric lighting technology was perfected, the years after Pearl Street told a different story. On both sides of the Atlantic men struggled to improve the performance of lighting equipment before the fragile utility industry sank under a sea of debts. Meanwhile the gas lighting companies lowered their prices and prepared to beat off the interloper. Despite the gleaming islands of electric light in the business districts of many cities, Edison's world was still gas lit and horse drawn. In 1887 the great cities of the industrialized West were still dark with fog and smoke, the "opalescent reek" described by Edison's contemporary Arthur Conan Doyle in the adventures of Sherlock Holmes.[19]

The West Orange laboratory was to be the moving force in the expansion of electric lighting during its second decade, for with no improvement electric light would remain a pinpoint of light in the nineteenth century world, a luxury for the very rich and a novelty for grand hotels and places of entertainment. Increasing the efficiency and lowering the cost of the incandescent lamp were major objectives. In 1887 he planned the new facility at West Orange, complete with a glassblower's room and a vacuum pump room, devoted to further research and testing of electric lights.

Building the Laboratory

Edison bought the land on which his laboratory was to stand in January 1887. The fourteen acres lay next to the main road of the Oranges, at the bottom of the hill that led up to Llewellyn Park. The site was conveniently situated close to Glenmont, and the nearby railroad gave easy access to New York and Newark. Henry

Holly, the architect of Glenmont, was retained to design the laboratory. He produced plans for a handsome three-story building 250 feet long and 50 feet wide. Charles Batchelor was given the task of turning these plans into bricks and mortar and spent the summer overseeing the construction. Edison was always around, either coming down from Glenmont or stopping on his way to the temporary laboratory at the Lamp Works in Harrison. The inventor exerted his powerful influence, firing Holly, denouncing bad workmanship, and constantly changing the plans. Yet Batchelor was accustomed to these kinds of distractions and worked on steadily.

Charles Batchelor had been at Edison's side since he joined the inventor's work force in 1871. Born in the north of England, he was one of the many machinists who left the factories of this industrialized Black Country for the textile mills of Massachusetts and New Jersey. Batchelor was sent to install machines in the Clark thread mills in Newark, met Edison, and never returned home. The dark-bearded Englishman had the craft skills that made Manchester the workshop of the world. A master of metal working, he could handle any job—from casting parts of a dynamo to fitting the fragile experimental filaments into incandescent bulbs at Menlo Park. His neat, well-organized notebooks reveal an orderly mind and precise hand. According to one of the laboratory staff, Batchelor was the "sometimes needed, conservative element of the combination," an ideal counterbalance to the mercurial Edison who often got so involved in an experimental project that he forgot everything else.[20] Batchelor was experienced in factory management and well suited to supervise the construction of the large laboratory complex planned by Edison.

First to be erected was the original laboratory designed by Holly. The west end of the laboratory faced Llewellyn Park and the road that cut through the Oranges. Here the graceful, arched windows of the library rose two stories to break the line of this sober industrial structure. The laboratory presented a dignified face to the world passing on Main Street. The library—always situated by the main entrance in all Edison's plans—gave an impression of intellectual effort within and lent support to the notion that learning was essential to inventing. Its magnificent wood paneling and large collection of books were the high point of any tour of the lab (fig. 1.3). At the other end of the laboratory stood the power house, which contained a boiler, steam engine, and several dynamos. This was the work end of the building where dynamos hummed to power the machines and light the complex.

Figure 1.3. The library in the main lab building. Edison's desk is in the foreground.

Edison had originally intended to crowd all the experimental areas, supply rooms, and machine shops on the three floors of this building, but the 37,500 square feet of floor space were not sufficient for his ambitious plans. Once the main laboratory was completed, work began on four smaller laboratory buildings, identical in size (100 ft. by 25 ft.) and positioned at right angles to the main building (see fig. 1.4). They were numbered 1 through 4 starting with the one closest to Main Street. Building 1 was devoted to electrical work; the galvanometer room in the early plans for the lab was now made into a separate building dedicated to electricity. It was constructed on a deep foundation of brick and cement to dampen any shocks to the delicate electrical measuring equipment inside. Great pains were taken to avoid metal fittings that might influence the electromagnetic devices used in experiments. This laboratory contained every type of electrical measuring apparatus, including galvanometers, Wheatstone bridges, and many experimental devices. A special rack and control panel was installed in

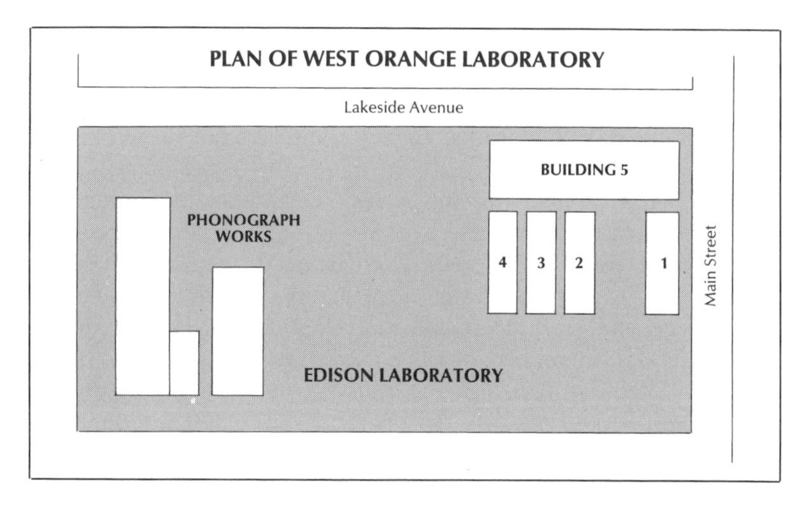

Figure 1.4. Plan of the West Orange laboratory.

this building to evaluate the performance of incandescent bulbs. This grew into a lamp testing room, which housed four or five experimenters. Several additions were made to the equipment to test more lamps under different operating conditions. The electrical laboratory in Building 1 provided essential testing services for the Edison lighting companies and, by virtue of its superior facilities, set the operating standards for them.

Building 2 was the chemical laboratory. It was fitted with a balance room and lined with shelves containing bottles of chemicals. The front part of Building 3 was used as a storeroom for the overflow of chemicals for Building 2, and the rear end was turned into a woodworking shop to make jigs and patterns. The last of the satellite buildings, number 4, was established as a metallurgical laboratory. It contained rock crushers and graders, assay furnaces, and containers for the many samples of ore that Edison had collected.

These four satellite laboratories indicate the experimental directions that Edison intended to follow: electricity, chemistry, and metallurgy. They complemented the work of the main building (now called Building 5) and completed the physical shape of the West Orange laboratory. The front of Building 1 formed an entrance gateway with the west end of the main building, and the space between it and Buildings 1 and 2 formed a courtyard, the same quadrangle effect envisaged by Edison in his early plans. The open area between buildings was not merely a meeting place; it was an important work space where experiments could be made on large equipment. The laboratory had an open architecture that was consistent with Edison's extravagant plans to add several more buildings. The haphazard addition of a new plant gave the lab an organic pattern of growth.

After the five main buildings were completed in September, Batchelor turned his attention to fitting out the laboratory. The power house was his first job. Then came the dynamo room to power the complex. This was soon equipped with a number of different generators, which could be evaluated while they produced electricity. The laboratory supplied current to Glenmont and other houses in Llewellyn Park and this small distribution system provided a useful test bed to simulate the operation of central stations.[21] It could be used to measure the efficiency of the distribution network of conductors and the accuracy of the meters that recorded output.

The laboratory was wired for electricity throughout and some of the experimental rooms were fitted with outlets of different voltages, yet electrically driven machine tools were still in the future. Batchelor followed current industrial practice in establishing a power house at one end of the main building and running shafts along the first two floors. Leather belts took the power from the drive shafts to the machines below.

The work of equipping the machine shops provided some diversion for Edison and Batchelor who scoured manufacturers' catalogs and purchased a wide selection of machine tools: lathes, drills, grinders, and presses. Batchelor divided them between two floors of the main laboratory building, with heavy lathes, drill presses, and milling machines on the first floor, and smaller lathes, grinders, and polishers on the second. The heavy machine shop on the ground floor was to produce large equipment, like dynamos, for central stations. It was fitted with a traveling crane. The second floor machine shop was called the precision room because the men who worked there were highly skilled mechanics and instrument makers who built experimental models.

At the end of September Batchelor began moving equipment from the temporary laboratory in the Lamp Works, where Edison and a skeleton crew of about ten experimenters continued the development of electric light, ore milling, and the phonograph. During October and November loads of "experimental stuff" were shipped by horse cart from Harrison. As soon as the experimental apparatus was installed, Edison and his men started work. The West

Orange lab opened for business in the first weeks of December 1887.[22]

During 1887 and 1888 a stream of consignments arrived in West Orange from all parts of America: machine tools, chemicals, electrical equipment, and loads of supplies—not only lengths of steel and pipe but rare and exotic materials such as seahorse teeth and cow hair. They were put into the main storeroom in Building 5, between the library and the machine shop. Experience had taught Edison that one could never anticipate what tools and materials would be required in an experimental project. It was his intention to have everything at hand in the new lab. He did not have to send out for rare books and new technical journals because they lined the shelves of the library, and he would not have to search the world for some unusual material because it was probably in the storeroom. His comment that "the most important part of an experimental laboratory is a big scrap heap" reveals his reliance on a well-stocked storeroom and a collection of apparatus and equipment left over from previous experiments.[23] The larger the scrap heap the better, for somewhere in the pile might be the all-important part.

Edison's interest in metallurgy was reflected in the great collection of different ores and minerals housed in the library and Building 4. The analysis of materials was one of his great strengths as an inventor; some of his important breakthroughs had been the identification of a material to perform an important task in an invention, such as the bamboo for the filament of his incandescent bulb or the chalk button for his loud-speaking telephone. Basic research into the nature of materials provided many happy hours of experiments for him. It also gave excellent value for the time and money expended, for he always sought knowledge that could be applied to more than one experimental project. In this way experiments were soon begun to examine the properties of hard materials that could be used both for the filament of the incandescent lamp and the stylus of the phonograph.[24]

Edison's goal to have everything at hand can be seen in the provision of a forge and furnace in the lab. Previously the construction of large metal parts, such as castings for dynamos, had been handled by outside contractors. Edison decided that the machine shops in the new lab would be completely self-sufficient. He believed that in industrial research bigger was indeed better. He had built a facility that could turn out "anything from a lady's watch to a locomotive" and stocked it with an extensive hoard of tools and supplies to achieve economies of scale by pursuing several experimental projects at the same time. There were real savings in reducing the expensive delays caused by waiting for supplies or a machining job. Edison noted that "inventions that formerly took months and cost a large sum can now be done in 2 or 3 days with very small expense."[25] Edison was not the kind of man who liked to wait; when he got an idea, it had to be immediately turned into an experimental model before the inspiration went. He hated to lose the momentum of an experimental project, preferring to stay up all night to complete it before the trail grew cold. In his words the laboratory was built for "rapid and cheap development of an invention."

Although the Menlo Park lab was the first invention factory, it was not custom-built for the job. The West Orange laboratory was built with the mass production of inventions in mind. Its facilities were purposely laid out to achieve efficient routing of experimental projects, the various elements flowing together from the library at one end of Building 5 through the experimental rooms on the second floor to the two machine shops. Edison's method of invention rested on cooperation between experimenters and machinists. The second floor of the laboratory was organized to facilitate this cooperation on a larger scale than ever before. The location of the precision machine shop next to the experimental rooms of the second floor established a floor plan built around the idea of speed. As ideas occurred to experimenters on the second floor they could quickly draw on the skilled mechanics of the precision room to build the models and devices they needed.

The experimental rooms, where the intellectual effort was concentrated, were the center of the laboratory. They were divided by wooden partitions and could be arranged to suit any project. Edison wanted a flexible floor plan to change experimental directions as he saw fit. The rooms were on each side of the main corridor that led to the precision shop. At the end of the row of experimental rooms on the right side of the corridor was room 5, the photographic studio. This room was kept locked but had an opening in the door to pass through supplies. The historic development of the motion picture camera was carried out in this room. Edison's room was number 12, to the left of the stairs and facing the laboratory's courtyard.[26] Across the hall was the room of Fred Ott, whose precision lathe often gave life to Edison's ideas. It was a short walk down the hall to take advantage of the skilled men and specialized machines of the precision shop. At any time, several experimenters would probably be in this shop, watching the construction of apparatus and swapping ideas and stories with the machinists.

The Context of Industrial Research

Although Edison claimed that "there is no similar institution in existence," his new laboratory was by no means the only industrial research facility in the United States.[27] In addition to the works laboratories and the testing facilities run by professional chemists, there were also more advanced laboratories to develop the technologies of the Second Industrial Revolution. Alexander Graham Bell established Volta Laboratory Associates in Washington in 1881, which included his cousin Chichester Bell and Charles Tainter.[28] Once the various Bell companies began operation of telephone networks, they discovered a need for engineering services to remedy faults in the system and improve equipment. The laboratories they set up were also charged with testing and the evaluation of new technology. One of the laboratory managers was Ezra Gilliland, a self-taught electrician and an old friend of Edison's from their days as telegraphers. Gilliland soon left the organization to join up with Edison.

The most striking application of science to industry was electricity, and every electrical pioneer had his own laboratory. A few miles down Main Street from the Oranges stood the laboratory of Edward Weston, a two-story building that contained a machine shop, an electrical room with extensive testing facilities, and a chemistry laboratory. This duplicated (on a smaller scale) the facilities Edison erected at West Orange.[29] Despite their association with scientific research, the laboratories of the Second Industrial Revolution were all based on machine shops. Charles Brush, whose arc lights had heralded the electrical age in both London and New York, had laboratories in the machine shops of the telegraph industry in Cleveland.[30] The inventor Elihu Thomson carried out experiments in a "model room" in his factory in Lynn, Massachusetts. He formed the Thomson-Houston company with his colleague Edwin Houston, which later became a serious competitor to the Edison enterprise in the development of electric lighting. Thomson's laboratory was called the model room because its function was to produce experimental devices and patent models. Precision machine tools occupied the bulk of the room, which also contained electrical instruments. With "a few tools and perhaps one or two workmen,"

Thomson hoped to produce "devices and new appliances . . . to be refined and immediately put into manufacture."[31]

This was also Edison's objective but on a much greater scale. In size and diversity of operations, Edison's West Orange laboratory stood alone in the nineteenth century. None of its contemporaries had a larger work force: the laboratory of the Bell Company employed twenty-nine men in about 1885, the staff of Charles Dudley's laboratory at the Pennsylvania Railroad was about thirty chemists, Thomson's model room and Weston's laboratory employed less than twenty people. In comparison Edison's new laboratory employed about 100 men and the physical facilities dwarfed all other laboratories. As one observer noted: "It is not a factory, as you might suppose, where steam engines or locomotives are made, it is a laboratory and private workshop of a private man."[32] This comment underlines the two essential differences between the invention factory at West Orange and other industrial research facilities: its size and its mission to translate the inspiration of one man into new technology. It was to develop both the practical and the fantastic products of Edison's imagination, from an improved electric meter to the pyrogenerator—a machine to produce electricity directly from coal that would "change the entire motive power and lighting of the world."[33] There was no other facility in the world with comparable resources.

The stated objective of the West Orange laboratory was to conduct experiments and "scientific investigations . . . with a view of making inventions useful in various arts."[34] Despite his many pronouncements to the contrary, Edison was not opposed to basic scientific research provided that it was directed toward some practical goal, however remote. His disdain of what he called "theo-retical" science came from his lack of confidence in the theory and his belief that research in academe was pointless and slow, the kind of research that Edison, the self-confessed hustler, would not tolerate in his laboratories. This prejudice, molded into his self-promotion as an ingenious tinkerer, has earned Edison an undeserved reputation for trial-and-error methods instead of systematic, scientific research, and denied his laboratories their rightful place as originators of industrial research. By defining industrial research as a science-based activity, some historians have excluded Edison's laboratories from the story of corporate innovation.[35]

Deciding the question of Edison's attitude to science and categorizing the work of his laboratory as scientific or nonscientific are essentially matters of semantics.[36] He used the term liberally and so did his contemporaries. "The keynote of the Victorian Era," announced the London *Times* in 1887, "is the development of scientific research, the concomitant growth of practical invention, and the expansion of industry which these have brought about."[37] Edison would have agreed with this. He would also have thought that this quote described the work of his laboratory.

At the time the West Orange laboratory was built, there was no clear dividing line between science and engineering, between basic research and applied research. There were few professional electricians and hardly any formally trained electrical engineers. The cutting edge of electrical knowledge was in the operation of lighting systems rather than in the few research laboratories in higher education. In view of the problems of running electric lighting systems, it was unavoidable that Edison's new laboratory had to conduct basic research into phenomena such as electromagnetism. This was carried out with no immediate payoff other than the realization that until more was known about electromagnetism, there was little room to improve the low output and poor efficiency of dynamos. As the eminent electrical engineer Silvanus Thompson noted in 1886, "Until we know the true law of the electromagnet, there can be no true or complete theory of the dynamo."[38]

Edison's invention factories were the pioneers of industrial research because they carried out organized, systematic research directed toward practical goals. Their work encompassed a broad range of activities. If Edison had planned to confine himself to trial-and-error, he would never have insisted on recording every experiment in the laboratory notebooks that were left on every work bench. The master of cut-and-try had indeed made his job easier by assembling a massive storehouse, yet he had also provided the most advanced testing and measuring equipment that money could buy—evidence that the work of this laboratory would not be only trial-and-error experimenting.[39]

The staff of the laboratory used the scientific method in their experiments, were well versed in theory (such as it was), and kept an exact record of their work. The laboratory notebooks kept at West Orange provide evidence of Edison and his leading experimenters theorizing about fundamental principles, making deductions from these principles, and testing the results by experiment. Some historians have interpreted Edison's activities as borrowing from science rather than creating scientific knowledge.[40] In this he was not alone; many engineers were doing the same in the last decades of the nineteenth century. Edwin Layton has put forward

Figure 1.5. Dickson's photograph of the newly completed lab, taken from Crookes Pond. The top two stories of Building 5 can be seen, as well as all four satellite buildings with number four closest to the camera. Some of the fine houses of Llewellyn Park are visible in the right background.

the useful concept of two communities of engineers and scientists coexisting. Instead of one community carrying out separate functions of science and technology, Layton sees two communities that dealt with the same knowledge but had different goals and systems of values. Engineers adopted the theoretical and experimental methods of science to solve technological problems. They expressed scientific knowledge in a different way and put it to a different use than scientists.[41]

The one value of the scientific community that was not in the invention factory was freedom of information. Although Edison planned to benefit from scientific knowledge and encouraged open communication in his labs, he did not allow the results of his work to be freely disseminated. His laboratory staff sometimes carried out research that they wanted to publish as a contribution to the scientific literature. Edison resisted all such entreaties because he feared that the findings might help his competitors.[42] At the same time Edison did not object to using his laboratory as an educational center. He hosted meetings for the scientific community such as a conference on electrical measurement for the American Institute in 1888.[43]

Such inconsistencies make it difficult for scholars to agree on Edison's place in the history of industrial research. The debate whether Edison's method was based on scientific research or cut-and-try will probably never be resolved. Nor will there be agreement on which lab should have the title of the world's first industrial research laboratory. There was no doubt in the minds of the men making their way to West Orange in the fall of 1887. To them Edison's genius had no limit and his ambition no boundaries. Stocked with every conceivable supply and led by the world's greatest inventor the new laboratory was "Heaven . . . certainly one of the finest in the world and the finest in the States"[44] (fig. 1.5).

CHAPTER TWO

The Machine Shop Culture

The worldwide publicity given to the laboratory emerging in rural West Orange ensured that there would be no lack of young men willing to work there. As soon as it opened a flood of applications arrived at Edison's desk: engineers with prior experience in the testing laboratories of the railroads or chemical companies; college men with degrees from schools like Rutgers College in New Brunswick, New Jersey; machinists from the large German community around Newark; and the sons of the wealthy who were eager to work for a man as famous as the Wizard of Menlo Park. Some were so desperate to be part of this famous laboratory that they offered to come for no wages; the honor of working with the Wizard was payment enough.[1]

Men journeyed to West Orange from all parts of the globe and from every strata of society. Dignified chemists from Germany, with distinguished credentials and appearance, worked next to itinerant machinists who journeyed back to the urban slums of New York and Newark after work. The work force, as one experimenter recalled, was a mixed bunch of "learned men, cranks, enthusiasts, plain muckers and absolutely insane men, as ever forgathered under one roof."[2] The boss of this unruly outfit presented no less an eccentric figure as the director of the largest industrial research facility in the United States. One observer gave this picture of Edison at West Orange: "As we looked down the alley . . . the well known figure popped out of a far door. Bare-headed in his shirt sleeves, vest flying open, trousers baggy and unpressed, he looked like nothing so much as a country store keeper hurrying to fill an order of prunes."[3]

Unkempt and usually unshaven, the great inventor was rarely seen outside the company of his men whom he treated as peers (fig.

Figure 2.1. Edison and the boys outside the Menlo Park lab. Batchelor is sitting on his left, Francis Upton on his right.

2.1). In workman's clothing, "he was as dirty as any of the other workers and not much better dressed than a tramp."[4] Experimenting at the first invention factory at Menlo Park was punctuated by gaming, practical jokes, and rowdy singsongs at the large organ that filled one end of the building (fig. 2.2). The all-night experimenting sessions, with their midnight feasts and hours of storytelling, were becoming as important a part of the Edison myth as the inventions themselves. Far from being sedate, intellectual environments with library quiet, Edison's laboratories were noisy, crowded places that often seemed on the verge of uproar. Informal and democratic, work there was far from the discipline and order found in the factories of the Industrial Revolution, the "dark satanic mills" where the work force felt like prisoners. The list of work rules and regulations—almost obligatory in factories—was absent in the machine

Figure 2.2. The famous invention factory of the Wizard of Menlo Park, taken in 1880 on the second floor of the lab. The Old Man sits in front of the organ.

shops run by Edison and his contemporaries. Like the skilled mechanics who formed the foundation of his labor force, Edison had a contempt for posted rules.[5] The story was often told of the new employee asking about the rules of the lab and receiving the following reply: "Hell, there aint no rules here! We're trying to accomplish something!"[6]

Edison seemed to be creating great inventions out of total disorder, yet underneath the apparent chaos was both organization and method. Molding a diverse group of men into a creative but controllable work force was the problem facing the manager of an invention factory. Edison solved it not so much by employing an advanced system of management as by recreating the craft culture that he had absorbed in the telegraph industry. This set of values and practices reflected a pre-industrial tradition that valued the skill of the worker and preserved the dignity and independence of his work.[7] I have called the work culture of Edison's labs the machine shop culture because it owed much to the machine shops where Edison had learned the skills of the inventor-entrepreneur. Although sparsely documented, and never referred to by Edison, the machine shop culture framed his method of inventing and gave work in his laboratories its distinct character. It was a unique craft culture subtly adapted to the needs of running an invention factory.

Edison first came into contact with craft practices in his youthful days as a telegrapher. He began work as a telegraph operator in 1863 in his home town of Port Huron, Michigan. After he had learned his trade, the seventeen-year-old Edison tramped the Midwest taking jobs in several telegraph offices. Although telegrapher and machinist were occupations that had come out of the Industrial Revolution, each identified itself with a collective work experience and shared the values and practices that made up a work culture.[8] Like the craftsmen of early modern Europe, these workers took pride in their work and believed in the social utility of their skills.[9]

Craftsmen were proud of their calling, whether it was hatmaking, puddling iron, or machining. Many had a distinctive dress or marks that identified them as members of a craft, such as the black hand of the hatmaker or the round hat of the glassblower worn by Edison in figure 2.2. They were devoted to their traditions, such as the "blue Mondays" that followed a weekend of drinking and the "blow out" that marked the passage of apprentice to journeyman. Each craft had its own vocabulary that helped bring its members together into a community of skills and interests.

Edison was introduced to the work culture of machine shops when he visited Charles Williams' establishment in Court Street, Boston, in 1868. He had taken a job in the Western Union office in Boston but was devoting much of his time to inventing. Williams' shop was one of the best equipped for electrical work in the country. Occupying the third floor and attic of the building, it contained several lathes of different sizes, a forge, and a store of metal parts and electrical apparatus. Williams made a living producing telegraph equipment and fire alarms. He also did custom work for a group of amateur inventors who could often be found in his shop. "Individual initiative was the rule" in Williams' shop and it fostered creativity, as seen in the experiments of Edison on the duplex telegraph, Moses Farmer on the dynamo, and Alexander Graham Bell on the telephone.[10] Some years after Edison had left Boston for New York, Bell invented the telephone in the attic room of the shop with the assistance of Thomas Watson, one of Williams' staff of mechanics.

Although the machine shop was primarily a manufacturing facility, it was also a place where an idea could be turned into an invention and an experimental model developed into a commercial product that could be manufactured on the spot. The small machine shop was the haunt of "independent, poverty stricken inventors."[11] Charles Kleinsteuber's shop in Milwaukee was the birthplace of another invention that was to figure in Edison's career. Like Charles Williams, Kleinsteuber made up experimental devices for inventors in addition to general machining, model making, and foundry work. In the same way that Williams' shop acted as a forum for creative minds, a happy fusion of talents brought about the idea of a writing machine. An amateur inventor, Carlos Glidden, was a regular visitor to Kleinsteuber's shop. Glidden saw a numbering device (to mark pages and bills) made for Christopher Sholes and the two collaborated on the development of a typewriter. Sholes patented the first device in 1867.[12] After some years of refinement, the machine was ready to be presented to capitalists who might promote it. Glidden and Sholes approached George Harrington of the Automatic Telegraph Company in 1870. Harrington showed it to his technical expert, Thomas Edison, who had just become Harrington's partner in the American Telegraph Works, a large telegraph factory in Newark.

Edison had arrived in New York in 1869, a twenty-two-year-old telegrapher with plans to make some valuable inventions. The telegraph industry was the high-tech field of the post-Civil War economy, providing the communications required in growing businesses such as the railroads and banking. The technology was advancing as fast as young men like Edison could devise improvements and put them on the market. The telegraph business was the main avenue of opportunity for ambitious inventors, and soon after his arrival in the financial center of the country Edison set himself up in a small machine shop across the river in Newark, New Jersey. His association with George Harrington was one of many partnerships and business deals that made him a successful industrialist—"a bloated Eastern manufacturer" as he described himself to his family back in Michigan.[13] He established several machine shops in Newark, some large facilities to manufacture equipment, some small experimental shops. The Newark shops had the most impact on Edison's development as an inventor and industrialist; the years spent in Newark were the formative period in his career.

Long an industrial city with a special pride in the skill of its metal workers, Newark had grown to industrial prominence in the first part of the nineteenth century when its position as a great Atlantic port made it accessible to the new ideas coming from western Europe. By 1860 it was one of the leading manufacturing centers in the United States.[14] Although its work force was not as large as those in other cities, it had a reputation for precision work. Newark businessmen claimed that "our people are mainly handicraftsmen, the others are mainly machinery workers."[15]

The ports of Newark and New York were a fulcrum of the Atlantic economy, and the stream of immigrants (although not yet the flood of the 1890s) brought skilled men from northwestern Europe. Newark had a heavy concentration of Irish, English, and German workmen in its immigrant population, which in 1860 constituted 37 percent of its total population.[16] Their skills were quickly employed in workshops and factories along narrow passages called Railroad Avenue or Mechanic Street.[17] They held on to their European identity and work practices. Many of the immigrant mechanics had served traditional apprenticeships in machine shops in Europe, moving from apprentice to journeyman (a worker who has mastered his trade and was allowed to work at it in another shop). In the early phase of industrialization, these craftsmen were the vehicle to transplant the technical knowledge of a new system of production. They also brought with them some of the customs and attitudes of a craft era that preceded the Industrial Revolution.

The production of textiles was the first new industry to be transplanted to North America's eastern seaboard. Several Scottish and English concerns built factories in Newark and the surrounding towns, filling them with laborers and skilled mechanics from the home country.[18] As the new factories were steam powered, some of the first machine shops in Newark were set up to assemble steam engines brought over in the early years of the nineteenth century. This expertise in steam power engineering was soon expanded to cover a variety of industrial technology, including locomotives and textile machinery.

The many textile mills of north New Jersey attracted machinists to keep the looms in running order. It was the practice to set up small machine shops at the back of mills where repairs could be carried out and new parts made. The ambitious machinists often left these shops to found their own. Their skills were much in demand as new fields were conquered by mechanization: hatmaking, printing, furniture making, and so on. As machine builders, the owners of the

Newark shops designed and manufactured the tools that transformed these activities. They were more than skilled machinists; they invented the new machine tools in their shops. It was commonplace for a shop to advertise its own patented devices in addition to the precision machining it offered to the trade. Edison's neighbors in the Newark shops could claim long experience in their craft (which normally went back to Germany or Great Britain) and a string of patents.[19] This was a fruitful environment for the young inventor.

Unlike other craftsmen who lost their skills with the introduction of machinery, skilled mechanics benefited from industrialization, which gave them many opportunities to advance. Edison achieved the upward mobility that many mechanics in Newark expected to enjoy in the practice of their craft. This was one of the ideals of the machine shop culture; the machinists had expectations of self-employment, and many who labored in the shops were driven by the goal of one day becoming master of their own shop. One machinist told a factory inspector in 1883: There are "so many men here who want to be independent that it is impossible to form a union."[20]

The Newark shops grew with the industrial development of the area, employing over a thousand men in the 1870s and 1880s.[21] These metal-working concerns carried out precision machining to order, in addition to manufacturing machine tools and complex products such as locks. In 1872 there were thirty-eight iron and machinery manufacturers and seventeen establishments in the hardware and tool trade. Included in this group was the shop of Edison & Unger, makers of telegraph apparatus and electrical machinery and experts in the application of magnetism to machinery (fig. 2.3).[22]

The Newark shops ranged from small establishments employing two or three men to complete works occupying several stories of large industrial buildings. In general, the machine shops were small factories devoted to the manufacture of metal goods. Power was brought from a distant steam engine by a system of drive shafts and pulleys, the leather belts running down to the lathes, drill presses, sanders, and other tools arranged along the lines of overhead shafts (fig. 2.4). The shops used general-purpose machine tools that could be employed for a wide range of manufacturing and repair jobs. At one end of the typical shop was a storeroom where all the supplies, such as wire, tubes, and sheet metal, were kept. This stock was the property of the owner of the shop. The hand tools, including the all-important cutting tools, remained the property of the workmen,

Figure 2.3. The shop of Edison & Unger on Ward Street, Newark, 1873.

who stamped them with their initials and kept them in their own tool chests.

The shops offered great flexibility in production. A shop could be hired to manufacture a wide range of tools and parts for a variety of industrial customers. Its owner or owners negotiated the price with the customer and sometimes carried out the work themselves. In the small shops a number of mechanics could be partners in the operation, each partner contributing tools, capital, or his skill. This is how Edison started in the business.

In larger shops the proprietor employed a staff of mechanics who worked for hourly or daily wages. Piece-work—an anathema to skilled craftsmen—was rarely used in the 1870s, but subcontracting of work was common in machine shops. It was widely used in mining, construction, and iron making; groups of workers would collectively decide the rates for each part of the contract.[23] The system in machine shops and armories was known as "inside contracting." After making an agreement with the customer to make the goods, the craftsman would subcontract with individual workers to make specific parts of the job or help him fill the order.[24] Although this practice was open to great abuse, in the machine shops

Figure 2.4. The main machine shop of the West Orange lab, as currently preserved in the Edison National Historic Site. The overhead belts are still connected, but the main drive shaft is powered by a large electric motor, center background.

operated by Edison it gave the individual craftsman control of the job and rewarded him for his skill. He was free to do the work as he wished with whatever help he required. The use of supplies and tools was recorded so that allowance could be made for the costs of production. This practice gave the machinist control over the pace of work and in effect gave him the opportunity to be his own master. It also offered some opportunity for entrepreneurship. Expertise in making a complex product, such as an automatic telegraph or a stock ticker, could be turned into a small business, as Edison himself had done. Capitalism was the force that powered the growth of the machine shops of Newark.

In contrast to the factories and textile mills of Newark, the machine shops gave individuals the opportunity to hire out the means of production. The shops contained all that was needed to design and make up a product—the power, machines, tools, supplies, and skilled labor—and all its resources were available for lease or rent. Young men like Edison were able to rent space to carry out their inventive work and contract with labor to make up their models. In some cases impecunious inventors did not even have to pay cash up front; the machinist or master of the shop did the work for a percentage of the profits.[25] Manufacture and invention in the telegraph shops were characterized by ease of entry, sharing of costs, and good communication.

The shops were a source of innovation because they brought together talented men and supported their initiative. The openness of the shops permitted the ready spread of information, not necessarily the written knowledge of science, but ways of doing things, techniques, and prior experience. An important part of the machinist's life was the "tramp" during which he wandered from town to town looking for work. The tramp lifted the young mechanic out of "mental ruts formed by a long apprenticeship and a narrow circle of acquaintances."[26] It was deemed to be an important part of any artisan's education, providing the experience required for a well-rounded craftsman.[27] In the world of metal working, the tramp was a vehicle for the diffusion of new ideas and techniques. The spread of the American system of interchangeable parts owed much to the skilled machinists who left the armories and took to the road.[28]

There was no factory fence or guarded entrance in the machine shops. A constant procession of men came through the shops: tramp mechanics, footloose and independent, called by to look up acquaintances and perhaps make some ready cash; craftsmen fresh from Europe came looking for work; and "wild eyed inventors, with big ideas in their heads and little money in their pockets," came to make up their experimental apparatus or watch the work of others.[29] They crowded around the long benches, which often ran the length of the machine shop underneath the windows. On the benches machines were assembled, tested, and disassembled; parts were taken out and altered; new arrangements were tried and old ones improved. Men lounged around—talking, smoking, and drinking. Tobacco smoke mixed with the odor of burning oil. Drinking and card games were accompanied by the steady banter of a close-knit group of men. This was the kind of scene that might confront a visitor to one of Edison's laboratories: the informality, the conviviality of master and workers, and the general air of work unbounded by rules or timetables.

When Edison moved from Newark to Menlo Park in 1876 he took with him his best workers, his equipment, and the work practices of the machine shop culture. The invention factory he set up in rural New Jersey was based on the shops he had operated in Newark. The machine shops of the telegraph industry proved to be an excellent base for innovation. They became a seedbed for a generation of electrical inventors and entrepreneurs, among them

Edison and Charles Brush. They and their machinists had little difficulty moving from the electromechanical devices of telegraphy to the larger equipment of electric lighting.[30]

One of Edison's great achievements at Menlo Park was to create a unique work environment that fitted the needs of an invention factory. Edison invented his own shop culture by duplicating the camaraderie that he had enjoyed in the machine shops of Newark. Like other American craftsmen, he adapted elements of European craft culture to suit his own ends.[31] His skill as a manager was to take some of the desirable elements of the machine shop culture—motivation, initiative, and flexibility—and graft them onto an organization that he controlled.

Like the journeymen of Europe, the work force of the Menlo Park laboratory lived a communal life, and not everyone could become part of their fraternity. They were a close-knit group, "a little community of kindred spirits, all in young manhood, enthusiastic about their work, expectant of great results," who worked, played, and lived together.[32] Like the other skilled workers in New Jersey, such as the hatters of Orange or the potters of Trenton, Edison's men had a strong sense of identity with their craft—in this case producing a stream of inventions—and with their unique work practices. They were bonded together by their pride in their craft and the special conditions in which they worked.

They had their own vocabulary, which was based on English working-class slang. "Muckers" (the name given to anyone in charge of experiments) came from the verbs to muck in—to pitch in, eating and working together, and to muck about—to fool about with little purpose other than amusement. Changing the first letter of the word would not be the only joke here; muck also signified dirt, and mucking up something meant a fine bungle. Mucking in was an important part of craft culture. Each man was expected to participate in the group's activities, maintaining the camaraderie and manliness that were essential to preserving unity and tradition.[33]

The muckers developed their own traditions, many of which came from Edison's fertile mind and quick wit. The pranks and practical jokes (which often involved scientific equipment) were a regular part of working in the lab, and only the uninitiated or the naive would drop their guard when assisting Edison. The shop culture did not frown on practical jokes because they underlined the essential freedom of artisans in the workplace. But any self-respecting factory master would have been appalled at the loss of work time and risk of accident, and many were. One executive of Brown and Sharpe, the machine toolmakers, lamented the time wasted on pranks and wondered how they ever got to be called "practical" jokes.[34]

Edison was more concerned with harnessing the creative powers of his men. That the invention factory was a pleasurable place to work was partly the heritage of the shop culture and partly Edison's intention to have it that way. As a research manager he achieved a delicate balance, controlling a group of talented, highly skilled men without hindering their initiative. Edison was no remote authority figure for his men; he was one of the boys who took a leading part in every activity on the shop floor—from experimenting to horse play. He acted like the master of the machine shop, the Old Man who was on personal terms with his journeymen and a father figure to the apprentices, working with them on the bench and sharing food with them on his table.[35]

In giving his workers a sense of shared identification with his goals, he got superhuman effort out of them. Working in the invention factory was a "strenuous but joyous life for all, physically, mentally and emotionally. We worked long night hours . . . frequently to the limit of human endurance."[36] Edison's well-known disregard for the nine-to-five discipline of work was an important part of the machine shop culture. The eccentric hours of work at the laboratory derive from the pre-industrial tradition in which craftsmen controlled the pace and timing of their work. This tradition clashed with an industrial order that attempted to maximize an investment in machines and buildings by organizing the workday of its employees.[37] The struggle between craftsmen and industrialists began with the issue of time, the freedom of deciding when to work. The first skilled craftsmen who came into contact with the American system, the machinists of the armories, revolted when timekeeping was imposed on them.[38]

Work at the invention factory took no heed of the clock. The experimental campaigns that went on for days made good stories for the newspapers; they gave the impression that work at the lab went on continuously. Edison had the ability to go without sleep for long periods, and his men often stayed up all night with him. What the public did not know was that after the all-night sessions they all went to bed for the rest of the day. By adjusting the hours of work to suit his needs, Edison made the most of the energy of his workers by keeping them at it when an experimental project was on the verge of a breakthrough and "enthusiasm ran high."[39] Thus, Edison forged an effective management style from the artisan culture of the machine shops.

The machine shop culture was built on the assumption that a craftsman who had mastered a body of knowledge became the best judge of its application. This emphasis on the individual ran counter to the values of a new industrial order that stressed conformity, discipline, and the strict observance of factory time. In contrast, the tempo of work in the machine shop alternated between bursts of activity and periods of levity and general mayhem. Yet the work somehow got done, and in the case of the Menlo Park laboratory the output of inventions was unsurpassed by any other research facility.

At the core of the invention factory idea lies a paradox: How can a creative process which is essentially uncontrollable and unpredictable be regularized? How can the mysterious act of invention be fitted into a system of mass production? The successful development of an electric lighting system at Menlo Park demonstrated that it could be done. A man of contradictions, Edison straddled the craft culture of the pre-industrial age and the industrial capitalism of the late nineteenth century. He encouraged a measure of freedom and independence in the workplace but never let control slip from his grasp. The machine shops of Newark blended craft skills with a strong vein of capitalism: the venture capital that played such an important part in Edison's career made its mark in the organization of his laboratories. The invention factories at Menlo Park and West Orange were unique. They perfectly represented the great transition in American history from the rural life of Edison's youth to the industrialized urban communities that used his inventions.

Machine Shop Practice

When Edison moved to West Orange he maintained the organization and the practices of the Menlo Park laboratory. At the new laboratory, amateur inventors and his friends were permitted to carry out their own experiments and use all the resources of the lab. They paid only for the supplies and labor they used.[40] He continued the practice of subcontracting work out to his men. As usual the lab was open to visitors and tramp mechanics. The openness and informality of the lab were part of Edison's conscious effort to create good internal communications; he knew the value of information sharing in the work of invention. One experimenter remembered that "we were all interested in what we were doing and what the others were doing."[41] Edison built the West Orange laboratory to facilitate internal communication, especially on the second floor of the main building where the experimental rooms were grouped together. He told one worker: "This answer is so good that I wish you would expand it a little, as I want to pass it along to the other boys."[42]

Another value that was transferred to the new lab was the insistence on the highest standards of craftsmanship. Precision was everything in machine shops. It was the key to a manufacturer's reputation and the success of his business. Precision could not be achieved with machinery; it rested in the skills of the artisan. It was these skills that put the craftsman on the same plane as the master and contributed to the egalitarianism of the shop.[43] The master depended on his men, for without their accuracy and workmanship he had nothing. Edison was equally dependent on the skilled machinists in his laboratory; he could not invent without them. Inventing in the 1870s and 1880s was practical rather than theoretical. Ideas had to be made into working models, if not for the enlightenment of the inventor, then for the requirements of patent law. Like many other inventors of his time—including Elihu Thomson, Charles Brush, and Edward Weston—Edison based his laboratory operation on a machine shop. In the years after he left Menlo Park in 1881, his base was never far from a shop, first at Bergmann's factory, and then at the Lamp Works in Harrison.

As Edison was not particularly blessed with good hand–eye coordination, he always surrounded himself with skilled machinists, such as Charles Batchelor and John Ott, who could interpret his rough sketches and produce experimental models in a short time. Edison would jot down the idea for an invention and have it made up as an experimental model in a machine shop. A hurried sketch on a scrap of paper with the legend "John Ott—make this" was the starting point of many of Edison's inventions.

John Ott joined the inventor's staff in 1871, when the twenty-one-year-old youth sauntered into Edison & Unger's shop on Ward Street, Newark. Edison had heard of Ott's reputation as a machinist through his partner Unger, who was Ott's cousin. As usual in a machine shop, disassembled machinery was strewn around the place. The story is told that Edison asked Ott if he could put a stock printer back together and make it work. Ott's confident reply went something like: "You don't have to pay me if I can't."[44] Ott had three qualities that Edison admired: "light fingers," inexhaustible patience, and an unswerving devotion to the Old Man. Such was Ott's skill that he was given the most delicate machining jobs in the laboratory, becoming Edison's "confidential experimental instrument and model maker." Their friendly, lifelong relationship embodied the unity of master and craftsman in the shop.

Edison worked in close cooperation with his machinists, overseeing the work and changing it as it took shape. It was in the process of altering an experimental model as it was being assembled that the germ of an invention sometimes emerged. In the inventor's own words: "Sometimes I get an idea and jot it down in the book and sometimes I would get it while the machine was being made and change it and then note it in the book."[45]

The book in question was the experimental notebook that played an important part in the work of the lab. Edison had begun using notebooks in the formative days in Newark, envisaging them as "a daily record containing ideas previously formed, some of which have been tried . . . and some that have never been sketched, tried, or described."[46] At Menlo Park and West Orange the notebooks would be left on the tables in the experimental rooms to be picked up whenever an idea or the results of an experiment had to be noted. Sometimes they were used by Edison to quickly get down an idea, usually in the form of a rough sketch; at other times they were used to record the progress and outcome of an experiment after it had been completed. These books also contain lists of experiments to be tried or jobs to be carried out by the laboratory staff. Although invaluable to historians and patent lawyers, the laboratory notebooks do not fully document the work of the lab; we have no record of the fruitful cooperation of experimenter and machinist as they pored over experimental devices in their efforts to make them work.

The tradition of shop culture was learning by doing; in Edison's lab it was inventing by doing, altering the experimental model over and over again to try out new ideas. It would be incorrect to think of the machinists in the laboratory as mere shapers of metal. This they did, and with great care, but their major contribution to the invention factory was as assemblers of complex experimental machinery. It was in putting these devices together—and making them work—that their skill and experience came into play. Edison was convinced that nothing that was any good would work by itself. His comment that "you got to make the damn thing work," is perhaps a reflection of the effort required to make experimental devices run.[47] It took a special skill to make a working model from a sketch roughed out on a page torn from an experimental notebook.

The machinists did more than act as Edison's hands; they filled in the details, making important decisions about how a job was to be machined and what materials to use. They applied their expertise in the struggle to get the thing to work, taking the initiative to modify devices and experiment with them freely. Machinist and experimenter were partners in invention. True to the democracy of the shop culture, Edison was always open to the suggestions and ideas of his men. Innovation in his laboratory was a cooperative affair; he rarely worked alone, despite his image as the lone, heroic inventor. A visitor to Menlo Park could often find him at a workbench, shoulder to shoulder with Charles Batchelor, both intently working on a device, the sound of metal against metal punctuated by a stream of instructions and suggestions from Edison.[48]

Edison made the basic partnership of experimenter and craftsman the nucleus of a team or gang. As he told an associate: "The way to do it is to organize a gang of one good experimenter and two or three assistants, appropriate a definite sum yearly to keep it going . . . have every patent sent to them and let them experiment continuously."[49] The basic team comprised an experimenter who provided the leadership and machinists and other craftsmen who assisted the team leader. In most cases machinists would divide their time between experimental groups, moving from team to team as required. The teams were free to draw on all the resources of the laboratory: supplies from the storeroom, scientific information from the library, advice from other muckers, and the hands-on experience of the machinists.

The chief experimenter ran the group with a minimum of interference from the Old Man. He would outline the task and give some pointers, but he normally relied on the initiative of the experimenter. When one man asked him what to try next, he said, "Don't ask me. If I knew I would try it myself."[50] Edison maintained the machine shop tradition of personal leadership. He liked to be among people and the people whose company he enjoyed the most were his fellow experimenters. He had a work space in nearly every part of his West Orange laboratory and made a habit of working alongside his men. He would also regularly walk about the lab, stopping in each room to check on his experimenters' progress. He started this technique at Menlo Park where each afternoon he would tour the laboratory, going up to each man at the workbench, questioning him about what he had done, discussing the results, and deciding what to do next.[51] An extremely observant man with an excellent memory, Edison never lost track of an experimental project and could remember previous experiments that might provide useful information. His mastery of his own "mental junk yard" was an asset to his men.[52] If a mucker was stymied by a problem, Edison could be relied upon to suggest another experiment or new approach. His enthusiasm for the job helped maintain the high morale of the work force.

The success of Edison's management style was based on his judgment of ability and character. He had to find bright, ingenious young men to lead the experimental teams with a minimum of supervision. His success as an inventor was built on their brains and his skill in framing the problem and picking the right men for the job. The choice of the team leader was critical. Edison was quick to remove leaders if they did not come up to his expectations. He fired one man because he was lazy and "demoralized the gang."[53]

The machine shop culture put more emphasis on practical than theoretical training, and Edison reflected this viewpoint in his choice of team leaders. Although he employed many formally trained experimenters, he did not always give them leadership roles, especially if he thought that they were not "practical men."[54] He was typical of the masters of machine shops in his lack of respect for academic credentials. In the old machine shop days anyone could call himself an engineer and many did. A degree from a university meant nothing to him, yet a man who walked in from the street and demonstrated extraordinary skill at the workbench could be sure of finding a job. Much has been made of Edison's anti-intellectualism and his disdain for theoretical science, yet his scorn for "professors" (as the college trained men were often called in the lab) came more from his belief that a college education was overrated than a hostility toward formal training. His comment on the career plans of Mr. G. C. Yee, a chemical experimenter in the lab in 1920, is revealing: "I cannot understand why he wants to attend college as I find he is really a fine chemist."[55] To somebody immersed in the shop culture, hard work and skill were to be turned into money and one's own shop, not advanced degrees.

Whatever qualifications a newcomer might possess, he still had to prove himself to Edison and the boys. Charles Clarke, who had a degree from Bowdoin College in civil engineering, described his "initiation into the brotherhood" at the Menlo Park laboratory. After arriving at the lab he kept to himself, but one night he joined the rest for a midnight feast and the usual joking, story telling, and teasing. Only after he had passed the verbal hazing was he accepted as a fellow mucker. Soon after the fun was over Edison stood up and hitched up his trousers—this was the signal to restart work.[56]

Edison thought that practical training was the best. His cronies in the telegraph and machine shops kept an eye open for exceptional machinists who were suited for experimental work, and referred them to him.[57] In the early days of electric lighting, a good man could expect rapid promotion as he mastered the technology on the workbench. James Bradley, the foreman of the Weston factory in Newark, began as a laborer in Edison's American Telegraph Works and went on to become a skilled machinist in his shops, rising to the exalted position of master machinist of the Edison Lamp Works.[58]

Many laboratory workers served a kind of apprenticeship in the shops of the Edison lighting organization before coming to the West Orange laboratory. The career of Reginald Fessenden is typical. His first application for a job at the lab was unsuccessful because after several years of higher education he did not know anything practical about electricity, and as Edison said, "I have enough men now who do not know anything about electricity."[59] Fessenden managed to get a job as an assistant tester of electrical cables for the Machine Works, which involved dirty work in the streets of New York City. He rose quickly in the organization, first as chief tester, then as inspecting engineer at the Schenectady Machine Works. As an engineer of the Machine Works, Fessenden got the opportunity to go to the main laboratory at West Orange. The Edison Machine Works kept some of its men at the lab to work on experiments related to its product line, paying the men's wages. On arriving at West Orange Fessenden was given several odd jobs around the site, such as wiring up new rooms. All the laboratory staff were expected to be able to work with their hands, and Fessenden's first jobs could have been a test of his abilities. He was then assigned as an assistant to the chief researcher on insulation, and it was under him that Fessenden learned the ropes. When all attempts at producing a satisfactory chemical substance to insulate wires failed, Edison decided to assign Fessenden to the job. When Fessenden protested that he was an electrician who knew nothing about chemistry, Edison retorted, "Then I want you to be a chemist. I have had a lot of chemists . . . but none of them can get results."[60]

Results and Rewards

Results were what the Old Man expected, and if they were not forthcoming, the unlucky experimenter could expect a sudden discharge from the laboratory. In early 1889 Edison terminated the job of Fessenden's superior, a Dr. Wuntz, on the insulation project, commenting, "I can't make his work pay me." Fessenden took over the leadership of the team, but a short time later Edison decided to end the project and disband the experimental team. Several men were transferred to ore milling work, and some left the lab permanently, including Fessenden. He received a good recommendation as a chemical (rather than electrical) experimenter, with the explanation that "his line of work has come to an end."[61]

Believing that "the real measure of success is the number of experiments that can be crowded into twenty-four hours," Edison made his men work as hard and long as he did.[62] He led by example. He was a demanding employer who did not suffer fools or poor workmanship gladly. He followed the dictum laid down by another master of a shop: "Be king, be a good king, deserve loyalty and remove all disloyal influences."[63] The press found him to be a charming man, but his men feared his temper and his sharp tongue. He "could wither one with his biting sarcasm or ridicule one into extinction."[64] At any time in the lab some teams were "on the ragged edge" and some were "the favorites."[65] The men all knew the score and strived to be in Edison's favor, especially when his displeasure might lead to dismissal. Some of his employees could not stand Edison's dictatorial nature and biting tongue, and at least one of them quit on the spot after he could no longer endure the verbal harassment. As one employee concluded: "Some of the men liked him. Many of them feared him. A few even disliked him."[66]

Although many muckers remembered their days at West Orange fondly, it is unlikely that they " were a happy family at the lab," as one experimenter later claimed.[67] In addition to the ragged edge of Edison's displeasure, there were feuds among experimenters competing for his favor. Working in the lab was not always pleasant or light hearted, and conditions were diplomatically described as rather exacting.[68] Tense and impatient during experimental campaigns, Edison kept his experimenters under constant pressure, submerging them with work and often criticizing the results. The muckers were often set impossible goals and were always working to a demanding timetable. Some of them suffered nervous breakdowns, others left in disgust. The hours of storytelling provided an essential safety valve to the stress of work. The occasional day off for the muckers at Menlo Park, which they spent fishing or playing on the electric railway they had constructed around the site, played an important part in maintaining morale.

For those experimenters who could stand the rigorous work schedule and survive the practical jokes, work at Edison's labs was exciting and rewarding. Those who got results were quickly promoted and given more responsibility. Each man in the laboratory received a different wage according to his skill and experience. At West Orange the wage of the machinists was set from twenty to forty-five cents an hour and was graduated in fractions of cents. Experimenters were paid by the week and their rates ranged from seven dollars to thirty dollars. Although the leading experimenters earned, on average, more than the skilled machinists, there were still plenty of the latter who earned more than experimenters by putting in long hours. It was possible for skilled machinists to earn more than twenty dollars a week—this was more than a college-educated experimenter could expect to receive and more than the average earnings of a skilled craftsman in Newark. The blacksmith and draftsman at West Orange each earned more than Fessenden, who was a talented experimenter with some useful experience. A good carpenter could expect higher wages than most of the experimenters. Craft skills were still in demand in Edison's laboratory and were well rewarded.[69]

The muckers were paid low wages and expected to work long hours. They were driven by the will to make their experiments succeed and the expectation of moving upward in the hierarchy of the laboratory. As an experimenter ascended the ladder he could expect higher pay and more important projects. The lab was full of ambitious young men who saw their future in invention. Edison commented, "Its not the money they want, but a chance for their ambition to work."[70] The world of the shop culture was arranged in a complex hierarchy of skills, reflecting several levels of accomplishment in the society of machinists. The invention factory had the same hierarchy. At the top stood Edison and the pantheon of the boys—a position here could lead to fame and fortune.

As members of the inner circle around Edison, the boys could expect a share of the profits of an invention. He had promised to repay those who assisted him in the electric light project and did so in the form of royalties (a percentage on the profits of an invention) and part shares in the manufacturing shops. Men like Batchelor and Kruesi were well rewarded. Samuel Insull used his connection with Edison to rise to power and wealth in the Chicago Edison organization. Some of the boys were appointed to the boards of Edison companies; others were given lucrative concessions to market new inventions. Edison's career reflected the American belief that hard work and initiative could take a man from rags to riches. Fred Devonald, newly arrived in the United States from Wales, took his first job as a laborer at the West Orange lab in 1888. By 1900 he was in charge of engineering at the laboratory. Some of the unruly youths who haunted the lab at night went on to have distinguished careers in the enterprise. Edison was the same as other masters in his employment of children, but child labor at his laboratory was not the same as the exploitation of youth in the textile mills. At West Orange they carried out light work, ran messages, and had the opportunity to learn the trade of the invention factory. Walter Miller entered the lab as a youth and became the expert on phonograph recording techniques, rising to the enviable position of being one of the boys.[71]

Apprenticeship played a vital part in craft culture. It transmitted knowledge and values and preserved the traditions of the workplace. Edison's laboratory played this role, serving as an institute of higher education, a training ground for many employees. The daily challenges, the freedom to experiment and to learn, the participation in the social life of the muckers, made work in the laboratory an exhilarating experience. One mucker commented that "my labor does not seem like work, but like study and I enjoy it."[72] Edison exerted a powerful influence on the young men around him: "The privilege which I had of being with this great man for six years was the greatest inspiration of my life." This sentiment was shared by many ex-muckers.[73] Working in the laboratory formed part of a basic training that several inventors and captains of industry claimed had been the turning point in their lives.[74] To those muckers who went on to hold management jobs in the growing Edison enterprises, the machine shop culture was an important reference point in their new tasks. Yet as the business organization grew larger in the wake of Edison's successful manufacturing strategy, the work culture of his laboratory would come into conflict with a new, corporate culture.

CHAPTER THREE

The Business of Innovation

Innovation is a term that Edison did not use. He described himself as an inventor and the work he did in the laboratory as invention. Yet to label Edison a mere inventor does not do justice to his genius, nor does it account for the enormous impact he had on American life. Inventing was the idea stage, the first step in a long process. Its formal ending came when a patent was filed. Edison considered getting ideas for an invention the easy part; the hard part was "the long laborious trouble of working them out and producing apparatus which is commercial."[1] Innovation defines Edison's work, taking it from the laboratory into the commercial world. Innovation covers the setting up of a commercial enterprise on an idea. Edison's record number of U.S. patents should not obscure his even greater achievement of founding several industries.

In Edison's view a patent was hardly worth the trouble of inventing something. He knew from experience that selling patents to businessmen often left the inventor shortchanged. More often than not the returns from a new idea went to the financier or manufacturer, while the inventor struggled to protect his patent in the courts and obtain his share of the profits. A patent alone was not enough, nor was an invention. The original idea had to be developed into something more tangible than a patent; it had to be transformed, or "perfected," into a working model or a prototype—something a businessman could see and touch rather than imagine. This was essential to obtain financial support. In Edison's words, the "money people" had to see money in an invention before they would invest in it.[2] Perfecting an invention included finding and remedying the bugs—the defects and design problems—that inevitably cropped up in the development of an idea into a working model or process. This stage of innovation ended when the invention was translated into a factory-ready prototype. The idea was now embodied in a technology, an amalgamation of ideas, knowledge, and hardware all directed toward a practical goal. Its value was much greater than a patent. The final step was "pioneering" a technology by putting it into production and proving its commercial feasibility. This meant financing and administering a manufacturing operation until it could be sold to entrepreneurs.

Innovation covers what Edison called inventing, perfecting, and pioneering a new technology. The business of innovation encompasses decision making, from establishing the technical goals of a research program to devising a marketing strategy for a new product. It also covers the management of the research and development effort, and the financing of the whole operation. Inventors in the nineteenth century often ignored the business of innovation, preferring to remain in the technical domain. This was fine for the individual who did not mind a life of poverty and obscurity, but for the operator of an invention factory the management of his resources was of primary importance.

Edison knew that "the most difficult thing for an inventor is to get hold of people who will back him up," but at this task he was an expert.[3] The Wizard of Menlo Park was known to work a special magic on impressionable financiers, and few of them got the better of him. His profitable association with some of the most notorious Robber Barons of the Gilded Age was responsible for his rapid rise to fame and fortune. Entrepreneurs, such as George Harrington and Jay Gould, had given him the capital to move from machine shop operator to professional inventor in the 1870s and the opportunity to play fast and loose in the booming telegraph industry. Research contracts with large companies, such as Western Union, and cash payments from entrepreneurs had supported the large laboratory at Menlo Park. If the telegraph and stock ticker had provided Edison's entree into the Wall Street financial community, then the electric light had consolidated his position as a commercial inventor. The unprecedented expense of developing the incandescent bulb had brought him into contact with some of the leading merchant bankers in America, including the partners of Drexel, Morgan and Company. They had participated in the formation of the Edison Electric Light Company, which raised capital "for the purpose of supplying Edison with funds for carrying out experiments in connection with the development of the electric lighting system."[4] This involvement in introducing a new technology marked a step away from their traditional business of selling government and railroad securities; it gave Edison hope that this quarter might be willing to finance his grand strategy of constructing the West Orange laboratory.[5]

He canvassed his contacts on Wall Street as soon as the work began on the laboratory buildings. The first to be approached were capitalists involved in electric lighting. J. Hood Wright, one of the most active partners in Drexel, Morgan, was invited by Edison in 1887 to join in financing the West Orange operation, but Hood Wright declined, probably because he was overextended in the electrical industry.[6] Next Edison approached the German entrepreneur, Henry Villard, an early believer in the potential of electricity and one of Edison's most consistent financial backers.[7] Villard had amassed a fortune in railroads and steamships. As president of the Great Northern Railroad he was convinced that electricity had commercial applications in traction as well as light. His financial and moral support helped maintain Edison's experiments in electric traction from the first trial run at Menlo Park to the working system constructed at West Orange. Villard advised and encouraged, giving Edison the benefit of his commercial experience and expectations for the bright future of electricity.

The venture capitalists around Edison provided more than encouragement and finance; their business networks were enlisted to bring his innovations to the marketplace. Villard energetically promoted the Edison lighting system in the midwest in the early 1880s, using his money and influence to set up several central stations.[8] This active involvement in the promotion of the new industry was cut short by the crash of his railroad empire in 1884. Villard had to leave the country for his native Germany. On his return to the United States, he was immediately invited to come and view the "fine new" laboratory under construction at West Orange.[9]

Edison had considered the finance of the new lab before the first brick was laid. The business of innovation started with his estimation of the market for the research and development services of the new facility. Edison planned to carry out contract research for companies and individuals, joint ventures with other entrepreneurs, and development work on his own projects. He settled on three major research areas: electricity, ore milling, and the phonograph.

The growing electrical industry provided the rationale for building the West Orange laboratory in 1887. Its financial planning was based on Edison's expectations of the research and testing needs of the lighting industry. His laboratory was to be the technical resource center for the many business organizations involved in electric light and power. Edison was already bound by contract to provide his experimental services to several of these companies, including the Edison Electric Light Company, and was confident that there would be many more opportunities for contract research. He hoped that the operating costs of his laboratory would be covered by income from research.

By the time he moved to Menlo Park, Edison's experience in the business world had produced sophisticated methods of supporting his laboratories. Instead of negotiating a research contract with an entrepreneur or a group of businessmen, Edison joined with them in forming companies to exploit a new technology under development. Edison usually received cash or stock for his valuable patents and a share of the profits, while the cost of experimenting was billed to a research contract. As the sole owner and operator of his laboratories, Edison merged his private income into the finances of the laboratory. The payments for contract research were funneled into the lab's account, as were his royalties from patents, dividends from the companies he formed, and the cash payments made under agreements with venture capitalists. Although he claimed that it was not the inventor's role to pay for research, Edison often financed the lab by moving payments from his personal account into the laboratory account. In this way, several sources of income could be directed into the financial support of the lab. In some cases he paid himself for contract research. This was the case with the manufacturing concerns he organized to provide the equipment with which the local illuminating companies were electrifying America, including the Edison Machine Works and the Edison Lamp Company. These manufacturing "shops," as they were called, were important customers for the laboratory, which was kept busy testing and improving all parts of the lighting system.

The intimate relationship between Edison and the electrical industry was revealed in his plan to make the new laboratory the one research and development arm of the many Edison lighting companies. Each part of the Edison electrical empire was to send its technical men to West Orange to become integrated into one organization.[10] This centralized laboratory, under Edison's personal direction, was to direct its efforts into experimental projects—such as the development of better insulation—that had tangible benefits to all concerned. The lighting companies expected the laboratory to produce new technology that would improve their competitive position, and Edison encouraged them to think that great things were to come. This arrangement had advantages for all sides: the companies hoped to get more for their money by centralizing technical services, and Edison knew that his decisions in the laboratory would influence the development of the electrical industry.

Edison was far too experienced in the business of innovation to restrict himself to one type of research or to sacrifice his freedom of action. In addition to electricity, he had chosen two other major technologies to be developed in the new laboratory. The phonograph and ore-milling projects were based (respectively) on new products and processes, and both were still in the development phase and not perfected. His invention of the phonograph in 1877 had brought him great fame but little in the way of profits. He organized the Edison Speaking Phonograph Company in 1878 to support the development of his fragile tinfoil instrument and to market its novelty value. The company put up $10,000 to support further experiments to improve the invention but nothing was done because his time was taken up with electricity. Once the electric light system was completed, Edison planned to return to the phonograph.[11]

The Edison Ore Milling Company was formed in 1880 to exploit his patented magnetic ore separator. The machine used electromagnetic force to extract metals from ore. The company was reorganized in August 1887, its capital increased, and a new research contract signed. The company provided $25,000 to cover the costs of research and Edison promised to "construct a special laboratory for the conduct and prosecution of said experiments."[12]

With these agreements in hand, Edison could be sure that there would be sufficient work for the laboratory when it was built. His calculations of the yearly income of the lab were based on contract research with the following companies:

Edison Lamp Co.	$10,000
Edison Electric Light Co.	$5,000
Edison Machine Works	$6,000
Edison Speaking Phonograph Co.	$3,000
Edison Ore Milling Co.	$3,000

Although this total of $27,000 is much less than the average annual income of the laboratory, it gives an idea of the respective weightings of contract research by product, indicating that electricity was by far the most important area of research.[13] Edison's claim that only half the expense of operating his laboratory was supported by contract research is hard to substantiate and could well have been exaggerated in the negotiations of research contracts. His policy was to cover his lab expenses with the proceeds of contract research, even billing his own private experiments to outside parties. Throughout his years as purveyor of contract research, Edison quietly blurred the line between the experiments he did for others and those he did for himself. Who was to know if a result from contract research was applied to another project or if experimental equipment built for one customer was used in work for another? Here were the advantages of contract research for an entrepreneur whose special talent was to see commercial applications for the results of experiments. The extensive contract research carried out by the laboratory staff opened up new areas of investigation and offered valuable spillovers of information that Edison was waiting to exploit.

The fundamental financial strategy of all Edison's laboratories was to provide the means for him to experiment. In his own words, he invented to raise the money to keep on inventing—the object of the exercise was experimenting rather than making money. The overriding goal of the business of innovation was to provide the financial support for Edison's life of experimenting.[14] The contract research, the deals with venture capitalists, and the many new companies were means to the same end. The West Orange operation was the private laboratory of one man. Its purpose was to carry out his private experiments. He boasted that "I have always got something new underway," such as a method of preserving fruit and meat or his scheme for talking pictures.[15] Some of these experiments were carried out to investigate totally new technologies, others were done for the fun of it. If one of these stunts struck gold, Edison was ready to drop everything and perfect it as a commercial technology. If it provided the key to another problem in a totally different project, he was prepared to quickly apply it. The new lab was built with this kind of flexible innovation in mind.

Edison could not have picked a better time to establish an industrial research facility. The flow of information on which his research operation was based had turned into a torrent, providing numerous opportunities for inventions. The accelerated development of the communications and electrical industries had led to advances on a broad front of technologies. Edison knew that this scientific and technical information could be applied to the work in hand at the laboratory and made sure that he kept abreast of research on both sides of the Atlantic. Although he was not to know that in Germany in 1887 Heinrich Hertz was undertaking some pathbreaking experiments on high frequency electromagnetic waves (radio waves as they are now called), he did know that it was well worth the effort to monitor the scientific literature from Germany.[16] The West Orange laboratory stood at the convergence point of an international communications network. Its purpose was to bring together flows of information at the right moment, providing the basic raw material for the invention factory.

The business of innovation was not based on technical information alone—trade papers and personal contacts brought in the commercial intelligence that framed the work on the experimental benches. Most of the projects undertaken at the laboratory had roots in both technical and business considerations. Take Edison's interest in storage batteries for example. In the 1880s the increase in knowledge about electrochemical reactions in cells made an improved battery possible. He knew the relevant scientific literature and had considerable practical experience with batteries. His belief that the existing lead acid cell could be improved was firmly based on theoretical knowledge and personal experience. He had also gathered information about the market for batteries. He estimated that the annual sales of storage batteries in the United States was about $1 million and expected that sales would triple "if a good battery could be obtained."[17] This type of thinking, which brought together technical know-how and commercial information, was the foundation of the business of innovation.

Edison the Manager

Edison liked to depict himself as an inventor who was drawn into the task of pioneering his inventions because of the timidity of investors. He let it be known that he would have preferred to stay in the laboratory, but somebody had to raise the money and supervise the factories. Edison's reputation as a businessman has never been great. His friend Henry Ford summed it up when he said that Edison was the world's greatest inventor and worst businessman.[18] This view has persisted until today, when it is still fashionable to explain the downfall of innovative companies in terms of poor management rather than poor technology. In his latest book, management expert Peter Drucker depicts Edison as a disastrously bad manager who ruined the companies formed to develop his inventions. Many Edison companies did go bankrupt, but not all these business failures can be ascribed to bad management. Drucker is doubly wrong when he concludes that "much, if not most, high tech is being managed, or more accurately mismanaged, Edison's way," for few of today's high-tech companies can duplicate the diversity of Edison's operations.[19] His strategy of basing the research operation on at least three different technologies gave his business a buffer against technical failure or economic depression. The history of Edison's business enterprise is not one of uninterrupted success, but it is also not a catalogue of failure. The fact that it survived the dog-eat-dog business world of the Gilded Age and the Great Depression is surely a point in his favor.

Peter Drucker has accepted the part of the Edison myth that depicts him as the eccentric inventor too concerned with his experiments to bother with business affairs. This view was bolstered by his well-publicized aversion to bookkeepers and record keeping. The story goes that he took the bills he received at his Newark shops and stuck them on a nail until his creditors took legal action against him.[20] He liked to be seen as a preoccupied inventor with no time for business. This image was useful in the constant litigation that threatened to take up all his time. Called to the stand, Edison could claim ignorance, and therefore, innocence, of the unsavory wheeling and dealing of the business world. This stance helped foster the myth of the great inventor and poor businessman. Edison took pains not to alter this perception, and for good reason.

His business associates knew a different Edison, a shrewd, calculating man who exercised fine judgment of the marketplace. The same energy and ingenuity that he brought to his experiments were applied to his business affairs. By the time he moved to West Orange, Edison had considerable experience in running a large business enterprise, for he had operated factories employing up to 100 men in Newark in the 1870s. One of his Newark shops was said to be among the largest telegraph manufacturing plants in the United States in 1874.[21] As he reminded one venture capitalist, "You are aware from your long acquaintance with me . . . that the works which I control are well managed . . . I know how a shop should be run & also how to select men to manage them."[22] Edison proved that he could pick good subordinates. Samuel Insull, Charles Batchelor, and Edward H. Johnson fully justified the Old Man's faith in them; all were capable managers who would have been an asset to any industrial enterprise. They were typical of the "honest, young, ambitious" men whom he picked to manage the new companies formed in the wake of his inventions. Edison might not have invented the incandescent bulb first, but he created the business organization and found managers with both technical and administrative skills to bring a new industry into being. The task ahead of these young men was daunting; it ranged from running a factory making incandescent bulbs to setting up a lighting company overseas. Many of them had only their experience in Edison's laboratory to prepare them for the challenge. Edward H. Johnson was a telegrapher friend of Edison who took on the job of establishing the English Edison Company and promoting electric lighting there in the first years of the 1880s. His performance gained the respect of the editors of the London *Electrician*, who called attention to Edison's skill in finding first class men, commenting: "We have always been sure that his business agents were equal if not superior to him in ingenuity."[23]

It took more than ingenuity to run a large organization like the West Orange laboratory. He set the business strategy and directed the research effort of the lab, making the important decisions about what experimental projects were to be undertaken and who should be assigned to them. He also decided when a project should be terminated, either because it had failed to produce results or because its work had reached a point where a prototype could be handed over for production. Although Edison gave his experimenters the maximum leeway in pursuing an experimental project, he wanted to decide the technical direction an idea should take and reserved the right to change it if he saw fit.

Several of Edison's experimenters who went on to be inventors in their own right, such as Frank Sprague and Nikola Tesla, claimed that working in his laboratory stifled their creative spirit. Both were impatient to promote their ideas for new technology and were guided by enthusiasm rather than diplomacy in their dealings with the Old Man. Sprague and Tesla later accused Edison of being blind to the commercial potential of the two technologies with which they later became associated, Tesla with alternating current and Sprague with electric traction. Both split with Edison when he refused to promote their ideas at the expense of his own work.[24]

Edison insisted on deciding what technologies should be developed and in what form they should be introduced. The business of innovation hinged on his appreciation of the commercial possibilities of a device on the workbench and an anticipation of its potential market. He encouraged initiative but frowned on too much of it, especially if it resulted in a complete commercial prototype. His relationship with Ezra T. Gilliland deteriorated for this reason. Gilliland and Edison were close friends, old telegraph cronies, who worked together on telegraph and telephone experiments after Gilliland left the Bell organization. At first they experimented in Bergmann's Works in New York City, each paying his own experimental expenses. When Edison set up production of phonographs on a small scale at Bloomfield, New Jersey, Gilliland was made factory manager. Gilliland kept on experimenting at the factory and produced a prototype with several improvements. He claimed that his new model was designed with cheap manufacture in mind, but Edison would have none of it and quickly scrapped Gilliland's innovation without even considering it.[25] Only one person was going to decide the form of the commercial prototype, and that was Edison.

He exerted great influence over the direction of an experimental project with his initial allotment of laboratory staff. In this task he was guided by his estimation of the experimenters' abilities and his sense of the importance of the work. Although he tried to match a man's aptitude to the problem, the history of his laboratory shows that he was not bound by credentials or experience. His men had to be flexible. The shop culture valued generalists, craftsmen who could turn their hand to any job on the bench and who were not afraid of taking on something new. The flexibility of the Newark shops was maintained in Edison's laboratories, where both staff and facilities were switched from project to project at the Old Man's command. The West Orange complex had been designed as a research facility that could be applied to any problem, its rooms ready to be reorganized for new experiments. Muckers were moved from room to room and were expected to pick up the new line of experiments and brief their replacements.

The invention factory was an operation where several different experimental projects were carried on at the same time. Edison kept many irons in the fire and several ready for it. The business of innovation involved monitoring the experimental work underway and then applying the resources of the lab at the right moment. Edison made the critical decision when to increase the effort. It could come in response to a perceived technical breakthrough or as an emergency measure when things went wrong. He would quickly marshall his troops, shifting in men from other jobs, forming larger teams (but always maintaining the same leader), and drawing on the large labor pool of the machine shops. When faced with problems in the storage battery project, Edison rapidly fitted up two rooms in the chemical lab and put twelve testers to work there night and day.[26]

The flexibility of the laboratory work force gave Edison the speed he valued in the process of innovation. It was one of the more concrete inheritances of the shop culture. Yet to use the tool he had shaped at West Orange, Edison had to develop some techniques that were more applicable to modern industrial management than craft work in the shops.

Good decision making required accurate and timely information about the progress of an experimental campaign, not just technical information but data on its cost and (most important) information to estimate the future cost of technological alternatives. Far from being disdainful of bookkeeping, Edison was a stickler for accurate records of the laboratory's work. Even when experimenters worked all night, they still had to sign out for the supplies and tools they used.[27] This preoccupation with record keeping began when the young inventor carried out contract research for telegraph companies and venture capitalists in the 1870s. In addition to results, his customers wanted a proper accounting of the money spent. Accurate record keeping was essential in maintaining financial support for the laboratory, which Edison knew was one of the most difficult tasks in the business of innovation.[28]

During his years in Newark, Edison made a bewildering series of contracts with different parties, several of whom were in direct competition. His partners in the machine shops wanted to see the outlay on personal experiments distinguished from manufacturing costs. This tricky situation required an exact breakdown of the cost of each job. The first records of experimental projects simply added up stores used and the time spent on the job by the machinists. As lab workers were usually paid by the hour, it was easy to figure the time spent on each job. It was a short step to make up a weekly list of jobs completed and distribute the time of each worker over them. This gave a good idea of the cost of each project. (At this time labor costs were by far the largest expense; the amount of supplies used in telegraph experiments was not great and overhead was of little concern.) At Menlo Park this system of accounting was used to allot labor time over each experimental project. Edison was not interested in the leather bound general ledgers or the accounts payable and receivable; he wanted all the relevant figures to be on one easily read page.[29] He received a weekly breakdown of the labor cost over each project on one sheet. Later his bookkeepers began to distribute the cost of materials over each experiment. In devising an accounting system that was probably more advanced than any other in use in machine shops or laboratories, Edison had a means to keep track of the expenses of each experimental project. This was used as information in decision making rather than for reducing the expense of research. He counted the cost but did not shirk from paying out enormous sums if he thought the experiments might turn up something valuable. Edison carried out a systematic review of the laboratory's work at least once a week, a considerable achievement in view of the large scale of his operations. By monitoring the muckers' work at the bench and consulting the weekly accounts, he could check on the vital signs of an experimental project before making a decision.

Bookkeepers were, therefore, an important and permanent part of the laboratory work force. In the early days at Newark and Menlo Park, the general administration of the laboratory had been one of the duties of Edison's personal secretary. Samuel Insull had defined the role of private secretary to Edison as masterminding his finances and controlling access to him. Insull's skill was to raise money in a hurry by selling Edison's notes to banks and investors. He also formed the companies, arranged for their finance, and sat on their boards. He made the job of private secretary one of the key positions in the Edison enterprise. An experienced manager, Insull was soon fully involved in the electrical industry after the success of the Pearl Street station. He was succeeded as secretary by a protege, Alfred O. Tate, who came to the Edison organization from the railroad industry. He began as stenographer to Insull but rose through the administrative ranks as an accountant. "I had the mind and inclinations of an engineer," said Tate, and like all who worked in Edison's laboratory, he took part in experiments.[30] As private secretary, he attended to the bookkeeping duties and had power of attorney to deal with financial affairs. In addition to dealing with all company business, Tate took charge of Edison's voluminous correspondence and his personal finances. He served on the boards of many Edison companies. His staff included a stenographer and a number of youths who did light office work. Edison's secretaries became the link between the laboratory and the companies he formed. By serving on the boards of the companies and attending on him in the laboratory, the private secretary could bridge the gap between laboratory master and corporate servant.

The Manufacturing Strategy

The chronic lack of operating funds for the new lab in West Orange brought discord between Edison and his financial backers. To build and equip the laboratory cost about $180,000, a sum that represents over $2 million 1987 dollars. In August 1887 Tate confided to Insull that the laboratory "was going to require a good deal more money than at first anticipated."[31] The capital hungry electric light industry had absorbed millions of dollars by 1887 and had exhausted many of Edison's financial backers. The industrialized West was in the middle of the Great Depression of the nineteenth century, prices were falling, and deflation gripped the American economy. This was not a good time to raise money.[32]

The expensive operating costs of the new laboratory, estimated by Edison to be from $60,000 to $80,000 annually, forced him to broaden the base of his operations to bring in as much money as possible.[33] The lab performed a variety of services under contract to outside companies. These ranged from continued work on an existing technology such as telegraphy, financed in part by Western Union, to surveys into completely new areas, such as copper smelting with electricity, commissioned by the Parrot Copper Company. Much of the contract work of the West Orange facility was a carryover from the Newark and Menlo Park laboratories. The lab still carried out research work for A. B. Dick on the electric pen (a stencil system), and the American Bell Telephone Company on telephony. The extensive publicity given to the West Orange laboratory brought in a stream of enquiries asking Edison to solve technical problems. A tobacco company asked him to develop a substance to bleach tobacco to the desired rich yellow hue without destroying its flavor. The Old Man was probably only too happy to experiment with one of his favorite habits.[34]

The policy of contract research meant that the lab had to be all things to all people. In these early years it carried out a wide range of tasks and was truly a multipurpose facility. The resources of the laboratory were not restricted to research and development; they were often applied to manufacture and repair. At the time that Edison was involved in the magnetic separation of ores, his laboratory could be hired to erect an ore milling plant with a capacity of 1,000 tons a day.[35] The two large machine shops in the main laboratory building at West Orange were used as a factory to make money from manufacturing telegraph instruments, such as the phonoplex. This was a successful invention that enabled several telegraph and telephone messages to be sent over the same wire. The railroad industry bought the phonoplex sets, which were assembled in Building 5.[36]

The great pioneering venture into electric light and power provided an interruption in a business career that had been dominated by communications. Edison returned to telegraphy and telephony at the West Orange laboratory. It was the major focus of research when the laboratory opened; in December 1887 he made a list of experiments to be started that included the phonoplex and "grasshopper" telegraph.[37] The latter was a wireless telegraph system that utilized the phenomenon of induction—when an electric current in one circuit produces a current in another close by—to communicate between moving trains and a fixed telegraph station. An electrostatic charge in the induction coil of the sending apparatus induced a similar charge in the induction coil of the receiver that sent a current in its own circuit to cause a telegraph "click." Edison anticipated a market in wireless communication over land and sea, and (along with many other inventors and entrepreneurs in the field) saw one way to achieve this in electromagnetic waves generated by sparks. He built apparatus that sent a high voltage electric charge over a "spark gap" between the points of two conductors. The spark generated in this way caused an electromagnetic disturbance that could be detected some distance away.[38] As the discoverer of "etheric force" (the name he gave to these high-frequency electromagnetic waves he had observed but did not understand), he had begun to think of electricity as a wave-like form that could be manipulated to provide light, power, and wireless communication. James Clerk Maxwell, the great theorist of electricity, had predicted that electromagnetic radiation could travel through space with the velocity of light. Edison recognized a commercial technology in the high-frequency oscillations caused by the disturbance of an electric spark, these invisible, speedy waves of etheric force traveling through the ether with messages from earthbound businessmen. Several experiments to transmit Morse code by interrupted sparks were made at West Orange. Although Edison claimed to have successfully transmitted messages with both induction and etheric force, his attempts to develop a wireless telegraph came to naught.[39] Nevertheless, these projects show that he had not lost his touch in identifying profitable new technologies in the laboratory phenomena he discovered in the course of experimenting. Edison was thinking of a compact signaling system that would not have taxed the manufacturing facilities at West Orange.

Edison's plan to manufacture new products at West Orange can be seen clearly in his idea of the Edison Industrial Company, "one organized avenue for the manufacture of all his inventions." This was not another large-scale enterprise like the great systems of electrical lighting, but a focused manufacturing operation to bring consumer durables to a mass market. With the enormous costs of introducing the electric light still fresh in his mind, Edison wanted to avoid "cumbersome inventions like the electric light" and concentrate instead on small products with a high profit potential and low capital requirement; he planned to supply small items of commerce—"useful things that every man, woman and child wants"—to be sold through a network of jobbers.[40] As a manufacturer in the late 1880s, Edison could use the telegraph and railroad to reach a national market. Perhaps influenced by the success of the sewing machine, he saw opportunities for consumer goods opening up in the 1880s. American cities were growing at a furious pace. Between 1880 and 1900 the urban population of the United States moved from 28 percent to 40 percent of the total. With the great boom in railroad construction over, cities were becoming the most important market for manufactured goods.[41]

Edison looked for new products in the further application of electricity, especially electric power that he thought could drive a variety of new products, from hand tools to tricycles. Electricity as a source of heat also provided some opportunities for innovation. He planned to build an improved snow sweeper on this principle. The gradual spread of electrical supply networks gave him the idea of producing machines powered by house current for use in homes and offices. Electric fans and heating elements could easily be devised and then made up at West Orange.

His confidence in his abilities as a machine builder was unbounded. He tried to interest capitalists in the development of a mechanical cotton picker to do for cotton what the combined harvester was doing for cereal production.[42] This experimental project was promoted despite his complete ignorance of cotton growing and only a passing acquaintance with reaping machines! After years in the telegraph industry, Edison was certain that he could master any electromechanical machine, and if it gave any trouble, then a night's work with the boys would certainly fix it. Such was the case with the typewriter. Edison claimed that the Sholes model was brought to his laboratory where he perfected it as a commercial product, replacing Sholes' clumsy wooden apparatus with a metal typewriter suitable for mass production. Sholes' biographer disputes this claim, and it is much more likely that this important work was carried out by master machinists at the Remington works.[43] Nevertheless, this was exactly the type of work that Edison intended to carry out at the West Orange laboratory: improving existing products, redesigning machines with commercial potential, and preparing them for manufacture.

The far-reaching impact of revolutionary new technologies such as the electric light should not obscure the fact that the invention factory was often directed into less dramatic but more profitable pursuits. The role of industrial research in the 1880s, as conceived by Edison, was not solely framed in brave new technologies and great new industries. The West Orange laboratory was a place where "old devices and articles" could be quickly and easily improved. The research of his laboratory was to be directed into finding more efficient—and therefore cheaper—ways of doing things, especially reducing the cost of manufacture of items as different as metal wire and plate glass. Edison was fascinated with all kinds of materials and was confident that he could find, or make, cheaper alternatives to many materials in everyday use. The large investment in the chemical facilities of the laboratory was expected to generate returns in the search for artificial substitutes for silk, ivory, and rubber.[44]

The story of the typewriter is testament to Edison's belief that the inventor was a victim rather than a benefactor of the innovative process. Sholes had little to show for his invention other than years of hardship and worry. The financial returns went to the Remington company, which manufactured thousands of typewriters in the years before World War One. Edison thought that it was not enough to develop new products; he was determined to control all stages of innovation, especially the manufacturing stage, in which he could employ the resources of his laboratory to continually reduce the cost of production. He claimed that any money he made was from manufacturing the invention, not selling the patent.[45] Rather than being a transition between the Menlo Park and West Orange laboratories, the short time at the Lamp Works in Harrison had a major bearing on Edison's career, especially in his thinking about the business of innovation. There he discovered the major article of faith of American manufacturing in the nineteenth century. The cost of making a product could be successively lowered as the manufacturer moved along the learning curve of mass production technology, and at each cost reduction more customers would purchase the product. The invention factory was to be used to lower progressively the cost of production. The miracles of cost reduction that had been achieved at the Lamp Works could be applied to other products. Edison formulated a grand strategy for the new lab, envisaging a large industrial undertaking that would manufacture the many new products devised in the laboratory: "My ambition is to build up a great industrial works in the Orange Valley starting in a small way and gradually working up—The Laboratory supplying the perfected invention(s) models pattern(s) and fitting up necessary special machinery in the factory for each invention."[46] Charles Batchelor was told of this scheme while the laboratory was being built. He wrote in his diary: "Edison's idea now for the future is to get up processes for manufacture and start factories . . . as soon as the new laboratory is finished this will be commenced in earnest."[47]

Edison's plans for his West Orange complex reflect the aspirations of a decade in which the United States became the world's leading manufacturer. A few years after the lab was built, the output of steel in American mills surpassed the production of the two leading industrial powers, Germany and Great Britain. By 1900 the United States was the world's leading industrial power. It might not have been his original intention to become a capitalist, but Edison played the role at West Orange, where he constructed many factories, formed numerous companies, and carried out business strategies similar to his industrial peers. Andrew Carnegie, John D. Rockefeller, Henry Ford, and other captains of industry promoted vertical integration of their operations, sought self-sufficiency in raw materials, and had a paternalistic, and often hostile, attitude toward labor. All benefited from technological innovation and all built successful business organizations on it.

While other industrialists were mass-producing sewing machines, clocks, agricultural machinery, and furniture, Edison was thinking of making office equipment. Complicated new products, such as the electric light, telephone, and typewriter, were slowly changing the American office, facilitating the task of managing the larger business organizations of the late nineteenth century. Edison's electric light had found the most rapid acceptance in offices (the first to be lit were the offices of Drexel, Morgan on Wall Street), and businessmen had purchased his stock tickers and electric pens in great numbers. Confident in his abilities to design business machines, Edison set about making office work easier and more efficient. He saw commercial potential in the typewriter and contemplated building a factory in the Orange Valley to manufacture an improved version. In 1887 there were several other typewriter companies in operation, and this threat of competition drew Edison toward an entirely new technology, an even more complex business machine that had the potential to revolutionize business communications—the phonograph.

When used as a dictating machine, the phonograph promised to ease the burden of business correspondence. He thought of the phonograph as a complement to the telephones and typewriters slowly coming into use in American offices. Incoming messages could be stored on the machine until they were typed, and outgoing correspondence could be dictated onto the phonograph and then given to typists. While Edison was experimenting with various forms of waxed paper to be placed around the cylinder he got the idea that a compact recording medium could replace the business letter completely; the cylinder covering, or phonogram, could be sent through the post. Edison visualized an office where paper was replaced by the wax cylinders of the phonograph; thus he was the pioneer in the paperless office.[48]

Edison thought that if the phonograph could be made as easy to use as the telephone and typewriter, it would replace the stenographer and make businessmen more productive, and he felt sure that the dictating machine would someday become commonplace in the office.[49] His business connections, such as A. B. Dick (the company marketing the electric pen), had shown interest in the phonograph and were eager to see the new machine as soon as it was completed. The phonograph was to be the first product to go into mass production at West Orange, and its success or failure would decide the future of the Edison enterprise.

The ground was broken for the factory, called the Edison Phonograph Works, in May 1888, and the buildings were completed by the end of the year. Located to the east of the laboratory, the Works centered on two large buildings: one was about 75 feet wide and 200 feet long, and the other was the same width but 400 feet long (fig. 3.1). The shorter building was located next to the laboratory and contained the many tools used to make the parts of the phonograph. A photograph shows that its interior was left open, allowing belts and pulleys to take power from the overhead main shaft to each machine (fig. 3.2). The larger building was divided into rooms separated by a central corridor. Here were offices, the assembly and testing rooms, and the japanning ovens for the phonograph's varnished finish. The wax cylinders and phonograph batteries were made in this building, which also contained the finishing and packing rooms. Between these two long structures were three smaller buildings, one of which contained the powerhouse for the Works.

Figure 3.1. The Phonograph Works soon after completion in 1888 or 1889.

Fully equipped with a large foundry and illuminated throughout by electric light, the Works could carry out all aspects of phonograph manufacture from toolmaking to testing.[50] It also had every facility to manufacture a wide range of precision products, especially those requiring electrical parts.

The completion of the Edison Phonograph Works in 1888 rounded out the original plan for Edison's West Orange operation—a large manufacturing facility adjoining a laboratory complex, all under the control of one man. In controlling manufacture, and ultimately the supply of raw materials, Edison was covering ground that is normally reserved for the large, integrated corporation. Yet at the same time, he had constructed a research laboratory that was equipped to move quickly into new technology, taking the larger organization with it. The diversity of experimental projects undertaken at the laboratory gave the Edison enterprise the means to ride

Figure 3.2. Interior of the main machine shop in the Phonograph Works.

out competition and depressions in one business. It was also the path for continued growth and expansion. The facilities erected at West Orange were the basis for the continual, systematic entrepreneurship that Peter Drucker praises in his books. At West Orange, Edison constructed the ultimate competitive tool in an age of competition: a facility that could move rapidly into new businesses—"those industries that offer the most promising field for invention and experiment"—manufacturing the products that emerged from the largest industrial research laboratory in the world.[51]

CHAPTER FOUR

The Phonograph: A Case Study in Research and Development

dison thought of the phonograph as his baby, a special discovery, an "invention pure and simple" that he had chanced upon in the development of another idea. It was one of the few inventions that had not been visualized before it was made. In Edison's argot the phonograph was more a *discovery* than an *invention.*[1] He stumbled upon it in his research on the automatic telegraph—a device that recorded incoming Morse messages on a moving strip of paper. In 1877 he patented an automatic telegraph that recorded the dots and dashes by embossing them on revolving discs of paper. He then used this idea in a device to record incoming telephone messages. It was in the process of carrying out these experiments that Edison discovered that the indentation of dots and dashes recreated a sound that bore some resemblance to the human voice.[2]

Edison's work on telephony and the acoustic speaking telegraph had drawn him into the study of sound and its wave-like forms. He was aware of the German scientist Hermann Helmholtz's research on sound waves and knew about devices like the phonoautograph, which used a rod attached to a membrane diaphragm to trace out the undulations of sound waves. As usual in the Menlo Park laboratory, the work on the automatic telegraph was sandwiched between research on several other projects, including the development of telephone repeaters (amplifiers) and recorders. The research on the telephone familiarized Edison with the thin discs of metal or animal membrane that acted like a diaphragm by vibrating to produce sound waves. By attaching a needle to a diaphragm and running a strip of paper underneath it, Edison was able to indent the sound waves of his shouts on the paraffin covered paper. The success of this experiment led to another device that imprinted sound waves on a moving cylinder covered with tinfoil. This prototype tinfoil phonograph had two diaphragms: one to indent the sound and one to recreate it to the astonishment of the listener.

The 1877 tinfoil phonograph was no more than a demonstration model of the idea, a scientific curiosity, or an amusing toy as *Scientific American* had noted.[3] Edison recalled that only a small number of tinfoil machines were actually manufactured (many in the shops of his associate Sigmund Bergmann), and that "no commercial instrument was made."[4] After the curiosity market for the phonograph had worn off, the tinfoil and the company formed to exploit it languished while Edison worked on his incandescent lighting system. Meanwhile, other inventors equally knowledgeable about sound waves and telephone technology saw commercial potential in the reproduction of sound. Alexander Graham Bell had almost stumbled on the phonograph idea in 1874, when he speculated that the phonoautograph format could be used to recreate the original sound. But Edison beat him to it, as he also beat Bell to the carbon button transmitter that dramatically improved the telephone. In 1881 Bell and his two associates in the Volta laboratory, Chichester Bell and Charles Tainter, thought they could beat Edison to a commercial phonograph that would support the costs of their experiments.[5]

Charles Tainter was a precision instrument maker who worked with Bell in the same way that Charles Batchelor assisted Edison. Tainter had begun his career as a maker of telegraph instruments and had developed his skills to include the manufacture of astronomical and other scientific equipment. His mechanical talents were his contribution to the laboratory.[6] Bell naturally gave the lab the benefit of his understanding of acoustics, and Chichester Bell contributed electrical and chemical knowledge. Together they improved Edison's machine by substituting wax for the tinfoil on the cylinder. While the tinfoil's needle displaced the recording medium, Tainter's idea was to cut or engrave an impression on it. This produced a more durable recording with clearer reproduction. Tainter also devised a separate reproducer arm in which a floating stylus kept a gentle, steady pressure on the needle in the groove, enabling it to ride out the bumps on the wax surface.[7] The recording cutter and reproducing arm were moved precisely along the wax cylinder by a lead screw assembly that ran parallel to the turning axis of the cylinder. The machine, named the graphophone, was a vast improvement on the phonograph. It had the fidelity of sound reproduction and simplicity of construction required for a commercial

machine. In 1885 Edison's English patents on the phonograph lapsed, leaving the field wide open on both sides of the Atlantic. The three Volta associates filed for several patents in 1885 and formed the Volta Graphophone Company the following year. They had the financial support of Sigmund Bergmann who manufactured their graphophones. Bell and Tainter exhibited their machine in 1886, bringing them to the attention of the press and a showdown with Edison.

Representatives of the Volta laboratory visited Edison at his temporary quarters in the Harrison Lamp Works to suggest that both parties come together to develop and manufacture an improved phonograph. He was offered a generous financial arrangement: all experimental costs paid by the new company and all capital requirements to manufacture the machines to be met, in return for a half share in the new enterprise. Normally, Edison would have jumped at such a proposition, but this time he would have none of it. His competitive instincts were aroused, and he feared that his rivals were going "to deprive me of the honor of the invention."[8] Edison could not be persuaded to come to an accommodation with Bell and Tainter, whom he referred to as pirates, because he was planning the construction of a laboratory that could catch up and overtake them in developing a commercial phonograph. Although Bell and Tainter had a considerable lead, Edison refused to work with them, claiming, "I have assumed all along that the business was open (to competition), and that with my experience and facilities I should have an advantage in competition."[9]

Edison saw his competitive advantage not only in the superior experimental facilities he had constructed at West Orange but also in his manufacturing strategy. He planned to perfect the phonograph in the laboratory and then quickly move into mass production. Tainter might be a good inventor, but that was only half the game. As Edison telegrammed his English representative, Colonel George Gouraud, "Tainter not practical man; don't know how to make cheap."[10] Edison was confident in his abilities to set up production in a few months and then continually reduce the price of phonographs in the same way he had cut the costs of incandescent lamps. The problem was that he did not have the factory. In the summer of 1887, while Batchelor was supervising the construction of the laboratory at West Orange, Edison was desperately trying to convince businessmen interested in the phonograph that the machine in his mind was more valuable than the one that Tainter had in his hand. He claimed in July 1887, "I have a much better apparatus (than Tainter) and am already building the factory." There was a grain of truth in this. A small factory had been rented in nearby Bloomfield to start making tools for phonograph manufacture; the inventive trio of Edison, Ezra Gilliland, and Charles Batchelor, based in the temporary lab in the Lamp Works at Harrison, had examined the graphophone; and Batchelor had done some experimenting on the wax compound of the cylinder.[11] But no progress had been made on improving the tinfoil machine, and Edison was playing for time while the new laboratory was under construction.

The experimental team in the Harrison Works was probably shaken by the advances made by Tainter, and there is no doubt that the graphophone, a sturdy, belt-driven machine, which was powered by a foot treadle in the same fashion as the sewing machine, had commercial potential. Edison had nothing to match this. The busy fall of experimenting in the temporary lab in the Lamp Works brought forward only a stopgap design that was not a commercial prototype.

The experimental work was transferred from Harrison to West Orange as the new lab reached completion. The new phonograph was unveiled at the West Orange laboratory as soon as construction was finished in November 1887. Its design borrowed heavily from the graphophone. It had a wax cylinder that turned on a threaded spindle, and the reproducer assembly was attached to a slide with an arm connecting it to the feed screw on the spindle, an arrangement that was inspired by the screw-turning lathes in both the West Orange and Volta laboratories. In contrast to the Tainter machine, the phonograph was powered by an electric motor positioned under the base plate. It drove the spindle through a system of beveled wheels. As the cylinder turned, the lead screw moved the traveler arm that pushed the reproducer along the cylinder. Control was achieved through a stop and start switch next to the motor. The reproducer and recorder diaphragms were mounted together in the form of a pair of spectacles. A needle in the recorder cut a groove into the soft wax cylinder and, after recording, the spectacle frame was then turned to bring the reproducer stylus onto the surface of the cylinder. The fragile reproducer consisted of a diaphragm connected to a curved steel wire stylus.[12] Several adjustment screws enabled the user to maintain the right pressure and angle of reproducer on the cylinder (fig. 4.1).

The general appearance of the phonograph introduced in 1887 indicates its makeshift nature. The lack of a cover for the machine

Figure 4.1. The 1887 Edison phonograph, as depicted in *Scientific American*. The recording stylus is engaged on the cylinder as the user dictates into the small horn attached to the spectacle diaphragm. The needle attached to the spectacle marks the length of the dictation on the numbered scale below it (*Scientific American* 7 [31 Dec. 1887]: 415).

and a casing for the motor point to a laboratory model rather than a commercial machine. The shape of the motor hints that it was originally designed for a round casing, such as the one used on the stock ticker. The phonograph of 1887 was certainly nowhere near as complete as the graphophone. Gilliland's frank admission of its inferiority probably played a part in his alienation from Edison.[13]

The technical work might have been going slowly, but the work of promoting the new phonograph was a great success. It was successfully publicized on both sides of the Atlantic, and this probably kept Tainter at bay until the business community saw what Edison had to offer. The publicity value of the world's largest industrial laboratory was not lost on Edison. The large, open space in the middle of the third floor of the main building was used as a demonstration area and lecture room, and well before the machines were running in the shops below, Edison had brought visitors there to demonstrate his phonograph.[14] By 1888 the lab was famous. The groups of visitors, many of them celebrities, who were given tours of the new facility served to bring wide publicity to Edison and his work. The propaganda offensive of the winter of 1887/88 gave way to a full experimental effort in the spring of 1888, when Edison marshaled the forces of the laboratory for its first great campaign.

The task that faced the laboratory staff in 1888 was far from easy. The original plan developed in 1886 was to develop a small, compact instrument that could reproduce the human voice at the same level of loudness and fidelity as the telephone. It was to be driven by a small motor that could be easily stopped and started.[15] The choice of an electric motor to power the machine was the critical element in deciding its form. Once the concept of the wax-covered cylinder was accepted, the major design problem in a talking machine was the power source. Tainter tested electric motors but decided against them because of the extra cost, complexity, and likelihood of breakdown. Edison, on the other hand, was confident that he could design an appropriate motor, and the simplicity of controlling the phonograph with a switch was an advantage over other power sources. Edison probably leaned toward an electric motor because he was already acquainted with it, and the experimenters in the electrical laboratory in Building 1 were busy with motor research. But the size and weight of the motor and the requirement of connecting it to a source of electricity worked counter to the development of a small, portable machine. The experimenters in the West Orange laboratory looked at clockwork motors as an alternative to the electric motor but found them to be lacking in power and too noisy. The electric motor solved the problem of noise but created several others, not the least of which was the difficulty of maintaining a steady turning motion—surges and drops in power in the motor distorted reproduction.[16] Edison's confidence in producing a constant speed motor was misplaced; it took many years of experiments and false starts before an adequate electric power source for the phonograph was developed. Another serious problem was the high consumption of current that quickly exhausted the chemical charge of the primary, rechargeable cell of the battery and contributed to the excessive number of breakdowns of the battery. This was more than an inconvenience for it increased the cost and reduced the reliability of the machine. In the race to catch up with the graphophone, cost and reliability were not the top priorities at West Orange.

The inferior reproduction of the tinfoil machine had been underlined by Tainter's improvement of the wax-covered cylinder that gave a much better sound. Moving to a wax cylinder gave Edison's men the opportunity to improve the fidelity of reproduction but presented an entirely new range of technical challenges, for the diaphragm and reproducer assembly had to be altered to work with a different recording medium. The wax compound had to be soft enough to take the indentation but hard enough to retain it over several plays. The first compound tried was a soft, soap based wax. The small pieces of wax that were dislodged as the recording was cut fell from the cylinder and gummed up the works of the machine. A further disadvantage of the soft wax was its inability to maintain good reproduction over several plays. Overcoming these difficulties involved considerable experimentation into the composition of the wax cylinder and the action of the stylus in the groove.

The number and complexity of the problems forced Edison to move men and resources into the phonograph campaign. Much of the third floor of the main building was taken up with the continuing development of the phonograph. A newly hired chemist, Jonas W. Aylsworth, investigated various wax substances for the cylinders; an English mucker called Gladestone tried to improve the output of the battery; a formally trained German experimenter and accomplished pianist, Theodore Wangemann, experimented on recording techniques; and Edison, assisted by the chemist, Dr. Franz Schulze-berge, applied himself to the problems of reproduction of sound and duplication of cylinders. In Building 1, the chief electrician of the West Orange laboratory, Arthur Kennelly, tested several types of electric motors under consideration for the phonograph.

Edison immersed himself in the problems of sound recording, which required further research in acoustics. The experiments in reproducers were paralleled by the continuing work on telephone transmitters and receivers; as usual he was hoping that one series of experiments might turn up some information useful in the other. His research began with a thorough examination of the operation of the diaphragm in receiving and producing sound waves. He then tested many organic and inorganic materials for the diaphragm, including ivory, horn, porcelain, rubber, shellac, sheet steel, and carbonized paper. This work has all the characteristics of the trial-and-error style of research that Edison had made famous, yet this was not blind, random testing of all the many different materials that had been collected in the large storeroom of the lab. Edison had studied the human ear to develop analogies between it and the reproducer of the talking machine, and this knowledge guided the search for what was to be a mechanical substitute for the small bones in the ear. Edison correctly saw that the linkage between the stylus and the diaphragm was the critical element in the quality and loudness of reproduction and directed his research effort accordingly.[17] As in the electric light filament project, Edison knew what he was looking for and roughly where to look. The experimental work included testing every likely solution, keeping exact records, and not stopping until all alternatives had been tried. This type of approach, blending theory and systematic investigation of a range of likely solutions, is similar to many research projects carried out in modern laboratories.[18]

A few feet from Edison another experimenter carried out the laborious testing of a multitude of substances for the wax compound of the cylinder. During 1888 the chemist, Jonas Aylsworth, labored on mixing, heating, and testing hundreds of wax compounds. He tried many different combinations of oils and fats to find a substance hard enough to take and hold the impression made by the phonograph stylus and soft enough to be shaved off when a new recording was made on the cylinder. Again it would be incorrect to label this research trial-and-error. Aylsworth had a successful career as an industrial chemist behind him and knew the likely parameters of the compound he was searching for. Starting with the beeswax used by Tainter, there were several other obvious choices, including paraffin and soap-based mixtures, that offered a good chance of success. Aylsworth was not blindly searching through a maze of different chemicals; his work took him along a small number of well-defined avenues. The job was to find the best alternative that fitted both technical and economic criteria. He finally settled on stearic acid and its salt, sodium stearate, a common fatty acid used in the manufacture of soap and candles. The mixture was heated to about 480 degrees fahrenheit and then filtered through fine muslin. On cooling, it congealed into a thick molten wax into which metal forms were dipped to make cylinders.[19]

As an industrial chemist, Aylsworth was concerned with ease of manufacture and quality assurance. Many of the operating problems of the phonograph were traced back to imperfections in the wax cylinder. The smallest crack or bubble in the wax compound—invisible to the eye—caused a crackle that could be distinctly heard as the record played. The wax was also sensitive to heat: too much caused softening and too little made the cylinder brittle and liable to crack. Much of Aylsworth's work was, therefore, directed at examining the manufacturing process and testing wax compounds under different operating conditions—two areas where theory was of little use. This project was one of the most important carried out at the lab. Experiments on the composition of the wax cylinder began as soon as the West Orange laboratory opened and continued into the twentieth century. The exact composition of the wax was one of the most valuable, and closely guarded, secrets of the Edison enterprise.

The development of the phonograph covered many different areas of research—chemistry, physics, electricity, and acoustics—and was a true test of Edison's team approach to innovation. Each of the different problems to be solved interfaced with the other and required close cooperation between the experimental teams. Edison had broken the task down into its component parts and assigned a gang of men to each; their close proximity on the third floor of the lab helped the internal communication that was vital for success. During the summer of 1888 Edison put more men to work on the phonograph, hoping to generate the synergy to finally overcome the interrelated problems that stood between him and a commercial machine. His secretary, Alfred O. Tate, reported that he had over a hundred men at work on about sixty different experiments. Less than a year after its opening, the West Orange laboratory was operating at full capacity, the phonograph campaign was in full swing and the effort in electricity and ore milling was maintained.[20]

Perfecting the Phonograph

One of the most famous photographs of Edison was taken in the summer of 1888. It depicts him as the Napoleon of invention, slumped back in his chair, exhausted and bleary-eyed after sleepless nights of experimenting (fig. 4.2). On his right hand sits Fred Ott, his personal machinist and close friend. Standing to his right is Charles Batchelor and to his left is John Ott. Fred Ott and Batchelor are more interested in the machine on the table than the camera. William Laurie Dickson, who stands between them, gazes intently into the lens and strikes a foppish pose. Sitting on the far right of the table is George Gouraud, whose stylish cravat and clean shoes immediately mark him as an outsider to the fraternity of the laboratory. His dress shows that he was not one of the boys. In front of them all, between the white soft wax cylinders and the glass battery is the so-called "perfected" phonograph, the result of their labor and the first major innovation to come out of the West Orange laboratory.

Edison's phonograph now appeared in the classic rectangular form, with its electric motor fully enclosed in a wooden casing. In much the same way as the precise electromechanical works of telegraph equipment were mounted on a metal casting that served as a base plate, the cylinder and reproducer assemblies of this phono-

Figure 4.2. A very important publicity picture for Edison, which announced the perfection of the talking machine, the new "improved" phonograph of 1888.

graph were placed on a metal plate so that they were easy to access and adjust. The overhanging mandrel on which the cylinder was mounted gave this design its characteristic look. The other prominent feature of this machine is the exposed ball governor at the opposite end of the top plate. Most attention had been paid to the drive of the machine, which was now completely redesigned. After experimenting with several forms of direct gear drive from the motor, Edison had returned to the belt drive and governor system employed by Tainter. A new electric motor had been developed for the phonograph which was much smaller than the motor of the previous model and consumed half the battery power. A governor with a speed control had been added to the drive gear to help maintain speed. The user decided on the speed of the motor and could precisely adjust the turning of the cylinder to achieve optimum fidelity. Controlling recording and reproduction was essentially the same as the 1887 design. The user adjusted the reproducer spectacle with small screws to ensure that it sat correctly in the groove and followed a scale underneath the cylinder to gauge how

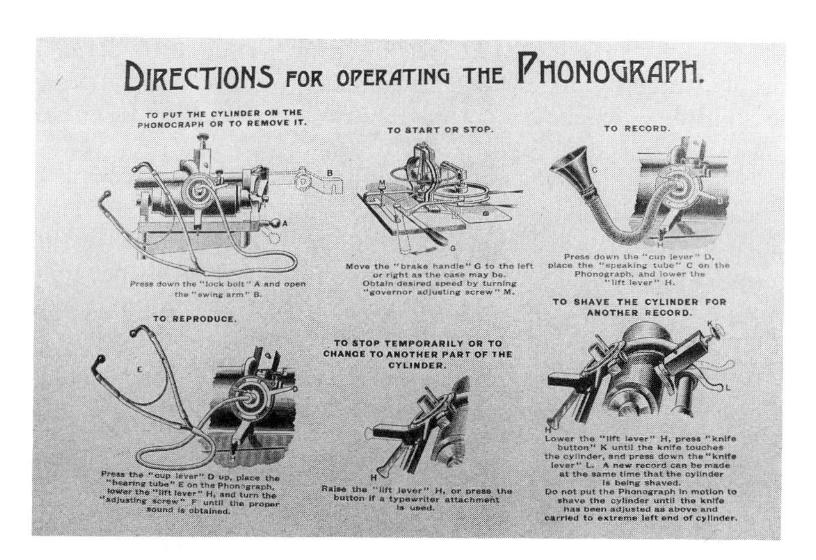

Figure 4.3. Directions showing how to load the cylinder on the mandrel, adjust the speed of rotation, and lower the diaphragm onto the cylinder to record or reproduce.

much playing time was left. A flexible speaking tube made it more convenient to dictate when it was attached to the recording diaphragm and permitted the listener to hear more clearly when it was attached to the reproducing diaphragm, (fig. 4.3).[21]

The wooden casing of this model and its cleaner lines indicate that Edison had at last designed a commercial machine. The 1888 perfected phonograph showed the attention to detail that came after a careful consideration of its use. The wax shavings that clogged the works of earlier machines now fell clear of all moving parts. Immediately after the photograph was taken, Colonel Gouraud took the first prototype to England to open up a European market for the machine. Edison was convinced that this product, the latest sensation coming from his laboratory, would have worldwide appeal. Once perfected, the phonograph should have entered the pioneering stage of production engineering and manufacture. The expectation among the jubilant muckers was that it was going to be put on the market in a few weeks. They could look out from their experimental rooms and see the factory buildings of the Phonograph Works rising up on the farmland next to the laboratory.[22] The photograph taken in June 1888 was supposed to mark the end of the phonograph campaign in the laboratory. Instead, it caught a short lull in the action. Expenditure on phonograph development rose rapidly in the last half of 1888 as its perfecting reached a new level of intensity. In fall 1888 experiments on the phonograph were costing about $3,000 a month—a sum that far exceeded the outlay on other projects.[23]

The laboratory staff worked feverishly through 1888 to complete the project and send a prototype to the Works. Edison exploited the high morale of the machine shop culture to keep his men at the task. A member of an experimental team remembered that they worked night and day during this time. One document shows that during one experimental campaign in 1888, over forty improvements were made, ranging from the fundamental change of altering the number of threads on the cylinder to the minor one of moving the location of small screws on the base plate. Many modifications were to make the machine easier to manufacture.[24] Edison personally took on the task of improving the poor sound reproduction of the phonograph. Not only was the voice of the talking machine faint and indistinct, but it could be lost through a mechanical fault in the main shaft and bearings or an imperfection in the wax of the cylinder.[25]

By October Edison was sure that he had finally perfected the instrument and wired Gouraud that he would drop dead when he heard the improvement in reproduction.[26] Nevertheless, the frantic activity continued into December, and Edison was so overburdened that he had to suspend the exhibitions and interviews in the laboratory, which were vital for marketing. No mention of the crisis within reached the public. The lab was flooded with inquiries about the phonograph, and the newspapers were full of articles about the wonderful new talking machine.

While Edison and his staff labored in West Orange, Tainter stole another march on them. In 1888 he rented an old sewing machine factory in Bridgeport, Connecticut, and began to manufacture graphophones. This realized Edison's worst fears, for not only did Tainter have a good machine—some said that the Edison perfected phonograph was a poor copy of the graphophone—but he now had the facilities to get to the market first.[27] While the Edison Phonograph Works was under construction, his Bloomfield factory was at full production, but this meant only a trickle of hand-assembled phonographs to exhibit to the press and prospective buyers. It was not until the fall of 1888 that the first assembly lines were completed at West Orange and the workers hired.[28] In the meantime, Edison could only hope that Tainter would not sweep the market with his improved graphophone.

Edison's plans for the manufacture of his phonograph were as ambitious as everything else he did at West Orange. He was going to make 200 phonographs a day by using a new system of manufacture that had first been developed for the production of firearms, claiming that "this machine can only be built on the American principle of interchangeability of parts like a gun or sewing machine."[29] The American system of manufacture, as it was called by suitably impressed Europeans, called for the breakdown of a product into parts, each of which was made to a precise standard. The final product was assembled by unskilled labor who bolted or screwed the parts together. This replaced the old system where a craftsman fashioned the product out of parts that were individually made for each job. American machine tool manufacturers had developed precision machines to mechanize the production of parts. Their lathes, milling machines, and universal grinders had been used successfully to mass produce watches and sewing machines. In the 1880s these machines were made available to manufacturers such as Edison and Tainter at lower prices than before—the happy result of longer production runs of standardized machine tools.[30] Those jobs that once took the time and skill of a trained craftsman, it was said, could now be done with a machine operated by a mere boy with only a rudimentary knowledge of the craft. The rule of thumb, the careful interplay of file and saw, could now be done with a universal milling machine to the same high level of precision every time.[31] Edison was part of a movement in American industry that sought to replace the skills of the work force with capital equipment and industrial organization. He claimed that much of the estimated $1 million expended on the Works went into special task machinery to produce the parts. The account records of January 1889 show an outlay of about $250,000 to establish the Phonograph Works, including $65,000 on machinery and tools and $20,000 on special phonograph tools.[32] Craft culture was still alive in the laboratory, but it was to be eliminated in the Phonograph Works.

The American system offered a high volume of output with lower unit cost. It was the basis of Edison's belief that he could manufacture much cheaper than any competitor. He was relying on his special methods of production at West Orange rather than his patents to achieve a competitive edge.[33] The West Orange complex encompassed a calculated division of labor that exploited the magnificent craft skills of the laboratory. The new manufacturing system required men of exceptional ability to set up the tools and gauges required in making standardized parts. As Edison explained in his original plan for the lab, its function was to provide the patterns and models that embodied the highest level of skilled work, while the Works executed the production and assembly of parts.

Charles Batchelor had begun the work of designing the tools to shape and drill parts of the phonograph in the temporary lab in Harrison in 1887. He worked out the arrangement of the drill press required to drill the base plate and made up the specifications to grind the cast metal parts of the upper works. He designed the holding attachments that held the cutting tools on the lathes. Edison and Batchelor kept in close contact with the makers of machine tools whose technical advances had made the American system possible. The grinding machines delivered by companies such as Brown and Sharpe gave Edison's men the precision needed in making interchangeable parts. Castings from the foundry required accurate finishing, and each part of the top works of the phonograph had to be ground to exact specifications. A slight wobble in the spindle of the mandrel that held the cylinder was sufficient to ruin reproduction. Subsequently the motor shaft and its bushes had to be ground to run true.[34]

The laboratory staff carried out the important task of designing the jigs and standard forms that guided the machine tools in the production of interchangeable parts. A carpenter's shop was established in Building 3 to make up the patterns used in production. Most of the special machine tools used to make phonograph parts were set up and first operated by the skilled men in the laboratory, and then transferred to the Phonograph Works.[35]

Increasing the scale of production from machine shop to mass production required some thought about the efficient layout of the machinery in the factory. In the days of the Newark shops, little attention was given to the placement of machines; the goal was to fill the space of the rented shops. This practice could not be continued in large factories with hundreds of machine tools.[36] In 1888 Batchelor joined Edison in working out the steps of production and arranging the machines in the Phonograph Works. Edison was omnipresent in the Works. It had been deliberately built next to the lab so he could personally supervise production.[37] Here was a strategy of control at work—Edison had only to step outside Building 5 and walk a few yards to the Works. He was in a position to oversee each part of the process—each step of drilling, grinding, and finishing—and he intended to keep improving it until he was satisfied.

One advantage of standardizing the parts of the phonograph was that it permitted constant development of the design. Edison hoped that each interchangeable part could be taken out, redesigned, and put back. He said that he wanted a design based on discrete, replaceable subassemblies "so that I can keep on improving."[38] This was a wise move because most of 1888 was spent in continually altering the design of the phonograph. Edison delayed delivering the prototype to the Works until the laboratory staff had eliminated all the bugs—the defects and minor design problems that caused malfunctions.

The factory started to produce phonograph parts in the fall of 1888, but by late November the superintendent estimated that daily output would reach only ten per day.[39] Edison and his staff were still working on the reproducer assembly, which was held back in the laboratory. The precision machining and assembly needed for the delicate diaphragms were done by the most skilled of Edison's men in the machine shop on the second floor of the lab. The factory assembled the top works of the phonograph and then waited until laboratory staff delivered the completed spectacle frame with the recorder and reproducer assemblies installed. Full-scale production of machines in the Works did not begin until the spring of 1889, and by June only a total of 1,200 had been completed. The daily output of the Works was less than fifty machines—much less than the original production target of 200 a day![40]

Edison had been highly optimistic in his plans to set up mass production of phonographs at West Orange. The American system of manufacture with interchangeable parts was easier said than done. He admitted later that "so called interchangeable parts have to be adjusted in place."[41] He found out that using unskilled labor was often a false economy. The boy operating the universal milling machine needed supervision from a man who knew the craft. In unskilled and inexperienced hands, machines broke down, tools were damaged, and parts were not interchangeable. The work force required an understanding of the product and the steps of its manufacture. Breaking in the workmen was responsible for many of the delays.[42]

Training the men took time and money. Practicing the new system of manufacture required more; discipline and order were required from both the men in the factories and the muckers in the lab. Edison wanted his laboratory to innovate with speed and flexibility. It had done so, but the sheer number of changes overwhelmed the men trying to establish order in the Phonograph Works. The machine shop culture of the lab, with its emphasis on experimenting until perfection was reached, was the opposite of establishing one standardized design and keeping to it. It was unrealistic to entertain thoughts of mass production when the prototype was being continually changed. Edison never did grasp the concept of good enough in the process of handing over a prototype to the Works. He was never satisfied, explaining to a financial backer that "we saw so many things that could be changed to make it more convenient to the customer" that the transfer was delayed.[43] The close proximity of the Works provided the temptation for him to walk over to the factory and make changes on the workbench, thus disrupting the efforts of the Works engineers to stabilize production. These practices, along with bad habits such as sending half-completed drawings to the factory, became commonplace in the race to get the phonograph into production. To other engineers designing equipment for manufacture, this amounted to suicide.[44] Charles Batchelor knew this, but who was he to dampen the enthusiasm of the Old Man in the middle of a frenzy of experimentation? The bad habits formed in the first years of the Edison Phonograph Works proved difficult to eradicate.

Marketing the Talking Machine

The failure to achieve mass production by the end of 1888 severely hurt Edison's finances and disrupted the time frame of his manufacturing strategy. The expenses of three years experimenting had exhausted his financial reserves and left him with little to show for it. The profits he expected from manufacturing phonographs had not materialized. He had paid for the Phonograph Works with his own money, liquidating much of his holdings in the electric light industry to do so. He also put up the money for the phonograph experiments and made loans to the Works to keep it operating. The alarming shortage of money forced him into emergency measures to raise cash. Samuel Insull was given the thankless task of peddling Edison's notes to businessmen and venture capitalists.[45] A search was mounted to find partners in the ruinously expensive effort to perfect the phonograph. In the spring of 1888, a group of prominent bankers and investors were invited to the laboratory to see the machine. The group was led by Jesse Seligman, senior partner in the banking house of J. and W. Seligman. Unfortunately, the phonograph would not work for the distinguished guests, and Edison lost a golden opportunity to finance the further development of the machine.[46]

Financial salvation was finally found in the form of Jesse Lippincott, a venture capitalist with an eye on a monopoly of talking machines. After purchasing the American Graphophone Company (the successor of Bell and Tainter's Volta Graphophone Company), Lippincott set out to acquire the Edison Company, paying $500,000 for Edison's interest in September 1888.[47] Lippincott's new organization was named the North American Phonograph Company. Its goal was to produce a commercial dictating machine to be leased to businesses. It franchised local companies and allotted an exclusive sales area to each one. Edison and Tainter manufactured their machines and supplied them to the companies. The customers could choose between the phonograph and the graphophone, and a royalty on each sale was paid to Edison and Tainter. The North American Company made a research contract with Edison to ensure further development of the technology. It gave him $15,000 in the first year and slightly less in each successive year. Lippincott had entered the phonograph fiasco and delivered Edison from what could have been total financial disaster.[48]

Several uses had been anticipated for the tinfoil phonograph. It could be employed in schools, courtrooms, and offices to record and reproduce the human voice. The perfected phonograph was to be used for correspondence. As early as October 1886, the experimental team developing the machine at the Harrison Works realized that the complexity of the phonograph precluded its general sale to the public. It was to be rented to businesses that would pay for its service and periodic adjustment.[49] This policy was maintained by the North American Company that expected its franchisees to provide repair service.

The fragility of the machine undermined all plans for its marketing; as more were put to use, more bugs emerged. One of the first business users of the perfected phonograph was Tate, who was pushed into employing it for his correspondence. He found that some of the chemicals in the wax cylinder tarnished the metal stylus of the recorder and rendered the machine useless after a few days of use.[50] The phonograph mechanism turned out to be highly sensitive to dirt and vibration, either of which could bring it to a halt. The laboratory staff had not been able to anticipate the abuse that the machine would receive in daily office work. Even the brushes of the electric motor had to be redesigned to avoid damage by people who turned the cylinder the wrong way.

Edison and Batchelor found out that mass production of a complex product required unrelenting attention to detail. The smallest mistake in manufacture meant that the machine would not work in the inexperienced hands of the franchisees of the North American Phonograph Company. There was also no lack of inexperienced hands in the Phonograph Works, for Batchelor found that training the new employees was a long and often frustrating task. Every stage of manufacture had to be closely monitored to ensure that bugs did not creep in; even the packing case had to be redesigned to prevent damage during shipment. Edison's men were locked into a cycle of constant innovation: product development, manufacture, delivery, feedback, and more development. The laboratory served as a conduit for information about the operating problems of the talking machine. All complaints, problems, and suggestions filtered through the North American organization to arrive at Building 5 where the harassed staff tried to keep up with them.

Many of the problems with the phonograph were traced to the power source. It soon became obvious that the battery and electric motor combination was totally unsuitable for office use. The primary battery that powered the motor was an electrochemical cell that required replacement of the chemicals once the electricity was discharged. Since the battery's charge ran down quickly, the user often faced the unpleasant job of filling the primary battery with malodorous, corrosive chemicals. Edison had to get more power from the primary batteries. This problem started him on a series of experiments that eventually led to a new type of battery. In the meantime, his customers struggled with leaky batteries, frequent breakdowns, and expensive replacement chemicals.

The other major complaint from users of the dictating machine was its poor reproduction. One described it as "but a parody of the human voice" that required "careful adjustment by a practiced hand" to be understood.[51] The feedback coming to the laboratory indicated that the phonograph was too complicated for office use; recording required a level of precision and manual dexterity that was usually lacking in an office worker. The staff of *Atlantic Monthly* magazine tried out a phonograph and concluded that it was far too delicate to be left to "the office boy or typewriter girl."[52] The spectacle assembly had to be handled carefully. Lowering it onto the cylinder needed a deft touch to avoid damaging both assembly and cylinder. The number of adjustments of the recorder/reproducer assembly was reduced, but the complaints about the difficulties of operation continued.[53] In May 1889 Edison acknowledged that there were still defects in his phonograph, but said they were being remedied "as fast as human flesh can stand it."[54] Each subsequent model of the phonograph had better reproduction, but it still did not satisfy enough customers to warrant the large-scale production planned in 1887. In the fall of 1889, orders were running at only a couple of hundred a month. As soon as the year lease ran out, customers were returning the machine to the North American Company.[55] The failure of the dictating machine idea forced Edison to look for other uses for the phonograph.

It was totally out of character for him to direct all his inventive effort into one product. He was always thinking about several projects at once, and the extensive facilities at West Orange continually tempted him to investigate tangents. In 1888 he told his lawyer, Richard N. Dyer, that he had many other ideas worth patenting in addition to the dictating machine: a toy phonograph for use in talking dolls, coin-slot amusement machines, and improvements in clockwork and hand-turning phonographs. He also experimented freely on the basic configuration of the phonograph, designing several disc phonographs, among them one with an articulated recorder/reproducer assembly. The laboratory staff developed several alternate methods of running the machine, including spring motors, electric motors wired to run on household current, motors powered by water, and a treadle system identical to the sewing machine and the graphophone.[56] "The amount of work to get phono out has been stupendous," Edison wrote in May 1889, and it was reflected in the enormous number of phonograph patents he obtained: thirty-one in 1888, twenty-two in 1889, and twenty-two in 1890.[57] The diversity of the research was both his undoing and salvation. A concentrated, focused effort on the business machine might have brought quicker results and speeded up the manufacturing process, but when the business model proved to be a failure, there were several other versions of the phonograph idea ready to be developed.

With the dictating machine in trouble, Edison looked to the doll to keep the business afloat. The doll was the first attempt to market the phonograph invention; experiments were begun in this area at the Menlo Park laboratory soon after the tinfoil phonograph was produced. These experiments were reestablished at the West Orange laboratory as early as December 1887.[58] Charles Batchelor, the father of two girls, took a special interest in the doll experiments, developing a small phonograph mechanism and fitting it into a doll that was less than two feet high. Batchelor took a one-half inch slice off the wax cylinder of the business model and attached it to a frame with a spring pushing against a drive wheel. A small diaphragm, with stylus attached, was placed into the groove of the pre-recorded wax cylinder, and this reproduced a childlike voice into a tiny metal horn. An important design criterion for the doll was cost—as usual Edison was hoping for a mass market—and an experimental project was soon begun on a "smaller cheap doll phono." In 1889 a group of businessmen from Boston established the Edison Toy Doll Company to market the doll. Edison and Batchelor received stock in the company in return for their inventive efforts, and the expenses of their experiments were billed to the company in Boston. The first doll patent was applied for in the summer of 1889, but it was not until the end of the year that the laboratory began to design tools and prepare for manufacture.[59]

While cautious businessmen might have had reservations about the reliability of a small doll mechanism, Edison was confident that he could produce a working doll phonograph. He planned to manufacture 500 a day in the Phonograph Works (fig. 4.4). There was nothing wrong in the marketing strategy; the talking doll proved to be so popular that orders soon outstripped supply. The problem was in the small phonograph movement inside the dolls. The diaphragm/stylus assembly simply would not stay in the fine groove of the wax record. Consequently most of the dolls failed to work properly, steadfastly refusing to talk for their owners. One dealer reported that 188 dolls were returned out of 200 sold.[60] Despite the strenuous efforts of the laboratory, the talking doll could not be made to work. This was an important project in the manufacturing strategy, for half the capacity of the Phonograph Works was devoted to dolls by 1890. Continual failure in the engineering effort led to the abandonment of the venture by the end of the year. Manufacture ceased, and Edison disassociated himself from the Talking Doll Company.

The doll mechanism was too fragile to be used as a toy. Reliability was also the chief problem with the business phonograph. Frequently, a phonograph that worked perfectly in the factory would grind to a halt after a few days of operation. The machine needed continual adjustment or it would break down. Similar problems occurred with the toy doll mechanisms—the dolls would talk perfectly when tested in the laboratory but would be out of order and useless by the time they got to the customer. Two years of intense effort at the West Orange lab had not been able to produce a technology equal to the demands of office workers or children.

While poor reproduction remained a constant gripe of business users, it did not seem to affect the general public's enjoyment of recorded music.[61] The fidelity of reproduction was not as important

Figure 4.4. Manufacturing the toy doll in the Phonograph Works about 1890. Racks of the phonograph mechanisms are placed in front of the workmen in the first two benches. The thin disc-like cylinders are attached by a vertical strip to the funnel-shaped amplifying horns made of sheet metal. Young men assemble and fit the mechanisms into the dolls, while the young women assemble and dress the dolls.

to an amusement listener, who (unlike the stenographer and typist) did not have to strain to hear each word. The entertainment value of the phonograph had been demonstrated by the showmen who attracted crowds to listen to the machine and pay for the privilege. Edison quickly developed a portable phonograph outfit for use by traveling phonograph exhibitors. This consisted of a basic perfected model with hearing tubes for up to fifteen customers and a collection of fifty pre-recorded cylinders of vocal, instrumental, and miscellaneous entertainment "of the finest quality."[62]

Penny arcades were set up by small businessmen to sell this entertainment on a larger scale. The laboratory staff designed an automatic player with a coin-operated mechanism that could be installed in arcades and other public places. The coin-slot, or nickel-in-the-slot, phonograph proved to be a highly successful product. It attracted attention, increased user familiarity with the machine, and maintained a steady level of sales. The coin-slot machine kept the phonograph alive in the mind of the public and gave much needed work to the laboratory and the Works.

Edison's commitment to the single-user dictation market in his 1887–89 patents is shown in his emphasis on speaking and listening tubes rather than horns that provided a means of acoustic amplification.[63] His staff had recorded music in his laboratory for experimental purposes rather than in anticipation of a market for this use of the talking machine. A user could record music or speech on the phonograph and then listen to it with a hearing tube, but this hardly constituted entertainment. As soon as the feedback from the local companies indicated that there was a market for an amusement phonograph, Edison responded. He began to consider the design of a phonograph for entertainment in 1889, speculating about a handsome wooden cabinet to make the machine look like a piece of furniture and developing a recorder/reproducer assembly suitable for music. The latter required a much louder play back than the dictating machine, and consequently Edison experimented with methods to amplify the sound emitted from the diaphragm. In November 1889 he applied for a patent that incorporated the fruits of the many experiments made during the year. This new design, an improved perfected phonograph, incorporated several changes in the top works, making the machine easier to operate. The spectacle with two diaphragms was replaced by one basic holder that could carry a variety of reproducer assemblies, including a single dual-purpose recorder/reproducer for the business market, and separate reproducers and recorders for the amusement market. This arrangement gave flexibility to the laboratory staff who could now design reproducers and recorders of different weights and composition. The reproduction of music (rather than the voice alone) required more volume and sensitivity over a wider range of sound frequencies. These were conflicting goals while the spectacle format was used

Figure 4.5. Theodore Wangemann recording piano music in the studio on the third floor of the lab. Recording horns of different shapes and sizes were tried. In the background a workman services a large recording machine.

with the soft wax cylinders—the heavier the weight of the reproducer, the greater the volume, but heavier tracking pressure marred fidelity and damaged the wax surface. Moving to separate devices for recording and reproducing functions opened the way for better sound reproduction.[64]

The unexpected popularity of the coin-slot phonograph created an urgent need for a supply of pre-recorded cylinders and changed the development path of the talking machine. The first step was to convert part of the top floor of the laboratory into a recording studio. Theodore Wangemann could often be seen there, sitting at his piano, surrounded by several different phonographs, sometimes accompanied by a four-piece orchestra. An accomplished pianist, he often recorded classical music on the phonograph. When it reproduced his piano playing, filling the room with beautiful music, all who heard it were most impressed, especially the Old Man. For the first time Edison realized that his machine could be used to recreate what he called "good" music, played by an artiste, and fit for something more refined than amusement parlors and bawdy houses[65] (fig. 4.5).

The initial concept of a talking machine used for amusement was that it would record the musical entertainment enjoyed in Victorian sitting rooms; it was to complement rather than replace the piano. The insatiable demand for studio recordings completely altered this plan. The new field of opportunity brought with it a demanding technical problem. A cheap method of duplicating these prerecorded cylinders had to be found before the amusement phonograph had a commercial future. The only way to do this in the 1880s was either to repeat the performance several times or copy the cut on the cylinder by a pantograph process. These two options were expensive and time consuming and produced poor quality recordings. Here Edison demonstrated his mastery of the business of innovation. He had lost his chance to monopolize production of the machines, and he anticipated increased competition in this area as more phonographs were sold. Yet when the entertainment potential of talking machines became evident, he recognized a future market for cheap, mass-produced recorded cylinders that would make this business much more profitable than selling the machines. He, therefore, set an experimental project in motion to duplicate recordings by making molds of cylinders and made it clear to the North American company that "What I want is the manufacture of duplicates."[66]

Edison was so enthusiastic about the future of the amusement phonograph that he began construction of new buildings at West Orange: one to produce wax cylinders, and one to construct the wooden cabinets to make the phonograph look like a piece of furniture.[67] This expense was added to the cost of several experimental projects that continued throughout the 1890s. In spite of the meager returns on four years of intensive work (1886–90), Edison was still prepared to invest in the amusement phonograph.

The future of the business phonograph and the North American Phonograph Company looked dim by the end of 1890. Edison ordered the Phonograph Works to finish all machines under contract and then concentrate on coin-slot machines to be sold directly to entrepreneurs (fig. 4.6).[68] By 1891 only nineteen local companies remained out of the original thirty-three franchised by the North American Phonograph Company, and the press was wondering about the disappearance of the Edison phonograph.[69] Edison was wondering about the estimated $540,000 of his own money sunk into the venture.[70]

The strategy of becoming an independent manufacturer of new products had met with disaster. His own ambition and the powerful influence of the machine shop culture had prevented him from

Figure 4.6. Assembling the coin-slot phonograph in the Works about 1890. A completed model M phonograph can be seen left foreground in front of the finished product. The customer put the coin in the slot on the right side of the machine and listened through the ear tube.

developing a commercial prototype that could be mass-produced in the Phonograph Works. The total failure of the business phonograph idea ruined his plan to make the new laboratory the center of a vast manufacturing complex. It forced him into more contract research to support the costs of operating the laboratory. This financial necessity fixed the electric light industry as the major customer of the West Orange laboratory and made electricity the central concern of the laboratory staff. Edison's business career was still dependent on the success of his electric lighting system.

CHAPTER FIVE

Edison's Laboratory and the Electrical Industry

he reduction of the cost of electricity was an integral part of Edison's dream of universal electric light and power. In the years after the opening of the Pearl Street station in 1882, electricity use grew slowly. The pioneer system did not make a profit in its first five years of operation, and the first reduction in the price of its current did not occur until the end of the decade.[1] Supply systems remained in the heart of great cities, serving businesses, places of entertainment, and the homes of the wealthy. The high price of electric lighting prevented its rapid diffusion and gave the gas companies some cause for optimism. As Edison expected, electricity was not cheaper than gas in most cities. The future of the electrical industry lay in reducing the cost of generation and distribution. Edison had this strategy in mind when he planned the construction of his laboratory in West Orange.

Another threat to his lighting companies had appeared by 1887; after spending the middle years of the 1880s in the doldrums, the electrical industry was growing and with it competition. Several large manufacturing companies, such as Brush and Thomson-Houston, had laboratories to develop their own technology. The incandescent lamp business was becoming very crowded. The Edison Lamp Company estimated that it was producing around 4,000 bulbs a day in 1888, compared with the 11,000 manufactured by other companies.[2] Edison had no illusions about the competitive nature of the industry he had created, telling one associate: "Try everything you can towards economy. No one is safe in this cold commercial world that can't produce as low as his greatest competitor. No matter how much money you are making never for an instant let up on economizing."[3]

Guided by this philosophy, Edison successfully reduced the cost of his incandescent lamp. In 1880 each lamp cost $1.21 to produce. By 1883 this figure had dropped to 30¢ and was still falling to 28¢ in 1889, and 22¢ by the end of the decade. He hoped to bring the cost down to 15¢ in the 1890s.[4] The mass production of incandescent bulbs was a severe challenge to Edison's application of the "American system" of manufacture. He rose to it magnificently, refining each step of the complex process that included blowing the glass bulb; fabricating the carbon filament, fixing it into the base, and then sealing it in the vacuum inside the bulb. He examined rejected bulbs to locate problems in the production process. He enthusiastically started on the design of machinery to take the place of men on the production lines. In 1889 a complicated machine was introduced into the Lamp Works to seal the filament support into the glass bulb. The first filaments had been prepared by hand out of natural fibers. They were replaced in the 1880s by nonstructured filaments of cellulose, which offered improvements in candle power and ease of manufacture. Edison designed a machine to mass produce filaments by squirting cellulose through a die "like macaroni."[5]

These machines reduced the skill required for manufacture and gave him some comfort with the rising demands of organized labor. In the 1880s labor had begun to find a voice in America. The eight-hour day was a heated issue at the time the West Orange laboratory was built and violent strikes were becoming commonplace. Labor unrest had caused disruptions at the Edison Machine Works and the Lamp Works.[6] Cuts in pay were no longer tolerated. As an industrialist of the Gilded Age, Edison's attitude to organized labor was typically harsh. Only in the friendly confines of the laboratory did a more enlightened atmosphere exist.

The muckers in the laboratory made their contribution to lowering the costs of electricity by making the lamp burn brighter and last longer. The first bulbs suffered from poor reliability and short life.[7] This was especially expensive for those lighting companies that offered free replacements of burnt out bulbs. Another problem was the blackening of the inside of the bulb by a carbon deposit that led to a loss of candle power. Edison found that the carbon particles were carried by an electrical current from the filament to the glass. (The "Edison effect" was caused by a flow of electrons moving without wires in the vacuum of the lamp.) The significance of this discovery was lost in the many other experiments to increase the resistance of the filament to bring more light with less current consumed. Characteristically, Edison did not stay within the confines of natural carbon in his search for a better filament. In addition to testing the carbonizing properties of several hundred materials, he experimented with a wide range of inorganic chemical compounds that might serve as filaments, including several incandescent oxides[8] (fig. 5.1).

The most dramatic improvement in electrical equipment occurred in the area of generation, where new knowledge of electromagnetism was applied to dynamo design. The first dynamos for the Edison system, the "long-legged Mary Anns" with their tall vertical field magnets, were designed at Menlo Park "in the days when guessing was a substitute for mathematics" and size was thought to increase the magnetic field.[9] A deeper understanding of the magnetic circuit, inspired in part by the theories of James Clerk Maxwell, enabled electrical engineers to obtain more powerful magnetic fields. The greatest success was scored by John Hopkinson, a scientist retained by the English Edison companies who was influenced by Maxwell's mathematical model of the electromagnetic field. Hopkinson shortened the field magnets of the Edison dynamo and did away with the multiple cores, improving the magnetic circuit and bringing an increase of efficiency of around 95 percent.[10]

The staff of the West Orange laboratory kept abreast of these developments in their program of evaluating equipment used by Edison companies and their competitors. When commissioned to design a new line of central station dynamos, Arthur Kennelly produced a machine that incorporated theoretical advances and the practical knowledge gained in operating supply systems. He joined the movement away from the format of two magnetic poles (north and south) to larger multipolar dynamos with four or more poles arranged alternately north and south. Larger prime movers were being introduced into power stations, and the new multipolar dynamo was found to have twice as much output per pound weight as any other comparable generator.[11] The dynamo developed at West Orange also met Edison's goal of increased efficiency and lower manufacturing costs. He informed Henry Villard in 1889 that the price of "all electric lighting machinery is entirely too high now."[12] The West Orange facility was to redesign each part of the system, from socket to central station generator, for greater economy in manufacture.

To reduce the cost of electricity, most attention was on the production of current until the end of the 1880s when it was realized that economic operation of supply systems required exact knowledge of the costs of production. Looking back, it is almost inconceivable that engineers (to use the word loosely) ran central stations with

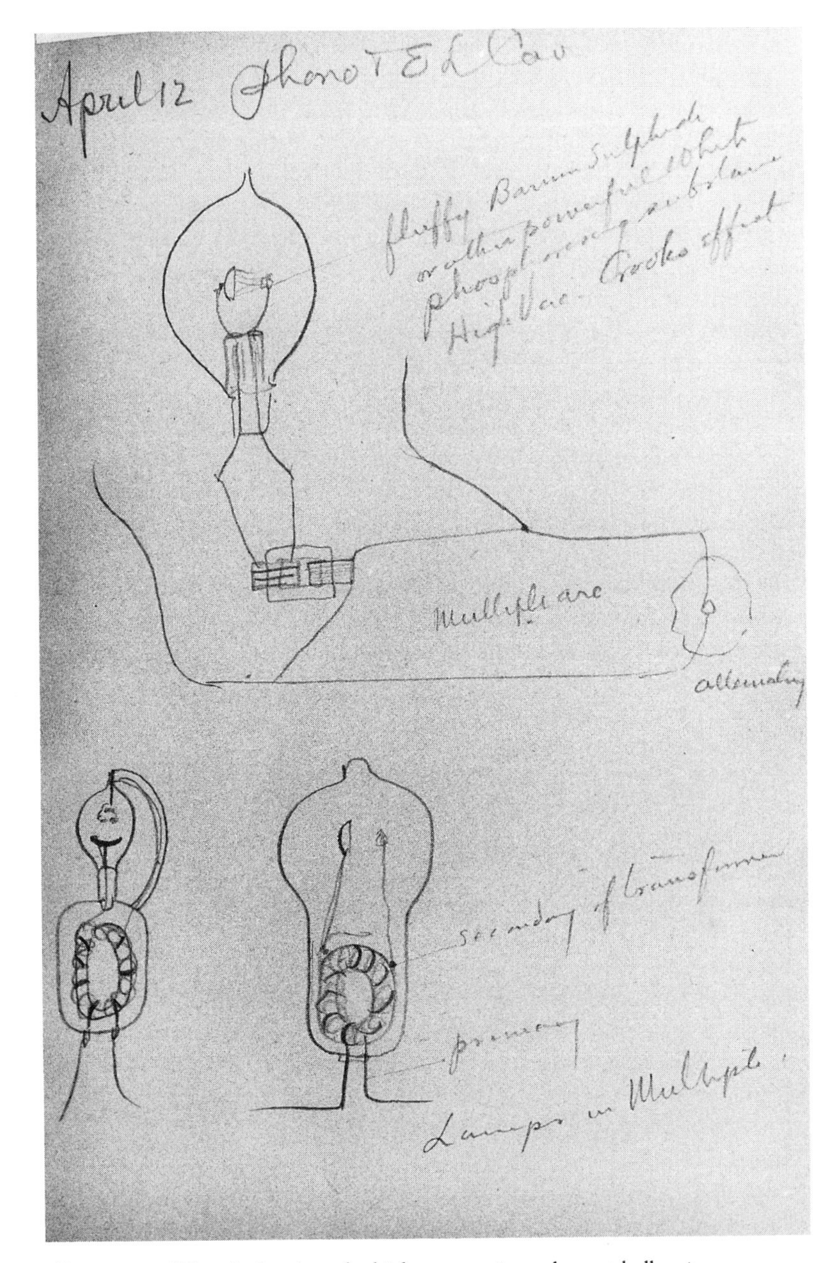

Figure 5.1. Edison's drawing of a high vacuum incandescent bulb using fluorescent salts on the filament (Notebook N880202, ENHS).

only a rough idea of how much electricity they were generating. Technological development alone was not the key to growth in the 1880s; the electrical press and engineering proceedings were also concerned with establishing the cost of generating electricity, down to the last ton of coal and pint of lubricating oil. A profitable central station operation required the skills of a cost accountant as much as an electrical engineer. It was not until the 1890s that the concept of a balanced load, which employed the dynamos of the central station to the fullest extent, was appreciated as the critical element in staying in business. Electrical measurement was the starting point of the new art of creating a profitable load and controlling the costs of generating electricity.[13]

The West Orange laboratory played an important part in the advance of measuring equipment from the fragile laboratory instruments like the tangent galvanometer to the direct reading meters. Many laboratories in the late 1880s were working on this problem, including those of Weston and Thomson. Weston was frustrated by the slow and inaccurate methods of electrical measurement in 1886, when it took a week to measure a dynamo's efficiency, and "an accurate reading was almost an act of God."[14]

Measuring the electricity consumed by the customer was an important element in figuring load. Edison had designed an ingenious electrochemical meter that consisted of two sheets of zinc immersed in zinc sulphate. The passage of current transferred zinc from the positive to the negative sheet. The sheets were then weighed and the difference in weight provided the means to calculate how much current had been used. This meter was typical of the instruments of the early 1880s; it was accurate and probably worked well enough in the laboratory, but under normal operating conditions it was cumbersome, expensive, and inaccurate.[15] Advances in the understanding of electromagnetism and in the manufacture of small springs and magnetic materials made a mechanical meter possible in the late 1880s. Edison was not the only one to see that electromagnetism could be used to drive small motors in the meter to record the passage of current. Elihu Thomson produced such a meter in 1889, and the race was on to bring out a mechanical meter for general use. Although Edison and Kennelly patented several meter designs, the Thomson-Houston meter won, achieving widespread adoption within the Edison system.[16]

One of the top priorities of Edison central stations in the late 1880s was instruments to measure their output. The first central stations were often crude installations of dynamos in rented premises. Hastily planned and equipped, the typical station had a primitive switchboard (with a wooden back that often caught fire) and few, if any, meters. In 1888 when a committee met to discuss some general engineering problems confronting the Edison companies, at the top of the agenda were pressure meters (to record voltage or the electromotive force of electricity) and ampere meters to show the strength of the current.[17] Arthur Kennelly led the way in the more precise measurement of electricity and the improvement of meters used in the Edison system. As professor of electrical engineering at Harvard and president of the Institute of Electrical Engineers, he worked on standardization and made a major contribution to electrical development.[18]

The presence of experts like Kennelly made the laboratory the technological nerve center of the Edison lighting industry. The main laboratory building housed the impressive library of technical literature and journals, the blueprints of many central stations built by the Edison Central Station Construction Department, and the results of tests of all parts of the Edison system. The laboratory served as a repository of operating experience that was of great value to the individual lighting companies. The staff of the galvanometer room (Building 1) uncovered new electrical knowledge that was of common interest to all members of the Edison lighting industry (fig. 5.2). Kennelly estimated the conductivity of cables and determined the safe carrying capacity of the wiring in use. The lab staff examined the physiological effects of electricity on the human body in order to reduce the injuries suffered by line men. The West Orange laboratory was a technical resource center where the lighting companies' operating problems, ranging from balancing their load to finding out the causes of explosions in junction boxes, were brought for solution.[19]

The most important advice dispensed by the laboratory staff concerned the extension of the distribution network of central stations. Edison had produced a mains and feeder system for the Pearl Street installation that economized on the amount of conductors used. The enlargement of the system took careful planning of the position of the mains, which brought electricity from the station, and the feeders, which took small amounts of current into local networks. The lab staff also estimated the weight of copper in the conductors, which determined the cost of expansion.[20] This service was invaluable at a time when the enlargment of the supply area was under consideration by all utilities. The experience of the Edison Illuminating Company of New York was typical of the early lighting

Figure 5.2. Dickson's photograph of the electrical laboratory (Building 1), taken soon after the West Orange laboratory opened.

companies. By 1886 it was at last enjoying an increase in earnings, but the profits lay in extending its distribution network to reach more customers. The company was inundated with requests for service from residences and businesses uptown from the Pearl Street station, but its generating capacity was too small to meet this demand. This was a barrier to the growth of both the Illuminating Company and the electrical manufacturers; the demand for lamps, motors, and fittings would be low until electricity was available uptown.[21]

The Pearl Street system of 1882 had been based on low voltage direct current that gave it a supply radius of one square mile around the central station. The small supply area was appropriate for a pilot system where corrections and repairs had to be carried out continually, but it became a liability as the popularity of electricity grew. The problem with low voltage direct current was that the size of the copper conductor had to be increased with the supply distance, and the high price of copper made the Edison system uneconomical over a supply area greater than two square miles. This was a serious obstacle to the future growth of the system.

The sharp increase in the price of copper had not gone unnoticed by Edison, nor had the commercial potential of expanding the distribution area. Immediately after the Pearl Street station was completed, he developed his three-wire system that used a common return wire to reduce the amount of conductors needed.[22] He went on to produce more economical systems using five and seven wires. His response to the long-distance problem was the "municipal" lighting system, which used higher voltages to serve large areas. This came with a high voltage (200v) lamp, and a cutout device that would maintain current if one of the lamps blew.[23]

Alternating Current

Edison and his staff examined all possible solutions to the long-distance transmission problem, taking an early interest in alternating current (a/c). While direct current flows from negative to positive much like water or gas along a pipe, alternating current reverses its flow at regular intervals. Several historians of the electrical industry have argued that Edison had an aversion to this technology because it was more complex than direct current. The records of the West Orange laboratory show this view to be unfounded; alternating current was one of the first experimental projects undertaken when the lab opened for business in 1887.[24] Alternating current can be easily transformed from low to high voltages, and higher voltages require a smaller diameter copper conductor. This made it economically feasible to transmit over long distances. After transmission, alternating current is stepped down at the point of use. Edison developed his own alternating current system and applied for a patent for this "system of electrical distribution" in December 1886. It employed high voltage current that was transformed down to produce low voltage electricity for business or home use.[25]

One method of transforming alternating current used the principle of induction, a variation in the current in one coil of wire that causes a current to be induced in a second coil placed nearby. The great advantage of alternating current in supply systems was the ease with which it could be transformed. In 1882 the first commercial transformers were produced in Europe by Lucien Gaulard and John Gibbs.[26] This development was followed by several other a/c lighting systems based on transformers. Edison was able to use his contacts in the European electrical industry to get an initial evaluation of the new technology. In 1886 the firm of Siemens & Halske (the licensee of the Edison system in Germany) sent him a detailed report.[27] The Edison Electric Light Company also took note of these developments in Europe and bought an option on the alternating current transformer system of Zipernowski, Blathy, and Deri (ZBD). Edison had all the information to assess the practicality of the new technology. He criticized the high voltage alternating current equipment of the late 1880s for its complexity, unreliability, and threat to public safety. The high voltages used in these systems greatly increased the risk of injury and fire. He maintained an unrelenting opposition to a/c, insisting that the Edison Electric Light Company drop its option on the ZBD patents—the easiest access to alternating current technology.[28]

As a practical inventor, Edison was concerned by the energy wasted in transforming current. As a businessman, he knew that the difficulties of metering alternating current made its commercial value slight. The lack of a motor to run on the new current was an even greater disadvantage; Edison considered that the motor load was the essential part of a profitable central station operation. He decided to build the first central station in a commercial district in downtown Manhattan (instead of the more thickly populated uptown), because he expected that electric lighting could not beat the competition of gas "but hoped to meet it by selling power."[29]

He told his friend and financier, Henry Villard, that alternating current was "not worth the attention of practical men"—meaning that it had no commercial potential.[30] Edison believed that large alternating current systems would cost too much money to construct and that any profit would be absorbed by interest charges on the high investment. His evaluation of all electrical systems was based primarily on commercial considerations—the best designed central station system was the one that returned the highest profit on the smallest investment. This viewpoint was shared by the capitalists running the electrical utility industry.

Research and Business

Edison's status as a great inventor and his informal influence in the world of electricity made his West Orange laboratory the symbolic center of the Edison lighting industry. The business of innovation was carried out in the main laboratory building, where board meetings were held and important decisions made on the development of technology. All technical issues were automatically referred to the laboratory. One of the first meetings held there was to discuss a technical report on the Westinghouse alternating current system.[31] Edison and the managers of the lighting companies had no difficulty reaching a consensus that long-distance transmission of electricity was the most pressing problem facing them. All were agreed that it was essential to develop a technology "useful for bringing in power at a distance."[32] The unresolved issues concerned the form of the technical response and the cost of the research program.

The gradual reduction of Edison's financial stake in the various companies that bore his name placed him in the uncomfortable position of being both master and servant of the lighting industry. On one hand he was the founder of the enterprise and the font of all electrical wisdom; on the other he carried out research, under contract, for managers who had to integrate the work of his lab into their own business strategies and financial planning. While the Edison companies were controlled by the Old Man and their management was carried out by the boys, there was little that Edison could not do; but as these companies grew into large public corporations, their management had to establish their own priorities.

The relationship between the companies and their founder was strained by shortage of money on both sides. The local lighting companies did not have the money to buy the system outright; consequently the Edison Electric Light Company and the manufacturing companies often had to accept stock in payment for equipment and patent rights. In assuming an ever larger financial stake in the growth of electric lighting, the companies had little left over to support research. Edison had overextended himself building the Phonograph Works and perfecting the machine. The ruinously expensive phonograph venture had eroded his credit standing and made it harder for all Edison companies to borrow money. Edison had to squeeze every cent from his customers, even resorting to some of his old tricks to cover expenses, such as charging his experimental supplies to other companies. When Samuel Insull took over the operation of the Machine Works he found that it was "constantly being asked to pay bills of the laboratory."[33]

The development of better insulating material illustrates the growing tension between laboratory and electrical companies. Insulation of the copper wires conducting the electricity was an unglamorous, but vital, part of the system—its performance under harsh operating conditions helped determine the reliability and economy of supply. The first electrical wires were uninsulated; they were simply attached to wooden carriers or even to gas fittings! Conductors inside houses and offices were later covered with paper, cotton, shellac, rosin, or paraffin wax. The underground mains of the Pearl Street station were insulated by a mixture of asphalt and paraffin that was pumped into a metal pipe containing the rope-covered copper conductors.[34] Edison had chosen to place the distribution system underground, and although this made for a safer and more reliable network, it increased costs and made it vulnerable to corrosion and decay. By far the largest number of breakdowns could be traced to problems with conductors. The lighting companies soon asked for a better insulating material. In August 1887 John Kruesi of the Edison Machine Works commissioned the development work, specifying that the insulation be waterproof, fire retardant, and flexible. Although this presented a difficult problem, Edison assured Kruesi that it would be solved in the new West Orange laboratory, which had facilities "to settle anything in a short time."[35]

The search proved to be long and difficult. Numerous materials were tried but few could successfully fit all three criteria. Mixtures that insulated often turned brittle over time, and most of the compounds were easily set on fire. The unfortunate experimenters also had to keep an eye on the cost of the materials under test, because an expensive insulator would never be accepted by the lighting companies. The longer the search for a better insulation went on, the more it stressed the working relationship between laboratory and Machine Works. Edison owned 90 percent of the stock of the Edison Machine Works, but his friend and associate, Samuel Insull (now manager of the Works), had to insist on prompt and accurate billing for experiments carried out by the laboratory. Insull knew all the Old Man's financial ploys, having devised many of them while he was Edison's secretary. He knew that the accounting techniques used at West Orange favored the lab and that the longer the delay in billing, the more likelihood of Edison adding nonrelated experiments to the bill. Insull complained about the high prices, which were about twice the normal cost for materials, and argued that "some of the charges are simply outrageous."[36] To reduce the costs to the Machine Works, Insull asked the lab to return all scrap to the Works. He also insisted on bills being presented promptly and at regular periods and argued that no other concern in the United States would dare to do business in this manner. His frustration is evident in this comment to Edison: "We have spent a great deal of money on experimental work on insulation and we ought to be getting some result out of it."[37]

This emphasis on results is evidence of a subtle shift in the attitude toward research in the Edison organization. The business of innovation was moving from the laboratory to the offices of the Edison companies and the boardrooms of banks. Managers had to be convinced that the benefits of experiments would warrant their cost before authorizing any expenditure. The first question asked was about "the nature and extent of the advantages to be derived to this Company . . . and to whom will the resultant benefits accrue?"[38] These were relevant questions to ask before embarking on an expensive experimental project.

Edison had little respect for the management of the lighting companies, whom he thought were amateurs in the business of innovation. He argued that their caution restrained him from putting his ideas into practice. He complained to Villard that "with the leadened collar of the Edison Electric Light Company around me, I have never been able to show what can be done."[39] Yet the costs of developing the system had made the banks that supported the electrical industry extremely nervous about embarking on ambitious experimental campaigns. Drexel, Morgan were cautious, conservative bankers who had moved out of their traditional business in railroads to enter into a highly speculative venture. Bankers and managers had to keep their attention on the rising debt. What Edison saw as lack of vision and courage, they saw as rational decisions about limited resources.

In contrast, Edison could not forget that the companies had been founded by him to support his research. He believed that they had reneged on the initial plan to make the West Orange laboratory the center of research and development. He criticized the directors of the Edison Electric Light Company for their unwillingness to use the competitive tool he had constructed for them at West Orange and blamed them for the ground lost to the competition.[40] Edison maintained that he had to fight the battles of the lighting companies with money from his own pocket, telling Villard that half the cost of meeting the technological challenge of the competition could not be billed to the Edison lighting companies.

The role of Edison's laboratory moved from the vanguard of the electric industry to its allotted place in the many engineering services available to the companies. The rapid expansion of the industry led to the creation of separate engineering departments in illuminating companies and electrical manufacturers. This duplication of engineering services on each level of the organization was wasteful, and management began to rationalize the engineering effort. In 1888 the Edison Electric Light Company formed a Standardization Bureau to oversee the standardization of each component part of the lighting system. This led to close supervision of the many engineering and test departments that were working individually on electrical problems. After six months such bad feeling had arisen between the Bureau and the engineering departments that the demarcation lines between them had to be redefined. The Bureau was given the power to initiate all engineering changes.

The Bureau took over many functions that had once been carried out at Edison's laboratory, such as the dissemination of operating experience and the collection of information about breakdowns. The central role of the West Orange laboratory as the source of innovation remained unchanged. It was to be called in when "entirely new forms are required which the factories are not prepared to originate."[41] Yet its work was closely monitored and all costs had to be justified. The Bureau approved the expenses of experimental work and insisted on prompt delivery of all bills from the laboratory. Any delay in remitting bills was subject to cancellation of the experimental work.[42] The move to a more formal relationship between laboratory and corporate customers defined the limits of the contract research carried out at the West Orange laboratory.

The Battle of the Systems

While the Edison interests remained aloof to alternating current, other lighting companies did not hesitate to bring out commercial systems. George Westinghouse, a self-educated inventor and engineer with the same entrepreneurial drive as Edison, thought alternating current had great potential and purchased the rights to the Gaulard and Gibbs system. He commissioned the inventor William Stanley to perfect a commercial transformer system. Working in his private laboratory in Great Barrington, Massachusetts, Stanley produced a transformer with parallel (rather than series) connections in the primary coils and used it in a small pilot system. In the same way that Edison used his influence in his business organization to banish high voltage alternating current, Westinghouse overruled opposition to it within his company and set about manufacturing the transformer system. In 1886 the first installation was carried out in Buffalo, and (if we are to believe the Westinghouse Company) many more followed.[43]

Tests on the Westinghouse system carried out in the West Orange laboratory confirmed Edison's risk assessment that high voltage currents were uneconomical and dangerous, yet the rapid acceptance of this new technology made the Edison companies reevaluate their earlier rejection of alternating current. They were now in a position to commission research at the West Orange laboratory. The steps they took led to one of the most colorful episodes in the history of electricity.

The "battle of the systems" was fought between the proponents of the Edison low voltage direct current (d/c) and high voltage alternating current (a/c). It took the form of a propaganda war as each side claimed superiority for its technology. Lasting from about 1887 to 1892, the battle of the systems was also a struggle to dominate the next round of electrical development. Winning the battle of words was a prerequisite to selling the larger systems that were to replace the old Pearl Street technology. Edison's spirited defense of direct current earned his West Orange laboratory a place in history as the beginning of electrocution; the small animals dispatched by powerful surges of alternating current became a mute testament to the risks of new technology. Yet in proving that high voltages were dangerous, Edison forfeited his place as the most progressive man of his age. Harold Passer, the great historian of the electrical industry, concluded: "In 1879, Edison was a brave and courageous inventor, in 1889 he was a cautious and conservative defender of the status quo."[44]

Edison's position on the risks of new technology had been formed in his days as an electrical pioneer, when he realized that large lighting systems would always have to operate in a regulated environment. It was no accident that the aldermen of New York were invited to Menlo Park to approve the lights being developed for the Pearl Street system. The issue of the safety of electricity was first raised by the gas companies in their efforts to discredit the new entrant in the lighting industry. The lessons learned during the first uncertain years of electric lighting stayed with Edison for the rest of his career. The first was that a new technology had to have the public's confidence to achieve commercial success. The second was that one well publicized accident could seriously hinder the further diffusion of a technology. The close calls in the first years of electric lighting confirmed his commitment to the engineering criteria of safety and simplicity. The fuse, or safety wire, he invented for the Pearl Street station became standard in all electrical installations thereafter.[45] Edison believed that the manufacturer of electrical equipment should have a craftsman's pride in the quality of his work. His electrical lighting system was marketed on the basis of its safety and reliability. Edison, and his laboratory, stood behind the equipment bearing his name—the "safest, most convenient, most efficient" lighting system, and "the only one that is not absolutely injurious to the health of the people using it."[46] The safety issue could also be used as a marketing tool to use against competition.

The many fires and deaths caused by accidents with arc lights and their high voltage supply in the 1870s were an indication of the special risks of electricity, and it was then that Edison began his lifelong opposition to high voltages. He knew that the notoriously poor workmanship of the electrical contractors, compounded by wear and tear on the equipment, created hazards with even the safest system. The risk of injury and fire became a major issue in the evaluation of alternating current technology, because as voltages increased so did the potency of electric shock. Alternating current was the most dangerous technology of its time not only because of the high voltages used but also because it enabled the creation of larger electrical systems that put more people at risk. Edison foresaw a time when whole cities would be threatened.[47] There can be little doubt that the a/c system introduced by Westinghouse was complex, experimental, and dangerous. Edison anticipated dire results: "Just as certain as death Westinghouse will kill a customer within six months after he puts in a system of any size. He has got a new thing and it will take a great deal of experimenting to get it working practically. It will never be free of danger."[48]

The great fear of the Edison interests was that a major disaster involving alternating current would harm the whole electrical industry and produce a backlash of legislation and fear that would hinder, or even stop, the spread of electrical technology. The demise of the English electrical industry was blamed on one piece of legislation, the Electric Lighting Act of 1882. The horrifying specter of municipal ownership of utilities was rising on both sides of the Atlantic, and with it were fears that all the capital invested in electric lighting might become public property.[49] Edison had more than money at stake; no other person was as closely associated with electricity in the public mind, and any accident would reflect on him and all the Edison companies.

The Westinghouse Company had no such reservations about the danger of high voltage current and began to market a/c equipment vigorously in 1886. It also used the Sawyer-Mann lighting patents to challenge the Edison monopoly of direct current incandescent lighting. By the late 1880s the two companies were locked in a bitter struggle. In some markets, such as New York and New Orleans, the Westinghouse licensee faced a well-entrenched Edison operation. In other cities the two systems were represented by well financed lighting companies that fought for domination of the market. This was the case in Boston where the Edison Electricity Illuminating Company (established in 1886) faced the Boston Electric Light Company. The competition was intense; large customers were offered discounts up to 60 percent. It soon became evident at the headquarters of the Edison Electric Light Company that the opposition was prepared to sell electricity at a loss to win lighting contracts; many of the Edison franchisees reported to the parent company that Westinghouse agents were pricing their current well below cost and these "ridiculously low prices" were winning customers.[50] Like Edison before him, Westinghouse was ready to take a loss to promote a new electrical technology.

The Westinghouse organization claimed that its system had a lower first cost than Edison's, bringing an outraged response from the West Orange laboratory, where a detailed costing of the Westinghouse system had been undertaken. The two parent companies fought a war of words, each attempting to promote its own system and discredit the competition. The Edison Electric Light Company under the presidency of Edward H. Johnson made every effort to refute the "extravagant and fictitious" claims made by Westinghouse. Johnson went into battle under the banner of "Patents, Investment, Futility of Guarantees, Danger and Moral"—stressing the strong patent position of Edison and the higher standards of his engineering.[51] Edison, who loved a fight and considered the Westinghouse Company to be the enemy, enthusiastically joined the fray.

The long-distance transmission capabilities of high voltage current gave Westinghouse the edge because he could claim the ability to reach any customer. The Edison companies believed that Westinghouse was recklessly pushing a system that was technologically and economically unsound, but they feared that his army of sales people would finally overcome them. Their response was to discredit the claims made by the competition; Edison himself was convinced that the "shyster" Westinghouse was in fact marketing an unsafe technology.[52] The series of experiments designed to prove this point became the best known episode in the battle of the systems and put the West Orange laboratory on center stage.

Edison did not originate the investigation into the killing powers of alternating current; he was drawn into it by a New York state commission investigating humane methods of capital punishment. In December 1887 the commission polled the leading figures in the electrical world on this issue, and Edison was one of the respondents. According to Arthur Kennelly, the chief electrician at the laboratory and the leader of the team investigating the effects of electricity, they had not distinguished between the two types of current in their experiments up to this point. A reporter from the New York *Herald* asked how much alternating current it would take to kill a dog. Edison told Kennelly to find out.[53]

Edison was not the only one interested in the effect of electricity on the human body. The therapeutic use of electricity had begun well before the incandescent bulb and its popularity grew steadily as electricity became more available. Among the people carrying out research in this area was Harold P. Brown, a self-styled consulting engineer and a rabid opponent of Westinghouse's alternating current system. Brown had no connection with the Edison enterprise other than his employment by an electric pen company in Chicago. In the spring of 1888 Brown began a publicity campaign to draw attention to the "large number of deaths caused by careless methods . . . on the part of many electric lighting companies." Harold P. Brown's "movement in favor of greater care for public safety" was aimed directly at alternating current.[54] He was steered toward Edison by F. S. Hastings, the secretary of the Edison Electric Light Company, who asked if Brown might use the West Orange laboratory for his experiments. Edison had no objection; many private individuals were welcome to work in his lab in the tradition of the old machine shops. Brown's colleague Dr. Petersen, a nerve specialist from Vanderbilt Hospital, was also welcome. Neurology fascinated Edison and he probably saw the electrocution experiments as an interesting diversion from the arduous work of perfecting the phonograph. Edison could have given Brown's work only a passing glance because he was fully committed to the phonograph. While Brown carried on his notorious electrocutions at West Orange, the

Figure 5.3. The dynamo room of the West Orange laboratory, where the infamous electrocution experiments of the "battle of the systems" were carried out.

resources of the laboratory were being thrown into one last desperate attempt to finish the phonograph campaign by Christmas 1888.

The Edison Electric Light Company had more than a passing interest in Brown's experiments, considering them to be of very great importance. Hastings mobilized the facilities of Edison's laboratory throughout 1888 to enable Brown and Petersen to prove that alternating current was dangerous. Arthur Kennelly reported that the experiments were done in a hurry, with makeshift equipment, and the whole process normally took less than an hour. They were carried out at night to prevent the curious public from attending and to spare Kennelly's valuable time, which during the day was taken up with more important work on batteries, motors, incandescent lamps, meters, insulation, and dynamos.[55] The climax to this series of experiments came on 5 December 1888, when several large animals were dispatched by Brown in the dynamo room (fig. 5.3). A Siemens alternator was used to generate the current, electrodes were attached to the unfortunate animals, and the current was increased in steps until death occurred. Kennelly insisted that there was no bias in the experiments and his notebooks testify to the scientific method used and the accuracy of the recording of results. Petersen dissected the dead animals and examined nerve tissue and blood cells under a microscope to look for the effect of the current.[56] Edison forbade any public comment by the laboratory staff on the grisly events of the fall of 1888. The results of the experiments would speak for themselves. The Edison Electric Light Company covered all costs, which were billed by the lab as "made for Mr. Hastings" with "full reports . . . made to the Company."[57]

To a curious public, and a generation of historians, Edison's opposition to alternating current hinted of pique. The great inventor seemed to be attacking the new technology that threatened his masterpiece. Why else should "the most progressive man of his age" cry halt to progress?[58] The answer was often seen in the obvious limitations of a self-educated genius: Edison opposed alternating current because he did not understand it. This view has led some historians to conclude that Edison vilified the new technology because he had no answer to it, that "only the unfair fight . . . was waged in earnest."[59] This comment ignores the fact that Edison's laboratory was leading a three-pronged counterattack on the threat of alternating current: development of high voltage d/c, development of a competing a/c system, and a campaign to discredit the "death current" of Westinghouse. Edison might not have had the mathematical education to understand the complexities of alternating current theory, and this could have deterred him from working with it. But he did have a special aptitude for mathematics, and the alternating current technology of the 1880s was far from scientifically complex. Moreover, his laboratory had all the apparatus and trained staff to develop alternating current equipment. Arthur Kennelly, for example, was a highly competent electrical engineer who went on to teach the subject and make several important contributions to high voltage technology.

As chief electrician of the West Orange laboratory, Kennelly also played a leading part in the other campaigns of the battle of the systems. The account records of the laboratory show that from 1888 to 1892 many experiments were carried out in alternating currents, high voltage direct currents, and devices like rotary convertors to join the two types of current into one system. The tempo of research into alternating current at the West Orange laboratory quickened as the competition with Westinghouse and Thomson-Houston grew more intense. By 1889 the Edison franchisees were alarmed at the inroads made in their business. Edison was told that "the call for some sort of direct current transformer is getting louder and louder every day."[60] Later that year the annual meeting of the Edison utility companies ended with an appeal for a high voltage technology to reach customers distant from the central station. The development of a high voltage direct current system (the municipal system) was underway at the laboratory and Edison was reported to be "not only willing but anxious that something should be done as to an arc light as well as an alternating current system."[61] The laboratory was carrying out its job to provide the new technology needed by the Edison companies.

At the very time when the battle of the systems reached a climax, the structure of the Edison lighting organization was radically changed. The many companies created by Edison to exploit his inventions were gradually consolidated into larger business organizations that were better equipped to compete in a growing industry. The formation of Edison General Electric (EGE) in 1889 amalgamated the Edison Electric Light Company with the various Edison manufacturing companies (the Machine Works, Lamp Works, and Bergmann & Co.) and the Sprague Electric Railway and Motor Company. Henry Villard mobilized German capital and joined with Drexel, Morgan in forming this large company, which was capitalized at $12 million. Edward H. Johnson believed that at last the Edison enterprise had the financial strength to meet the competitive challenge and excitedly predicted, "goodbye Westinghouse!"[62] Villard and Morgan thought differently; EGE was the first step toward a powerful international trust. Villard had dreams of combining the leading German and American electrical companies into a giant multinational corporation.[63]

The prospectus of EGE stressed the links between the West Orange laboratory and the technical staff of the Siemens & Halske Company (the holders of Edison's lighting patents in Germany), claiming that the exchange of technical knowledge gave the new company "a vast advantage not heretofore possessed."[64] The Edison electrical companies had always made the most of their special relationship with the famous inventor and the prestige of the largest industrial research center in the United States. Deputations from potential customers were given a tour of the facilities "for the sole purpose of impressing upon them that whatever improvements are made in the Edison system that their company will receive them."[65]

The improvements were the products of commissioned research. The loss of income from the manufacturing shops made Edison more dependent on EGE for income for his laboratory. The shops had been important customers for the services of the lab but now their research needs were integrated into a corporate research and development policy. EGE was now the major contractor for the services of the laboratory. It recognized "the importance of still further perfecting and cheapening the manufacture installation and maintenance" of the Edison system and entered into a contract with the West Orange laboratory to acquire all further improvements.[66] But it retained the right to choose what technologies it wanted. The acquisition of the Sprague Electric Railway Company, for example, meant that it had little interest in the traction system under development at the West Orange laboratory. The leadership of the new company was conservative and so were the banks that stood behind the consolidation. Their goal was to maximize the return on the great investment in electric lighting. Any new technological direction had to be judged against this standard. The management of Edison General Electric quickly moved to reduce costs, including the costs of research and development. All research was to be carried out under contract, and bills had to be presented promptly for payment. The company reserved the right to dispute bills and refuse to pay for research they had not commissioned. The "leadened collar" was being pulled tighter.

Edison's growing dissatisfaction with the management of EGE was centered on their inability, or refusal, to introduce the improved lighting technology developed in his laboratory. He noted caustically in 1890 that they asked for a system that could compete with Westinghouse, but they did not employ his five-wire system because "they were unwilling—although the system was a success—to enter into the competition when using so many wires."[67] Edison criticized the company for not energetically promoting his high voltage direct current system. They too were afraid of the hazards of high voltage current and thought it was too risky to introduce it commercially. When EGE finally offered an a/c system to its franchisees, it did so with a caveat that the technology was "dangerous and inefficient at best, but procurable on demand."[68] The management of EGE shared some of Edison's fears about the dangers of this technology, but as businessmen they had to have it available if their customers wanted it.

Both Edison and the parent company promised the franchisees that high voltage a/c and d/c systems would be available to them by the end of 1889. This development program was plagued with delays caused by poor liaison between the laboratory, Machine Works at Schenectady, and company headquarters in New York. The problems of coordinating the research effort at West Orange with the manufacturing effort at Schenectady proved the wisdom of setting up Edison's own manufacturing facilities next to his laboratory. All the projects which required cooperation between the laboratory and Machine Works suffered from long delays; even the high voltage municipal lamp, an invention that Edison favored and one that he had repeatedly promised to the utilities, was not delivered on time.[69]

The delays in the production of this equipment have reinforced the view that Edison and his laboratory did not have the inclination or ability to work with high voltage technology. Edison's emphasis on practicality and profit obviously swayed him toward the high voltage direct current solution, but this did not stop him from exploring all avenues of long-distance distribution, including alternating current. The Edison General Electric Company was probably not too concerned with the delays in developing high voltage technology because it wanted to avoid expensive investments in new systems and was wary of new equipment that would make its inventory obsolete.

The formation of Edison General Electric was the first step in a gradual consolidation of the electrical industry that reflected a general trend in American business. This was the era of big business and the trust. The pursuit of monopoly power replaced competition with price-fixing and patent pools. Edison was opposed to this movement in the electrical industry, explaining, "I can only invent under a powerful incentive. No competition means no invention."[70] Edison was ready, and probably eager, to take on Westinghouse in the same way that he had met the challenge of those "pirates" Bell and Tainter. Yet it was pointless to face the competition without the resources of Edison General Electric behind him.

The question of alternating current had become a business matter rather than a technical issue. An impending union with Thomson-Houston removed the need to bring out an a/c system because Thomson-Houston already had a proven system on the market. In 1890 when Edison General Electric finally prepared to introduce an alternating current system, Insull advised Edison that they should withdraw from the battle of the systems before this was done.[71] While EGE wavered, development work on high voltage a/c technology continued through 1891 at the West Orange laboratory, and Edison himself worked on this project.[72] A small test installation was made, but the system was never formally introduced because the consolidation with Thomson-Houston was clearly on the horizon.

In 1887 the West Orange laboratory had been the focal point of Edison's electrical empire. He had the power to decide on experimental projects and to suppress technologies that he did not like, such as the ZBD alternating current system. Two years later the laboratory was just one part of a large organization and most of its electrical work was done as commissioned experiments. Edison no longer had the power to set the direction of research for the companies that bore his name. He did not have the freedom of action to be the courageous and pioneering inventor he had been in 1879. The business of innovation was now firmly in the hands of management, and it was carried out in the offices of the company, not in the laboratory. The company did not want to bear the costs of a major assault on the long-distance transmission problem and Edison knew this. The leadened collar of EGE prevented him from fully committing his laboratory to the battle of the systems.

CHAPTER SIX

Diversification in the 1890s

t the beginning of 1889 Edison reassured his associates in the electrical industry that he was about to finish the work of perfecting the phonograph. Then he would take a "fresh pull" at electricity.[1] Unfortunately for the Edison companies, the development of the phonograph was far from finished and consequently the promised fresh pull was lost in a bewildering array of other experiments. The furious pace of work did not let up at the West Orange laboratory in 1889. Edison's time was spread thinly over many different experiments; over seventy jobs were on the laboratory's books.[2] The Old Man's desire to be part of every experimental campaign worked against the completion of several important projects, such as the design of a high voltage alternating current system, because work was stopped while he supervised the completion of the drawings.[3]

The failure to get the phonograph right the first time delayed the transfer of resources to other projects, creating a backlog of work at the lab. Edison was like a juggler with several balls in the air at once. He repeatedly promised the ore-milling syndicate that he would soon perfect his system of magnetic ore separation, but as long as the phonograph refused to work properly there was little time for anything else. Tate assured them in September 1888 that the talking machine was at last perfected and out of the inventor's hands, but in reality it still took up most of his time.[4]

In 1889 some new developments in the ore separation project did attract Edison's attention away from the phonograph and the fresh pull in electricity. His experiments in magnetic separation of ores had started as a tangent to the work on the lighting system at Menlo Park. He was looking for a way of extracting metals from ore for use as filaments in incandescent lamps. At the same time the urgent need to get more power from dynamos led the lab staff to investigate methods of increasing the force of magnetic fields. Characteristically, an increase in knowledge in one area presented an opportunity in another. This was the kind of innovation at which Edison excelled: magnetic separation of ore emerged from the convergence of two different experimental projects. The idea was to use powerful electromagnets to divert iron particles from a falling stream of ore. The iron ore fell in one bin and the rest in another. It was simple and inexpensive and it appealed to Edison enormously. Experiments were carried from the Menlo Park lab to Harrison and then to the new metallurgical laboratory (Building 4) at West Orange. The easy success of these first experiments convinced Edison that he had found a commercial technology. "It's going to be immense," he excitedly told Batchelor.[5]

Edison's agents were ordered to scour the country for deposits of iron ore, gold, and other precious metals that could be separated by this process. An ideal site for magnetic ore separation was found in the iron mines of Pennsylvania and New Jersey, which were exhausted of high-grade ore. A pilot plant was established at Bechtelville, Pennsylvania, in mid 1889. Edison was soon convinced that he could make money by extracting iron from the low-grade ore that mines threw away as too lean to be worth processing. By the summer of 1889 he was said to be intoxicated with the profit potential of ore milling.[6]

The comparative advantage of a laboratory built for flexibility and speed was that it could move quickly into new technologies. Consequently Edison began to shift his experimental staff from the phonograph and electricity into ore milling. He continued to flit from one experimental project to another, spending half his time at the bench and the other half keeping the customers for contract research at bay. The great appeal of ore milling was that it gave him an opportunity to escape from the pressures of the laboratory and the business of innovation into the fresh air of the mountains where he prospected for magnetic ore and other metals. But a return to the lab was inevitable, as were the enquiries about the availability of the new phonograph, the status of the alternating current system, and the fate of the Edison typewriter.

Alfred Tate's innocent remark, "I often have wondered if Edison might not have accomplished more if he had attempted less" strikes at the heart of the drawback of the largest laboratory extant.[7] With the resources at hand to settle any experimental problem, Edison could never resist starting something new before a project was completed. It was only shortage of money that forced him to narrow the focus of the laboratory's work. At the beginning of 1889, the expenses of developing the phonograph led to a cutback of the experimental effort. Work was ended on the cotton picker and the hearing aid. The pyromagnetic generator project was abandoned as hopeless. There was no longer the time or money to find a substitute for ivory, rubber, silk, or mother-of-pearl. The development of an improved typewriter was shelved after more than $1,500 had been spent on experiments. Preliminary experiments on a flying machine were written off, as was basic research on etheric force, induction, and light waves.[8]

Lack of money might have cut back on experiments, but it could not stem the flood of ideas coming from Edison's fertile mind. His interest in flying machines, for example, continued unabated through the 1890s. He had several ideas for powered flight, including a disk of shutters that was moved up and down by a perpendicular piston. In trying to find a mechanical duplicate of the flapping of a bird's wings, he was following the false path taken by many inventors of flying machines. Nevertheless, a shop order was opened in 1892, a model built, and Edison spoke enthusiastically about his experiments to the press.[9] The account books of the laboratory are littered with speculative ideas, called stunts by the lab staff, that did not get beyond the experimental model stage. Although Edison liked to think of himself as a practical inventor, a businessman who judged a project by the size of the financial return, the records of the laboratory show a man happily investing his fortune into scores of experiments carried out more for the fun of doing it than the expectation of profit.

The last ten years of the nineteenth century offered so many new fields for Edison's attention, from aviation to X rays, that he found it difficult to concentrate his resources on any one project. It sometimes took a crisis, such as a mutiny of financial backers or a competing invention from a rival, to focus his mighty energies. If Bell and Tainter had not brought out their graphophone in 1886, the story of perfecting the phonograph might have been completely different. If George Westinghouse had listened to reason and not rushed alternating current technology to market in 1886, the pace of electrical development might not have been so intense in the decade that followed.

Even with the battle of the systems raging, Edison found it difficult to keep to his program of perfecting the lighting system. Sometime in 1889 or 1890 Edison drew up a list that illuminates the fate of his alternating current system. It begins with the objective of constructing a cheap dynamo "for competitive purposes." Next come improvements in the incandescent lamp: eliminating the loss of candle power, increasing its output of light, and decreasing its cost. Second from last is a plan to develop a lamp for use on Edison's high voltage (200 volt) systems. Finally, at the bottom of the list, and hurriedly crossed out: "To devise a complete system of alternating . . . ."[10]

In addition to the high priorities of electric lighting and distribution, there were several other areas of electrical research that beckoned Edison. Advances in the manipulation of magnetic fields had led to improved dynamos and motors with increased power output and power-to-weight ratios. Smaller, more powerful motors gave Edison new hope for electric traction, an application of electricity in which he was a pioneer. In 1880 with the assistance of Charles Batchelor, he had set up an electric railroad at his Menlo Park laboratory. They mounted a "long-waisted Mary Ann" dynamo on a truck and used brushes to pick up power through the iron wheels. The dynamo acted as a motor, and its drive was a leather belt connecting it to the wheels. America's first electric railway was soon forgotten in the struggle to complete the Pearl Street system, but Edison kept it in mind. He picked up the experiments in the early part of 1887 and planned to continue them at the large laboratory then under construction.

While the attention of the press and the electrical world had been on Edison's new laboratory in 1887, a former employee of the Edison enterprise was making electrical history in Richmond, Virginia. Frank Sprague constructed the first commercial electric traction system in 1887, in a heroic engineering campaign that would have done credit to the Old Man. Admitting that his ideas were "largely underdeveloped and untried," Sprague contracted to install a complete streetcar system in Richmond, comprising generating station, track, forty cars, and accessory equipment.[11] A formally trained engineer of great talent, Sprague left Edison's employ to establish his own company in 1884. He had the support of E. H. Johnson who used his influence to promote Sprague's motors in the electric lighting industry, much to the chagrin of Edison who considered Johnson's support of Sprague treachery. Although Edison thought of Sprague as a dangerous rival, he found it difficult to find fault with Sprague's motors, which were quickly put to use in industry and offices.[12]

The Sprague streetcar motor was the foundation of his traction system and an important factor in its success. Sprague fitted his compact motor to the driving axle of the vehicle and used gearing to keep the armature far away from the spur reduction drive. While one part of the motor was meshed into the gears on the axle, the other was sprung from the car frame. This widely copied design reduced the damage from continual vibration and made the motor more resistant to the dust and dirt that were the bane of the streetcar engineer.[13] Sprague solved the critical problem of current pickup by refining the overhead pickup first developed by the Belgian inventor, Charles Van DePoele. After bearing the brunt of heavy criticism from the engineering fraternity, Sprague proved that electric traction was a practical form of transportation. The Richmond installation had the same impact as the Pearl Street station; its successful opening in 1888 led to contracts for 200 more traction systems to be installed in American cities, 90 percent of them with the Sprague system.[14]

A boom in streetcar building on both sides of the Atlantic followed Sprague's success in Richmond. The growth in the electrical industry during the years 1889 to 1893 was in traction rather than in large distribution systems for lighting. The great demonstrations of long-distance transmission of electricity, such as the Frankfurt–Lauffen scheme of 1891, made the headlines, but the day-to-day profits of the utilities were in providing light and power. The latter became increasingly important in the creation of a profitable load factor. In the ten years after the construction of the West Orange laboratory, the business of the electrical industry was power. Power from direct current electricity made its mark in industry, business, and tall buildings. Alternating current might have great potential for large supply systems, but the lack of a motor made it an impractical and noncommercial technology.

The electric streetcar appealed to investment bankers and city fathers alike; it was faster, cheaper, and less trouble than horse-drawn transportation. The electrical manufacturers rushed into electric traction. Edison General Electric acquired the Sprague company and marketed his overhead pickup system, while Thomson-Houston bought out DePoele and manufactured his equipment. The two companies competed furiously for the large new market for electric traction, and both claimed to have 100 systems in operation by 1890. Five years later there were 1,300 miles of electrified tramway in the United States.[15] The streetcar became the most potent force of change in the growth of urban communities. Electric traction

Figure 6.1. The West Orange laboratory early in the 1890s at the apex of its powers and the height of its fame. Some of the Edison Phonograph Works can be seen behind the satellite buildings on the far left.

promoted the spread of suburbs and gave workingmen cheap transportation from home to work. Few cities were untouched by the streetcar. A line was laid from Newark through the Oranges in 1892. It was one of the 400 electric streetcar lines in the United States at that time. The cars that left Newark every twelve minutes became the lifeline of Edison's factories, bringing workers from the sprawling urban mass of northeastern New Jersey to work in West Orange.[16]

While streetcars glided past on Main Street, the laboratory staff labored inside to make up the ground lost to Sprague; Edison was not willing to allow the glory of one of his inventions to go to somebody else (fig. 6.1). In contrast to the almost universal use of the overhead pickup, Edison concentrated on a third rail to deliver low-voltage current to the cars. Sprague powered his cars by two fifteen HP motors, but Edison chose only one four HP motor for his streetcar. EGE had a large stake in the Sprague system and was naturally reluctant to support a competing technology, especially when the scale of traction equipment made research costs extremely high. Therefore, Edison looked for financial support from his old friend (and railroad magnate) Henry Villard. They embarked on an ambitious project to develop Edison's traction system at the West Orange laboratory. Motors and drive gear were constructed in the machine shops. A test track was built from the lab to the streetcar line running along Main Street. The quarter mile of track was dotted with transformers which reduced the 1,000 volt transmission to a safe twenty volts for the live third rail.[17]

Edison's choice of a third rail pickup did not solely reflect technical criteria; he thought the overhead wire was too dangerous to be used in a commercial system. These wires were used extensively by the telephone, telegraph, and utility companies, which strung them by the hundreds from poles and roofs. In city centers across the United States, overhead wires were an unsightly reminder of the risks of electricity. The first regulations to protect the public from the dangers of electricity were aimed at overhead wires, which were considered by many as agents of death. Alarmed by the growing number of fires and accidents, city governments had begun to ban the wires from urban centers by 1888. The great blizzard of that year did the job for them in many northeastern cities; thousands of wires came down, demonstrating beyond all doubt their fragility and the danger to the public. Yet the utility companies and the communications industry continued to resist efforts to eliminate the wires. This struggle was seen by some of the press as a confrontation between the "trusts" of big business and the interests of the people.[18] At the time his laboratory opened, Edison was confident that overhead wires in city centers would be completely prohibited and subsequently designed his traction system with very low voltages and a third rail pickup. In acknowledging that this absolutely safe form of rapid transit would not see general use, Edison argued that the extra cost of his system would be justified by the density of the traffic in city centers and the absence of the dangerous wires.[19]

The problems with using such low voltage were to ensure a clean pickup from the third rail and obtain sufficient torque from the motors. The experimental results were not promising because mud or moisture on the pickups interrupted the flow of current. Edison

Figure 6.2. The Ott brothers (John on the left, Fred on the right) working on some traction equipment outside the main machine shop of Building 5, West Orange laboratory.

worked through 1889 and 1890 on this problem while the staff of the laboratory designed the other parts of a complete system of electric traction. The work on motor drives, controllers, and car brakes was added to the many other electrical experiments undertaken at the laboratory (fig. 6.2).

The Paris Exposition

In the middle of the crowded summer of 1889, with the phonograph, alternating current, and ore-milling experiments at critical points, Edison left West Orange and joined hundreds of American engineers and businessmen who were crossing the Atlantic to visit a large industrial exhibition in Paris.[20] The series of international meetings that had begun with the Great Exhibition at the Crystal Palace in 1851 were the major showcases for machinery builders and inventors. The "American system" had been unveiled at the Crystal Palace, the great Philadelphia Centennial Exposition of 1876 had marked the sensational debut of Bell's telephone, and the Paris Exhibition of 1881 had been a triumph for Edison's electric light.

The centenary of the French Revolution provided an excuse to celebrate the social and economic changes wrought by an even more powerful revolution. In 1889 the technological elite of the industrial West met at Paris, in the shadow of the newly constructed Eiffel Tower, to pay tribute to the advances of the Second Industrial Revolution. Edison had prepared an impressive display of his inventions to exhibit, hoping no doubt to attract investors.

That he should take a trip to Europe in the middle of several important experimental projects is a measure of his confidence in his staff. It is also an indication of the international reach of his business enterprise. Edison had thought of international markets since he had crossed the Atlantic in 1873 as a telegraph engineer. Many of his innovations had also crossed the ocean. The duplex telegraph, phonograph, and electric light had all been quickly diffused throughout Europe. Edison's reputation on the Continent had never been higher, and his business connections there assured him that his new inventions would receive a favorable reception. Colonel George Gouraud had the franchise to market the phonograph in Europe. An agent for Edison's inventions since the 1870s, Gouraud fancied himself an expert on new technology and the business of innovation. His expectations for the perfected phonograph in Europe were great, as was the size of his order—1,000 machines, the largest single order placed at the Phonograph Works. Edison was still not sure how much of Gouraud's enthusiasm was justified and how much was bluster. He intended to find out for himself in Europe.

While Edison was feted in Paris, his laboratory workers maintained the effort on several important projects. One of the office staff noted that everything went on in a "hum drum kind of way" while the Old Man was away, each experimenter with his allotted task and team leaders sending regular reports to the boss.[21] Although it needed Edison's mercurial presence to galvanize the lab into action, it could function without him. Charles Batchelor was placed in charge of the lab in Edison's absence. He supervised research, made all the business decisions, and managed the Phonograph Works. He assured Edison in Paris that production of wax cylinders was underway at a new building in the Works. The manufacture of primary batteries was going so well that Batchelor had plans to sell the surplus to outside customers.

Batchelor personified the duality of manager and experimenter that the Old Man required from his chief lieutenants. Batchelor took an active part in experiments on ore milling, phonographs, and electricity, working on them simultaneously as his notebooks show. He supervised the experimenters at work in Building 5 on various aspects of the phonograph. In addition to the continued effort to perfect the talking machine for business use, Batchelor reported to Edison that progress was being made on the toy doll and long-playing phonograph.[22] Work continued on the Edison traction system with Batchelor supervising experiments on a car brake and motor.

In Building 1, Arthur Kennelly led the development work on an alternating current machine and the high voltage municipal system. Both were eagerly awaited by the Edison Machine Works and the utility companies. Reginald Fessenden directed a team of experimenters in the chemical lab (Building 2) in the search for a better insulation material for electrical conductors. Finally, W. K. L. Dickson carried out research into magnetic separation in Building 4, in which he had installed several ore separators.

Dickson divided his time between Building 4 and the photographic room on the second floor of the main laboratory where he worked on the motion picture camera. Batchelor joined him in experiments on strips of film and the machines to show them. With Edison in Paris, Dickson persuaded Batchelor to erect a new building specially designed for the experiments on moving pictures. A two-story photographic studio was quickly erected adjacent to Building 4.[23] This was one of many new structures added to the laboratory complex. The lean-tos, shacks, and other temporary buildings, each designed to suit an experimental project, formed a congested suburb around the lab.

The wide range of experimental activities carried out in the laboratory at West Orange was illustrated by the size and variety of the Edison exhibit at the Paris Exposition. In addition to the electrical, phonograph, and ore-milling displays, Edison showed many different types of telegraph apparatus and some unique inventions like the Edison-Sims torpedo and the pyromagnetic generator. The exhibit for the Paris Exposition was a major project for the laboratory. In December 1888 work began on the design of the display, which was constructed in the machine shops and Phonograph Works. It was then crated up and shipped to Europe, accompanied by some of Edison's laboratory staff under the leadership of William Hammer, a master of electrical promotion who had successfully installed lighting displays at numerous exhibitions. Edison's electricity exhibit was one of the centerpieces of the exhibition. It featured thousands of electric light bulbs, a complete electric power station, and several electric signs and displays. It took up an acre of fair grounds.[24]

Edison's enthusiastic reception in Paris underlined his international status as a great inventor. A constant flow of visitors arrived at his suite in the Hotel du Rhin. Edison was delighted at the warm welcome he received in Europe. He was honored by many professional and scientific groups, including the French Academy of Sciences and throughout his trip was able to gather information that was relevant to the work being done at West Orange. The collection of scientific information from all parts of the world was the foundation of Edison's method of innovation. Wherever he went, he exchanged information with the European scientific community and often got an opportunity to see the latest advances for himself. At a dinner to honor the pioneer of photography, Louis Daguerre, he talked with Etienne Marey, a physiologist with considerable experience in photographing animals in motion. His photographic "gun" had caught the action of birds in flight. The visit to Marey's laboratory in Paris was the turning point in the development of motion pictures.[25]

Western Europe was a hotbed of electrical development at this time, especially in high voltage and alternating current schemes. Edison's travels through Europe brought him to several important electrical installations. In London he visited the Deptford central station under construction by the brilliant electrical engineer, Sebastian Ferranti. Because of its size and position (well outside the city center on a site easy to supply with coal), the Deptford installation was the central station of the future. Ferranti's plan to supply a million incandescent lamps in the metropolis and send the current in from Deptford at the unheard-of pressure of 10,000 volts made this the most ambitious supply scheme of the nineteenth century. Despite his opposition to high voltage alternating currents in the United States, Edison did not condemn the plant he saw at Deptford. He did warn against the danger of high voltage currents, but the engineering work impressed him, and certainly Ferranti's standards were higher than those of the contractors installing Westinghouse systems in the United States. There was much in the Deptford scheme that appealed to Edison; an ambitious venture carried out on an unprecedented scale, it promised to bring cheap electricity to the masses. Ferranti was an engineer in the Edison mold, talented, ambitious, and fired by a vision of universal electric light. Edison later described the young man as "the brightest electrician and engineer abroad."[26]

On an extended trip through Germany, Edison was given a tour of the great Siemens Works in Berlin where his dynamos and incandescent lamps were made under license. Siemens & Halske were the leading electrical manufacturers on the Continent. Their international organization and highly respected engineering resources were the heritage of the days when the firm dominated telegraphic communication across the Continent. The formation of EGE united the engineering staff of Siemens and the West Orange laboratory in the effort to increase the efficiency of the original lighting technology. Edison had strong links with Germany. Newark's large German community was the source of many of his laboratory staff, some of whom, like Johann Schuckert and Sigmund Bergmann, had gone on to build electrical manufacturing concerns in Germany. His many friends and business connections in central Europe kept him informed of the research work of companies and individual inventors. German industrial laboratories were supported by the best technical education system in the world, and Edison made it his business to find out what was on the horizon in electrical development. Research on electric lighting, especially on new incandescent filaments, was well advanced in Germany. He probably listened to the research plans of Siemens & Halske with special interest.[27]

The business of innovation made Edison more than a spectator of the European scene; he used his time in Europe to promote his inventions and monitor the companies set up to market the products developed back home at West Orange. The perfected phonograph was the most popular part of the Edison exhibit at Paris, drawing the largest number of visitors after the Eiffel Tower. With interest in the machine at a high point, Edison took this opportunity to reorganize the company responsible for marketing and distribution in Europe. Gouraud had a flair for promotion but little financial clout; he was quietly moved aside to make way for the New York banking house of Seligman and Seligman.[28] The reorganized Edison United Phonograph Company was an important customer for phonographs in the 1890s, providing considerable business to the Works in slack times.

The most important business transaction of the European trip was Edison's purchase of the rights to the Lalande battery. This primary cell provided a reliable source of electricity that was desperately needed for the phonograph. It also became the basis of a profitable new business of manufacturing primary batteries for use in industry and the railroads. As soon as word got out that Edison was making his own batteries at West Orange, the laboratory received many inquiries from people interested in securing a reliable battery with long life. It was understood that any battery produced by the Wizard of Menlo Park would be superior to anything else on the market. The Wizard took note and began to consider the mass production of an Edison battery.[29]

Prior to the construction of the West Orange laboratory most of Edison's international business activities had been restricted to the sale of rights to inventions. This was a quick and simple way of raising money. The agency of these transactions was the bankers and venture capitalists who financed his work in North America. Their international networks facilitated the sale of franchises and the formation of foreign companies. The franchisees usually agreed to purchase equipment, such as phonographs or dynamos, made by Edison in the United States. The successful overseas operations often progressed to manufacturing Edison products under license. In the case of companies with their own engineering resources, such as Siemens & Halske, foreign affiliates became partners in the further development of the technology. From this point they could easily become competitors. The answer to this threat was the overseas transfer of the manufacturing strategy begun at West Orange. Edison wanted to control the profits of manufacturing overseas and considered setting up plants in Europe, making his enterprise truly multinational. He planned to acquire a factory in Belgium to make his toy doll and use it as the base of a sales network that would reach all parts of the Continent.[30] This project was never realized because the toy doll failed as a commercial product.

At the Paris Exposition of 1889, only the phonograph was a success, and the rapid growth of the amusement phonograph industry gave Edison another opportunity to develop manufacturing facilities in Europe. The shortage of money during the 1890s restricted his role to that of a partner rather than the owner of these enterprises, yet the policy had been set. Edison was determined to exert more control of the commercial development of his ideas, at home and abroad. By the turn of the century, he had established two new industries in the industrial core of the West. Films and recorded music were produced through the cooperation of business organizations in Europe and North America, their products moving easily from one continent to the other along with the creative people who made them and the money that financed them. The development of these two major industries in the 1890s put the West Orange laboratory at the center of an international business enterprise.

Edison's Folly: The Ore-Milling Venture

On his return to West Orange, Edison was greeted with good news about the experiments in hand. Batchelor reported progress on the phonograph and electrical projects; Dickson had made significant advances with moving pictures; and the trial run at the Bechtelville separation plant had been a success. Encouraged by this promising result from the ore-milling experiment, Edison bought a mine at Ogden in the mountains of northwest New Jersey and began to equip it with his ore separators. In the early 1880s Ogden had been (with the adjacent Lehigh Valley) one of the leading iron ore producers in the United States.[31] Now much of the magnetite iron ore had been extracted and abandoned mines dotted the sides of the Musconetcong Mountain. Edison thought that his Ogden plant could successfully process the low-grade ore and in doing so demonstrate the practicality of his magnetic ore-milling technology in exactly the same way that the Pearl Street central station had shown the viability of the electric light.

Up to this point the ore-milling project had developed along familiar lines: research led to patent applications, a model separator was constructed, and attempts were made to sell the Edison ore separators to mines. Edison found to his disgust that the mine owners were a conservative group with little interest in a new processing technology. The pilot plant at Ogden was intended to attract groups of investors into the business and to convince mining companies of the commercial potential of the new technology. It was only when this effort failed that Edison took over the ore-milling operation. He decided that it was a waste of time persuading mine owners to adopt the idea of magnetic ore separation. Instead he was going to run the mines himself and then negotiate with the mine owners from a position of strength, once his method had been proven successful. He moved from selling separators to going directly into the mining and concentrating business.[32]

The appeal of ore milling to Edison was its profit. The man who prided himself on the practicality of his innovations claimed in 1892 that ore milling was the most commercial thing he had ever done. He had convinced himself that this technology could play an important role in the iron industry by revitalizing eastern blast furnaces, which were in steady decline because of the exhaustion of iron ores.

Shortage of ore had forced iron masters to import iron ore from Cuba and Spain. New Jersey still had large deposits of low-grade ore, and Edison thought his separating machines could make a paying proposition out of these dead mines. His ore-milling company would be his Standard Oil—a large, fully integrated operation with few competitors and incredible profits. At the height of the project he envisaged running eight large mills in New Jersey with a total yearly output of more than $10 million worth of ore, bringing an annual profit of $3 million.[33]

Edison was not the only entrepreneur-inventor trying to magnetically separate iron from low-grade ore. The originality of Edison's idea was the sheer scale on which he intended to extract usable iron from the ore. He argued that mining profits came not from the richness of the ore (the prevailing business strategy of the mining industry), but from the highly efficient processing of large amounts of ore. While no one thought that very low-grade ore (20 percent magnetite content of the ore) could be economically processed, Edison believed that the economies of large-scale processing could make it a paying proposition. His ore-milling technology was striking in its originality. The editors of *Iron Age* concluded: "He sweeps aside accepted practice in this particular field and attains results not hitherto approached. He pursues methods in ore dressing at which those who are trained may well stand aghast."[34] His grand plan was to change the supply of raw material to the iron industry with his ore separators. The historian Harold Passer's description of Edison as a "conservative defender of the status quo" does not fit his activities during the 1890s; instead we see a courageous inventor and reckless industrialist who went against the conventional wisdom of the iron industry and poured money "like water" into a radical technology.[35]

Edison could embark on an original technological path because he had the resources of the West Orange laboratory behind him. Conventional ore-crushing and ore-grading equipment was not big enough or good enough for him; the ore separation plants were custom built to his own design. He had the largest private laboratory in the world to make up and test new ideas. Nearby was the Phonograph Works to manufacture the massive ore-crushing machines. Edison's plans for magnetic ore separation were based on widespread use of electric power and automatic machinery—the ore was not going to touch human hands from extraction to final loading. As a pioneer in continuous flow processing, he had to devise the special equipment, which was all produced at West Orange, down to the telephone network that was essential to manage this gigantic operation.[36] Edison planned to use his laboratory to take iron mining from primary processing to an efficient, modern industry, embarrassing the conservative mine owners while taking their profits.

Everything connected with the ore-milling project was massive, from the size of the machinery to the amount of ore processed each day. The ore seams at Ogden ran in a straight line, so Edison used giant earth movers to cut open a huge strip mine in the side of the mountain, creating an enormous artificial canyon. He bought large steam shovels (one was the biggest in America) and several traveling cranes to carry out this work. After the ground had been torn apart, the huge chunks of rock were removed and carried on railroad skips to the large rock-crushing machines where they were broken up into successively smaller lumps. The job of breaking up the rocks was done by giant cornish rolls that rotated in opposite directions to crush the rocks as they fell between them. These machines were designed at the West Orange laboratory and were probably the largest ever built, each weighing thirty-five tons. The rock fell from these rolls to a conveyor belt that took it to an intermediate crusher, and then to Edison's "three high" rolls which produced a fine powder out of the crushed ore. The powder was dried and lifted fifty feet to the top of the separation tower that dominated the skyline of the Ogden plant. The ore fell through screens and past hundreds of successively more powerful magnets. These deflected the stream of iron into waiting bins. The tailings were also collected and given several more trips down the separation tower. The final concentrate was about 68 percent iron.[37]

Conveyance of ore was one of the important engineering problems to be solved because it offered the greatest opportunity for cost reduction. Edison operated his own railroad at Ogden, and although it gave him the greatest pleasure, it did not solve the difficulties of moving materials cheaply and efficiently about the site. A new answer was required for this old problem. In collaboration with a rubber-goods salesman, Thomas Robbins, Edison built rubber conveyor belts to move the ore through the processing stages. The rubber belts ran on rollers that were specially designed to resist dust and dirt. This automatic, self-lubricating system anticipated the assembly line of the twentieth century. Henry Ford later claimed that it was Edison's example at Ogden that inspired him to devise his assembly line for the Model T.[38]

Edison worked hard to reduce the costs of producing his ore because he knew that transportation costs would increase its price substantially. He aimed to deliver ore by rail to the eastern blast furnaces at a lower price than ore from the Midwest, enabling the eastern furnaces to compete with those in the South and Midwest. The plan reflects the nineteenth century confidence in economies of scale; Edison thought he could produce cheaper iron concentrate by processing more of it.

Edison built his own community around the mill, transforming the deserted hamlet into an industrial complex where hundreds of men worked. To a man fascinated with machines, the plant at Ogden was a paradise. Edison loved to watch the great shovels and cranes at work and felt at home in the uproar of the mill: "On all sides the roar and whistle of machinery, the whir of conveyors and the choking white dust . . . Big wheels revolve in the engine house . . . little narrow gauge locomotives puff their way in and out between the buildings . . . with shrieking and whistling wheel and brakes."[39] He reveled in the primitive life at the mill and the world of miners and mountainmen in the remote New Jersey highlands. The spartan existence also invigorated Charles Batchelor who was soon recruited to join in the fun. Talking with the boys all night, sleeping on floors, clambering in and out of the giant machinery, and surrounded by dirt and equipment on all sides, Edison was in his element. He wrote to John Kruesi: "Come with Batch and I into the mining business where you may be dirty but very happy."[40] The original plan of making the ore milling project a joint venture with investors or mill owners was now forgotten. Edison intended to run the operation with his own funds, telling the press that the profits were not going to be shared with outside parties, "except the boys here who own the thing with me."[41]

At Ogden Edison was able to recreate the close-knit comradeship that characterized the machine shop culture. Working at the mill, as it was called, was just like the good old days at Newark and Menlo Park, where he and the boys ate, slept, and worked together. Edison naturally brought several muckers with him from West Orange. After a communal dinner at the end of the day, Edison would lead a discussion of the experimental and engineering work on hand. This was called "going to school," and every man was expected to talk about the areas in which he was involved. This emphasis on communication was the basis of Edison's team approach to innovation. It was also an opportunity to test the mettle of the boys. Edison "loved to nag his men, haze them with questions, just to see if they knew what they were doing."[42] When they were not hard at work, the boys occupied themselves with practical jokes, cockfights, rattlesnake fights, and boxing matches. For Edison, this was the life!

The Ogden plant was completed at the end of 1891 and the first order for iron concentrate was received. Edison now began to devote more of his time to producing the first commercial concentrate. He spent most of each week at the mill, returning to West Orange on Saturday. He spent Monday at the lab reviewing the weeks's experimental work before traveling back to the mill. Edison was an absentee manager of the laboratory. His absence from the experimental rooms of West Orange sharply cut back his output of inventions. A decade that started with an average of over thirty patents a year ended at a much lower level: five patents a year in 1892 and 1893, no patents in 1894 and 1895, and only one in 1896.

Edison was obsessed with the ore-milling project on which his own personal fortune rode. He claimed to have invested every cent he had in it. The first year of operation at the mine was marred by continual breakdowns as machines failed under the tremendous strain of crushing rocks. The engineering problems of running the giant rollers soon overcame the workers at Ogden and the financial resources of the Edison enterprise. The cost of failure was high. Samuel Insull calculated that the ore-milling venture was running a loss of $6,000 a week in the summer of 1891![43] Convinced that the engineering problems could be solved and confident that the project would eventually be successful, Edison was determined to get the ore mill working whatever the cost. He admitted that it was quite a problem but commented that "when we get in working order we shall easily solve it," revealing an unswerving faith in the existence of a technical solution to every problem.[44]

Once again, he had underestimated the number of technical difficulties in perfecting an innovation. His attention was focused on the output of concentrate, and he assumed that it would perform satisfactorily in the blast furnace and that the furnace men would make any small adjustments required in processing a new form of ore. Yet the powdered concentrate he delivered to the iron masters was an unwelcome novelty to them. Their dissatisfaction with the way it acted in their furnaces threatened to end the ore-milling project. Edison responded by returning to the laboratory to devise a method of molding the concentrate into briquettes. This improved the handling characteristics of the ore but did nothing to reduce its high phosphorus content—a major disadvantage because phosphorus in the ore made the final product brittle. Edison became so absorbed in these chemical problems that he gradually lost sight of the economic parameters of the project. It had become more a technical challenge than a business venture.

Electricity in the 1890s

With Edison fully occupied with ore milling and the phonograph project dead, the laboratory's work was concentrated on electricity. The Edison enterprise began the 1890s on the foundation of three main businesses: phonographs, ore milling, and electricity. In 1890 the lab billed about $80,000 of research expenditure. Roughly $20,000 went to the phonograph, slightly more went to ore milling, and the remainder was divided between the electrical research for Edison and the contract work for the electrical companies.[45]

Experiments in electricity increased during the first half of the decade as work on the phonograph declined. A new contract was made with Edison General Electric in 1890 in which the company agreed to support the costs of half of Edison's experimental time for the next five years. Three-quarters of the laboratory's work time was to be devoted to electrical development, the company agreeing to pay expenses up to a weekly limit of $1,200. Its annual report justified this outlay by pointing out the importance of further development of incandescent lamps. It concluded: "In view of the value of Mr Edison's services, the Company has a new contract with him, under which he devotes a greater part of his services to the Company."[46] This generous support made up for the withdrawal of the company's men from the West Orange laboratory. The plan to make the laboratory the center of engineering and innovation in the lighting industry had been replaced by a formal relationship between an independent laboratory and the company.

The facilities of the laboratory, which by contract were "at the disposal of the company," were applied to the production of a diverse range of electrical equipment for use with both direct and alternating currents. The West Orange laboratory remained an important resource for the local Edison lighting companies that continued their policy of expansion during the 1890s. Much of the planning for these larger systems was done in the lab, and in some cases Edison himself contributed to the design of central stations and distribution networks. The Chicago Edison system, under the direction of F. S. Gorton, was typical in that it quickly ran out of capacity to meet demand and Gorton was impatient to extend service. Edison and Kennelly devised a distribution system for Chicago, employing high voltage direct current and the municipal lamp that operated on 200 volts. In 1891 Edison reassured Gorton that "We . . . are planning for the future so that what we do put in will match the extended system 10 years from now."[47] (The large maps used to plan the spread of mains from Chicago's city center are still in the laboratory.)

Edison General Electric even took up Edison's diverse research into traction after he had given up on his low voltage system. They signed a new accord in 1890. The staff of Building 1 examined the Sprague motor, altering its design and improving the method of its manufacture.[48] Traction motors were designed at the laboratory and built at the Schenectady Machine Works, which had the large-scale facilities to develop the technology, including an indoor test track. Edison and Kennelly often traveled to Schenectady to test traction and high-voltage distribution systems.[49] This work had important technical applications in dynamo construction and manufacture.

Edison's special expertise in the manufacture of lamps continued to be a valuable asset to the company. In 1891 and 1892 he was called in to reorganize production at the Lamp Works in Harrison, spending more than 500 hours getting the Works "out of the difficulties caused by failure of the employees to do their duties."[50] The court's decision of 1889 had confirmed Edison as the inventor of the electric light, and Edison General Electric wanted the world to know that they had his patents and genius at their disposal.

The most important research done for EGE at the laboratory was on the incandescent lamp. Edison's room 12 on the second floor of the laboratory was fitted out to work on filaments. Nearby were the glassblower's room and the vacuum pump room that took their power from the overhead shaft running from the precision machine shop. In the 1890s the search for a better filament went on with more financial support than ever. The cellulose filament represented the peak of lamp technology in the ten years after Edison's 1879 invention. Although the life of the lamp had been significantly increased, and its output of light was much improved, most of the electrical energy was transformed into heat rather than light. A new decade of lamp development was beginning in which several new filament technologies were pursued, including incandescent oxides, rare earths, and metals with high melting points. The directors of Edison General Electric and its successor, General Electric, knew that a radically new lamp technology could decimate their major business. They were determined to acquire or control any important improvements in lamps.

Although Edison could hardly have been pleased that his name was dropped from General Electric when it was formed in 1892, he did not sever relations with the company, as some have claimed. His secretary, Tate, attributed Edison's annoyance to the loss of his name from the industry he created, but the inventor was probably more upset at the high valuation put on the assets of Thomson-Houston in the terms of the amalgamation. As expected, Henry Villard had watered the stock of the smaller Thomson-Houston Company to bring it up to par with that of Edison General Electric. The formation of GE is often interpreted as a watershed in Edison's association with electricity. The press saw it as the trust versus the lone inventor, with the latter frozen out of the industry he had created. Many text books and television programs depict Edison as the founder of General Electric, when in fact it was the work of Henry Villard and J. P. Morgan.[51] Edison had no great desire to maintain a financial interest in the electrical supply and manufacturing industries, which he considered were too crowded. His whole purpose in founding the West Orange laboratory was to move out of highly competitive markets into businesses founded on new products. Edison saw few prospects in the reorganized electrical industry.[52]

Edison was made a director of General Electric but rarely attended meetings. He did, however, continue to carry out research for the company, including the development of alternating current systems. The laboratory staff continued the development of rotary converters and rectifiers that performed the important task of converting alternating current into direct current and vice versa. Edison experimented with an alternating current motor—a critical part of a commercial supply system that was still unavailable in the early 1890s.[53]

Despite the bad feeling caused by the formation of General Electric, the company still needed the resources of Edison's West Orange laboratory to keep pace with dangerous competitors like Westinghouse. The ties between Edison and the electrical industry were not broken in the 1890s; in fact they grew stronger. Over the five years from 1890 to 1895, EGE and GE paid the laboratory over $121,000 for its work on improving and developing electrical equipment. This work formed the largest single project for the Edison laboratory in the early 1890s, as table 6–1 shows. Electricity was the important thread that ran through all the experimental projects: electromedicine, phonographs, moving pictures, and ore milling.

The Great Depression

The financial strain of running the ore-milling plant led Edison to cut back expenditure at West Orange. His first step was to reduce

Table 6.1. West Orange Laboratory Experimental Expenses, 1891–1895

	Billed			
Year	Edison General Electric and General Electric	Thomas Edison Personal	Others	Total
1891	$ 31,944	$15,722	$20,729	$68,395
1892	36,992	9,050	8,956	54,998
1893	23,302	5,182	14,175	42,659
1894	14,382	6,024	14,076	34,482
1895	14,541	1,918	65	16,524
Total	121,161	37,896	58,001	

Source: Edison accounts, Billbook # 13, ENHS.

Note: The sales year ran from October to October.

the labor force. In 1891 the lab employed about sixty-five men and only 160 employees remained in the Phonograph Works—a significant decrease from 1890 when 500 worked there.[54] In 1892 the Works was manufacturing four different types of phonographs but the failure of the business phonograph as a commercial product prevented the long production runs that were essential for mass production. After a baptism of fire during the campaign to perfect the phonograph, the Works had been reorganized in 1889. A new superintendent with experience in precision manufacture was hired and an engineering department created. As these resources could not be wasted, Edison instructed the Works management to obtain outside work to keep the machines running. The Bates numbering machine company was purchased in 1892 so that the numbering stamps could be manufactured in the phonograph factories. Much of 1892 was taken up with efforts to win a large contract from the Thomson-Houston company to manufacture small motors, a project that was especially suited to the facilities of the Works and vital for the financial health of the Edison enterprise.[55] The amalgamation of Thomson-Houston into General Electric brought an end to all hopes to manufacture electric motors, but the Works continued to manufacture meters and instruments for General Electric.

At the end of 1892 Edison was back at Ogden, installing bigger and better equipment and confidently looking forward to success. His plan was formulated on the rapid conquest of the engineering problems and an increase in scale. The cost of installing colossal machines was immense, and he could only afford it by a gradual liquidation of his stock in GE. He did this because he needed money quickly, not because he was peeved over the formation of General Electric. The large investment in the Ogden operation was consistent with his strategy of becoming a manufacturer and was not a reaction to his exclusion from the electrical industry.

The "new and magnificent industry" that impressed many foreign visitors to the United States was built on weak foundations—its forced expansion produced a facade of impressive development that hid its low returns.[56] The intense competition between electrical manufacturers had produced many supply schemes where there was little demand for electric light. This was especially true of high voltage systems; profits in electric lighting lay in supplying dense concentrations of customers and not in distributing electricity over wide areas to the lucky few. One executive of General Electric remembered, "Customers did not exist; they had to be created."[57] Yet the practice of creating demand for electricity by financing supply organizations was not without pitfalls for the electrical manufacturers. In 1890 Edison General Electric held $4 million of illiquid stock in local supply companies. The burden of debt grew steadily during the decade. By 1893 its successor General Electric held an estimated $16 million of stock. At the same time the excessive cost of developing a radically new technology had begun to take its toll on its major rival. Westinghouse nearly went bankrupt after pouring money into its alternating current system.[58]

The financial panic of 1893 administered the shock that almost brought down the American electrical industry. Each local lighting company that went out of business eroded the financial reserves of the electrical manufacturer that had accepted its stock in payment for equipment. The depression continued into 1894. From 1892 to 1894 output of electrical equipment declined by 37 percent, and only the steady growth of electric traction prevented an even worse decline. There were rumors on Wall Street that GE was about to declare bankruptcy. The electrical industry teetered on the brink; it was "perfectly dead" according to Tate.[59]

While the electrical industry declined, Edison was comforted by the money he was going to make in ore milling. Here was a business that he thought would "outrank the incandescent lamp as a commercial venture."[60] It was only a matter of processing a raw material and delivering it to the iron masters. There was surely no need for expensive technical support and years of research to improve the product.

Edison's optimism in his ore-milling venture counterbalanced the great depression in manufacturing that soon made itself felt in the West Orange complex. Demand for consumer goods was tumbling and with it went the amusement phonograph business. There was no work for the men in the phonograph factories. Many were discharged, further reducing the work force that had suffered cutbacks since the beginning of the decade. Edison also reduced the lab staff. By 1896 the factory inspectors of the state of New Jersey found only ten men at work in the laboratory and just over 100 in the Phonograph Works. The next year there were twenty-eight workers in the lab and 144 in the factories around it.[61] Table 6.1 shows the decline of research work at the West Orange lab.

A few miles down Main Street stood the workshops of the hat makers of Orange. The hatters had their own craft culture that reflected many of the values of Edison's muckers. The hatters were proud of their skills, fiercely independent, and closely bonded in their work culture. Yet they were unable to prevent the introduction of automatic machinery and the ravages of the depression. From June to October 1893 eight finishing shops in Orange discharged more than half their workers.[62] These were bleak years for workingmen. Edison had no difficulty in recruiting them by the hundreds for his ore mill at Ogden, and soon the population of the mine swamped the local community. A village of wooden houses was built to house workmen; a post office was added later; and the name of the community was changed to Edison.

The depression ruined Edison's ore-milling venture because it led to reductions in the price of iron ore from the Midwest. Edison's original business plan was based on undercutting the competition's price of about $6.00 per ton. His own ore was priced between $4.00 and $5.00 per ton, and he expected to produce briquettes as low as $3.50 per ton as the Ogden plant grew more efficient. But when the price of Minnesota ore dropped to $3.00 a ton, the chances of a successful conclusion to the ore-milling venture grew more distant. The Mesabi ranges in Minnesota could provide the cheap ore required in the steel mills of the Midwest. In 1893, 600,000 tons were shipped; by 1894 the figure was 2,000,000.[63] The more efficient exploitation of the Minnesota ore beds brought the price down to $2.50 per ton by the end of the decade. Yet Edison still persevered at what was called "the most extensive experimental research on record."[64]

The discovery of the ore-rich Mesabi ranges in 1890 should have brought an end to the work at Ogden. The ore was easy to mine and perfectly suited to the production of Bessemer steel in Carnegie's giant mills. When Edison heard the news, he is supposed to have said, "You might as well tell them blow the whistle" at the plant.[65] Yet he continued to pour money into the operation, rebuilding the plant in 1893 in preparation for another test run. Undaunted by the technical problems and the disinterest of iron masters, he labored through 1894 to increase the size of the ore-milling plant. He might have known in his heart that the cause of magnetic separation was lost, but he never admitted it in public. His failure to get out of ore separation, to cut his losses in 1890 or 1891 and go on to the next project, was a mistake of heroic proportions. Why did Edison break the cardinal rule of the business of innovation and attempt to perfect a technology that was not commercial? He chose the perfect time to embark on the campaign. Labor was cheap, and iron ore was a basic raw material that would not suffer fluctuating demand that made the production of consumer durables such as the phonograph an uncertain business. There were also several technical reasons to keep at ore milling—the automatic conveyor system and special crushing machines were successful innovations. There was one good economic reason—Edison ran ore milling as his own business and would reap any profit, however small. He had lost his influence on the direction of electrical technology, sold out the phonograph to Lippincott and his company, but had managed (mainly by default) to stay in charge of the ore-milling venture. He told reporters that "in my future campaigns I expect to control absolutely such inventions as I shall make. Whatever the profits are I shall control them."[66] This was a prophetic remark in view of his return to the manufacturing strategy, a move precipitated by a new phonograph product and the unexpected success of one of Edison's stunts.

CHAPTER SEVEN

Moving Pictures

dison's greatest experimental success at the West Orange laboratory ran concurrently with his greatest failure. The motion picture was a truly original innovation that created its own history as it went along. The novelty that became the first entertainment of the masses began as a simple peep-hole machine that showed the viewer a few seconds of moving pictures. Its development in the laboratories of West Orange took second place to the ore-milling venture in the 1890s, yet years after the magnetic separation of ore was forgotten the movies stood out in the public's mind as another piece of magic from the Wizard. They reestablished Edison as the world's greatest inventor after the loss of face in the electrical industry. It was no less a triumph for his business enterprise; the first viewing machine, called the Kinetoscope, was the vanguard of a new industry.

The circumstances of this innovation provide ample proof of the advantages of the invention factory idea. Motion pictures were conjured out of the many flows of information that converged on the laboratory. The idea became a stunt. It was possible, it just might work, but it could only be fitted in the schedule when the ore-milling project was not in operation. The original idea was turned into an experimental model, a complicated assembly of metal, wire, and photographs that was made to work by the experimenters and mechanics in the laboratory. To them it was just another job, another shop order to be executed, and another challenge to their skills and ingenuity. Once the machine left the shop for the experimental rooms on the second floor of the lab, it was up to greater minds to find a commercial technology in the machine and create the value of the machinists' work. The great failure of ore milling was essentially a failure of the business of innovation rather than the result of poor workmanship. Given sound business judgments and workable designs, Edison's men could be relied upon to produce the experimental machines that were the intermediate products of innovation. Their job was to put them together and make them work. They employed their knowledge of similar machines to make the small adjustments and improvements required in the rapid development of a new technology.

The debate about the true inventor of motion pictures is therefore academic: several men made vital contributions to this technology within the fluid organization of Edison's laboratory. Many more working outside the laboratory contributed the results of their experiments via the scientific press or through personal contacts with Edison and his staff. The moot point of the history of this innovation was that it could never have been the work of one man alone. Like the electric light, motion picture development was international in scope. Edison's achievement was to be in the right place at the right time. Although the original idea was his, the successful invention of the Kinetoscope was the result of the various skills of his men who turned his vision into metal. He might have been stretching a point to claim the credit for the invention of motion pictures, yet without the fruitful cooperation of the men in the invention factory—a idea that was his alone—there would never have been moving pictures in the nineteenth century.

The project began as one of Edison's many scientific interests. The dramatic development of photography in the years after the Civil War encouraged him to believe that there were commercial opportunities in this new technology. He had already employed photographic techniques in his study of electric light, and the new West Orange laboratory was equipped with a dark room. His original plan for the new lab was to span the three disciplines of electricity, chemistry, and metallurgy. Always alert to the opportunities of combining new technologies into an innovation, Edison quickly added two more areas of investigation: acoustics and photography. It was a small expense to purchase the equipment and books to explore these new fields, and Edison was ever hopeful that some small piece of information discovered in these areas might have important application in another.

After reading the scientific literature about electricity, he joined many other practitioners in conceptualizing it as lines of molecules moving in waves. He saw a connection between the waves of electromagnetic force, the spiral waves that held the recorded sound of the phonograph, and waves of light.[1] He started to experiment on this relationship at the Harrison laboratory, attempting to produce a photographic image of sound by means of a reflecting disc attached to a phonograph that projected the lines of sound waves onto a long strip of paper. This work was taken up as soon as the West Orange laboratory opened in December 1887. One experiment focused light onto his electromotograph—a chalk drum relay used in telegraphy—to test the changes in electrical resistance and produce a sound that could be picked up by a stylus in the wave pattern burnt into the chalk.[2]

On a more practical level, Edison saw that an immediate use for photography was to translate the waves of the phonograph cylinder into sight as well as sound. He thought it might be possible to "devise an instrument that should do for the eye what the phonograph does for the ear," storing visual images that could be reproduced at a later time by the user.[3] Edison had collected several toys that gave the visual impression of movement, such as the zoetrope. This device was the work of the English mathematician W. G. Horner. It used the illusion of persistence of vision—when the eye retains an image a short time after it has disappeared. The zoetrope was composed of pictures viewed through slots in a cylinder; as the cylinder turned, the pictures appeared to move. Edison also followed the progress of experimenters such as the Frenchman, Etienne Marey, who had manipulated still photographs to produce moving pictures.[4] By the time the laboratory opened in 1887, Edison had the information to develop a mechanical means to exploit the persistence of vision phenomenon by moving images quickly past the viewer's line of sight.

The inspiration to begin this experimental project came from one of the many visitors to the famous laboratory in West Orange. In February 1888 the pioneer photographer Eadweard Muybridge visited Orange to give a speech and afterwards came over to the lab to demonstrate his sequential photographs of animals in motion. Muybridge had mounted these pictures on a zoetrope and successfully recreated movement. Edison had been in communication with Muybridge before the new lab was constructed and had acquired some of his photographs, but seeing Muybridge's work at first hand inspired Edison to duplicate Muybridge's approach in a phonograph format.[5]

The experimenter given this task at the West Orange laboratory was not an ordinary member of Edison's work force. William Laurie Dickson was no rough mechanic; his photograph shows him as something of a dandy, a man of genteel background and some education who felt a cut above the grease-stained muckers and tramp mechanics who worked alongside him in the laboratory. Dickson had emigrated from Scotland to the United States with one object in mind: to learn the profession of inventing from the master. His letter to Edison was typical of the many sent by ambitious young men: "If you would try me, I will take the lowest place in your employment, until you find me worthy of something higher, so passionately do I love your profession."[6] Dickson was not successful in gaining immediate entrance to the lab but worked his way through the Edison lighting organization until he joined the great inventor at the temporary lab in Harrison. An enthusiastic amateur photographer, Dickson became the official photographer of the laboratory—no small job considering Edison's worldwide fame. It is likely that Dickson discussed the feasibility of moving pictures with the Old Man while they experimented at the Harrison Lamp Works.[7] Now he had the resources of the large new laboratory, including a photographic studio, to develop the idea. Yet as chief of the metallurgical laboratory in Building 4, Dickson had to divide his time between ore milling and motion pictures. This was not unusual. Dickson and Edison worked closely on the development of ore-milling machinery, and at the same time they pondered a means to record sight and sound.

Edison naturally hoped to achieve visual reproduction in the same mechanical form as he reproduced sound, using the phonograph analogy in his first experimental apparatus. The revolving cylinder—a recurring motif in Edison's inventions—was chosen as the format of the experimental model.[8] Consequently the first experiments at West Orange consisted of laying very small photographs on a cylinder, slightly larger than the tinfoil phonograph but of the same general arrangement, and viewing them through a microscope. Both light and sound needed amplification as they emerged from the turning cylinder. Dickson described the first apparatus as having the sound and picture cylinders running on the same axis and powered by the same electric motor. The motor came from the phonograph. Edison first tried the obvious answer to the problem of synchronizing sound and pictures. The cylinder had to move forward one frame for each exposure and then hold the image steady for a fraction of a second so that the viewer's eye could record the image.

The first months of experiments constituted the exploratory work to find out if there was any merit to Edison's basic idea that images could be manipulated to show movement. By October 1888 Dickson and Edison were convinced that it would work and a caveat—a preliminary declaration of an intent to patent an idea—was filed for a "moving view" apparatus that could recreate things in motion. The basis of this invention was the cylinder format in which the rapid movement of many small photographs per second gave the illusion of movement. The series of pictures were photographed on the cylinder "in the same manner as sound is recorded on the phonograph."[9] The viewer looked through a microscope device to see the moving pictures and listened to the accompanying sound track through tubes attached to a phonograph.

The experiments had moved out of the realm of speculation; a shop order was opened for the project, which enabled Dickson to draw on the resources of the lab. All supplies, machining, and experimental time was billed to the order, which provided a means to account for the expenses of the venture. During 1889 equipment to take microscopic photographs was acquired, and a special studio was constructed next to Building 4. Dickson was assigned an assistant, a mechanic called Charles Brown, and the two worked behind a locked door in room 5 of the main laboratory building.[10] This experimental team was later assisted by several other mechanics and experimenters who were shifted into the project as the need arose.

Although the basic idea was simple, there were many problems to be solved before a commercial machine could be contemplated. The cylinder format had the disadvantages of requiring a very small image (impractical with the primitive film then available) and focusing on a curved plane which produced images with a fuzzy surround. The difficulty of synchronizing sight and sound was that the former required intermittent motion while the latter needed continuous motion. This was a mechanical problem that was handed over to the mechanics for solution. Their experience in working on telegraph instruments obviously was helpful. They experimented with the same break wheels, tuning forks, ratchets, and electromagnetic devices that they had used many times before.

Dickson soon found that the microscopic photographs affixed to the drum were impracticable and began to increase the size and improve the definition of the image. He moved from fixing small photographs onto the drum to coating it with a photographic emulsion of gelatine containing silver salts. By the end of 1889 his efforts had brought forth the critical breakthrough in the development of motion pictures: the use of a perforated filmstrip in place of the photographs on a revolving drum. The most important contribution to Dickson's work at West Orange came from the manufacturers of photographic film: John Carbutt in Philadelphia, the Blair Company in Boston, and the Eastman Company in Rochester, New York. As soon as Dickson found out about their innovation of strips of celluloid film coated with sensitive emulsion, he obtained some samples from John Carbutt. Here was the advantage of a wide information net that could give the workers at West Orange the benefit of technological advances elsewhere. The thick sheets of film were divided into strips with the top side of each strip cut into the form of teeth. A ratchet engaged the teeth and moved the film steadily past the lens. It was only a short step to using a filmstrip with perforated edges.

Celluloid filmstrip was also obtained from Eastman's company, and the specifications of the first order for this improved film, made by Dickson in September 1889, defined the standard 35mm width of film used today[11] (fig. 7.1). Dickson's progress was dependent on the filmmakers producing finer emulsion and faster film that enabled him to make clearer images. Eastman's film had the advantage of being thinner than Carbutt's, permitting the design of a camera that worked with flexible filmstrip instead of thick sheets. Edison had considered the use of a "long, tape-like sensitive film" in his early sketches of 1887, and the weight of the testimony given by Edison and the laboratory employees involved in the invention points to the development of a filmstrip with perforated edges before the trip to the Paris Exhibition. The film historian, Gordon Hendricks, is skeptical of this claim, believing that it was made to improve Edison's patent position. Hendricks argues that Edison got the idea for a perforated strip from Etienne Marey. Hendricks' persuasive version of the invention hinges on Edison's trip to Paris where he saw the continuous filmstrip in Marey's laboratory. Edison is supposed to have said, "I knew instantly that Marey had the right idea."[12]

On his return to West Orange, he incorporated this idea into another caveat filed in late 1889. This final piece of information completed the first phase of this experimental project, in which the important innovations came from outside the laboratory: Muybridge's photographs showing motion, Marey's filmstrip, and Eastman's thin celluloid film. The contribution made by Edison's laboratory was to bring all these advances together into one innovation. The store of expertise in telegraph technology at the lab was Edison's trump card in the successful construction of a film camera; the mechanics of moving a filmstrip through a camera was much the

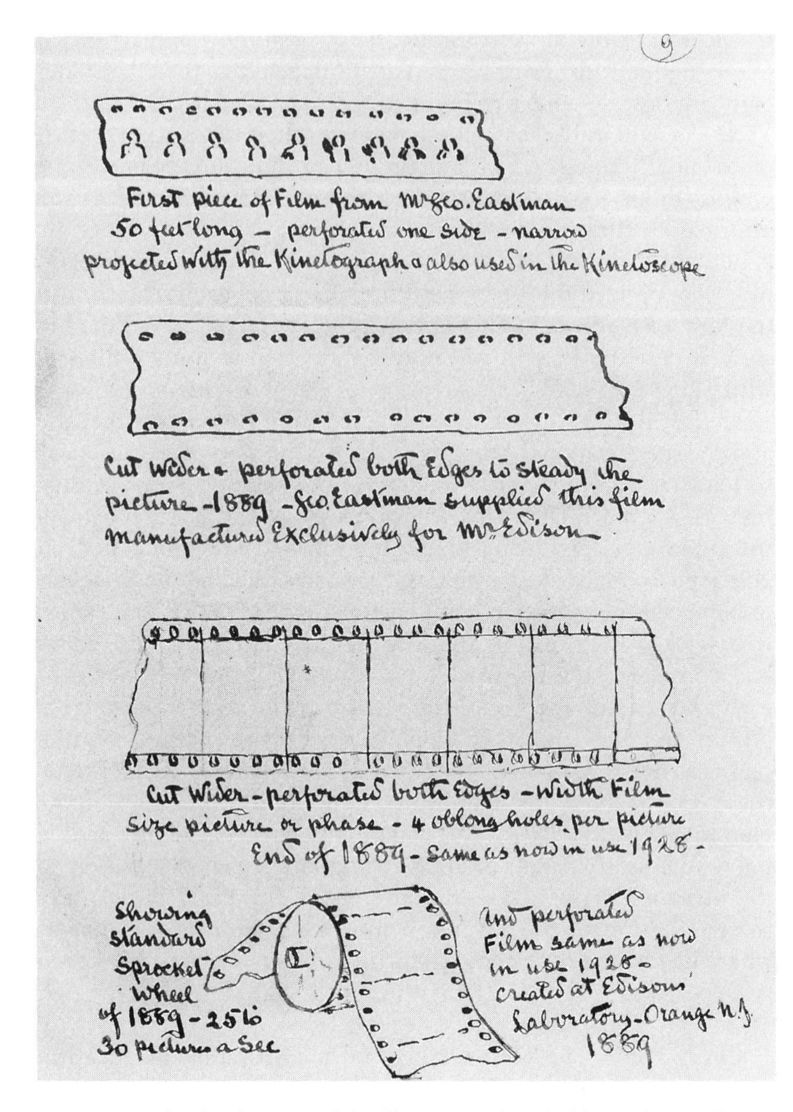

Figure 7.1. The development of the filmstrip, as described by Dickson in his own account of the invention.

same as the movement of perforated tape through an automatic telegraph.

The development work on the camera, later called the kinetograph, continued through 1890 and 1891, aided by better filmstrip and mechanical improvements in the machine. The laboratory's staff found that the celluloid film produced by the Blair Company was best suited to moving pictures. New models of cameras were built that employed larger images as Dickson experimented on the format of the film. Although he slowly improved his method of taking pictures on the filmstrip, he was faced with problems in the film feed mechanism that caused an unsteady, flickering image. The inherent elasticity of the celluloid strips and the lack of a uniform thickness compounded these difficulties.[13] The mechanics who were called in to help Dickson, chiefly William Heise, Charles Kayser, and the Ott brothers, had years of experience in techniques of moving a tape rapidly and exactly past an aperture. Some of them remembered the lessons gained from developing Edison's stock printers in the early 1870s. They altered the feed to provide a more positive movement that would not shift the film around at high speeds. The addition, and subsequent modification, of a Geneva stop mechanism was a major step in developing a reliable camera and projector.

Figure 7.2 shows Charles Kayser at work on the motion picture camera. The device has the temporary look of an experimental model, with fragile belts crisscrossing the top works. The model sits on a horizontal wooden base in exactly the same way as the telegraph and phonograph. The strip kinetograph used two electric motors, one to drive the shutter and the other to advance the strip by means of perforations on each side of the strip (fig. 7.3). One motor came from a phonograph and the other was probably made for an electric fan. The strips of film, which were from twenty- to fifty-feet long, were held in two drums situated horizontally across the camera mechanism. The film was held steady by teeth that engaged the perforations while the exposure was made and then moved forward by an escapement mechanism for the next exposure. The rate of exposure could be taken as high as forty-per-second although twenty-per-second was sufficient to give the illusion of movement.[14]

An application for a patent for this model was finally made in 1891 when Edison filed a description of two machines: the Kinetoscope for viewing pictures and the kinetograph for taking them. By this time the camera had only one motor and a simpler, more efficient drive mechanism. It came complete in a lightproof box with an eye piece for the operator. The camera was conceived by Edison as the manufacturing part of the system that produced the film. It was a totally different machine from the one people associate with today's motion picture camera. Housed in a large rectangular

Figure 7.2. Charles Kayser at work on the first motion picture camera in 1889 or 1890. "Junk," or old equipment and experimental apparatus, is strewn about the room.

box, it was a bulky device intended to remain in the Edison Works. It was designed around the electric motor which powered the shutter and moved the film rapidly past it. The shutter and motor mechanism were assembled as one unit and positioned directly behind the single lens. The machine was fitted with a start and stop switch and a counter that showed amount of film used. The camera was attached to a cast iron base (no doubt to dampen the vibration) and connected to a source of electricity.[15]

The Kinetoscope was a pine box with an eye piece at the top for viewing. The upper part of the cabinet contained an electric motor, shutter, and incandescent lamp to illuminate the images. The electricity for the motor was provided by a primary battery, like the phonograph. The lower part of the box held the fifty-foot loop of

Figure 7.3. A close-up photograph of the device shown in front of Kayser in Figure 7.2. The strip kinetograph of 1889 had two electric motors on either side of the two drums holding the film and the shutter and lens assembly.

film, consisting of the positive images strung together in ribbon form. The film was arranged over rollers and rapidly moved to the top of the machine for viewing. A rotating slotted disc exposed each image to the viewer at rates of about thirty to forty-five frames a second, giving a viewing time of about half a minute[16] (fig. 7.4).

Marketing Motion Pictures

Designed as an arcade machine in which the customer paid to see a filmstrip, the Kinetoscope was a hurried first attempt to get moving pictures into a form suitable for commercial exploitation. It was closely modeled on the coin-slot phonograph, which was the only part of Edison's talking machine business to experience any growth of sales in the gloomy years after 1892. The traveling showmen who had taken "portable phonograph outfits" on the road were now being replaced by small businessmen who purchased "stationary exhibit outfits" and installed them in arcades or phonograph parlors.[17] If the size of their operations is any guide, their business remained good during the years of the Great Depression, for in some big cities there were large parlors with over seventy machines. An

Figure 7.4. The Edison peephole Kinetoscope, showing the interior of the machine with its loops of film. The peephole can be seen on the underside of the raised top piece.

arcade machine was, therefore, the logical starting point for marketing motion pictures. Beginning in 1893 the two types of coin-slot machines were produced concurrently in the laboratory and then installed side-by-side in amusement parlors (fig. 7.5). Edison's plan was to exploit the novelty appeal of moving pictures by selling Kinetoscopes to amusement parlors, departments stores, hotels, and bars.

Naturally the first entrepreneurs attracted to the Kinetoscope

Figure 7.5. The interior of Peter Barigalupi's arcade, Market Street, San Francisco. Coin-slot phonographs are lined up on the left, with ear tubes dangling down; Kinetoscopes are on the right.

came from the phonograph business. Edison gave his secretary A. O. Tate the rights to market the machine. Tate was joined in the Kinetoscope Company by Thomas Lombard of the North American Phonograph Company and E. A. Benson of its Chicago franchisee. Edison made a contract with one of his mechanics to manufacture the first batch of Kinetoscopes, an indication that he was not fully convinced that this technology was ready for mass production in the Works.

Tate and his associates saw 1893 the Columbian Exposition in

Chicago as an appropriate place to launch the invention, for this event was to dwarf the recent celebrations in Paris by the size of its crowds and the extravagance of its spectacle. But to Tate's growing frustration, the mechanic, James Egan, did not hurry to complete the contract. Edison could not be prevailed upon to push him along, perhaps a relic of the old machine shop tradition of allowing a craftsman to work at his own pace.[18] Edison missed an opportunity to steal the thunder from his great rival in the electrical industry who scored a major success with his exhibit at Chicago. The Westinghouse display of a complete alternating current system was a milestone in electrical development, an engineering triumph that opened the way to widespread use of this technology in all its forms. Although the West Orange laboratory prepared a small display for the Exposition, there was no great wonder to show and little money to spare on exhibits. The name of Edison still blazed at Chicago, however, thanks to the indefatigable efforts of General Electric and the utility companies, which saw the wisdom of maintaining their symbolic ties with the great inventor.[19]

The first batch of Kinetoscopes was shipped from West Orange in April 1894. Tate crossed over from New Jersey to supervise the delivery to the Holland Brothers' amusement parlor on Broadway near 27th Street in New York City, ignorant of the excitement that these plain pine boxes would arouse. His description of the first night of business gives an impression of their appeal and the profits that could be made:

> I was kept too busy passing out tickets and taking in money . . . If we had wanted to close the place at six o'clock it would have been necessary to engage a squad of policemen. We got no dinner. At one o'clock in the morning I locked the door and we went to an all-night restaurant to regale ourselves on broiled lobsters, enriched by the sum of about one hundred and twenty dollars.[20]

This amount for one night's work was more than most men made in a month at the laboratory! The impression made by a few seconds of flickering image was irresistible and soon crowds gathered at the first moving picture parlors. The Kinetoscope quickly became the sight to see, providing a handsome return on the estimated $24,000 spent on experiments. As Gordon Hendricks has pointed out, the $7,940 for machines and $369 for films paid to Edison by the Kinetoscope Company might seem insignificant compared with costs of today's movies, but the result caused no less excitement among its audience.[21] The silent pictures of Kinetoscope had a universal appeal. A few months after the first Kinetoscopes left the West Orange Works for amusement parlors in Manhattan, fifty machines were sent to Europe. London, Paris, Rome, and Copenhagen soon had their own Kinetoscope parlors.[22] Edison was sowing the seeds of another vibrant business in the same way that he had founded the electric light and phonograph industries in Europe.

There was little time for Edison to rest on his laurels after patenting the technology and introducing a commercial product. The technical limitations of the kinetograph soon became obvious. Its single lens and static view could only record whatever was placed in front of it, and the limitations of the mechanical film feed fixed the length of film at about fifty feet—any longer caused stress that broke the film. Film subjects were fixed at about thirty-five to forty seconds, and the public soon tired of the same short filmstrips. As the demand for fresh film subjects rapidly outstripped supply, new facilities were constructed at West Orange to mass produce film. The most visible of these additions to the plant was the special studio constructed on the grounds of the laboratory. Although it was described as a "revolving photographic building," it has become famous under the nickname coined for it at the lab—the Black Maria. The building was about forty-feet long and ten-feet wide. Made of wood and covered with black tar paper, it moved on a circular track to follow the movement of the sun. Artificial light was used in this studio and part of its roof could be opened to let in natural light (see fig. 7.6). This was the first motion picture studio built expressly for this purpose, replacing the film shed erected by Building 4. Its construction, which began in 1893, marked the point when Edison saw the commercial potential of motion pictures and made a commitment to move wholeheartedly into their manufacture and sale. The building cost $650 and earned its place in film history.[23]

At one end of the Black Maria was the room in which film was loaded in and out of the camera. The heavy box-like machine was moved along rails that ran the length of the studio; its rigid stability was essential to achieve a clear picture with the slow film of the 1890s. The interior of the Black Maria was made completely black to create high contrast and give a sharp, well-defined image for the Kinetoscope strips. The descriptions provided by Dickson show that this studio could accommodate talking pictures; a phonograph was connected to the camera during filming and attempts were made at synchronization by adjusting the electric motors that powered both machines.

Figure 7.6. The Black Maria, 1893 or 1894. Fred Devonald is standing by the entrance. Another employee looks down from the opened-up roof section, which allowed the sunlight into the main studio, center. The kinetograph was housed on rails in the low section of the Black Maria on the right.

The subjects filmed in the Black Maria came from vaudeville, the most popular form of public entertainment at the time and a favorite of the staff of the West Orange laboratory. Edison and the boys often took an evening off to wander the streets of Manhattan, visiting bars and music halls. The self-contained vaudeville skit could easily be transferred to a thirty-five-second Kinetoscope film. During 1894 and 1895 a procession of vaudeville performers made their way from Manhattan to West Orange to perform in the Black Maria: animal acts, dancing girls, strong men, and famous figures such as Annie Oakley and Buffalo Bill. Edison took a personal interest in finding new talent; after a performance of the *Gaiety Girl* in New York, he went backstage and persuaded three of the ensemble to visit the laboratory and dance before his camera.[24] For the mechanics and laborers of the laboratory, who doubtless could not have resisted the temptation to peek, this must have been a rare treat!

The Kinetoscope did not have a long life as public entertainment, but as the pioneer in the field it had an eventful one. Although it suffered from constant neglect, the motion picture project was taken from an experimental device to a commercial product in about five years (1889–94). This was not as fast as some in the Edison organization would have liked, but it was a considerable achievement for the laboratory. The rapid development of a new line of business, carried out without a major technical hitch, proves that the West Orange work force did have the speed and flexibility Edison expected of them.

Electrical Research at the End of the Century

The two motors that powered the strip kinetograph made it a typical product of Edison's laboratory in the 1890s. The film camera was another motor-operated machine like the phonograph, Kinetoscope, and electric fan. An early believer in electric power, Edison had devised motors for transportation, industrial use, and office equipment. The motor load played an important part in Edison's conception of a commercial electrical supply system. He knew that lighting alone would not fully employ the power plant of central stations and expected that demand for electric power would provide the business to make electricity supply profitable. The history of the utility industry in the 1890s proved him correct. The exponential increase in the use of motors was responsible for improving the load factors of central stations, saving many from extinction. Of all the electrical technology that passed over the experimental tables at West Orange, Edison chose to develop the fractional horsepower motor, beginning with designs for one-half and one-sixth horsepower motors. These were best suited to power the consumer durables, the useful things that every man, woman and child wants, that Edison saw as the future of the supply industry.

The electric fan was the first of a long line of motorized products to come out of the West Orange laboratory. It was a simple machine—a motor, base, fan blades, and wire guard—that was easy to assemble in the Phonograph Works. The staff of the electrical laboratory in Building 1 designed motors to be used on all kinds of house current and a shop order was quickly executed to design a fan motor to run off the Edison Lalande cell. The fan became universally popular. By the end of the century the electric fan was as indispensable to office work as the telephone and electric elevator.

Where Edison led the way, many were soon to follow. It was not difficult to copy the fan. The great number sold by the Edison Manufacturing Company was soon surpassed by thousands of electric fans manufactured by small companies eager to enter the market for electrical goods. Edison was not so naive as to think that competition in this field would not come quickly. As he said, "When an invention by means of competition fails to pay a satisfactory profit we can drop it and substitute something else."[25] The moving picture was exactly the kind of technology to move into; it rested on a firm base of the laboratory's prior experience and was much harder to copy than the electric fan. The 1890s saw Edison move from crowded industries into new areas where he had the competitive advantage, such as electromedicine.

The investigation of the effect of electric currents on the human body was not merely an episode in the battle of the systems to be forgotten after General Electric introduced an alternating current system. Arthur Kennelly with the neurologist Dr. Petersen continued to experiment on the medical applications of electrical currents. Their work was directed toward anesthesia, an important new area of medical research that was vital to the development of surgical techniques in American hospitals. Ironically they worked with alternating currents, the form of electricity that Edison is thought to have wanted outlawed. In 1893 the two experimenters successfully used low-voltage alternating current to produce an anesthetic effect, demonstrating the basic paradox of modern technology: the currents that killed could also be those that healed.[26]

Edison also experimented on methods of electrical therapy, a form of home health care that had achieved great popularity after its introduction in midcentury. The direct application of electricity to the body, in the form of magnetism or low voltage current, was widely used to treat a multitude of ailments. In 1889 Edison presented a paper to the International Medical Congress in Berlin about his experiments on electrical endosmosis to reduce gouty swellings.[27] Although the popularity of patent medicine and quack doctors reached a high point in the Gilded Age, Edison resisted the temptation to introduce a cure-all therapeutic device. His experiments, which had the financial support of Henry Villard, were directed toward producing medical equipment that could be sold under his name. The Edison Manufacturing Company produced apparatus for physicians and dentists, such as dental drills (employing his small motors), and cauterizing equipment.

These products from the West Orange laboratory offer a vivid contrast to the heavy electrical equipment developed at Menlo Park. Relatively inexpensive to develop, the innovations of the 1890s suited the facilities erected at West Orange and answered Edison's need to raise money quickly in the wake of the disastrous expense of the ore-milling venture. When shortages of money brought work at the Ogden mine to a halt, he returned to the electrical, phonograph, or motion picture projects in the lab. In his words, "I have no money . . . but am back in the lab making it."[28]

Typical of this strategy was the rapid development of X rays as a commercial technology. The German physicist Wilhelm Roentgen discovered X (for unknown) rays at the end of 1895. This type of electromagnetic force has a wavelength similar to light that is capable of penetrating solids. Edison claimed that he started experimenting on them only ten hours after he saw the cable announcing Roentgen's discovery. This project took advantage of the lab's expertise in electricity, photography, and electromedicine. Little time was wasted in assembling all the necessary equipment, including the apparatus to generate the high voltages needed to produce the rays. A typical burst of activity in the West Orange laboratory, with Edison and his team working night and day, duplicated Roentgen's experiments and produced a working device, called the fluoroscope, in less than two weeks.[29] It consisted of a screen coated with fluorescent emulsion and a vacuum tube with two electrodes. A charge of electricity bounced off a metal target produced the rays, which were deflected through the patient to the screen where the image was viewed. Edison sent a working fluoroscope to Michael Pupin of Columbia University at the end of March 1896. By July he had had completed a fluoroscope manufacturing plant at the West Orange complex, proving again that his laboratory had lost none of the speed that had characterized operations in the 1880s[30] (fig. 7.7).

At the time of this experimental project, the work at Ogden had reached a climax. He admitted to the press that "it will either work or bust," and unfortunately it failed.[31] Failure at Ogden brought him back to the laboratory to raise money. In October 1895 he wrote to F. Fish, general counsel of General Electric: "I am getting close to the end of my mill biz and will soon be able to come back to work."[32]

The work was the electrical research of the lab, most of which

Figure 7.7. The Edison fluoroscope, the first X-ray machine. The operator is Charles Dally, brother of the ill-fated Clarence Dally.

was financed by General Electric. In Edison's absence, Arthur Kennelly and the staff of the electrical laboratory in Building 1 had continued their development of an alternating current system: alternators, motors, meters, transformers, and measuring instruments.[33] This work was probably undertaken to match the research effort of GE's great competitor, the Westinghouse Company. After spending much of the 1890s in the doldrums, the market for alternating current equipment was slowly growing in the latter years of the decade, especially after a suitable a/c motor had been introduced. Both companies wanted to supply the equipment needs of utility companies and industry, which they expected would grow significantly as more alternating current supply systems were installed. Yet as Edison had predicted, consolidation did replace competition in the electrical industry. In 1896 GE and Westinghouse came to a patent agreement that replaced strife with profitable coexistence.

The relationship of GE and Edison's West Orange laboratory changed after 1896 when a new agreement with the company gave

Edison $1,000 a month to pay for the experiments on finding improved filaments for the incandescent light. Instead of a wide ranging experimental effort into all aspects of electricity, the company now wanted a focused program in Edison's special area of expertise. The work on transformers and meters was continued at the lab, but now it was supported with Edison's money. Although he could only have experimented on filaments in the winter months when the Ogden mill was closed down, he kept up a stream of enthusiastic reports to General Electric. In December 1896 he wrote to Fish, "I have reached a very curious stage in the lamp filament work . . . evidently I have a perfect filament."[34] There are few other details of this filament, which was probably an experimental proposition rather than an innovation. As Edison told General Electric in 1897, "I hope to either succeed in the filament or come to an understanding with you to continue." Although he did continue experiments on incandescent filaments and fluorescent lamps into the twentieth century, the lack of results brought a gradual decrease of the company's financial support.[35]

The policy of the electrical manufacturers during the 1890s was to engage individual inventors under contract to carry out research in their areas of expertise. Westinghouse supported William Stanley's laboratory at Great Barrington, Massachusetts, and paid him a small monthly retainer.[36] Similarly, General Electric retained the services of Elihu Thomson as a consultant. The development of some new lighting technologies in Germany helped change this strategy of contract research. German scientists, some employees of electrical companies and some working on their own, were exploring the incandescent effect of certain rare earths and metals acting as filaments. Walter Nernst, an electrochemist at the University of Göttingen, followed the path of Edison and several other inventors in developing refractory oxides that became incandescent at high temperatures. In 1897 he patented a new filament that consisted of a short rod of various oxides of metals. More efficient than the carbon filament lamps in use, his lamp was an important breakthrough in lighting technology. It had an especially long life when used with alternating current, and Westinghouse quickly acquired the patents and introduced the Nernst lamp in North America, the only important incandescent lamp not controlled by General Electric.[37]

A greater threat to GE's dominance of the lamp market came in the form of metallic filaments with very high melting points. German scientists were experimenting with rare metals, such as osmium and tungsten, which promised to totally outperform the carbon filament lamp in luminosity, efficiency, and useful life.

The important patents in electric lighting upon which the electrical industry was founded were all to expire by the end of the 1890s, opening the field to all comers. Foreign electrical manufacturers, especially the giant Siemens company of Berlin, saw their opportunity to move into the North American market. This gloomy prospect, and the growing vulnerability of the existing lighting technology, forced General Electric into action. Under constant pressure from its brilliant electrical engineer, Charles Steinmetz, the management of the company decided to establish a research laboratory, staffed by formally trained scientists, to investigate new areas of electricity. The objective was to prevent the high ground of lighting technology from falling into the hands of the Germans. The company was concerned about the high cost of this establishment but appreciated that failure to control the new metallic filament lamps would be more expensive in the long run.[38] While General Electric built its own laboratory at Schenectady, New York, in the first years of the twentieth century, the financial support given to Edison's laboratory decreased to approximately $25 per week, which was no more than a token payment. The years of contract research at West Orange had come to an end.

The new lab at Schenectady is often taken as the progenitor of modern industrial research, as different from Edison's invention factory as trial and error is from basic scientific research. Its focus was narrow; it looked into metallic filaments, mercury vapor lamps, and the Nernst lamp. Steinmetz's original idea was to concentrate on these critical areas and then carry out general chemical research in any spare time that remained.[39] In contrast, the West Orange laboratory covered a much wider range of electrical technology spanning light, power, storage batteries, measuring instruments, etheric force, electromedicine, and anything else that came into Edison's mind. While General Electric's laboratory moved methodically and cautiously toward predetermined goals, the invention factory was set up to quickly exploit any technological opportunity that presented itself.

General Electric's development of the X ray provides a significant contrast to the operation of Edison's laboratory. The company moved slowly to introduce a machine developed by Elihu Thomson but soon removed it from the market. The problem with the X ray was that it was impossible to control. The unpredictable behavior of the first X-ray tubes put the utility of the technology into doubt, and its inherent danger was well demonstrated by injuries and deaths following exposure to the rays. One of Edison's assistants, Clarence Dally, died from radioactivity poisoning—an unwelcomed first for the West Orange laboratory.[40] GE's caution was as much tempered by the small market for this technology as its danger. A greatly improved tube, employing the same material used in the new lamps, was the important factor in convincing the management of GE to move into the manufacture and sale of X-ray machines. The company was considering many paths for diversification after suffering a depression in its main business in 1914. The demand for X-ray equipment during the First World War ensured its future as a commercial technology, as it also did for radio. GE had established itself in the lucrative manufacture of the vacuum tubes used in both these twentieth century marvels.[41]

Edison's involvement with X rays produced less profit and no prestige. After quickly exploiting the original idea, even producing a peep-hole machine for the 1896 Electrical Exposition in New York, he moved on to the next project. No more experimental work was carried out on X rays. Many half-hearted attempts were made to utilize the idea, but the sale of X-ray sets by the Edison Manufacturing Company to hospitals ended well before 1914. The difference between these two development paths, one corporate the other chaotic, reflects two different approaches to the business of innovation. Edison had none of the patience and caution that brought GE success.

This episode also serves to underline the difference between the laboratories at Schenectady and at West Orange. It is easy to imagine the chemists and physicists of GE's lab carrying out pure scientific research while the muckers of West Orange did their work with hammers and screw drivers. The formally trained staff of the General Electric laboratory are usually portrayed as carrying out their work apart from the commercial demands of the parent organization. GE told its stockholders that the laboratory was to be "devoted exclusively to original research."[42] This was the theory of the corporate laboratories; the practice was much different.

Both laboratories carried out manufacturing. GE's facility made X-ray tubes by the thousand, for example. It was not unusual to see craft work at the Schenectady lab. On one occasion a blacksmith from the machine works was enlisted in the struggle to make malleable tungsten for filaments.[43] By the same token it is not difficult to see Kennelly carrying out basic scientific research at West Orange, aided by a body of theoretical knowledge, precise measuring instruments, and the systematic records of all prior experiments. The corporate laboratories and the invention factory had a duality of purpose that covered both theoretical investigation in the library and engineering work at the bench. The staff of the West Orange laboratory were equally at home undertaking basic research or troubleshooting in the Works. Improving, altering, and repairing existing products were all in a day's work. Edison's men took a hands-on approach to the problems that came up in manufacturing, modifying machines in the Works just like the Old Man had done in the 1880s. Similarly many of the corporate labs established around the turn of the century had to divide their time between basic research and engineering problems.

The industrial research laboratories had to prove their worth to a cost-conscious management, in much the same way that Edison, Elihu Thomson, and William Stanley had had to satisfy their corporate customers in the 1890s. The expense of research had to be justified, and this could only be done by showing results. Willis Whitney, the manager of General Electric's laboratory, moved in this direction very quickly after the lab opened in 1900. By 1901 the laboratory was solving engineering problems brought to it by other parts of the company. Much the same can be said for the Bell Company's pioneer industrial research laboratory. Hammond V. Haynes, the director, had begun projects to examine the fundamental science of telephone transmission in his laboratory. But he soon gave this up, abandoning theoretical work for "the practical development of instruments and apparatus."[44] The basic scientific work, he decided, was best carried out by institutions like the Massachusetts Institute of Technology (MIT) and Harvard.

If pure science is defined as basic research to discover new knowledge without the motivation of future profit, there was little pure science in industrial research. As one chief engineer of AT&T put it, "the practical question is, 'Does this kind of scientific research pay?'"[45] The Marxist historian Harry Braverman puts it more bluntly: "Corporate magnates, still impatient of free and undirected research and anxious for nuts-and-bolts engineering innovations, hardly bothered to conceal beneath their new commitment to science a contempt for its most fundamental forms."[46] In corporate laboratories intellectual freedom was often reined in by budgetary requirements and scientific freedom undermined by the need to keep the advantage gained by research. The work of their staff was primarily dictated by corporate strategies and not by scientific considerations. Research meant working towards specific technological goals.[47]

The important difference between Edison's laboratory and the facilities of GE or AT&T lay in the relationship of the research effort and the manufacturing organization. Edison's plan was not for a laboratory to support manufacturing or to provide defensive technology to ward off the competition. His manufacturing strategy placed the West Orange laboratory in the position of both developing new products and supervising their manufacture. The laboratory initiated the work of the greater organization rather than responded to its needs. The difference between the status of this independent laboratory and the corporate R&D department is evident to the eye. The West Orange lab is an imposing, free-standing structure that expresses pride in the work of research. The industrial research laboratories of the early twentieth century were unimposing structures, such as the wooden shack that housed the GE laboratory in Schenectady, offices tucked away in corporate headquarters, and machine shops hidden in corners of large factories. As the driving force of a great international business, the West Orange laboratory stood at the center of an industrial empire.

CHAPTER EIGHT

An Industrial Empire

hile extravagant stunts like the X ray kept his laboratory in the public eye, more mundane experimental projects, such as the phonograph, had a greater influence on the Edison enterprise in the 1890s. The development of a low-cost phonograph for amusement use had begun about 1888. The development cycle for the phonograph took about ten years for each major innovation. Ten years after inventing the tinfoil machine in 1877 Edison built a laboratory that would perfect the phonograph as a commercial product. The enormous effort of 1888 led to an improved talking machine but it had little commercial potential. Ten years' more work was required to identify a market and develop the right technology for it.

A meeting of the franchisees of the North American Phonograph Company in 1890 underlined the need for a lighter, more compact talking machine with better reproduction. The delegates were given a tour of the West Orange laboratory where they were shown a new, more sensitive recorder and given an option to purchase pre-recorded cylinders that had been duplicated by Edison's experimental molding process. By 1890 the development path of the amusement phonograph was set: a reproducer that could recreate music loudly and clearly, a more compact mechanism, and a method to duplicate pre-recorded cylinders.[1]

In his original concept of a phonograph industry, Edison had expected profits to come from the sales of machines. The unexpected success of the coin-slot format alerted him to the money to be made in selling pre-recorded cylinders. Although recording was a difficult, time-consuming operation that required expensive facilities in the studio, good recordings were at a premium. In the slow year of 1891, for example, the New York Phonograph Company spent $15,000 on "original" masters.[2] Although the West Orange laboratory had the best recording facilities in the industry, its output was limited by antiquated methods of duplicating the masters by multiple recording or the pantograph method. Production was low—only 160 a day in 1892—and would remain that way until Edison's engineers perfected the duplicating process.[3]

While an experimental team worked on this problem, Edison and Walter Miller (the recording expert of the laboratory) improved the special machine used to make recordings. The introduction of a hard, jeweled stylus to cut the groove in the cylinder instead of a sliver of steel was a major innovation in phonograph technology; it gave improved response in the higher frequencies and was less liable to inflict damage on the wax surface.[4] The more sensitive recorder was soon matched by a new design for the phonograph reproducer. Miller collaborated with experts on sound and hearing in his search for a reproducer that could recreate the clarity and brilliance of recorded music without wearing out the wave cut into the groove.[5] The "automatic" reproducer was a feather-light device that had a diaphragm made of a microscopic sliver of glass. Its stylus assembly was balanced to track in the groove without riding out or repeating. Unlike the standard recorder-reproducer used on the business phonograph, it did not require adjustment.[6]

This innovation was incorporated into the new phonographs introduced in 1893. Called the M types, they were a direct descendent of the "perfected" phonograph of 1888—the top works remained essentially the same except for the new reproducer. These machines were solid, weighing at least fifty lbs., and quite expensive. Although the price range (from $150 to $200) was less than the $250 price tag of earlier models, the phonograph was still too expensive for a mass market.[7]

The research work on reproduction continued at West Orange to Edison's satisfaction. He designed a horn to amplify the sound coming from the diaphragm, and by 1894 he was convinced that his experimental phonographs reproduced music "almost perfectly."[8] There remained only the engineering problem of the power source. Here Edison made an abrupt break with past practice by rejecting the electric motor that had been at the center of his perfected phonographs since 1887. The adoption of the spring motor for the phonograph in 1896 was the critical step in reducing the size and weight of the machine.

The phonograph of 1896 was designed to play back pre-recorded cylinders. The distinctive governor assembly of the old models was replaced with a simple belt drive connected to the spring motor underneath the casing. The new model came equipped with the automatic reproducer and a recorder could be bought as an extra. A simple horn was attached to the reproducer to amplify the sound. There were two main controls: a start-stop switch to activate the drive and a screw to retard or accelerate the speed of cylinder rotation. The move away from the business market, with its emphasis on recording, was the key to simplify the operation of the talking machine; the user only had to place the reproducer arm on the cylinder. In one stroke many of the operating problems of the earlier models had been removed. More information about the customer's use of the machine led to this important development.[9] The spring motor model was simple, sturdy, and solid. As the cost of manufacture was reduced by using many components of the old M type, the price came down to about $100.

The immediate commercial success of the spring motor phonograph resurrected Edison's plans for an industrial complex in the Orange valley. Just ten years after the tremendous failure of the business machine, he now had the product to mass-produce at the West Orange complex. He also had control of the phonograph business because he had delayed introducing the spring motor model until he had extricated himself from Lippincott's North American Company. The expense of perfecting the business machine in 1888 had forced Edison to sell his rights to the invention to Jesse Lippincott. The latter's death in 1891 produced a vacuum in the management of the North American Company and gave the graphophone interests an opportunity to overcome the miserable record of their machine by a counteroffensive in the courts. When they appeared to be in a position to gain control of the North American Company's patents, including all those produced at West Orange, Edison acted quickly. He became president of the company and made Tate vice-president. His first policy was to dismantle the franchise system and replace it with a sales network. The franchisees were to become sales agents who received a royalty on every phonograph sold in their areas.[10]

The price of developing a successful amusement machine had been high, and neither the liquidation of Lippincott's fortune nor the payments from the North American Company could cover the cost of years of experiments at West Orange.[11] Edison was sure that the phonograph was superior to the graphophone. It was no accident that he timed his takeover of the marketing network to coincide with the introduction of a commercial phonograph. He instructed Tate to put the North American Company into bankruptcy and arranged to sell phonographs directly from the factory to individual dealers, bypassing the franchised local companies. The National Phonograph Company was formed to manage the nationwide system of phonograph dealers, and only the Columbia, New York, and New England Companies survived from the original North American Company franchisees. Once this maneuver was completed, the new Edison spring motor phonographs were announced. The technology of the amusement phonograph neatly fitted his manufacturing strategy. The simpler operation of this machine removed the requirement for an extensive system of sales and service support. Edison realized that he did not need a large dealer network to reach the market; he could sell machines directly from the Phonograph Works to anyone who wanted to resell them.[12]

Edison was not alone in the musical phonograph market. The Columbia Phonograph Company—one of the franchisees of the defunct North American Phonograph Company—had acquired the rights to manufacture graphophones and already had a spring motor model priced below the Edison product. The West Orange laboratory staff was immediately put to work to reduce the size and cost of the spring motor phonograph. The "Home" model was built around a much smaller spring motor that reduced its weight by almost twenty lbs. Billed as "the machine for the millions" it sold for $40, bringing it into competition with the low-priced Columbia models.[13]

The race was on between the National and Columbia companies for the new market for cheap amusement machines. Further price reductions by Columbia forced Edison's engineers to redesign the basic spring motor model to bring its price down to $25. The new "Standard" model introduced in 1897 was stripped of the distinctive overhanging mandrel that had characterized all phonographs produced at West Orange. Its motor could play two or three records at one winding compared with the fourteen claimed for the sturdy spring motor of the original 1896 model. Yet it weighed only seventeen lbs. compared with the forty-three lbs. of the prototype, and its price of $20 undersold the popular $25 Columbias. The final shot of the West Orange lab in this price war was the $10 Gem introduced in 1899. A tiny (seven lb.) machine that could only play records, it was designed to compete with the cheap European machines entering the American market at that time.[14]

These new phonographs could not be called inventions and hardly justified the label of innovations. Edison commented that the 1896 model was the result of design changes rather than any patentable technology.[15] There was no flash of inspiration in the development of this machine and its successors, only the slow improvements that came after years of investigation into sound and a better understanding of what the user wanted. Once the new design was standardized, the staff of the lab quickly modified it to suit the competitive market. The 1896 phonograph was a stage in the process that turned the unreliable, overcomplicated machine of the 1880s into "a first class, plain, practical machine" with a mass market in the twentieth century.[16]

With the worst of the Great Depression over by 1896, there were plenty of customers for the cheap spring motor phonographs. The urban population was growing at a furious rate, and America's great cities became the market for the electrical appliances and phonographs manufactured at West Orange. With sales of machines booming, the need to mass-produce recorded music became even more acute as the nineteenth century drew to a close. A team had been working since 1888 at West Orange to produce molds of recorded cylinders that could be used to make numerous duplicates of the originals and now the end was in sight.[17]

This project centered on a process of making a metal negative—a matrix—of the master cylinder to mold duplicates. The process developed in the laboratory used a vacuum deposit technique to plate a thin layer of gold on the white wax master. This microscopic layer was built up by electroplating copper or nickel on it to form a metallic matrix—a process that took several days and the utmost care. After removing the original cylinder, the matrix master was cleaned with benzene and dipped into hot wax to make a submaster from which many copies of the original were molded.[18] By 1899 Charles Wurth's experimental team had finally produced acceptable duplicates of pre-recorded cylinders from their matrices, but the work went on to eliminate the bugs in the molding process. The problem that Wurth faced was that shrinkage of the grooves during the duplicating process adversely affected the sound quality of the recording, and this was the one area in which the Old Man would allow no compromise. Edison records had to have perfect fidelity. As most of the duplicating process could not be seen or controlled, a painstaking effort was made to gauge the amount of contraction and compensate for it by using fewer cuts per inch and a slower recording speed for the master, and then increasing them gradually through the submaster to copy stage.[19]

The hardness of the cylinder wax was a critical element in these experiments. Wurth and his team had to work closely with Aylsworth who was developing new compounds to bear the heavier tracking pressure of improved reproducers. The hardness of the wax was also a critical element in capturing a wider range of frequencies, especially the low, bass notes that cut a deeper groove in the record. The steady progress in reproducers and recording techniques in the 1890s made the work of Jonas Aylsworth critically important, for any improvement in these areas had to be matched by better wax compounds, or the benefits might be wasted. The criteria that bounded his work were demanding: the ideal substance did not only have to be very hard, but also had to be easy to mold into cylinders and predictable in the duplicating process. His improved wax compound was made up into the "new high-speed hard wax" molded records that Edison planned to make in the tens of thousands.[20]

It was not until 1901 that all the many difficulties had been overcome and the commercial production of records begun. The laboratory staff executed a shop order to double the capacity of Wurth's trial plant and began to design the automatic machinery for the large-scale duplicating facilities Edison planned.[21] The years of experiments in duplicating—carried on without interruption since 1888—had now paid off; the molded cylinder records that sold for fifty cents only cost seven cents to make.[22] In 1890 Edison had seen that the business of selling amusement phonographs was based on the availability of reasonably priced records. At that time records were made by anyone who could handle the delicate recording diaphragm of the perfected phonograph. Consequently, recorded cylinders were scarce, high priced, and low quality. Ten years later, when phonograph sales were soaring, the National Phonograph Company could deliver the right product at the right time. The successful conclusion of the cylinder duplication project became the foundation of Edison's phonograph business. Output of records was 25,000 a day in 1903 and rose rapidly.[23] Edison's dream of an electric light in every home had now been replaced by one of putting a phonograph in every sitting room in North America and Europe.

The record-buying public had a wide choice of recorded music—popular songs, dance tunes, and romantic ballads—but little opportunity to listen to serious music because the cylinder could only play for two minutes. The most that could be crammed onto the cylinders were solos from well-known pieces—Gilbert and Sullivan were especially popular—and short selections from overtures. Symphonic music was difficult to record because arranging the large number of instruments in the studio was a problem.[24]

Humorous monologues taken from vaudeville were easy to record and immensely popular with an audience who saw the talking machine more as a novelty than a source of music in the home. Irish humor—in monologues and songs—was a fixture in recordings, but the most popular of all humorous material in the Edison catalogue were the "coon" songs in which white vaudeville artists portrayed black stereotypes in comic monologues, in such forgettable recordings as "Who Dat Say 'Chicken' in Dis Crowd" and "De Coonville Cake Walk."[25] Not all the performers on Edison records were professional comedians; in some cases members of Miller's clerical staff made coon recordings, a reflection of the amateurism that marked the first recordings and the ease with which this kind of humor could be put onto a cylinder.[26] The same stereotypes were depicted in the comic films made by the Edison company. The humor of popular films, such as the Uncle Josh series, could also be enjoyed on pre-recorded cylinders.[27] Uncle Josh was a country bumpkin whose exploits in the big city were easy to predict. The humorous possibilities of country folk coming to terms with modern city life were well represented in commercial recordings, which at this time were purchased by city dwellers.

The talking machine was primarily sold in the growing urban market. Judging by the recording catalogs at the turn of the century, the musical taste of the record-buying public tended toward sentimental favorites that romanticized rural life, such as "Church Scene from the Old Homestead" or "When It's Moonlight on the Prairie." Recorded music reflected the cultural values of the men who made it and its affluent urban audience. It gave these Americans what they wanted to hear, "the dear old songs of heart and home" as one recording catalog put it.[28]

The first decade of the twentieth century marked the transformation of a novelty into an article of mass consumption, as the phonograph moved from amusement parlors to the living rooms of average Americans. Recorded music could now influence as well as reflect the taste of its audience, which grew larger every year. Phonograph sales per annum in the 1890s started from $25,000 and soon reached $250,000 by the end of the decade. In 1900 the National Phonograph Company sold over a million dollars worth of goods. Each successive year of the twentieth century brought an increase in sales.[29] The West Orange complex was producing a complete line of ten phonographs by 1903, ranging from the deluxe Concert model

Figure 8.1. A proud Edison next to a spring motor model phonograph. Its long amplifying horn required a stand to support it. This photograph was taken in 1906 at the height of the cylinder's popularity.

that used large five-inch diameter cylinders to the baby Gem. Edison was coming close to realizing his goal of one of his machines in every home, announcing that "There is no family so poor that it cannot buy a talking machine"[30] (fig. 8.1).

The dramatic growth of the phonograph business made its mark on the Edison enterprise in the form of a massive expansion program. The West Orange complex grew larger as several new buildings were erected to manufacture cylinders. The demand for recorded music also brought a factory complex at nearby Silver Lake to life to supply the stearic acid and other bulk chemicals required for cylinder production. Edison had acquired forty-seven acres of land at Silver Lake, New Jersey (now in the community of Belleville) as part of his Edison Industrial Works scheme of 1888, hoping that this would become a manufacturing plant for joint ventures with investors. When this plan came to nothing, he set up the Edison Manufacturing Company there with his own money in 1889 to produce the chemicals needed in phonograph records and batteries. The Manufacturing Company also managed the production of Edison-Lalande batteries, electric fans, and electromedical equipment. Once commercially introduced, the motion picture was handed over to the company, which formed a Kinetoscope department to oversee production of films and arrange for their distribution.

Edison's business strategy for both the motion picture and talking machine was to maintain a monopoly of the media. The technology was shaped to this end. The first film camera was designed to be operated in the Black Maria, within the West Orange complex. Known as the "dog house," the kinetograph was too heavy and cumbersome to be used on location. Edison maintained a strict policy of controlling the use of the movie camera, not letting it out of the hands of the cameraman during film production.[31] He wanted a monopoly of the film media and a captive market in the peep-show exhibitors who needed a constant supply of film. In the phonograph business he hoped that his patented duplication method would dominate the mass production of recordings. After years of costly experiments, he saw his opportunity to cover his expenses in a booming market. The first priority was revenue and as long as the motion-picture business kept the Manufacturing Company fully occupied, he was loath to push the technical development of the product.

Edison has been criticized for neglecting the development of film projection. Some historians of motion pictures portray him as uninterested in this technology and slow to develop a projector. Yet by 1895 he had made $75,000 from sales of Kinetoscopes.[32] This was reason enough to stay with the Kinetoscope until it lost its commercial appeal. He was working to advance the technology of motion pictures, but he chose to give projection second place to the development of talking pictures, which he thought would define the future of the industry. He believed that motion-picture technology would reach the point where grand opera could be reproduced by one of his machines "without any material change from the original."[33] This was a much harder technical problem than projection, and one to which he was committed from the first days of research on the Kinetoscope. This objective might seem too ambitious in light of the primitive state of the technology at the turn of the century and the difficulty of synchronizing sight and sound. A commercial system was another twenty-five years in the making, but Edison was not to know this. In his favor were the undeniable expertise of the West Orange laboratory staff in the technics of sound and several years of experiments.

The first step toward talking pictures was the kinetophone, a peep show machine introduced in 1895. This was a Kinetoscope with a cylinder phonograph installed in the casing. The viewer looked through the peep hole to see the image while listening to the sound track on eartubes (fig. 8.2). It is doubtful that sight and sound were synchronized in this machine despite Dickson's description of recording phonographs used in tandem with the kinetograph in the Black Maria film studio. The viewer probably heard a musical accompaniment to the filmstrip. Although produced in small numbers, and by no means a commercial success, this machine established the precedent of a self-contained sight-and-sound system that Edison saw as a consumer product that would go into the homes of Americans. His goal of providing "honest workmen with grand opera at a price he [sic] can afford to pay" might have been somewhat off target in his appreciation of the needs of the audience, but it was far ahead of its time.[34] His mind was set on producing talking pictures. This had been his original intent in the motion-picture project, and he was confident that this innovation could capture the attention of a large audience. Unfortunately the development of this idea was interrupted by the impact of film projection that completely changed the motion picture business and undermined the dominant position enjoyed by the Edison Manufacturing Company.

Film Projection

While the laboratory staff struggled with the problems of synchronizing sight and sound, several other inventors (many of whom were in Europe) were concentrating on the much simpler problem of projecting moving pictures. The transatlantic information network that Edison had so successfully exploited in the past now got the better of him. The Kinetoscope had stimulated interest in motion pictures and Edison's failure to obtain foreign patents for his motion

Figure 8.2. Enjoying the sight and sound of the kinetophone about 1895.

picture inventions gave European inventors a clear field to advance the technology.[35] The most revolutionary innovation came from the Lumiere brothers in Lyons, France; their cinematographe camera was portable and it could be adapted for use as a projector. Work on a combined camera-projector was underway at West Orange, but the Lumiere brothers beat Edison to the marketplace. Their camera was light, compact, and movable, providing many new possibilities for film subjects.

The introduction of new film cameras in North America gave entrepreneurs the technology to enter into film production and ruined Edison's hope for a film monopoly. The availability of cameras unleashed a wave of cameramen/producers, some with imported cameras, some with copies of Edison cameras, and some with machines of their own invention. Many of the new entrants were men of mechanical bent hoping to break into a growing business by working up cameras in makeshift laboratories or machine shops. If Edison had been born twenty years later, he would have been one of them, but in 1897 the fifty-year-old businessman was cast in the role of defender of a proprietary technology. His competitors came from all walks of life and from all parts of North America and Europe. The only thing they had in common was the notion of making money from motion pictures. A traveling showman called William Selig saw the Kinetoscope and was inspired to build his own camera in his small workshop in Chicago. The arrival of Lumiere machines in Chicago provided more ideas for Selig, who went on to produce a camera and a projecting machine. The Selig Polyscope Company was soon making films on location in the back streets of Chicago and showing them in vaudeville houses.[36] Independent inventors such as William Paley, a recent immigrant from England, took their cumbersome cameras on the road and made the short-lived, flickering films that were the raw material of the new business.

The Latham brothers were typical of the "pirates" that Edison now had to face. They were businessmen with considerable vision who also produced some important technical innovations with the help of William Dickson (who left the West Orange laboratory in 1895). They developed a projecting device, called the eidoloscope, in their laboratory in New York City and in 1895 exhibited their boxing films with it. As soon as this development was noticed at West Orange, Edison assigned Charles Kayser to the task of designing a projector superior to the Lathams'.[37] Kayser was a competent experimenter with considerable experience in motion-picture technology, but he was not in the same league as the man he replaced, who was now in the enemy's camp.

A much more dangerous competitor than the eidoloscope soon appeared in the form of the Biograph—the projection machine of the American Mutoscope Company. Formed in 1895, this company had the services of Dickson, its own laboratory and manufacturing facilities. Its studio revolved on the roof of its offices on Broadway, duplicating the format of the Black Maria at West Orange. The company quickly branched into projection. The Biograph projector was the work of Dickson, who used a larger film format (68 mm film width compared with 35 mm) to give it a superior image.[38] During 1896, the first year of film projection, the Biograph machine stood out by virtue of its technical excellence. The renamed American Mutoscope and Biograph Company became one of the most important producers in the history of silent films, aided by the genius of the director D. W. Griffith who made many Biograph films.[39]

The proliferation of projection machines in the United States accelerated the decline of the Kinetoscope business. In addition to a larger and much clearer image, the Lumiere and Biograph projectors had an intermittent motion to give an illusion of movement that was lacking in peep-show machines. The Kinetoscope Company was now in the hands of two entrepreneurs, Norman Raff and Frank Gammon, who were described by Edison's biographer, Matthew Josephson, as "speculators, backers of race track entries or theatrical shows." Their gamble on the Kinetoscope had failed by the end of 1895 when it became obvious that the novelty was wearing off. Raff admitted, "Our candid opinion is that the Kinetoscope will be a dead duck after this season."[40] Raff and Gammon had to take action to revive their flagging business. They saw the future of moving pictures in projection and consequently when they found a good projector, in the hands of C. Francis Jenkins and Thomas Armat, two amateur inventors in Washington, they quickly acquired the rights to it.

In January 1896 Raff and Gammon brought Armat's machine to the West Orange laboratory to show it to Edison. He agreed to manufacture it in the Phonograph Works and have it sold under the name of the Edison vitascope. There was good sense in this because Edison's name on the machine assured its acceptance. In a letter written to Armat some time later, Edison acknowledged that the Armat design was superior to the one he had at the time. He stopped Kayser's work to concentrate on Armat's machine, which was the first commercial projector to be mass produced for the American market.[41]

The debut of the Edison/Armat projector, at the famous Koster & Bial's music hall in New York City on April 23, 1896, was a turning point in the history of motion pictures. Visual novelties were commonplace in vaudeville, but this was something different. The vitascope soon found a place on vaudeville programs all over the United States, replacing the limited audience for peep shows. The projector was not limited to a fifty-foot loop of film and this gave exhibitors the opportunity to present longer programs. The ready acceptance of projected motion pictures dramatically increased revenues, for each filmstrip had an audience of hundreds instead of the individual peep-show customer. Businessmen bought franchises from Raff and Gammon and took the vitascope on the road, spreading the fame of the movies and attracting more entrepreneurs to the industry. The newspapers, unconcerned with its true provenance, hailed Edison's latest invention as another marvel of the West Orange laboratory. The movies as mass entertainment had begun.

The popularity of the vitascope grew so quickly that Edison had to take notice. While he amused himself by planning spectacular train crashes to be filmed, the Kinetoscope department of the Edison Manufacturing Company struggled to meet orders for films. At this early stage in the development of motion pictures, film subjects were limited to documentaries, travelogues, and short comedies. In the words of film historian Charles Musser, "cinema was used primarily as a visual newspaper in the late 1880s and early 1900s."[42] Despite having the largest production facilities in the industry, the Manufacturing Company could not keep up with the demand for films and had to turn to independent producers to supply them. William Paley was put under contract to supply films for the Edison enterprise.

The impending Spanish-American War had caught the imagination of the American public, with the help of the strident nationalism of the Hearst yellow press, and gave a considerable boost to the entertainment business. Biograph, now the chief competitor of the Edison company, had sent one of their cameramen to Cuba to film the historic events there and Edison followed suit by dispatching William Paley to the front to make war films. The motion picture camera influenced politics as well as entertainment. America's Manifest Destiny was glorified on film and in popular music, and Edison—like many of his generation—believed that his country deserved a place as a great power. The recording catalog of the National Phonograph Company was swollen with patriotic songs and military marches that reflected the current sentiments of the United States. War films had a strong appeal to audiences from all classes. Some theatres ran them continuously during the war. The great popularity of these films enabled many small producers to gain a foothold in the industry, either by filming their own subjects or illegally copying (duping) the Edison product. The American Vitagraph Company, for example, staged the Battle of Manila with model ships, advertising them as "original and exclusive war films."[43] For

American Vitagraph and many other small producers and film exhibitors, such as William Selig in Chicago, war was good for business.

The initial strategy of the major companies was to lease film exhibition services and retain ownership of both film and projector. But the boom in film projection attracted so many independent (and usually illegal) operations into the business that the industry leaders were forced into selling their products on the open market. All that a prospective film exhibitor needed (after acquiring the movie) was a rented hall and a good projector. As soon as the Armat machine was brought into the West Orange laboratory, Edison's mechanics began the job of simplifying it to reduce the cost of manufacture—a task to which they had become accustomed. At the same time several experimenters were developing an improved projector under a veil of secrecy; Edison's lab was not the only one racing to bring out a new machine.

The Edison Projecting Kinetoscope—a redesigned vitascope—was part of the transformation of another novelty to an important entertainment industry. The 1897 model Edison Projecting Kinetoscope came complete for $100. The 1898 model could be used to project slides to keep the attention of audiences while the films were changed. Its advertising maintained that the vibration problem, which "has been the principal defect in projecting machines," had been solved.[44] Edison's laboratory had improved the basic projecting machine, and now it was on the market at a competitive price. The author Terry Ramsaye, a contemporary of this period of movie history, remembered that anyone who could turn a meat grinder could operate the machine.[45] The exhibitor could also purchase "genuine Edison films" which were sold by the foot. The ready availability and cheapness of the Edison projector helped stimulate the motion-picture industry; a small investment was all that was needed to exhibit films. A showman could offer a complete program to the theatre manager: film, projection, and accompanying patter.

Two new industries were being created with the output of the Edison Phonograph Works, providing many with the chance to make money from Edison products and Edison with the profits from sales of equipment. The flood of cheap phonographs and the end of the franchise system meant that anyone could create a small retail operation with a storefront and a few phonographs. Men who had traveled door-to-door with the machine, or who had run the business from a corner of a hardware store, now took a step toward respectability by running a record shop—the wide choice of recordings made it imperative to offer the customer a place to listen to the latest songs.[46] Edison thought that selling the talking machine and recordings was the best business proposition at the turn of the century because it was easy to enter and profits were high, estimated by him to be from $100 to $200 a week.[47] The seemingly unlimited audience for moving pictures provided another opportunity for small businessmen and a market for the Edison Manufacturing Company. In the same way as he had bypassed the franchise holders of the North American Phonograph Company, Edison left Raff and Gammon high and dry and went into direct sales. He intended to sell his projectors to any showman and entrepreneur who could afford them.

The policy of gaining control of manufacturing had certainly paid off by the end of the century. The profits coming in from phonographs and motion pictures offset the gigantic losses of the ore-milling venture. Edison's greatest gamble of the nineteenth century ended with one last effort to turn the Ogden mine into a paying proposition. He was not a man to give up on an experimental project. A slight rise in the price of midwestern ore encouraged him to return to the mine in 1897. He spent most of that year in Ogden, leaving the operation at West Orange in the hands of a subordinate.[48] Edison was convinced that he had "almost turned the corner" on the ore-milling venture in 1898 and told his business associates that the work was finished. But the cold winter of 1898/99 forced him to stop work. He returned to West Orange, admitting defeat in private while maintaining a brave front to his business associates.[49] This was Edison's greatest failure both financially and technologically; he had been unable to make money from his magnetic separation idea and had lost a fortune in the process.

Had the mill at Ogden been a success, Edison's reputation as an inventor would have been superseded by his fame as an industrialist. The triumphant years of the electric light would have paled against the easy profits of his ore-milling business. The end of this dream might have broken a lesser man, but Edison took the disappointment with his customary good humor. He had other irons in the fire and was, as usual, full of ideas for new projects such as an improved storage battery and Portland cement. The failure at Ogden did not even tarnish his reputation—the vitascope had reaffirmed to the American public that Edison was still the Wizard capable of producing amazing new inventions. The decade that started with the grand industrial engineering of electric traction systems and ore separation ended with the cheap phonograph and moving pictures, two unlikely products for the creator of the electric lighting industry, but welcome nevertheless.

Expansion and Integration

In each successive year of the twentieth century, the factories of the West Orange complex produced more phonographs and motion pictures. By 1900 the labor force had increased to about 1,000 men in the Works and 150 in the laboratory—a significant increase over the work force of the 1890s.[50] The expansion of the motion-picture business led to the enlargement of the manufacturing and film-processing plant. Wherever Edison cameramen took their pictures—whether it was Edwin Porter's films of the America's Cup or Robert Bonine's views of Alaska—they were always brought back to West Orange for processing. The Works also manufactured projectors and accessories in great numbers in addition to an impressive number of phonographs. By 1906, a record-breaking year for sales, the Works was producing thousands of machines and hundreds of thousands of cylinders every month, but this still was not enough to meet demand. In 1907 over 20 million Edison records were sold.[51] The amusement phonograph was now the core industry of the Edison enterprise. Ten years after the introduction of the spring motor phonograph, this business had transformed the West Orange complex. The many shacks, lean-tos, and other temporary structures—the architectural heritage of contract research in the nineteenth century—were swept aside as several new buildings were constructed.

The first to be built were the facilities to produce records. Extensions were added to the back and sides of the Phonograph Works, and a separate building to make records was constructed parallel to the laboratory on its North side, adjacent to the ends of the satellite buildings (fig. 8.3). It was 240-feet long and by 1906 five stories high. Building 24 (as it was called) contained the machines to mold the records and rooms for testing and quality control. Edison sometimes carried out experiments in this building. A new front facade was erected to the Phonograph Works, crossing the T of the buildings that faced Lakeside Avenue. The power house was re-sited at the end of this new structure that provided a massive concrete surround to the old wooden Phonograph Works, greatly enlarging the floor space of the factory and accommodating a work force that had swollen to 4,000 by 1908.

Figure 8.3. Artist's impression of the West Orange complex, reproduced from the *Edison Phonograph Monthly,* 1903. Comparing this scene with Figure 6.1 shows how much the Phonograph Works had grown. The original two factory buildings still stand in the middle of the Works, with the railroad spur running between them. The Black Maria can be seen in the space between lab and Works.

The most impressive addition to the West Orange complex was the new office building completed in 1906. It stood next to the laboratory on Lakeside Avenue, an easy walk from Edison's desk in the library of Building 5. Four stories high and 137-feet long, the new building housed the office workers and management of the Edison companies.[52] The National Phonograph Company occupied most of the administration building. It controlled the flow of materials through the Phonograph Works and managed the distribution network of jobbers and local dealers.

The company sold to a worldwide market through its overseas affiliates. Its staff had to deal with information coming in from sales, manufacturing, and the laboratory. Their number increased with the volume of production, as did the tasks they performed. As the phonograph industry became more competitive, the company extended its control over its dealers, prohibiting them from discounting prices and eventually requiring them to sell Edison products exclusively. By the 1920s the U.S. distribution network exceeded 10,000 dealers. This large organization required constant supervision to maintain the morale and obedience of the retail outlets.

The Edison enterprise was moving from the personal endeavor of its founder to raise money for his laboratory to an organization of professional managers running an international business. The business of innovation that had once been carried out by the experimental bench or in the hallways of the second floor of the laboratory was slowly moving to offices in the administration building. The small-scale administration of the nineteenth century was evolving into a large organization comprised of specialized departments. This process had begun during the 1890s when the ore-milling project took Edison away from the laboratory for long periods. While he was away, the responsibility of running the West Orange operation was given to William Gilmore, one of Samuel Insull's proteges. The unprecedented growth of Edison's phonograph business soon after and the demands of serving a national market justified a large, white-collar work force that was organized into specialized departments. The advertising, accounting, legal, and purchasing departments administered a growing number of transactions in the first decade of the twentieth century. Each was run by a professional manager. The creation of specialized functions in the organization had the wholehearted approval of Edison, who viewed this as a more efficient way to do business. He stressed the economy of centralizing all purchasing, for example, into one department to administer all the needs of the enterprise.

The Legal Department grew out of Edison's close relations with his patent lawyers. The work of obtaining and defending Edison's patents became vitally important in the twentieth century. The "patent wars" dominated the motion-picture business until the creation of the Motion Picture Patents Company in 1908. The Legal Department relentlessly harried film producers for infringing Edison's camera patents, driving many of them from the business.

The patent position was equally important in the phonograph industry, a highly litigious business where the rights to several critical technologies were disputed. The failure to secure full patent coverage in the area of composition and duplication of phonograph records cost Edison dearly. The Legal Department also provided useful information for the laboratory staff by reviewing all relevant patents in the field and monitoring the technical advances of the competition. The complex legal arrangements of the companies formed to market phonographs in the nineteenth century provided plenty of headaches for the Legal Department in the twentieth.

The marketing of the consumer durables produced at West Orange brought Edison's name into millions of households. His face and signature were recognized all over North America and Europe. The Advertising Department coordinated the promotion of his products, supporting newspaper and magazine advertising on a national basis. Edison had been his own promoter in the nineteenth century and had taken the work very seriously. He astutely manipulated the print media to exploit the Wizard of Menlo Park myth. In the twentieth century the Edison myth was promoted for the same commercial ends but on a larger scale. The public had come to expect magic from the Wizard and high-quality products from his factories. The highly publicized West Orange laboratory was playing its part in molding the public's impression of Edison as the world's greatest inventor, laboring in his magnificent laboratory to produce new wonders.

The growing size of the operation, the diversity of products, and the sheer number of transactions carried out by the Edison enterprise required a centralized organization for bookkeeping and information gathering. The Accounting Department was given the herculean task of bringing all bookkeeping functions into one office. It provided the information needed to coordinate production and sales. It replaced the part-time activities of Edison's secretaries, foremen in the machine shops, and unruly young men who doubled as clerks when they were not running errands or mucking around in the experimental rooms. The Accounting Department was to become more important as the flow of information grew, yet it began as a small office with one professional accountant. Walter Eckert did practically all the work of producing balance sheets and statements.[53]

The supervisors of the specialized departments formed a slim layer of middle management. Many of them had started in lowly posts, gradually rising to positions of influence within the organization. Several promising young employees were assigned to head sales departments at the beginning of the century. Nelson Durand took over the Dictating Machine Department; Frank Dolbeer was appointed sales manager of phonographs in 1911; and Leonard McChesney started his career as a clerk in the Advertising Department and rose to the post of advertising manager for motion pictures. These men went on to play an important part in the affairs of the Edison enterprise in the twentieth century. They answered to William Gilmore and sat on the many committees that carried out the administration of the Edison enterprise. As early as 1908, the managers of the various departments were coordinating their actions into an organ of central control. An executive committee was formed in that year to manage the business while Gilmore was away. Each head of department was in attendance to deal with business that affected his department, and collective decisions were made on important questions.[54]

The Decline of the Shop Culture

The influx of professional administrators in the organization diminished the importance of the boys who had traditionally blended technical and management duties in the Edison enterprise. Although mechanics like the Otts were still at Edison's side, providing a link of continuity with the old machine shop era, there were many new faces in the laboratory by the end of the nineteenth century. The older generation of laboratory workers was spent and the new one was quite different. The ranks of the boys steadily diminished in the 1890s when several of them left West Orange to go their own ways. Edison had once commented that the boys worked with him for fame and fortune. Many of them saw their big chance in the 1890s and left the lab. Batchelor quietly disappeared from the scene and devoted himself to travel, his health, and his investments; Kennelly left to become professor of electrical engineering at Harvard; Samuel Insull took a job at Chicago Edison; and A. O. Tate resigned because he felt that Edison had not been fair in his dealings with the franchisees of the North American Phonograph Company.

In place of the trusted associates who were at home in both the laboratory and the boardroom, Edison hired professional administrators like William Gilmore. Although a powerful force in the management of the Edison companies, Gilmore was limited to financial and administrative affairs—his status did not equal that of Batchelor or Tate, and Edison thought nothing of overriding his decisions. Tate's place as Edison's secretary was taken by Johnny Randolph who had started as an office boy in the 1880s. Once elevated to the post closest to the top of the organization, Randolph found that the job had its disadvantages, notably the avalanche of paper involved in constructing the industrial empire and the impatience of the Old Man. Randolph's rise to power is evidence of Edison's continuing policy of promoting the faithful. Fred Devonald, for example, had worked his way through the hierarchy at West Orange and owed everything to Edison. He was finally made chief engineer but was only in nominal charge of the laboratory; Edison made all the important decisions. Gilmore, Randolph, and Devonald wielded considerable power within the organization, but they were not given the same broad responsibilities as the boys of old, nor did they receive the respect that was given to the machine shop fraternity. They were employees. They were constantly harried by the Old Man. When their usefulness was over they were discharged.

It is possible to explain these changes in terms of the gradual evolution of an administration in response to the requirements of a larger business organization, as another episode in the development of modern management. The machine shop culture that produced experimenters with excellent management skills had flourished when the enterprise had been relatively small and concentrated on research and development. But a great industrial enterprise that made products as diverse as primary batteries and motion pictures required an administration of specialists who had to confine their activities to one branch of management. The size of the business and the complexity of the technologies under development worked against the possibility of Edison finding another Charles Batchelor to accompany him in the twentieth century.

There were also irrational, emotional explanations to explain this separation of function in the Edison enterprise, and chief among them was Edison's growing disillusionment with the boys. Several of them had committed the cardinal sin of disloyalty to him. Some, like E. H. Johnson, threw in with rival inventors; others, like Ezra Gilliland, simply took the money and ran.

Edison's sorrows over the ungrateful boys came to a head with the defection of William Dickson, a close associate and protege, who left the laboratory in April 1895 to join the Latham brothers syndicate. Ironically Dickson's downfall came from the old practice of cooperative working in the laboratory. William Gilmore's uneasiness with the joint projects with the Lathams was upheld by Edison, and this led to accusations that Dickson was being too friendly with a potential competitor. Dickson left West Orange and took his experimental notebooks with him, providing the Latham company with his considerable experience in motion pictures.[55] He later joined the Biograph Company, which became a dangerous rival in the motion-picture industry.

Dickson was not the only experimenter to take his specialized knowledge to a competing company; the great competition within the phonograph and motion-picture businesses lured away several key employees, including Eugene Lauste (who worked with Dickson on motion pictures) and John English, the superintendent of the Phonograph Works. The two new technologies developed at the

West Orange laboratory in the 1890s had created turbulent, competitive industries that were driven by innovation. They were also highly profitable; in 1910 the phonograph and motion-picture businesses were each worth well over a million dollars in sales to the Edison enterprise. Ease of entry and high profits had attracted many small companies into these businesses, which in turn contributed to the pace of technical change. The laboratory staff were now pitted against the engineering departments of aggressive and innovative companies such as Columbia and Biograph. Columbia proved to be a constant thorn in Edison's side. It led in the technical improvement of talking machines, manufactured cheap machines for a variety of small companies in the business, and even branched out into film production.[56]

The technological battle was fought with both fair means and foul. The intense competition placed a premium on industrial espionage, which was practiced freely during the 1890s and 1900s when it was said that each company maintained spies in the enemy's camp. Each side carefully examined the opposition's machines to try to gain the advantage. Edison was not above instructing the laboratory staff to copy the successful engineering of the opposition or to lure away the skilled men of other companies.[57] Edison's efforts to match the improved phonograph of Bell and Tainter were assisted by one of their skilled mechanics who came to West Orange with valuable knowledge of their experiments.[58] From the beginning of the talking-machine business, a premium was placed on the skills and experience of the men who understood the workings of the new machine. The first "perfected" phonograph, it will be remembered, left West Orange for England shortly after the famous photograph of Edison and his team was taken in the summer of 1888. A laboratory technician called Hamilton was assigned to accompany Gouraud and the machine across the Atlantic.[59]

The value of specialized skills in these new technologies made the defection of a key experimenter or mechanic something to be avoided. One man could give the opposition the benefit of years of experience. Although the staff of the laboratory were expected to reflect the shop culture values of flexibility and adaptability, the fact that they had pioneered several new technologies in the 1890s made some of the muckers highly valuable. The leading recording engineers of the mature phonograph industry, such as Fred Gaisberg of Victor and Calvin G. Childs of Columbia, began their careers in small companies and moved to dominant positions in the recording industry.[60]

The industries of the Second Industrial Revolution are generally accepted to be based on scientific information, leading historians to coin the term "scientific industries." Although one usually thinks of the electrical, chemical, and steel industries in this context, the talking-machine business was no less scientific, and it was sometimes claimed that this field contained "the richest contribution science has made" to industry.[61] The production of pre-recorded cylinders, for example, was based on a curious mixture of scientific technique and craft skills. The surfaces of the blanks used to make the masters were machined to absolute smoothness by a special cutting machine adjusted by a micrometer. They were examined under a microscope to ensure no imperfections.[62] Yet success in recording music was generally accepted to be based on the skills of the recording engineer because most of the important decisions taken in the studio were judgments based on prior knowledge and experience. The man in charge of the recording had to match the composition of the recording diaphragm with the voice of the performer; the thinner the glass diaphragm the more sensitive it was to high notes. Choose too thin a diaphragm and the result was a blasting of the recreated sound, but a thicker, firmer diaphragm could fail to capture the highs notes of the performance. The shape, size, and elevation of the recording horn were also important considerations for the recording engineer. The placement of musicians in the recording studio was critical in mixing the different sounds in the recording. Although this task involved a great deal of trial and error, it depended on the technician's intuition for the way a voice or instrument would sound as it was reproduced from the soft wax of the cylinder. The recording staff had to continually adjust the recording parameters to suit the music and coach the performers to interpret it in a manner suitable for good recordings; "avoid singing with too much expression" was the advice of the National Phonograph Company.[63] These techniques could only be learned by doing. Formal education was of little use in placing instruments in a recording session or filming a moving train. The experimenter's understanding of and feel for the materials they used—the wax compound of phonograph cylinders and the celluloid of films—played an immeasurable part in the quality of the final product.

Craft skills were evident in the international diffusion of these technologies, for nobody knew the subtle intricacies of recording or duplicating phonograph records better than the West Orange laboratory staff. They followed the technology across the Atlantic, setting up recording studios, establishing plants to produce the wax compound in bulk, and installing the equipment to duplicate pre-recorded cylinders. They knew when the chemical compound to make cylinders looked and smelled right, and the exact temperatures to be used in molding duplicates. Their experience helped them identify and remedy bugs in the process. The formula for the wax compound of Edison records could not be written out and sent through the mails; it required an experienced eye to judge the quality of the bulk chemicals and successfully mix them. Charles Wurth and Walter Miller had mastered their respective crafts to the point where the lab staff named processes after them. They transferred their specialized knowledge in person, training the first technicians of the European phonograph industry.[64] In the same way as the technology of the first Industrial Revolution had been spread by skilled mechanics from the old world to the new, the technology of the second was diffused by experts from the new world to the old.

Edison was well aware of the advantages and disadvantages of the international diffusion of new technology. He knew that it could eliminate the economic benefits of industrial research in a very short time. His goal was to profit from overseas sales and not see his technical lead ebb away as it was taken up on the other side of the Atlantic. The pressure of maintaining security in new technologies was aggravated by a distrust of the patent system. His failure to quickly patent important advances in motion picture cameras, for example, reflected his lack of confidence in patent protection. He preferred secrecy to filing patents. The West Orange laboratory became a security zone in the Edison complex, with mysterious experimental projects going on behind locked doors.[65] This movement undermined the old shop culture that had flourished in an atmosphere of friendly communication and cooperation. In the late 1880s the pressure of work and the constant demands of an adoring public forced him to limit access to the experimental rooms. In the 1890s he still maintained the tradition of allowing other inventors to work in his laboratory but was more selective, only allowing allies, like Harold Brown, and business associates, like the Latham brothers, access to his experimental facilities.[66] His experiences with the latter proved that the machine shop tradition of cooperation had its disadvantages in the modern competitive world.

Members of the Latham syndicate worked with the staff of the West Orange laboratory in the improvement of film handling in cameras and projectors. One of them, Enoch Rector, successfully enlarged the spool capacity to take longer films that exceeded a running time of two minutes. This innovation, called the Latham loop, was soon widely diffused. Although this technology had begun in the West Orange lab, Edison could not keep it under his control.[67] This kind of reversal served to undermine the shop culture at the West Orange laboratory, forcing Edison into the role of defending proprietary technologies with tight security and strict control over the lab staff. Skilled men who had once been free to move from job to job were now bound to the company that paid their wages. In the twentieth century recording engineers had to sign contracts to ensure that their knowledge would be not be transferred to the competition.

The shop culture had thrived in an open laboratory where individual initiative was rewarded. The spirit of entrepreneurship had fired the innovations of the Newark and Menlo Park periods, with mechanics and experimenters enjoying an upward mobility and a share of the profits. The shift to specialization of function at the turn of the century reduced the opportunities available to the laboratory work force, making it harder to move up from the machine shop. The practice of contracting out manufacturing jobs with mechanics had been maintained in the 1890s when the shop orders show several occasions when special experimental models or one-off customized products were made by mechanics under contract.[68] The contracts signed between Edison and his skilled machinists allowed the latter to determine the pace of production and subcontract for additional help. Yet the tools produced by the craftsmen to carry out the contract remained the property of the company.[69] The day was gone when the machinist kept his tool box with the specialized cutting tools as part of his contribution to the creation of value in the machine shop. The practice of contracting could not survive in the twentieth century when competition in the key businesses acted as the catalyst to bring greater efficiency to manufacturing at West Orange. Initiative and independent action had to be sacrificed to bring discipline and order to the factory floor. Separating the functions of experimenting, manufacturing, and management reduced the influence of the shop culture, and the systematic restructuring of manufacturing administered the decisive blow to craft practices in the laboratory and Works.

CHAPTER NINE

Thomas A. Edison, Incorporated

Edison and his managers had every reason to be optimistic about the new century: their phonograph business had finally taken off and they were in a strong position to dominate the motion picture industry. The Edison Manufacturing Company had become the foremost manufacturer of films and projectors in Europe and the United States.[1] Taking a leaf out of the book of the Edison Electric Light Company, the Manufacturing Company energetically defended its patent rights. A court decision in 1901 upheld Edison's original patents and put him in an unassailable position to control the production of motion pictures. "We have won" crowed the Manufacturing Company, convinced that only it had the right to make films and that the patents could now be used to drive the competition from the field.[2] The audience for popular films and music appeared to be infinite. Edison had given the public light, and now he was giving them entertainment; films and music for the masses would surely be as profitable as electric light and much less trouble to perfect.

In addition to these two established businesses, Edison hoped that his storage battery experiments would lead to a new business for the West Orange complex. Edison had long thought that a better alternative to the existing lead-acid combination could be found. During the 1890s he had experimented with storage and primary batteries, primarily in his efforts to improve the power source of phonographs and Kinetoscopes.[3] With the ore-milling project over by 1900 and the phonograph business well established, he turned to an experimental search that he had been considering for some time. The muckers in the chemical laboratory looked for an electrochemical combination that was more efficient than the existing lead-acid cells, concentrating on lighter alternatives such as cadmium, magnesium, and nickel for the lead of the battery. Edison wanted a storage battery that delivered a high charge per pound weight so that it would become the standard power source for electric automobiles. His great expectations of the storage battery business were reflected in a rush program of establishing manufacturing facilities. The Edison Storage Battery Company acquired an old brass mill in Glen Ridge, New Jersey, to serve as a factory. The Works at Silver Lake were enlarged and a small laboratory established there to carry out research on chemical manufacture. As soon as the experiments were underway at West Orange, Edison began the construction of several large factories to make storage batteries across from the laboratory on Lakeside Avenue.[4]

The storage battery factories were only one addition to an industrial empire that was spreading out from West Orange. The Edison Portland Cement Company was set up to exploit the cement-making machinery developed in the aftermath of the ore-milling venture. The company acquired limestone deposits in west New Jersey and constructed one of the biggest cement-making plants in the world. The scale and layout of this plant owed a great deal to Edison's experience at Ogden; it contained automatic machines driven by individual electric motors, and belts for material conveyance.[5]

In both the motion picture and phonograph businesses it was desirable to site studios close to the artists rather than bringing them to West Orange to perform. A recording studio was built on Fifth Avenue in New York City, the cultural center of the nation. The growing popularity of the movies justified the construction of a film studio in 1901 on 21st Street to replace the facilities at West Orange.[6] Motion pictures were made year round in this new addition to the empire, under the direction of the cameraman who had complete control of artistic and technical matters. The longer length of films and more elaborate productions inevitably led to larger studios to replace the roof-top installations and crowded rooms of buildings in downtown Manhattan. The Edison company followed its chief competitors—Biograph and Vitagraph—into the suburbs of New York City where they all built large studios equipped with artificial light. In 1905 the Edison company began the construction of a studio in the Bronx.

By 1910 Edison had created a diversified manufacturing operation, based on phonographs, motion pictures, electrically powered products, and primary batteries. The center of this sprawling industrial empire was the main laboratory building at West Orange. The laboratory staff were not solely concerned with research; they planned the layout of new factories, designed the machinery, and often made the tools in the lab and installed them in the Works. They were particularly busy with fitting up the storage battery factories. The intricate, and often highly original, battery-making machines occupied them for many years. They also carried out general engineering work in all of Edison's factories and even cleaned buildings at Silver Lake![7] The flurry of activity at the West Orange lab also produced changes in its physical structure and, as always, the laboratory staff carried out the work of rearranging the rooms and installing new equipment. Building 2 was altered to suit the storage battery project. The metallurgical lab in Building 4 was refitted to carry out duplication of master cylinders for the phonograph business.

There had been little planning in the assembly of Edison's business enterprise. He pursued the policy of expansion without regard to the overall development of the organization, forming new companies and building factories as the need arose. Each new product led to a new company and often to a new manufacturing facility, as was the case with the storage battery and Portland Cement at the turn of the century. Edison also embraced the strategy of vertical integration, acquiring the facilities to control each stage of production in much the same way that Andrew Carnegie integrated his steel making operations and John D. Rockefeller brought all stages of oil production and refining under one organization. In addition to the current economic wisdom of securing the sources of raw materials and lowering costs, there were good technical reasons to take over the production of raw materials. The requirements for chemicals needed in the manufacture of batteries and phonograph records were unique to Edison's operations; the purity of raw materials played a major part in determining the quality of the final product. Even the slightest impurity in the chemicals used to make wax compounds, for example, could ruin the reproduction of sound from the phonograph. Edison's standards were the highest in the industry, and only his own raw materials were to be used in manufacture.

Edison envisaged controlling the storage battery operation from the nickel mine to the assembly of automobiles. Although problems with the battery experiments kept him fully occupied, he could not resist interfering with the construction of automobiles and electric delivery wagons.[8] In 1908 he finally acquired the Lansden Electric Car Company of Newark. This completed his integration of his storage battery business from processing of raw materials at the

Silver Lake complex, through assembly at West Orange, to making and selling the battery-powered automobile.

The same policy of integration was applied to the phonograph business. Edison bought out the companies manufacturing spring motors and eventually took over the fabrication of the wooden cabinets. Where possible, Edison attempted to gain control of companies providing parts or materials for his factories and acquired smaller competitors in the New Jersey area.[9] Edison also had an interest in factories in Europe making his talking machines and records. His business was in the front rank of the larger organizations formed in the era of consolidation in American industry.[10]

Now an industrialist of the first order, Edison could afford to indulge his fancy. Like many other Americans at the turn of the century, he was fascinated with the automobile, which he thought would end horse-drawn transportation in America's cities. Firmly convinced of the inevitable ascendancy of the automobile, he recognized it as the great commercial technology of the twentieth century. He hoped that an electric automobile, employing his battery, would become universal in the United States.[11] He embarked on a love affair with the horseless carriage, purchasing so many high performance automobiles that a large garage had to be built between Buildings 1 and 2 to house his collection. His leisure time was soon dominated by the automobile, tinkering with it, tuning it, and most of all, driving it full out on New Jersey's country roads. The elderly inventor enjoyed the thrill of speed: "The sport of kings, I call it—this automobiling at seventy miles an hour. Nothing on earth compares with it. In its newest development, electricity has once more come to the service of mankind."[12]

Edison being Edison, the electric automobile naturally had the greatest appeal to him. As soon as the first models were on the market, he sent for the manufacturers' catalogues and bought one made by the Pope Manufacturing Company of Hartford, Connecticut, the largest American manufacturer of automobiles at that time. Edison was delighted with the electric car. He had previously considered electrifying the bicycle, but now he saw a much more dynamic product.[13]

He was not the only electrical engineer who saw the potential of applying the technology of electric propulsion to automobiles. The inventor and engineer Elmer Sperry recognized electric automobiles as a field for his expertise in streetcar motors and controllers.[14] Most of the technology was already there; all that remained was to apply it to a car and provide the current. Electric power was ideal in

Edison's view. It was cheap, easy to control, and a great deal safer than gasoline or steam cars. The people with the money to purchase automobiles agreed with him; the silent, smokeless automobile was the easiest to drive and the most comfortable to ride in. In 1899 electric and steam cars accounted for the majority of sales.[15]

Edison began to consider building an automobile in 1899. The staff of the laboratory obtained plans of many different car bodies and gathered extensive literature about heavy-duty motors and controllers. By 1902 Edison could report to an expectant press that his boys had made up a battery-powered rig and were test running it.[16] He promised significant improvements in the duration of the battery's charge, which at that time restricted the range of the automobile to about twenty miles. His participation in several long-distance runs proved that touring was possible in an electric car, providing that utility companies could build recharging facilities along the route. The electricity supply industry played a vital part in the future of this technology because recharging determined the range and the economy of the electric car. The private automobile owners promised to be excellent customers by consuming great amounts of current at night when demand was low. Edison was not exaggerating when he predicted that this type of power usage would lead to "a new epoch in electricity."[17] He was particularly interested in electrifying delivery wagons because their needs were especially fitted to the convenience of electric transportation. Their work was restricted to short runs around the city which did not exhaust the charge of the battery, and they could be recharged at night when electricity was cheap. As usual Edison was highly optimistic about both the performance of the battery and the ease of building a drive train and chassis around an electric motor. The Wizard's prophecy was that electric automobiles would be light, cheap, and easy to operate. He correctly anticipated the gradual decrease in cost of the automobile that would bring it within the means of middle-class Americans. He predicted its general use in the years to come.[18] Edison looked forward to another brilliant innovation—the electrically powered car—to burnish his reputation in the new century.

The Formation of TAE, Inc.

This optimism and confidence were gone by 1910. A series of failures in experimental projects, damaging court decisions, and finally a depression in the main business brought Edison and his managers back to earth. By then his enterprise was on the verge of a financial crisis. A major reform of his business organization was undertaken. Each part of the organization underwent a clarification of function that marked the evolution of an industrial empire into a modern corporation. A policy of centralization concentrated power at the top.

Each of Edison's businesses suffered setbacks after the promising beginning of the twentieth century. The first disappointment was in motion pictures. The favorable court rulings were overturned, and even the efforts of the Legal Department could not hinder the activities of the many independent film makers who crowded into the field. By 1900 profits had fallen for the Manufacturing Company. Supply of films had increased but exhibition was not keeping pace with production. The public were tiring of the same film subjects—travelogues and topical films about natural disasters and current events—and the average exhibitor could not afford to purchase a great variety of film subjects. With demand low and excessive litigation adding to costs, William Gilmore admitted that the movies were now "a non-profitable field." Edison was beginning to appreciate the fundamental instability of the industry, harboring thoughts that this business might be just a fad.[19]

The same intense competition marked the cement business. Soon after the Edison Portland Cement Company started operations, overcapacity in the industry led to price-cutting that eliminated profits. The venture had begun with the good idea of exploiting Edison's patents and experience in materials handling, but it ended as a financial drain on the organization.

The storage battery project illustrated all the things that could go wrong with even a surefire experimental campaign. The decision to produce a better storage battery was not an act of haste as Edison rebounded from the ore-milling failure, but a rational decision based on the available resources of the laboratory. Its work force had considerable experience in battery technology and all the equipment and facilities were at hand. The experimental project was built on years of research into primary and storage batteries.[20] It moved across the familiar territory of primary battery electrochemistry which had been under investigation at the lab since the 1880s. It did not begin with a frantic search of every possible electrochemical combination for a storage battery, but with the customary study of scientific literature.[21] The experiments carried out by the Swede, Waldemar Jungner, on nickel-cadmium batteries were carefully analyzed. Edison also collected data on the Electric Storage Battery Company that held the patents of lead-acid storage batteries and

Figure 9.1. Edison in the chemical laboratory (Building 2, West Orange) in 1904.

marketed the widely used "exide" cell. After considering the available information, Edison was convinced that an alternative to the lead-acid cell could be found "if a real earnest hunt were made for it."[22]

In 1903 Edison thought he had found the right electrochemical combination and quickly moved into production. His new nickel-iron battery had a better performance than the lead-acid cell; this justified a higher price. It sold well to users of electric cars and delivery trucks. Unfortunately the charge of the new batteries deteriorated over time, and, even worse, the cells leaked acid and so were unsuitable for commercial use. By 1904 Edison was forced into an expensive recall of the 14,000 type E batteries sold. With characteristic determination, he set about finding the bugs that caused loss of charge and made the batteries leak (fig. 9.1). Although he remained optimistic in public, he revealed in private that he was not sure if he could solve these technical problems.[23] His grand plan to develop electric automobiles came to nothing, giving ammunition to his critics who remembered previous claims. "Mr Edison's bunk has become somewhat of a joke—a real joke," noted *Motor Age*.[24]

It took him five years of intensive work before he had a battery, the type A, ready to challenge the lead-acid cell. But by then the enormous costs of expansion and research swamped the paltry finances of the Edison Storage Battery Company, which became the financial weakling of Edison's companies. In the years up to 1909, it had no product to sell and was burdened with the heavy cost of research and development. By 1905 it was carrying $154,693 of experimental charges, which was about equal to the value of machinery and tools at both West Orange and the Glen Ridge factory.[25] Its financial statistics were so poor that Edison decided that it would be unwise to release them to credit companies. The failure to get the storage battery onto the market put an unbearable strain on Edison's finances. He supported much of the Storage Battery Company out of his own pocket and raided the coffers of the National Phonograph Company for the cash to support his experiments. It was estimated that he invested $1.9 million in developing the battery.[26]

The storage battery research had a detrimental effect on both Edison's finances and health. The desperate search to find the bugs in the battery wore him out as he drove himself to the limit to save the project. Experimenting was dangerous work; Edison had been burnt by explosions, gassed by toxic fumes, and exposed to radiation. The trauma of an unsuccessful campaign also took its toll. William Dickson was probably not the only mucker to suffer a nervous breakdown. The pressure of managing the Edison enterprise was too much for Johnny Randolph, who in 1908 took his own life.

In 1900 Edison was an energetic fifty-three-year-old, but his health had begun to deteriorate. He became seriously ill in 1900 and 1901. His hearing, which had been damaged in his youth, grew progressively worse. He underwent emergency surgery on ear infections in 1906 and 1908. Such was the danger of this condition that he decided to "retire from commercial business altogether" in 1907.[27] Edison's business associates and the general public were told that he had retired from commercial work and was now going to devote himself to general experimenting for his own personal enjoyment. In 1908 his secretary announced prematurely that "we no longer do commercial work at the lab."[28]

The failing health of the founder presented an immediate problem for the management of the Edison enterprise, for there was an urgent need to shape his sprawling industrial empire into an organization that could carry on without him. Edison had filled many positions within his companies: chief executive, inventor, engineer, and financial planner. Each of these functions had to be delineated and filled with a professional manager who could maintain day-to-day operations of his department. By 1907 William Gilmore was running the Edison enterprise through the managers of each department. His authority extended from the administration building into the Phonograph Works and beyond. He instructed his subordinates "to act intelligently and wisely without disturbing or bothering Mr. Edison."[29] The goal of the administration was to keep operations running smoothly while the Old Man busied himself in the laboratory.

Although the ranks of administrators had grown steadily during the years of the phonograph boom, the ramshackle construction of his business empire prevented a truly efficient administration. Many of the operations were widely dispersed in New Jersey and New York, and the management of far-flung units, like the newly constructed film studio in the Bronx, was free to become "a law unto itself."[30] The complicated legal and financial relationships between Edison's companies meant few clear lines of authority. This was not seen as a problem in the balmy days of the phonograph boom, when profits were high, growth was spectacular, and most attention was on research. The steady income from the phonograph business masked the structural weakness of Edison's empire.

The phonograph boom was stopped dead by the depression of 1907. After a summer in which all production records were broken, a financial panic in the fall led to a serious reduction in business by 1908. The National Phonograph Company was not alone in watching its market diminish alarmingly (see table 9.1). Sales of talking machines declined during hard times, yet the decline in sales of Edison's phonographs lasted much longer than any of its competitors. The panic of 1907 signalled the end of Edison's dominance of the industry. The National Phonograph Company had been challenged by several new companies and one powerful technology. The ascendancy of the disc was proof that Edison's once unquestioned technological leadership had ebbed away.

In 1900 Edison's phonograph business was based on the soft wax cylinder that could play for about two minutes. The record-buying public could not be expected to remain satisfied with the appeal of short recordings—the future of the industry lay in longer recordings of more serious music. In this development the National Phonograph Company lagged. The disc machines made by the Victor and Columbia companies had the advantage of longer playing records, ranging from four to seven minutes, which could go well beyond

Table 9.1. Sales of Phonograph Merchandise, 1899–1921

Sales Year*	Edison	Victor
1899	$ 755,922	
1900	1,120,143	
1901	1,052,771	
1902	1,865,538	$ 1,893,057
1903	2,629,144	1,715,419
1904	3,243,832	2,108,174
1905	4,762,115	2,835,337
1906	6,852,383	4,638,183
1907	7,104,628	6,015,240
1908	4,831,131	3,631,285
1909	3,989,961	4,961,546
1910	3,640,140	6,648,734
1911	2,627,098	8,245,009
1912	2,217,547	10,472,806
1913	3,688,472	15,308,460
1914	4,144,013	16,206,598
1915	6,759,844	22,269,792
1916	9,091,471	29,594,385
1917	10,841,260	
1918	11,160,983	
1919	19,903,861	
1920	22,689,289	
1921	7,271,416	

Source: Edison 1925 Phonograph folder, ENHS; Victor Historical File, box 11, AHC.

*The Edison sales year ended February 28; the Victor sales year ended December 31.

short popular songs. As the competition gradually increased the playing time of records—in both cylinder and disc formats—the West Orange laboratory had to follow suit.

A longer playing cylinder had more grooves cut into it per inch than the standard 100, and this necessitated changes in the wax of the cylinder (to make it harder) and in the design and weight of the reproducer (to avoid skipping from the smaller grooves). The laboratory staff had experimented freely in this area since the 1890s, even going up to 400 threads per inch in experimental conditions.[31] The main problem was that the tiny sapphire point of the reproducer stylus jumped out of the groove or broke down the thin walls of wax between the grooves. The closer grooves also cut down the frequency response because the reproducer stylus could not be adjusted to fully capture the wide undulations of the bass notes. Yet these technical problems had to be solved or the prospects for the National Company were dim. The introduction of four-minute cylinders, first for the Twentieth Century Graphophone in 1905 and then by the Indestructible Record and Columbia companies, not only threatened to cut into Edison's cylinder business but also to promote the sales of machines capable of playing them.[32]

In the years following the 1907 depression, sluggish sales and high inventories at West Orange indicated that customers were being lost to more innovative companies. Announcements of 200-thread records under development at West Orange could not stop the decline in sales or the restlessness in the ranks of dealers. Morale had deteriorated to such a low point by 1910 that the National Phonograph Company had to quell rumors that it was getting out of the business.[33] Sales of the Edison cylinder machine decreased rapidly while the disc machines marketed by its competitors were gaining acceptance and market share. Ahead lay fiercer competition and difficult technical challenges. In 1910 the National Phonograph Company was still the flagship of the Edison enterprise but its sales had decreased every year since 1907.

The National Phonograph Company had to support the ailing storage battery and cement companies that were hemorrhaging funds. By 1910 it was taking over $100,000 a month to keep the weaklings' heads above water, and this soon began to put a strain on the finances of the Phonograph Company, which was also meeting a large part of the laboratory payroll.[34] The financial strength of the Edison enterprise was sapped by the failure of the storage battery project. The improved type A battery had been successfully introduced in 1909, but unfortunately the market for electric automobiles had dried up. The internal combustion engine was now dominating the field, forcing Edison to look for other uses for the battery. In the meantime, the losses of the storage battery continued to mount. What was worrying the accountants of the National Phonograph Company was that "the earning power of the profitable companies here is not as great as it once was and the expenses of the unprofitable concerns are not diminishing."[35]

It was probably the impending financial crisis in 1910 that hastened the creation of Thomas A. Edison, Incorporated. The banks on whom the company depended were becoming nervous about extending more credit. The problem was to ensure that the profitable companies had enough funds to develop new products while avoiding financial disaster at the hands of the unprofitable cement, storage battery, and Lansden Electric Car companies. In addition to supporting the weak links of the Edison organization, the National Phonograph Company was also facing a highly competitive market in which technical leadership had been lost. The executives of the National Company shared Edison's faith in the ultimate profitability of his storage battery and Portland cement, but feared that lack of credit would stop the Edison enterprise completely if drastic action were not taken in 1910. They stressed the need for a sound financial base to support the weaklings and provide centralized management. They appealed to Edison to cut back on some of the unprofitable ventures, especially the cement company, and generally economize until the introduction of new products could replenish the treasury. The Edison organization had to be placed on a sound financial basis until new businesses such as the storage battery could return a profit.[36]

The development of a professional management structure in the Edison enterprise had opened up a window to information about other companies' financial strategies and organizational plans. An avenue of communication was established between the staff of the National Phonograph Company and its contacts in banking and accounting. Professional accountants were a particularly fruitful conduit of information; they saw a great number of different companies in the course of their work and their detailed financial knowledge gave them valuable insights into company organization.[37] The idea of centralizing the Edison enterprise around one business unit probably originated with the management of the National Phonograph Company who then sold the idea to the Old Man. In 1910 Edison made the decision to merge several of his companies into one corporation, and the National Phonograph Company was chosen to be the core of the new organization.[38] It had the largest staff and its accounting, legal, advertising, and production departments were to be the building blocks of a new administration.

The newly reorganized National Phonograph Company was capitalized at $12 million and renamed Thomas A. Edison, Incorporated (TAE, Inc.). The corporation's initial press release stressed that it was a reorganized, rather than a new company, and that it was the beginning of a movement to bring all Edison companies under one corporation for the purpose of "business convenience."[39] It absorbed the New Jersey Patents Company (which held many valuable Edison patents), the Edison Manufacturing Company, and the Business Phonograph company. Other (less profitable) companies were excluded at this time in order that the new organization present a healthy balance sheet to the financial community.[40]

The new administration added a layer of vice-presidents between the chief executive officer (Edison) and the middle managers. Executive management was carried out by two vice-presidents and a general manager who, with Edison, formed the Executive Committee that decided on all matters of business policy. The chief executive and general manager was Frank L. Dyer. Edison's chief counsel since 1897, Dyer replaced Gilmore in 1908 when the latter protested Edison's policy of taking cash from the Phonograph Company to finance the storage battery experiments.[41] Dyer was more suited to the legal maneuvering that played an important part in managing the Edison enterprise. The monopolistic Motion Picture Patents Company (MPPC) was in the planning stage at this time and the shrewd lawyer was a suitable replacement for the combative Gilmore who bossed the organization in the tradition of a nineteenth century factory manager. Legal struggles had become a decisive element in the phonograph and motion pictures businesses, in which patents and franchise arrangements were hotly contested. The franchisees of the defunct North American Phonograph Company had sued Edison for a share of the profits of phonographs sold in their areas, claiming that they had been illegally deprived of their franchises by his manipulation of various phonograph companies. A court case paralyzed Edison's talking machine business from 1908 to 1909, when the franchisees won $425,000 in damages. It was Dyer's job to prevent this kind of defeat in the future.

The guiding principle of TAE, Inc. was to reduce the cost of running the business. This process had begun when the bubble burst in 1907, forcing the National Phonograph Company to contemplate reduction and reorganization of the work force.[42] TAE, Inc. was to be the vehicle to bring this process to all of Edison's many business enterprises, and its early years were marked by vigorous reduction of the labor force, especially white-collar workers.[43] Its formation provided some immediate financial relief in that one organization could do away with the duplication of functions within several Edison companies and provide cost-effective management.[44] The consolidation of many functions into one compact body increased the productivity of a greatly reduced administration. According to one historian of the organization, TAE, Inc. was "a splendid example of what a vast amount of business could be accomplished with little or no office overhead."[45]

A system of management by committee was established to give the central administration oversight over the working of the organization. A departmental committee was set up for each product. This brought together representatives from the Works, laboratory, and sales organizations. Each of these three branches had its own committee that answered to the Executive Committee. Every decision requiring expenditure had to be supported in committee. The men at the top had the last say on large expenditures through a bylaw making any order or contract worth over $10,000 subject to approval and ratification by the board of directors.[46]

Changing the name of the National Phonograph Company to Thomas A. Edison, Incorporated, also had some publicity value, for Edison's personal reputation was making him an American folk hero in the twentieth century. Frank Dyer noted that the principal reason for changing the name was to "bring home more firmly to the public the fact that this is Mr. Edison's personal business and that his personality stands behind it."[47] There can be no doubt that he was the greatest asset of the new company. His patents were valued at $6 million in the financial statement of the new corporation, and this represented over half its book assets. TAE, Inc. was centered on the person, achievements, and future inventive potential of Thomas Edison.

The Engineering Department

The laboratory underwent a reorganization shortly after the formation of TAE, Inc. A new department was formed in the lab to streamline the process of making factory-ready prototypes and transferring them to the Works. The Engineering Department was to bring production engineering under the control of management. Here was another job to be bureaucratized and fitted into the work of planning, production, and sales. The new department was typical of the specialization of function that characterized the early years of TAE, Inc. Its formation marked the beginning of a sustained effort to reform manufacturing at West Orange.

The high price of Edison products was the motivating force behind these changes. Twenty years after Edison had articulated a manufacturing strategy based on systematic reduction of cost, the factories of West Orange had still not achieved the economies planned in 1887. Lowering the costs of production had become essential in the phonograph business because TAE, Inc. faced competition that invariably undersold it. Many of the new companies entering the talking machine business in the twentieth century put their hopes on low prices rather than quality goods. The efforts of the laboratory staff were not enough to keep pace with the price cuts of the opposition. Jonas Aylsworth and the cylinder reproduction team in the laboratory were successful in reducing the cost of making the "gold moulded" cylinder from 7 cents to 3 cents, yet by 1904 the price of an Edison record only dropped from 50 cents to 35 cents. Columbia had originally priced its records at 25 cents and kept reducing the price.[48]

The competition was growing at the low end of the market where sentimental attachment to the Wizard was often overcome by his higher prices. Edison had no illusions about the need to reduce prices of the popular models. His prescription for the electrical industry was equally applicable to the talking machine business—only those who could reduce their costs would survive.[49]

Edison considered that reducing the cost of manufacturing electromechanical machines was one of the strengths of his laboratory. It took the form of separating the machine into its component parts and working on the design of each part to make it cheaper to assemble the whole phonograph. The plan was to produce a standard design of the fewest possible parts and to make each part to this standard. The essence of mass production as practiced by Edison and his generation was to reduce the cost of making a product by standardizing its design and organizing long, uninterrupted manufacturing runs. Yet this goal was confounded by the growing line of Edison phonograph models and the practice of making continual design changes and special one-off models.

These practices can be traced back to the early 1890s when the machine shop culture flowered at West Orange, and failure of the dictating machine forced Edison to broaden the base of his manufacturing operation. The Works became a loosely organized facility that manufactured everything from Kinetoscopes to numbering machines. It was a production engineer's nightmare: a do-all general purpose factory, with no central task or stabilized engineering designs, and few long production runs. The proximity of the laboratory to the Works encouraged Edison and his muckers to redesign parts and then walk them over to the Works, making the alterations to the machine on the bench. Subsequently, the Works was inundated with design changes and often had to manufacture products before the laboratory finished the design.

This was the heritage of the shop culture in which experimenting took precedence over production engineering. The hierarchy of skills at West Orange gave the experimenter in the laboratory predominance over the foremen in the Works. Improving the product was much higher on the list of priorities than keeping to one, stable design, and it was also a lot more rewarding for the muckers and their boss. Although these values suited the shop culture, they were maintained at the price of high manufacturing costs because the Edison Phonograph Works was called upon to manufacture a wide range of machines. As a point of honor, Edison and his laboratory staff had to meet every improvement introduced by other competing phonograph companies. New shapes of the amplifying horn, different lengths of cylinders, and special reproducers came out of the laboratory faster than the Works could deal with them.

The men in the laboratory thought nothing of completely changing an existing design to meet a new technical requirement or to carry out a single promotional stunt. A special kinetograph, for example, was built in 1899 in a larger format to take special pictures of the Klondike gold fields. In a small manufacturing organization, such as a machine shop, these one-off jobs were easily incorporated into production. In a large operation, design changes took more time and had a greater chance of disrupting the complex balance of mass production.

The laboratory's response to extended-play records of the competition is a case in point. It naturally brought a crash program that produced the four-minute Amberol cylinder in 1908 and an entirely new machine to play it in 1909. The Amberola phonograph employed a moving mandrel and a fixed reproducer, instead of the earlier design of a reproducer moving along the fixed mandrel, standard in Edison phonographs since 1887. This was a premature debut of a new technology because the soft wax of the cylinder was too fragile to hold up the thinner walls between the grooves.[50] It was also too soft to retain the high notes after repeated play back. What was a disappointment for Edison's customers was a disaster for the

Phonograph Works. Edison had a policy that his customers should not be left with obsolete machinery, so each new model or design change was accompanied by a conversion kit to adapt old machines. The Amberolas with their totally different mechanism were accompanied through the Works by several mechanical devices to convert existing phonographs. This added more variants to a phonograph line that was already crowded with numerous models and types. To make matters worse, the conversion kits were difficult to make, adding to the frustration of the foremen and engineers in the phonograph factories.

The great number of design changes in Edison phonographs also made life difficult for the dealers, who were kept in a state of confusion by the fast-moving lab in West Orange. In many cases they were not told about changes—they had to find out for themselves by examining the machine after they had uncrated it! The dealers had plenty to complain about: modifications that did not work, changes in the machine that made its operation difficult, and poor assembly. In many cases dealers had to overhaul and repair machines before they could be sold.[51]

The more new designs were produced in the lab, the more production changes had to be made. The chaos in the Phonograph Works became steadily worse in the early years of the twentieth century as the laboratory pushed forward with a stream of new products in the phonograph and motion picture businesses. The result of this unhappy situation was serious delays in manufacturing schedules.[52] The absence of a routine for making design changes created an atmosphere in the Works where standardization of design and manufacture was virtually impossible. It was a case of every foreman for himself and few records were kept. One engineer recalled, "New models chased themselves through the factory without red tape—that is, if they could find their way through the maze of cubby holes and partitions that formed the factory floor."[53]

Many small design changes were made with the good intention of streamlining production; but their number, and the number of people making them in both laboratory and Phonograph Works, led to more difficulties. In some cases changes cancelled each other out; at other times conflicting modifications led to "expensive mistakes" when put into production.[54] This haphazard approach to production engineering, typical of the Edison lab and Phonograph Works in the nineteenth century, could not be tolerated in the highly competitive situation that existed after the 1907 depression. Independence and initiative had their disadvantages. It soon became evident that changes would have to be made to eliminate some of the abuses of manufacturing practice in the Edison Phonograph Works.

Edison was not slow to realize that production engineering was the key to maintain low costs. Yet this important job was widely dispersed about the laboratory and Works. Although several experimenters worked on phonographs, there was no permanent position in the laboratory devoted to improving the steps of its manufacture. The work was carried on by Edison himself (when the mood caught him) and the superintendent and foremen of the Phonograph Works. The Works superintendent had been in nominal charge of factory management and engineering since he was set up in his own department at the beginning of the 1890s.[55] In the twentieth century the superintendent carried out a wide range of design work and even experimented in the laboratory with Edison. Superintendent Peter Weber and his staff played their part in developing new products and modifying existing ones.[56]

The transformation of all these design changes into new blueprints, tools, and manufacturing processes was as important as the actual engineering of the changes. This task went, by default, into the hands of the foremen of the machine shops and Edison Phonograph Works. It was done in an informal, haphazard way; there was no proper set of drawings kept and each foreman made his own changes to suit himself and usually failed to inform anybody else. The tools used in production were designed and made by Otto Weber "largely as he saw fit."[57] No records were kept of this vital work. The internal politics within the Works often determined the changes made and the lack of reporting them. Strife between foremen was sometimes at the root of the lack of cooperation in manufacturing. The Works was dominated by strong wills and unbending personalities that the superintendent was hard put to control.

The problem was to organize these forces to be able to make changes quickly and to readjust all manufacturing processes to suit. What was lacking was one authority to coordinate all the parts of the process (research, design, drafting, toolmaking, and production) and control it from laboratory to Works. In 1910 Edison made the decision to form an "engineering and experimental department" in the lab that was to develop new machines from scratch and then send the models and blueprints to the Phonograph Works for manufacture. The Engineering Department was also given the authority to approve any change in an existing design of a machine.[58]

A more formal structure was created to replace the personal

relationships of the boys in the lab and the foremen in the Works. A chief engineer was appointed over the Engineering Department to administer production engineering. His first and most important duty was to redesign phonographs to make them cheaper to manufacture.[59] When the European affiliates of the National Phonograph Company were troubled with low-price competition, a number of very cheap machines were sent to the West Orange lab for assessment. Dyer reassured the European managers that the Engineering Department would design a low-price cylinder machine for them.[60]

An important function of the department was to work with the other engineering and Works departments—it was intended to be a permanent link in the chain of development that began in the laboratory and ended in the shipping department of the Works. It was accountable to the committees that supervised its work, providing a formal locus in the process of product development that was amenable to bureaucratic control.

The new department fitted easily into a system of centralized control. A new design, a working machine, needed the approval of the Executive Committee before it was turned over to the Drafting Department for a set of blueprints to be made up and sent to the Works. Changes in existing models were overseen by the Manufacturing Committee, a body comprised of executives of TAE, Inc., the chief engineer, and other Works staff. Formed in June 1910, the committee assessed the economic consequences of a proposed design change and decided if it was to be implemented.[61] All suggestions, requests, and proposals for design changes were submitted to the Manufacturing Committee. If it agreed to the change, it was the job of the chief engineer to implement it.

The chief engineer was the important link between the researchers in the lab and the staff of the Phonograph Works. After designing new parts and improvements for machines, he handed over all relevant data to Peter Weber in the Works, who then supervised the production of the tools. The chief engineer was responsible for recording this information and distributing it to the relevant production departments.[62] The Engineering Department was a part of the new bureaucratic organization of TAE, Inc., becoming a processing point in the ever increasing paperwork generated by the Works and laboratory staff. All changes in product design or manufacturing process were authorized by Engineering Notices that became the instrument for the chief engineer to supervise production engineering.

The Engineering Department was set up to work on phonographs, and was placed on the payroll of the National Phonograph Company. It was finally decided to put it onto the laboratory payroll because it was expected to produce patentable designs of phonograph improvements.[63] Its staff was initially drawn from the engineering office of the Works and the experimenters in the lab who were experienced in phonograph technology. They soon enlarged the scope of their work to cover primary batteries and motion pictures, producing a line of projectors, including one for use in the home.

While Edison and his helpers carried out newsworthy experiments, the Engineering Department had the important task of executing frequent changes in the product line. The demands of a highly competitive phonograph industry, with its rapid technological advance, had forced the Edison Phonograph Works into what historian David Hounshell calls "a new era" in mass production. Annual model changes required a system of "flexible mass production" and an organization to carry it out.[64]

CHAPTER TEN

The Diamond Disc

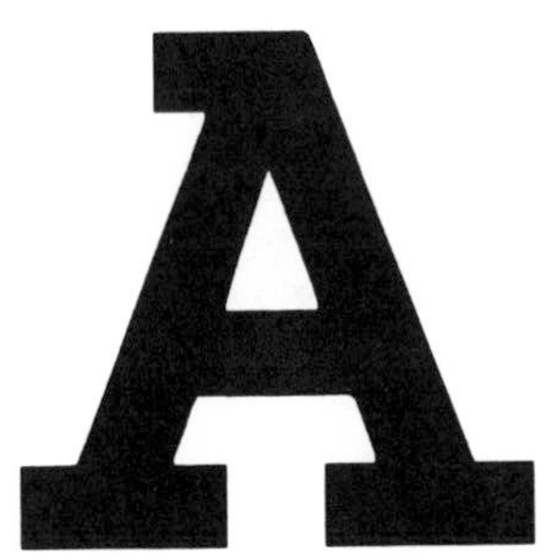

A test of the reorganized engineering services of the West Orange laboratory was not long in coming. Beginning in 1913 TAE, Inc. introduced a new line of disc phonographs and records that required concerted action from laboratory, Works, and central administration. At stake was the future of the phonograph business.

The steady decline in sales of cylinder phonographs after 1907 alerted TAE, Inc. to the realities of the phonograph market: the cylinder format was finished as a commercial technology and the future belonged to the disc record. In the financial year ending February 1912 a total of 61,320 cylinder machines were sold, compared with the 289,822 sold five years earlier. The Victor Company, TAE, Inc.'s greatest competitor in the talking machine business, sold 185,766 disc machines in 1912. Edison's company was one of the few remaining cylinder manufacturers. The two-minute "gold moulded" cylinder and the four-minute Blue Amberol could not compete with discs that could play up to seven minutes. TAE, Inc. began planning for a disc-playing phonograph in 1910, the new disc machine being viewed as the product that would "put new life into the business" and return the company to its former position of prosperity.[1]

After anticipating the disc format in his earliest patents, Edison went on to thoroughly investigate it. He discovered that the cylinder had several theoretical advantages over the disc: it could be turned at a more constant speed, and the needle was not pressed into the groove by centrifugal force at the end of the recording. The cylinder format enabled better tracking at a more constant speed and thus better fidelity, yet the phonograph-buying public were less interested in technical perfection than they were in longer playing time and ease of storage. Edison's inability to fully comprehend this reality was to be a fatal disadvantage in the phonograph business.

The disc-playing machine had a history almost as long as the cylinder phonograph. The pioneer disc instrument, called the gramophone, was invented by Emile Berliner at the same time that Edison opened his West Orange laboratory. Berliner's experiments in telephony brought him into the realm of sound recording, in the same way that Edison, Bell, and Tainter had followed this technology to the talking machine.[2] The gramophone incorporated several important innovations that were not found in the phonograph or graphophone. The reproducer assembly was housed in a sound box that was drawn across the record by the spiral movement of the groove, eliminating the need for a feed screw or other arrangement to move it. Berliner used a lateral cut in the groove instead of the hill-and-dale method used on the cylinders. The acoustic advantages of moving the stylus from side to side, rather than up and down in the groove, were hotly contested in the phonograph industry. The advocates of the lateral cut claimed better frequency response because the stylus did not have to fight gravity in its modulations. Opposing this viewpoint were the supporters of Edison's system who argued that the stylus had a greater freedom of movement in the groove and could therefore capture the whole tonal range of the human voice.[3]

One undeniable advantage of the Berliner system was the ease of duplication of disc records. Berliner perfected a method of making masters by recording onto a zinc plate covered with fatty film and then using acid to etch the groove marked in the film. These masters were electroplated in copper to make matrices that were used to stamp the impression onto hard rubber discs. This was much simpler than making a mold of a cylinder record and forcing a blank into it. Berliner was the first to use masters to duplicate an unlimited number of records, making him one of the pioneers of the talking-machine industry. After developing this process in 1893, he went into business as the United States Gramophone Company.

Within a short time of its introduction, the gramophone was being manufactured in Europe and the United States. Although sales of the machine were aided by its low price (the first cost $12), and the cheapness of the disc record, the gramophone was not a great commercial success because of its poor reproduction. The problem was maintaining constant speed with the hand crank and thus achieving constant pitch. The experimenters at the West Orange laboratory struggled with the power source of the phonograph for about ten years before they found the answer in the spring motor. Berliner had also looked at spring motors but, like everybody else in the industry, found them wanting.

In 1896, the year that the National Phonograph Company introduced its spring motor phonograph, the Gramophone Company made a contract with a machinist called Eldridge Johnson to develop a suitable motor for Berliner's machine. Johnson ran a small machine shop in Camden, New Jersey. Although his principal business was making wire-stitching machines, Johnson had experimented in other fields and had been granted several patents, which made him a fairly typical machine shop operator. Johnson produced a quiet, easily regulated spring motor that maintained constant speed to the end of the disc record, overcoming the increasing pull as the needle got to the end of the spiral. This major contribution to the development of disc technology brought him closer to the Gramophone Company. He was soon turning out motors, sound boxes, and other parts for the machine in great numbers, and before long complete gramophones were being made in Johnson's shop. An experimenter at heart, Johnson continued to improve Berliner's machine. He redesigned the sound box around a mica diaphragm dampened by rubber rings. This reproducer was far superior to the original and became the basis for further improvements.[4]

The Johnson-designed gramophone was sold as the Trade Mark model of 1896. Priced at $25, this product competed with the popular cylinder machines. A smaller, cheaper version of the gramophone was soon introduced to match Edison's Gem model. At the bottom of Johnson's line was a very small disc player, aptly named the Toy, which was introduced around 1900 to sell at less than $5.[5] The Columbia Company soon began to manufacture disc players after concluding a cross licensing agreement with the gramophone interests in 1901. Their Eagle model, which only played discs, sold for $10.

All these disc machines had the disadvantage of poor sound reproduction when compared with an Edison cylinder. The Berliner acid-etching process (to make masters) eradicated some of the sound waves recorded in the groove, and caused indistinct or muddy reproduction. Disc records had a hissing surface noise that was the result of the action of the steel stylus tracking along the wax recording media—rubber having been discarded as a medium when mass production of discs was undertaken.

Johnson knew that the gramophone did not have a commercial

future with inferior sound reproduction and set about improving the duplication process. He spent $50,000 and two years experimenting before he produced the critical innovation in recording. It is not too great an exaggeration to call it "an invention that was to revolutionize the talking machine industry." His master disc was engraved with the sound waves from a specially designed recording machine that was built around an Edison phonograph. The wax master was then electroplated to make a matrix—somewhat like the process developed in the West Orange laboratory.[6] The dramatic improvement in the sound reproduction enabled the disc to meet the high standard of tonal quality and articulation established by the Edison cylinder records. The improved gramophones and their disc records were accepted by the musical trade and sold well.

In 1901 the Victor Talking Machine Company was formed on the Berliner and Johnson patents. Johnson followed Edison's practice of selling franchises in Europe and set up companies in England and Germany. A record factory was established in Germany to mass produce discs that were recorded in Europe, and a fruitful exchange of recordings was begun in the same way that Edison companies on both sides of the Atlantic took advantage of each other's libraries of films and recordings.[7]

At this point the Victor record catalog was similar to those of the National Phonograph and Columbia Companies: comic songs, ballads, military marches, and instrumental solos. Many professional singers were unwilling to risk their reputations by recording their performances on talking machines with unpredictable, and often unsatisfactory, reproduction. This barrier to the development of recorded sound was removed by a series of historic recordings made by Victor recording engineer Fred Gaisberg in Europe. After learning his craft in the studios of several American companies, Gaisberg had gone to England to work for the British Gramophone Company. He disobeyed their instructions when he recorded ten songs by the Italian tenor Enrico Caruso for the unprecedented fee of $500. Caruso's voice was perfectly suited to the mechanical reproduction of talking machines at that time; it emerged clear and strong from the horn to fill the room with wonderful sound. His recordings were issued in the United States in 1903 on the new twelve-inch Red Label discs. They caused a sensation among music lovers, and Caruso went on to be the first performer to enjoy a million sales of a recording. This triumph bolstered his career, made millions for Victor, and opened the way for more classical recordings by established singers.

While Columbia followed Victor's lead by quickly introducing its own series of grand opera recordings, the National Phonograph Company stayed with its usual fare. Its most successful European imports were recordings of humorous songs by the Scottish music hall artist Harry Lauder. "I Love a Lassie" and "Roamin' in the Gloaming" were well-loved songs, but they did not compare with the great Caruso, at least in the minds of the middle class who purchased talking machines.

Sales of gramophones and disc records rose steadily during the first decade of the twentieth century. Eldridge Johnson's small shop was slowly engulfed by the many factory buildings erected to make machines and discs, as he followed the same strategy of expansion pursued in north New Jersey by another successful machine shop operator. The one-story brick building in Camden was soon dwarfed by a cement, steel, and glass complex of manufacturing plant and administrative offices, presenting a scene remarkably similar to the one at West Orange.[8] Victor also supported a laboratory where Johnson continued to perfect the gramophone, fitting ball bearings to ease the sound box's journey across the disc and developing a counterweight in the horn that decreased tracking pressure and damage to the groove. His most important innovation was the development of a tapered tone arm to amplify the sound from the sound box as it passed to the horn. An ingenious pivot and bracket for the horn meant that the sound box and tone arm moved across the disc without bearing down on the stylus in the groove. The innovation of using a tapered tone arm as a part of the amplifying horn was an important advance in the technology of talking machines, and a valuable patent for the Victor Company.[9] It gave excellent sound reproduction and more volume than the lightweight Edison reproducer that could not exert much tracking force on the soft wax cylinders.

The moving tone arm was at the heart of a new line of Victor machines introduced in 1903. These popular models became the foundation for the company's growth in the first decade of the twentieth century. Sales rose from 7,570 machines in 1901 to 94,557 in 1910. The 1907 depression hurt Victor's sales as much as it damaged those of the National Phonograph Company, but the effect of the depression was short lived in Camden because a new disc machine brought out that year became Victor's most successful product.

The Victrola was announced in 1906 and went into mass production shortly afterward. It was Eldridge Johnson's masterpiece, a brilliant combination of improved technology and inspired design.[10] The spring-driven Victrola had the tapered arm connected to the horn which gave it an uninterrupted air passage that increased its diameter at a constant rate, reproducing the mechanical amplification of sound in the same manner as a musical instrument. Instead of sitting on top of the machine, the horn was shaped to turn inward underneath the top plate—making it totally enclosed within the cabinet. Volume control was achieved by adjustment of the door or louvres at the front of the machine. The Victrola used the same mechanism as the exposed horn models that preceded it. In fact its smaller horn gave it an inferior sound to those machines, but this did not concern the thousands who bought it. The totally enclosed works gave the Victrola a completely different look because the horn—overbearing, unsightly, and prone to collect dust—had completely dominated the line of the phonograph. Unlike all the other talking machines on the market, the Victrola could be taken for a piece of fine furniture and many people bought it for that reason. Priced at $200, the Victrola was aimed at the luxury market and quickly hit the mark. It became the fashionable machine of the middle class, the phonograph of the twentieth century that influenced the format of talking machines for years, during which time many people came to call any talking machine a Victrola—to the disgust and anguish of Thomas Edison.

The West Orange laboratory's first response to the challenge of the Victrola was a cylinder machine designed to play the four-minute Amberol cylinder. The Amberola came enclosed in a cabinet that bore more than a passing resemblance to the Victrola, and its price of $200 brought them into direct competition. The Victor Company claimed (with good reason) that the Amberola was a copy of their machine, even to the point of a similar name. The Amberola was not the only Victrola copy on the market, which was soon crowded with machines such as Grafonola and the Musictrola. The Victor company believed that they held the key patents to the Victrola and were prepared to license them, but at a very high price. The staff of the West Orange laboratory had the job of working around Eldridge Johnson's important patents, especially the enclosed horn and tapered arm innovations, while the Victor Company continued to protest bitterly and threaten legal action.

The introduction of the Blue Amberol record in 1912 rescued the Amberola machine from commercial failure. These records had a bright blue celluloid surface laid around a plaster of paris core. Celluloid had originally been developed as a substitute for ivory, but its hardness and ease of molding made it suitable for records. Jonas Aylsworth had investigated the new material in the first years of the twentieth century, but unfortunately his employers had not secured patent protection for its use in records, blocking this promising avenue of development from the National Phonograph Company. Thomas B. Lambert's patents effectively deprived them of the use of this material, a fatal handicap in the phonograph industry.[11] Aylsworth spent years looking for a substitute without success, and while he labored in the chemical laboratory, the National Phonograph Company had to make do with soft wax cylinders. This material was totally unsuited to the longer playing time, and the increased number of grooves per inch of the Amberola cylinders—causing skips, scratches, and eradicated high notes. It pushed many Edison customers into the arms of the competition. Once a licensing agreement was concluded to allow the Edison company to employ celluloid, Aylsworth and the cylinder record teams at West Orange showed what they could do with a superior recording medium. The Blue Amberol recordings have a brilliance and clarity that stand out even in the age of digital recording. Over seventy-five years after its introduction, the Blue Amberol still has a devoted following who claim that no record produced since can beat its sound quality.[12]

The Blue Amberol records marked the apex of cylinder technology. Yet dramatically improved sound and a longer playing life could not win back the customers who had migrated to discs. Although the Amberola was a highly successful product, it was no match for the Victrola that was steadily gaining in popularity. By 1910 the affluent urban markets were "Victrola crazy" according to the Edison sales force. The introduction of two less expensive ($75 and $100) tabletop models in that year boosted Victor's sales considerably. TAE, Inc.'s phonograph organization was in a state of disarray: chaos in the Phonograph Works, dealers depressed, sales force demoralized, and steadily falling sales. Its line of phonographs, "in worse condition than at any time in the last three years," could not compete with those of Victor and Columbia. There was no doubt that "the Victrola is in the lead" and it was sweeping the Edison phonographs from the market.[13]

There was no time to be wasted in introducing a product that could make up the ground lost to the Victrola. The development of a disc machine was the most important experimental project carried out in the West Orange laboratory. It was probably not the most expensive campaign, but it was the most dramatic and best publicized. The Victor and Columbia companies joined the management of TAE, Inc. and its dealers in nervously waiting for the machine that was expected to bring the technology to new heights.[14]

The awful cost of failure discouraged any innovation; TAE, Inc. played it safe by making a copy of the Victrola that incorporated some small improvements in the reproducer assembly, which had previously been developed for the Amberola line.[15] The much-touted Diamond Disc phonograph was remarkable only for its lack of originality. Its greatest innovation, the diamond point of the stylus, was the logical next step after the sapphire point first introduced in 1893. Most of the research and development was done to circumvent the patents on the swinging arm and enclosed horn, the lab working closely with the Legal Department of TAE, Inc. Victor's engineers found the machine clumsily assembled but were impressed by the ingenuity with which Edison had circumvented Johnson's patents.[16] The Diamond Disc had a special feed device to move the reproducer across the record to avoid infringing on Victor's patents.

The one original part of the disc campaign was the work of Aylsworth, who was charged with finding a harder compound for the disc records which could stand up to the diamond stylus. Working at his house in Orange, where he had a fully equipped lab, Aylsworth experimented with compounds that could be molded into shape and would not melt with heat. He examined the properties of phenolic resins that could be molded by heat and pressure. A short distance from West Orange, Leo H. Baekeland, a Belgian inventor, was experimenting in his home laboratory with the same resins in his efforts to find a substitute for shellac. He found that phenol resins could be mixed with binding agents and molded into hard, heat-resistant shapes. Aylsworth improved the phenol resins first developed by Baekeland and produced a material called condensite, a purer resin with fewer by-products. (Condensite refers to the condensation of phenol and formaldehyde which produces the resin.) The development of condensite was a valuable contribution to TAE, Inc.'s phonograph business and a major chemical innovation. The Diamond Disc record was the only thing that impressed Victor and Columbia. Its exceptional clarity owed a great deal to the hardness of the condensite surface.[17]

The bulk of the research effort at West Orange was applied to recording and reproducing music in a disc format. The lesser job of designing the playing mechanism was given to the Engineering

Department. Edison was most concerned with the reproducer assembly and the problem of transferring the hill-and-dale cut to a flat surface. The recording cutter made a deep groove in the disc to accommodate the hill and dales of sound. Edison decided to use a smaller diameter stylus with a greater tracking weight to produce a loud, clear reproduction while keeping the stylus firmly in the groove. The improved reproducer owed much to the experience of developing the Amberola machine and what Edison called "years of empirical experiments" in which thousands of diaphragms were made up and tested.[18] The reproducer for Blue Amberol records had a diaphragm composed of layers of cork and compressed paper and a spring-loaded weight that kept the stylus in the groove and acted as a mechanical amplifier. Edison had first introduced floating weight diaphragms for the molded cylinders brought out in 1903. This design was refined in the years that followed. The diaphragm developed for the disc machine was slightly larger than the Amberola diaphragm, and came with a heavier floating weight.

Larger diaphragms and mechanical amplification were means to obtain a louder reproduction, but more volume produced more violent vibration of the diaphragm which tended to "blast" the sound, eradicating the transient and sibilant sounds—the S sounds that tax the reproducer. Edison found an answer to this problem in semi-viscous substances (or goo as they were known in the lab) that he had used as the agent to hold the diaphragm in position. The movement of a recording diaphragm suspended in this kind of fluid was effectively dampened. These advances were incorporated into the recording machines that made the disc masters, with a discernable improvement in the sound quality of reproduction. [19]

These advances in the technology of sound enabled TAE, Inc. to portray the Diamond Disc as the ultimate in sound reproduction, a talking machine that could capture the slightest nuance of sound, "the overtones," even in demanding classical music. Edison's own account of the introduction of the disc line (produced for advertising purposes) was that he had labored for years before he came up with a disc technology that met his high standards. His new machine had to be free of the distortions and "irritating scratchy tone" of disc players made by his competitors.[20] Edison's laboratory had normally set the technical standards for the competition to follow, but in this case the shoe was on the other foot. TAE, Inc. had to convince the public that its new disc phonograph was not just another Victrola but a completely different machine, for neither the lab nor the company wanted the Diamond Disc judged by Victrola standards. Edison's Advertising Department had to persuade customers that the Diamond Disc was an entirely new product, a technological breakthrough in talking machines. Described as "Edison's masterpiece," the Diamond Disc was marketed as a superior technology, as were most of the products manufactured at West Orange. This strategy worked well in the nineteenth and early twentieth centuries, but by 1914 the situation had changed in the talking-machine industry. As TAE, Inc.'s advertising manager pointed out in 1915: "It isn't sufficient for us to say that we have the best machine. Victor have spent 14–15 years making this market and they won't let us walk in and take it."[21]

As long as TAE, Inc. stayed with the cylinder format, the two leading companies in the talking machine industry split the market nicely: the Victrola was bought by the affluent middle class and city dwellers, while the Edison cylinder tended to be the favorite of rural America. Many music stores carried both lines. Both manufacturers stoutly opposed price cutting and lived together in what Johnson called this "wholesome and commendable relation" until 1913, when TAE, Inc. announced the introduction of a full line of disc machines.[22] They ranged from the $25 table model up to elaborately ornamented machines (including Louis XVI and Chippendale styles) with prices as high as $450. Such was the need to fill the gap left by declining cylinder sales that large-scale production of the disc was planned as soon as the prototype was sent over from the laboratory. The Phonograph Works was committed to manufacture 600 machines a day. This was a major undertaking done in a hurry, much like Edison's first venture into the phonograph business back in 1888.[23]

The introduction of the disc line was plagued with delays reminiscent of the 1888 debacle. It encompassed two periods of intensive work: from 1909 to 1912 to produce the first models and from 1912 to 1915 to eliminate the glaring defects in them. During these years Edison gained over seventy patents and his key researchers were granted several more. Yet this great effort did not produce immediate success. Although the larger, free-standing cabinet models were well received by Edison's faithful customers, the low-price table models were criticized as inferior. One jobber wrote, "It does not give satisfaction . . . is not a credit to the Edison disc line and should not be offered for sale."[24]

The first test of the new system of producing factory-ready prototypes and putting them into production resulted in failure. Although highly efficient in theory, guiding a new design through a maze of bureaucratic controls and checks took time. Research and development by committee did not have the speed that Edison valued so greatly. Engineering problems with the budget phonographs had also played a part in the delays in bringing out the disc line. The motor did not maintain constant speed to the end of the disc and thus could not maintain an even pitch—probably the result of the extra load on the motor caused by the mechanical feed. These models had so many bugs that the Engineering Department could not produce a reliable machine that could be sold for less than $50. In the wake of repeated breakdowns, a new design was made to sell at $60. Yet the motors still performed so poorly that the A 60s, as they were designated, had to be withdrawn from the market. Edison ordered the Engineering Department to start all over again. They hurriedly set about a new design that changed the motor and feed assembly, and a complete set of drawings was soon forthcoming from the Drafting Department. The B 60s were still not completely satisfactory—with the result that the laboratory continued the struggle to produce a Diamond Disc machine below $100.[25]

The Old Man, far from satisfied with the performance of the laboratory during the disc campaign, blamed the delays on the incompetence of his men, as he usually did when confronted by failure.[26] Although administrative reform promised improved efficiency on paper, it fell below expectations in practice. The bad blood among foremen in the Phonograph Works undermined the cooperation that was vital for a speedy development job. The Works was still "more disorganized than organized," and foremen worked for their own personal, political ends.[27] The formation of a new organization and the massive reductions that followed, had intensified the political struggles in the Edison enterprise, bringing anxiety and strife to the vulnerable middle managers. In 1912 one engineer in the laboratory wrote to Edison's son, Charles, a student at MIT, "Everybody is afraid that somebody else is going to beat them out of their job" and indicated that the atmosphere was not pleasant in the West Orange facility. He told Charles Edison that "your father knows this condition exists, and it disgusts him very much."[28]

Edison Back at the Helm

From the organizational point of view, Edison's likes and dislikes should have mattered less as professional managers took over decision making. After all, the Old Man was now well into his sixties and surely on the point of retirement. The new organization had been given all necessary powers (on paper) and stood ready to take command. All that was required was an orderly handing over of power. Nothing of the sort happened. The importance of the disc project kept Edison at the wheel of the enterprise, and despite the establishment of an Engineering Department and the appointment of a chief engineer, Edison still led the laboratory work force in person, telling a friend in 1911, "I have 250 experimenters who depend on me daily for directions and when I go away it greatly disturbs the organization."[29] The new bureaucratic organization of the lab functioned at his whim. Although the position of chief engineer had become part of the formal organization of the laboratory by 1910, it was filled at Edison's discretion. He directed the activities of the chief engineer and was not above interfering in his work, sometimes taking projects from him to work on personally.[30]

The new position of chief engineer was part of a general movement within TAE, Inc. to specialize functions. Yet he could not be a specialist because the work of his department—originally focused exclusively on the phonograph business—was soon expanded to cover many products, including a wide range of electrical and motion picture equipment. It was part of the machine shop culture that every man be a generalist, able to turn his hand to whatever job was placed on his bench. This worked counter to the corporate program of specialization and made life difficult for the chief engineer.

George Bliss had been brought to West Orange because his expertise in electric motors and regulators exactly fitted the direction charted by senior managers, who had just created an electric motor department in the Works. Yet Bliss was asked to undertake a broad range of work of which electrical machinery was only a small part. His tenure as chief engineer spanned production engineering on the disc and storage battery projects, and meant the design of several new machines. His department also carried out work on cylinder phonographs, primary batteries, motion pictures, and special jobs, such as designing a small electric motor for a prospective Edison lawn mower and building a complete electric delivery wagon. Edison's order, "if Bliss feels job is too big—give it to Weber," reveals that the pressure was becoming too great for Bliss.[31] The Old Man did not have confidence in Bliss and gradually eroded his power to the point of ordering that all work carried out in the machine shops needed his prior approval—proof of his total control of the lab. Bliss only lasted one year before Edison removed him in 1912.

The next chief engineer was personally appointed by Edison. He

Figure 10.1. Miller Reese Hutchison. This 1912 photograph was suitably inscribed to "My Big Chief from His Friend and Worshiper, Hutch."

was a prolific young inventor who had also demonstrated considerable skill as an entrepreneur. Born and educated in the South, Miller Reese Hutchison had produced a hearing aid, invented the electric klaxon horn, and was in the process of developing an automobile self-starter when he first came to visit the West Orange laboratory (fig. 10.1). His record as an inventor was supplemented by his success in self-promotion—a trait he shared with his idol.

Hutchison's rapid rise to prominence in the enterprise was more the result of his personal friendship with the chief executive than his considerable talents as an engineer. Hutch and the Old Man became inseparable. They relived the frantic activity and warm comradeship of the machine shop culture, experimenting all night, joking and reminiscing while they worked, and starting a multitude of new projects.[32] Their friendship rekindled some fond memories of the good old days for the elderly inventor; at last he had a successor to the faithful Batchelor. The emergence of Hutchison as an influential figure in the organization helped keep its center of gravity firmly in the laboratory. The framework of committees set up in the administrative changes of 1910 remained in the background. The executive power in TAE, Inc. was in the small group around Edison, who made the important decisions, overruled the actions of the committee system, and planned for the future.

Edison's decisions in the laboratory still had a profound effect on the operations of his industrial empire—the business of innovation had hardly been touched by the formation of TAE, Inc. One technical choice that was to be decisive in the future of the phonograph business was the design of the reproducer assembly for the Diamond Disc. The question of compatibility with the Victor discs already on the market was soon raised by the management of TAE, Inc. The strength of the Victor Company was its impressive catalog of recordings. It had set a course of signing the leading performers, such as Caruso, to record on its premium-priced labels. Music lovers flocked to buy these high-quality recordings and the record-buying public were impressed by the quality of Victor discs even if they didn't appreciate the music. The managers of TAE, Inc. wanted to make a reproducer that could play the lateral-cut Victor records to increase the market for the disc machine. This was obviously the right business strategy to follow, because the Diamond Disc would immediately have had an enormous selection of recordings available for its customers. Edison had little sympathy with these arguments and refused to develop a lateral cut reproducer. It was a matter of pride for him. He could not let the Victor Company set the standards for the phonograph industry to follow, especially for the inventor of the talking machine. He wanted his disc machine to stand on its own merits. The senior management in TAE, Inc. was faced with the task of persuading the Old Man to see reason before the forces of the marketplace did it for them.[33]

Despite the presence of several experts in the organization, Edison took over the vital area of music selection and recording. In the competition with Victor, this was probably more important than the actual machine. Many people bought Victor phonographs because they wanted to hear the famous singers who recorded on Victor records, and many deserted the phonograph to gain access to Victor's library. Edison took up the study of serious music and concluded that much of it was bunk.[34] He maintained that many of the famous artists recorded by his competitors were better at acting than singing. After acknowledging that the existing stock of Edison recordings was poor, the Old Man set out to improve both the technique of recording and the selections available on Diamond Discs, evaluating thousands of cylinder recordings before choosing the selections for the new disc. His strategy was the opposite of Victor's: "I propose to depend on the quality of the records and not on the reputation of the singers." Edison believed that music lovers would prefer a better recording to "a rotten scratchy record by a great singer."[35]

The experience with motion pictures had made Edison and his managers wary of dealing with first-class talent who invariably cost more money and were difficult to handle. The strategy for musical talent was to keep them as anonymous, and as low paid, as possible, even to the point of leaving the singer's name off the recording—a disastrous miscalculation that went counter to the public's increasing identification with the singer rather than the song! Edison's influence was decisive in putting the technology before the artist. The machine was to be the selling point of the Diamond Disc.

It was ironic that the musical offerings of a leading phonograph company were filtered through the prejudiced, damaged hearing of a sixty-year-old man. Edison enjoyed his job as music director, keeping tight control of recording and the signing of talent. His opinionated views about music were often pressed upon his subordinates, Walter Miller and George Werner, who were subjected to his prejudiced directives about recording music.[36] Although his own taste in music was mundane, one of Edison's great ambitions was to record classical music played by large orchestras. To permit the recording of larger groups of musicians, a new studio was constructed to Edison's specifications in Building 4 of the West Orange laboratory. He made thousands of experiments in the quest for the perfect, realistic reproduction of sound. Many remember Edison as the man who brought the wonders of electric light to the world. Yet in terms of the amount of time spent experimenting on the phonograph, Edison should be remembered as the man who labored for years to bring us

Figure 10.2. Edison and some muckers listening to a prototype of the Diamond Disc machine, with some apprehension, in the West Orange lab in 1912.

the clear, faithful reproduction of music. This was his life's work[37] (fig. 10.2).

Talking Pictures

Edison's quest for perfect reproduced sound also dominated the development of his motion picture product; talking pictures remained the primary goal of the research program. A new series of experiments on the kinetophone—the name given to the arcade machine—was begun in 1899. The tempo of research picked up in 1908 during the negotiations leading to the formation of the Motion Picture Patents Company (MPPC). At the same time that this agreement was signed in the library of the West Orange laboratory, an experimenter was working full time on talking pictures in the studio on the second floor.[38] Monthly expenditures on this project grew steadily in the years that followed. The success of the film projector shifted the focus of these experiments to a machine that could provide a sound accompaniment to a projected film—the new kinetophone was a film projector linked to a special phonograph.

The four-minute cylinder format was unsuitable for the longer films that followed such box office successes as *The Great Train Robbery* of 1903. Audiences in the first film theatres—the nickelodeons—wanted to see films that told a story. The first task in producing talking pictures was to develop a longer playing phonograph. It is significant that Edison chose to remain with the cylinder format for the kinetophone, even though the disc offered greater playing time. Experiments were undertaken at the lab to test the feasibility of discs, but the kinetophone was finally developed around cylinders of 4 1/4-inch diameter and 7 1/2-inch length. These outsize cylinders were the culmination of previous development work on longer cylinders for dictating machines and larger diameter cylinders for the Concert phonographs introduced in 1899. They offered slightly louder volume and slightly longer playing time (six minutes instead of the four of the Blue Amberols), but taken as a whole, the kinetophone was an unimaginative, unsuccessful response to the challenge of talking pictures. Edison was playing it safe, making the most of his previous cylinder experience and the existing resources of the lab. Here at last was the "conservative defender of the status quo," the title that Edison did not deserve in the battle of the systems.

The creation of the Motion Picture Patents Company (MPPC) in 1908 could be said to have removed any need for further development of motion picture technology, as it brought an end to the "unfair and ruinous competition" which had deprived inventors, like Edison, of a fair return on their motion picture patents.[39] The MPPC was a monopolistic organization formed on the basis of the major patents in motion picture cameras, projectors, and films. The ten members of the company, or trust as it was called, comprised the leading members of the American film picture industry. The MPPC claimed control over the production, distribution, and exhibition of films. Rather than sell a movie outright, it retained title and sold the right to exhibit it. The Trust issued licenses to filmmakers and collected the royalties from the exhibitors who showed the Trust's films. It was similar to many other patent pools and monopoly organizations in American industry and it finally achieved Edison's dream of a monopoly of his film camera patents. He received the lion's share of the income of the MPPC, which was about $1 million a year from 1909 to 1914.[40]

Film historians have often interpreted the MPPC as a conservative monopoly that retarded the development of film both as art and mass entertainment. The members of the Trust are sometimes depicted as conservative businessmen only concerned with mass-producing standard lengths of film for a captive audience. The two important developments in film subjects during the heyday of the MPPC (1908–13)—the development of longer feature films and the star system—are ascribed to the independent filmmakers who finally broke the Trust. It can be argued that the MPPC had no incentive to innovate. The MPPC claimed to account for seven-eighths of film production and distribution and struggled to meet the voracious demands of the film exhibitors. Longer and more elaborate productions, with highly paid stars, would merely add cost to the product.

Recent scholarship has revealed a different story. Far from being a retarding influence on American film, the MPPC brought order to chaos and created a strong American industry that could compete with European filmmakers. The Trust was concerned not only with making profits but with raising the quality of film subjects. The motion picture industry evolved in the same uncertain institutional environment as electric light. Several public groups had condemned motion pictures as immoral and had successfully banned them from their communities. The MPPC did its part to improve the quality of movies because respectability attracted a middle-class audience. The nickelodeons certainly started as places of bad repute but they were soon frequented by the middle class. Edison had plenty of experience operating in a regulated industry. He made it clear that his goal was to produce educational, family films, and to make sure that this business was not hampered by regulation. One of the important functions of the MPPC was to work with the authorities to achieve self-regulation of its output. As the leading member of the Trust (with Biograph), the Edison company developed longer, feature films and signed the stars of the Broadway stage.[41] Like the disc machine, Edison's motion picture business was increasingly aimed at the more affluent middle class.

When the film exhibition business began in the late 1890s, Edison was primarily concerned with reducing the price of the standard filmstrips made for exhibitors—the artistic content of the film was not considered important.[42] Films were treated as a commodity at West Orange, as another product to be economically manufactured and sold. As the business grew from a novelty to an industry of international scope, the managers of the Kinetograph Department, led by James White, paid more attention to the quality of the film product. By 1910, they were busy buying the rights to literature, including the work of Mark Twain. During the first decade of the twentieth century, the typical Edison film was a one reeler, running less than 1,000 feet. The length of films had increased substantially by 1914 when one multiple-reel (two- or three-reel epic) and four single-reel films were produced by TAE, Inc. each week. When the MPPC succumbed to competition and unfavorable court decisions about 1915, the Edison company was attempting the transition to the longer features and established stars that dominated the film industry.

Although no supporter of the star system, Edison played an important part in maintaining the quality of the films produced by his organization. He took the same personal interest in films as he did in the record catalog, passing judgment on the output of his studios, attending movie theatres to test audience reaction, and making his views known to all involved in film production.[43] He had a special interest in educational films that reflected his often publicized desire to see his inventions used for education. The National Phonograph Company offered a series of instructional cylinders to be played on the special school phonographs, and the Edison Manufacturing Company produced a series of "educational films" that covered geographical, agricultural, industrial, historical, scientific and military topics.[44] In the twentieth century, Edison filmmakers worked in conjunction with institutions such as the Bureau of Health and the National Association for the Study and Prevention of Tuberculosis.

In 1912 an ambitious plan to produce a library of films on basic science was undertaken at the West Orange laboratory. This was a major project planned over eight years to provide a visual resource for schools. Edison was going to revolutionize education. His sole aim, according to his secretary, was to "emancipate the children from . . . laborious and useless work in acquiring an education."[45] The Old Man took this project very seriously. He assigned one of his best experimenters, W. W. Dinwiddie, to supervise film production, and converted Building 1 (the old electrical laboratory) into a studio. Edison's motives were not purely altruistic; films on electromagnetism and chemistry were easily made in the West Orange laboratory where experimental apparatus could be used instead of the expensive sets required for cowboy and adventure movies.[46] There was also a pressing need to produce a library of films for the

Figure 10.3. A retouched publicity photograph showing a family watching the Edison home-projecting Kinetoscope. The manufacturer would have us believe that this machine is so simple it can be operated by a child.

home-projecting Kinetoscope. This machine was a small projector designed to be operated by the user in the comfort of his sitting room (fig. 10.3). Edison had again anticipated an important market and moved into it before the technology was ready. The high price of machines and slow growth of a film library doomed this foray into the consumer market.

The key to success in this business was the medium rather than the machine; it was pointless to market a talking picture machine without the films to go with it. Recording the sound track to accompany the film presented some difficult technical problems. The West Orange recording staff had plenty of experience recording musicians, but actors and dancers presented a more complex sonic environment and posed some tricky questions: How does one increase the sensitivity of the recording without also increasing the volume of background and reflected sound? How does one position the actors for best technical results without destroying the quality of the performance? The recording was done with a large horn on the phonograph which picked up the sound of the action on stage. It took an experienced man to keep the voices of the actors clear without holding the horn too near their faces. The requirements of good drama and good technical reproduction rarely coincided. When sound recording of films was later tried with electrical equipment, a generation of actors had to be retrained in new techniques of making movies.

Once the film had been made, there still remained the task of amplifying the volume of playback to ensure that the audience could hear the sound track in every corner of a large auditorium. This problem had been of utmost concern to the experimenters in the laboratory since the 1890s because their phonographs had a softer playback than the competition's. They found that increased volume could be obtained by putting more pressure (tracking force) on the needle as it made its way through the grooves, but this increased wear on the cylinder surface, which eradicated the high notes and obscured fidelity. Several inventors had achieved higher volume by mechanically amplifying the sound in the reproducer and horn assembly. One of them, Daniel Higham, had patented an over-sized reproducer with a friction valve device that acted as an amplifier. His services were acquired and he came to West Orange in 1908 to play a leading role in the development of the kinetophone.

In 1910 an experimental machine was demonstrated to the press. The laboratory was the scene of this demonstration, which included pictures of an announcer making various sounds and a dancer performing with musical accompaniment. Although Edison promised a perfected machine in a year, Hutchison commented, "We are up against a pretty stiff proposition" in getting it to work.[47] He tried using several horns to get a more balanced and louder sound track, but Edison told him that thousands of experiments with two or more horns had not produced acceptable results.[48]

Thousands of additional experiments were required to perfect the synchronization of sight and sound. The considerable distance between the film projector at the rear of the auditorium and the phonograph behind the screen added to the complications of adjusting the two machines to work in unison. A clutch mechanism was developed to alter the speed of the playback in order to synchronize it with the film but this was a very difficult mechanism to work. It was linked to the operator in the projection booth by a cord running on pulleys. The projectionist had to make continual adjustments to keep the sound in time with the image.

The kinetophone team began another year of experiments in 1912, extending one of the longest running projects on the laboratory's books.[49] Edison was determined to see the project through to the end. The business of innovation was still dominated by belief that a technical solution existed to every problem. In the spring of 1912 the kinetophone team finally succeeded in recording some cylinders that passed the Old Man's scrutiny. Preparations were made for the machine's introduction but Edison would not let it out of the laboratory until he was satisfied with it.[50] Once the experimental model had passed muster in the laboratory, there still remained the work of producing a commercial machine and its accessories.

Like the early phonograph, the kinetophone was too complicated to sell directly to the users; Edison planned to use his own men to manage its commercial introduction. They were picked from the West Orange work force and underwent extensive education in the laboratory. They had to pass stringent tests, including written exams, before they were sent out.[51]

The work on talking pictures reached a climax in the spring of 1913 when Hutchison noted in his diary that the machine was at last working properly.[52] The kinetophone was formally introduced with Edison's claim that he had perfected talking pictures. The first public performances in a vaudeville house in New York led the press to announce yet another marvel from the West Orange laboratory. Enthusiastic audiences all over the country thrilled to the kinetophone as it was taken on the road. The high novelty value brought in the crowds. The *New York Times* noted, "Gasps of astonishment could be heard in the audience" as the pictures on the screen spoke.[53] The talkies had begun well before *The Jazz Singer.*

The successful introduction of talking pictures added to the revenue coming from the motion-picture department of TAE, Inc., which was the most profitable part of the Edison enterprise in 1913, and made up for the losses incurred in the slow introduction of the Diamond Disc.[54] The income from motion pictures supported the enterprise through the difficult years of perfecting the disc machine and making up a record catalog. Yet after its brilliant start, the kinetophone quickly declined, never to recoup the great expenditure on experiments. As was to be expected, the novelty value soon wore off. The decline of the kinetophone is best explained by the problems of maintaining and operating a highly complex machine. One error in synchronization could turn a triumph into a disaster, and cheers into jeers. More than one kinetophone show was booed off the stage! It was too difficult to control the kinetophone with a string that reached from the machine (which was next to the screen) to the projection booth. The problems could be traced back to the underpaid operators, who lacked the motivation and skill required to work the machine, and the rats who ate through the control cords.[55]

Edison was embarrassed by the theatrical disasters of his new invention, but he was not going to let one failure stop him. In the beginning of 1914 he told Hutchison that they were going to "perfect the kinetophone to the limit." Instead of filming around the existing schedule of the Bronx and 21st-Street studios, Hutchison was to have his own kinetophone studio in West Orange, specially built to absorb the reflected sounds that ruined sound tracks. Edison was working flat out to perfect the kinetophone. He had filled a notebook with fresh ideas and told Hutchison, "We are the only people who can do it . . . show the theatrical people that scientific people can beat them at their own business."[56] Here Edison revealed the basic strategy of the phonograph and film operations; commercial success could be attained with technical perfection that overcame aesthetic or dramatic considerations.

This concern with the technical rather than the dramatic was an appropriate one for the West Orange laboratory. The attention of the laboratory staff was on technical problems; each man was concerned that his area of responsibility performed to the Old Man's satisfaction. The recording engineer had to dominate the proceedings or the result was a cloudy sound track.[57] The problem of reflected sound in the studio made it imperative that the actor speak directly into the horn. The short kinetophone subjects were directed by laboratory staff who placed the actors and staged dialogue to achieve this end. Even if the film was approved by the engineers at West Orange, it still could not compete with a professional entertainer who used other systems of sound motion pictures. Audiences wanted a show and not several unrelated technical demonstrations of the capabilities of the machine. One observer concluded that compared with a good showman with second-rate equipment, the superior Edison machine still looked amateurish.[58] The kinetophone failed to entertain audiences and soon vanished from the scene. No further effort was made to develop talking pictures at West Orange.

This failure hurt the Edison enterprise and must have been a severe disappointment to the Old Man. Technology had not triumphed over art. The kinetophone was the last of the highly ambitious experimental projects that had characterized the West Orange laboratory. Hundreds of thousands of dollars, every possible resource, and the best brains in the laboratory had not brought success. Years of costly research into color films were also abandoned. The great leap forward into talking pictures had not been achieved, dooming TAE, Inc. to compete in terms of the commercial appeal of its movies. There were to be no more ambitious forays into revolutionary new technologies at West Orange; the kinetophone was the last hurrah of the "largest lab extant." Innovation at the Edison laboratory was to be directed into less expensive and more predictable pursuits.

CHAPTER ELEVEN

The Rise of the Organization

he year 1914 was a watershed in the history of Edison's business enterprise. The outbreak of war had a deleterious effect on the operations of a multinational organization. Even more damaging was a serious fire that almost consumed the West Orange complex. The conflagration started in a shed containing inflammable motion-picture film stock, which stood in the middle of the old Phonograph Works. The fire began in the afternoon of 9 December 1914 and by nightfall the 1888 wooden factory buildings were burning furiously. It spread rapidly through the rest of the Works, creeping closer to the laboratory buildings and their precious experimental models and notebooks. The flames from the chemicals and other combustible materials used in manufacture could be seen seven miles away in Newark. After many hours of anxious suspense, the fire slowed down as it moved along the administration building toward the laboratory. In the early hours of the tenth, it appeared that the old brick buildings of the laboratory were going to survive, and with them the priceless masters for the Diamond Discs, but by the time the fire had been brought under control that morning, thirteen buildings had been badly damaged or destroyed at an estimated loss of a million dollars.[1]

The rapid recovery from this disaster is a testament to Edison's personal leadership of the organization. Despite his ill health, he led the clearing work that began the day after the fire. Experimenters, managers, and laborers returned to pick their way through the rubble.[2] Soon after the disaster a document entitled "Manufacturing Resources for a Quick Resumption" circulated the West Orange complex. At the top of the list, "Edison, himself—24 Hours a Day."[3] The great fire underlined that he was still the driving force of his industrial empire. The retirement of 1907 was forgotten. Increasingly poor health was ignored. The challenge of introducing the Diamond Disc line had now turned into a campaign to rebuild both the organization and the factories of the West Orange complex.

Charles Edison, the second child of Thomas and Mina and the first boy born to this union, was now at West Orange in training to take over the business. A graduate of MIT, Charles joined the company in 1913. As he showed his son the ropes, Edison was probably feeling the pressure to conserve his business enterprise in preparation for the eventual handing over of power. After the fire he gave his son a personal demonstration of the determination needed to run a large corporation. He did not have time to mourn the damage done; instead he reassured Charles, "We'll have a much better plant when we get through. We've swept away all the old shacks, and now we can have a good plant."[4]

The fire did indeed do more to reorganize the Phonograph Works than all the administrative reforms introduced from 1900 to 1914. By destroying all work in progress, the fire gave Edison's engineers the opportunity to start from scratch. New procedures could be instituted from the ground up. The fire also forced management to establish a strict set of priorities for manufacturing. In 1913 the Works was called upon to manufacture five types of outside horn phonographs, four models of Amberolas, five different disc machines, Opera phonographs with outsized cylinders, dictating and shaving machines for commercial use, special phonographs for language study, and "school outfits" for educational purposes. A wide range of motion picture equipment and accessories, Bates numbering machines, and special jobs for the laboratory were in the pipeline waiting to be made up in the Works.[5] When the Works reopened many of these products were quietly forgotten, never to be manufactured again.

The production of the Edison disc phonograph, especially the redesigned B 60 economy model, was given top priority when production resumed in 1915. The first disc was produced in the storage battery factory opposite the laboratory by the end of January 1915—a feat of which the management and labor force at West Orange were justifiably proud. Those phonograph lines that were continued after the fire were consolidated into a smaller number of models that would not tax the reduced manufacturing facilities. The popularity of the Blue Amberol records had given new life to Edison's cylinder business, extending the line of Amberola phonographs to at least twelve different models by 1914. Although plans had been made to consolidate this profusion of machines into a more manageable line, no progress was made until the fire intervened. The Amberola line that reemerged from the ashes was based on three models that all used the same mechanism, with an extra spring in the most expensive model. John Constable, the phonograph expert in the laboratory, took personal responsibility for standardizing the parts of the Amberola around the smallest number of components. He laid out the steps of production and supervised each step from the laboratory, providing an important precedent for the manufacturing reforms after 1914.[6]

The fire was a major turning point from the engineering and manufacturing standpoints. It was also the symbolic watershed between a period of product development and the years of manufacture that followed. TAE, Inc. told the Internal Revenue Service that the pre-1914 period was the development stage, when Edison, "being more interested in the perfecting of the product," directed the resources of his enterprise toward research.[7] After the fire, TAE, Inc. prepared to concentrate on manufacture. Edison had anticipated this shift as early as 1910 when development in the three major businesses—phonographs, batteries, and motion pictures—was in progress. He wrote to Sigmund Bergmann in Germany that "our pioneering and experimental period is now over and we are going for cost reduction."[8] The continuing problems with the disc machine spoiled this schedule, delaying the introduction of several new products. But by the end of 1914 the Engineering Department was hard at work producing drawings for the new model Amberolas, the line of disc phonographs, an improved super Kinetoscope, and a new dictating machine.[9] All these machines had been in the pipeline for some time. The design of the new dictating machine, for example, had been years in the laboratory before the Old Man finally gave it his approval. The drawings for the new business machine were completed at the end of 1914, and the tools were being made up in the Phonograph Works when the fire struck.[10]

Once the Works had been rebuilt, the task of the laboratory staff was production engineering rather than product development. This is not to say that experimenting ended at the laboratory, but that it gradually assumed secondary importance to the overriding goal of manufacturing more phonographs at lower cost.

Edison and the small group of muckers around him carried on with the improvement of the disc machine and record. Their first priority was to eliminate surface noise on the playback. After claiming superior reproduction, Edison could not be satisfied with anything less than perfection, and it came as a great shock to all in his laboratory when the disc records failed to live up to their expectations. The problem emerged after discs had been on the market for some time and users had played their records repeatedly. The diamond point of the Edison disc phonograph was claimed to be superior to the competition's steel stylus because it not only sounded better but also did not have to be replaced—the steel needle point grew dull over time. Unfortunately, the diamond stylus acted as an abrasive on the record surface, causing an annoying scratching sound while it wore away the groove.[11] Edison went back to the laboratory to solve the problem.

In the years after the fire, Edison supervised every major experimental campaign in the laboratory in addition to his work on improving the disc. No production model left the laboratory without his approval. The fire seemed to have injected a new energy, an urgency, into his work. With loans from banks and Henry Ford, he set about the task of rebuilding the West Orange complex. His secretary and biographer, William Meadowcroft, had been with him long enough to have seen many periods of frantic activity, yet he recounted that 1915 stood out as a year of massive effort at the West Orange laboratory.[12]

The Divisional Policy

Edison was contemplating more than rebuilding his industrial empire after the ravages of the fire; he was planning another major organizational change to suit the strategy of mass production. And true to the absolutism of the Old Man in the enterprise, these changes were planned, decided upon, and implemented from above.

Thomas A. Edison, Incorporated, had been a response to falling profits and overextended finances. It provided economical management to a broad range of operations. A level of middle management had been created and specialists hired, but their influence was circumscribed by Edison's absolute power and their numbers reduced by the cutbacks that followed the formation of TAE, Inc. Decision making remained in the hands of a few senior executives around Edison. Proximity to the Old Man was still the key to power in the organization, as the meteoric rise of Miller Reese Hutchison had shown. The centralized organization of TAE, Inc. tended to concentrate power in the upper level of management, distancing it from day-to-day activities. In the words of one manager, "We were suffering from having the brains all on top and the labor all underneath with nothing in between."[13]

The shift of emphasis to manufacturing and marketing naturally put new demands on the administration. Mass production demanded a system of mass distribution, but TAE, Inc. was still struggling on all fronts against a competition that was well-entrenched: Victor and Columbia dominated the phonograph business, the Dictaphone Company (a subsidiary of the Columbia Company) held sway in the dictating machine market, and the Exide batteries of Electric Storage Battery Company were the products to beat in this field. The motion-picture business had been taken up after the fire, but the competition in this industry reached new peaks after the demise of the Motion Picture Patents Company opened the floodgates for new entrants. The Portland Cement Company faced falling prices in an overcrowded business.

The policy of diversification had created many different businesses in the Edison enterprise, each one with its unique technology. Each of the Edison products had a different market: primary batteries were sold largely to the railroad industry, storage batteries went to industrial users and delivery companies, the dictating machine was marketed to the business community, and the phonograph and motion picture were articles of mass consumption that demanded a special knowledge of the world of entertainment. The top-heavy administration of TAE, Inc. was perceived as a disadvantage in selling a broad range of products in different markets. The men who were closest to the point of sale had the expertise but none of the initiative in the company.

The senior management of TAE, Inc. recognized that each of the diverse products they manufactured would be better marketed by those who were knowledgeable about the business. The rapid swings of public taste in phonographs and motion pictures had shown how important it was to react quickly to changes in the market. The divisional policy was framed with this goal in mind. It made sense that products as different as cement and primary batteries should be marketed by independent organizations. The divisional policy was intended to move decision making closer to the customer by giving middle managers the opportunity to exercise their special technical or marketing skills. It created channels for their input within the organization, allowing them to provide the timely information with which to better apply the engineering and manufacturing resources of TAE, Inc. to a changing market situation.

The evolution of the modern business organization has been chronicled by the historian Alfred Chandler. He has shown that the multidivisional, decentralized structure came as a response to problems of running the large diversified businesses created during the merger movement at the turn of the century. Companies diversified into new product lines to use their resources more profitably, but as they took on more types of products, they required a new organizational plan for the administration of their complex businesses. The divisional structure answered many of their needs and eventually became the standard of business organization in the twentieth century. Beginning with Du Pont and General Motors, Chandler has described how big business in America embraced this structure.[14]

As Edison's policy of diversification had come two decades before those of Du Pont or General Motors, his move to a divisional structure preceded theirs by several years. Yet there is some similarity between TAE, Inc. in 1914 and General Motors before it was reorganized by Alfred Sloan in the 1920s. Sloan described the chaotic empire built by the pioneer automobile entrepreneur William C. Durant thus: "We were short of cash. We had a confused product line. There was a lack of control and of any means of control in operations and finance, and a lack of adequate information about anything."[15]

This was the position of TAE, Inc. in 1915. There was the same lack of information about production, sales, and finance, but on a larger scale because Edison made many more different products than General Motors or Du Pont. The cost of rebuilding after the fire had substantially increased the debt of the company. The requirement for more precise financial accounting came with the burden of debt.

The divisional policy was laid out by Edison and announced on March 15, 1915. It assigned each major product in TAE, Inc. to a separate division, each the responsibility of a division manager. The divisional structure was modeled on the U.S. government. The central administration was the federal government, each division was a state, and the division manager was the state governor.[16] Each division stood on the foundation of its particular business and was expected to make its contribution to the profits of the enterprise. The central administration was to give the division managers "as wide latitude as possible," after setting general policy. Its function was to coordinate the activities of the divisions. At the top of the divisional structure stood Edison. He did not concern himself with the day-to-day affairs of the divisions, for as Meadowcroft wrote: "His forte does not lie in the lines of routine business methods."[17]

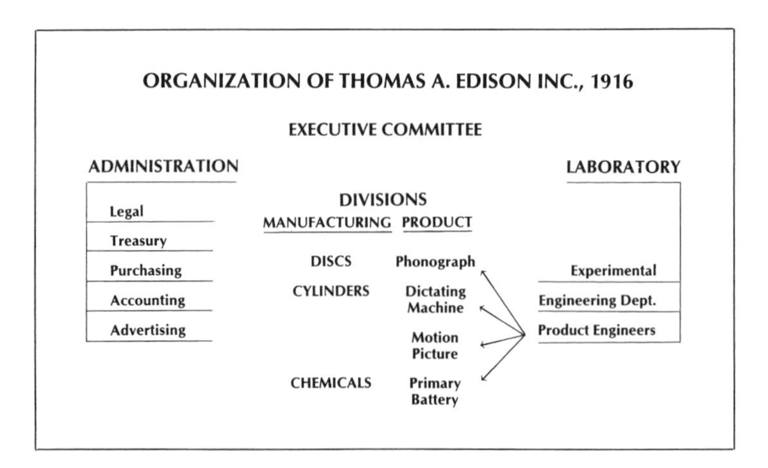

Figure 11.1. Organization of Thomas A. Edison, Inc.

His role was to carry out strategic planning, assisted by a few key vice-presidents.

The move toward an organization based on marketing divisions had begun after the creation of TAE, Inc. when the sales of each product were administered by a sales manager and a small department. Many of the division managers were sales orientated, including William Maxwell of the Musical Phonograph Division and Nelson Durand of the Dictating Machine Division. The divisional policy was first introduced in the primary battery business, because this product was clearly independent and its market was narrow. The other lines followed. The laboratory was classed along with the central administration as a service organization. Its work was divided into three basic functions: experimental, product engineering, and engineering services. Its status reflects the management philosophy that the laboratory provided a basic service to the divisions. It was not made into a separate division that had to justify its existence by showing a profit at the end of the year.

Figure 11.1 shows the organization of TAE, Inc. in 1916, in which the divisions represent the Edison product lines: musical and business phonographs, disc and cylinder records, numbering machines, chemicals, primary batteries, and motion pictures. Some of the divisions were oriented toward marketing, such as the Musical Phonograph Division that was conceived as a jobbing organization that bought the machines and records manufactured by the various manufacturing divisions and marketed them through a network of dealers. Other divisions were primarily in manufacturing, including the disc and cylinder divisions and the divisions that produced raw materials, such as the Phenol Resin Division.[18] Materials and labor moved through the organization via interdepartmental transfer.

The divisional policy accelerated the process of organizing each aspect of manufacturing into a separate unit, each under the control of a manager. In time the departments that carried an increased load of business were turned into divisions, such as the Reproducer Assembly Division of the Phonograph Works, established in April 1916.[19]

The divisional policy was not only to produce more informed and energetic marketing of each product but also to set up a costing structure to provide the information needed to run the business. The division was created to group together the costs of either making or selling the product. The division manager was made accountable for all costs and was expected to show a profit for his operation if he was selling, or keep within an existing cost structure if he was manufacturing. (TAE, Inc. followed current industrial practice by separating manufacturing costs from marketing costs.) The continuing departmentalization of manufacturing enabled better cost control by setting up more points in the process to estimate cost, and each division had to account for the resources and materials it used. It was expected that the division managers' concern with keeping costs down would bring efficiency to the administration and prevent it from becoming too large.

Under the old system at TAE, Inc. there was no responsibility for following the expenses of making the goods because they were all manufactured under the same roof and with the same costing system. The expenses of administration and sales were grouped together as overhead. A highly dispersed business, such as motion pictures, defied accurate estimation of the costs of production because there were many different processing points and a diverse number of overhead charges. During the balmy days of the phonograph boom, TAE, Inc. manufactured thousands of products without an exact knowledge of cost. "We never knew whether we were making or losing money," remembered one executive.[20] This situation could not be allowed to continue.

The central organ of the new divisional policy was the Financial Executive. This body monitored the financial transactions of the divisions and acted to reduce excessive expenditure. The divisional structure was to be responsive to the economy drive of the central management, who made it known that this was Edison's personal policy.[21] The head of the Financial Executive, Stephen Mambert, implemented a much tighter purchasing policy on all parts of TAE, Inc. Previously this policy was poorly defined and the purchasing department was small. In the new order all requisitions for materials were funneled through Mambert's organization. The enlarged purchasing department could not act until it received proper authorization from the Financial Executive. Mambert's power extended to approving the sale or disposal of by-products and waste. The divisions were told that their dealings with outside manufacturers should be confined to information that was needed to make "an intelligent request" for goods or supplies; they could no longer act as an independent purchasing agent.[22]

As the clearing house through which all orders for materials, tools, and services had to pass, the Financial Executive could exert a powerful influence on the activities of the divisions. Its manager exploited his position to move quickly up the administrative ladder. Stephen Mambert began his employment in TAE, Inc. in 1913 as a cost clerk. A year later he was made assistant to Charles Edison. Although he was being groomed to take over from his father, the twenty-four-year-old Charles found many distractions during his early years in the company, including the delights of Greenwich Village, which was only a few minutes from West Orange by automobile.[23] Mambert, in contrast, was both conscientious and ambitious and stayed close to the hub of the organization. He gained the ear of Edison and was soon promoted to the new position of Efficiency Engineer and placed within the laboratory organization—a sure sign of his favor with the Old Man.

The cause of efficiency in American business was led by Frederick W. Taylor and his associates who attempted to apply scientific principles to increase the productivity of the worker on the shop floor. Their goal was to find the best way of doing things and systematically apply this knowledge to all aspects of production, increasing the efficiency of the "human machines" of the organization.[24] Taylor began his career in a machine shop, where he found to his amazement that it was the workmen who really ran the shop.[25] He would probably have come to the same conclusion if he had visited the machine shops at the West Orange laboratory. Taylor set out to wrest control of the pace and character of work from the machinist and give it to the manager. Shop management in his view controlled every aspect of work in the shop to fit the specifications laid down by management, using various coercive measures to make workmen do things the right way.[26] Establishing the right way was, of course, the province of the manager rather than the artisan. It began with the analysis of the work process by time and motion studies and ended with its complete standardization. This process became popular in large companies with extensive manufacturing operations as the gospel of efficiency spread beyond the machine shop. The focus of the Taylor system was the planning department that directed and coordinated production through the work of professional engineers. "Scientific management" has been interpreted as the attempt by engineers to retain control of large business organizations that outgrew the shop culture of the machine shop.[27]

Although Edison was not impressed by the entreaties of management consultants, his enterprise was influenced by the efficiency craze of 1911, which followed publication of some of Taylor's important texts, but not in the area of manufacturing and not to the benefit of the engineers. Efficiency at West Orange was concentrated on management, specifically on identifying and controlling costs. Those who benefited from this movement were accountants and bureaucrats like Stephen Mambert. Engineers and their machine shop culture became a target for the efficiency movement. The central goal of this process was control—whether in the machine shop or in the operation of divisions.

The power and influence of Mambert's Financial Executive was built on a foundation of better record keeping within TAE, Inc. It relentlessly pressed each manager for the exact details of his operation. It scrutinized the payroll and hunted down discrepancies. It checked, and double checked, accounts payable and receivable.[28] Although theoretically independent of central management, the division managers were brought to heel by a collar of paperwork. They were instructed to send copious information about their activities to the central management. When Edison began the divisional policy, he instructed every division to supply him with daily reports of orders and shipments.[29] Although it is doubtful if many were read, the stream of memos arriving at his desk provided the means of monitoring the work of the divisions. This policy was continued by Charles Edison who requested monthly reports about cost reductions and new methods of production.[30]

A new purchasing policy created a massive amount of paperwork because even the smallest transaction between divisions and departments of TAE, Inc., which had been informal and unrecorded, now had to be carried out under a purchase order, including all the work of the laboratory. The lab staff found themselves filling out many

Figure 11.2. Stephen Mambert with the "Financial Cabinet," his staff of accountants, financial analysts, efficiency clerks, and yes-men.

new forms to document even the most trivial job. Shop orders, requisitions, and Engineering Notices formed the paper trail followed by engineers and muckers as they went about their work.

The flood of information within the organization promoted the influence of bureaucrats, from the lowly cost clerk in the Works to the professional accountants in the central administration.[31] Such was the increase in the production of statistics that the Accounting Department was enlarged, and a new department, the Efficiency Department, was created. This was Mambert's domain. He also controlled the work of the accountants, bookkeepers, and statisticians in the accounting and auditing departments. Mambert sent his men to every part of the organization; some acted as office managers, some operated the new mechanical accounting machines installed in all the divisions, and others simply carried out Mambert's dictum of "better financial practices," which was based on copious reporting of information to the Efficiency Department[32] (fig. 11.2).

While Taylor's disciples used the "decimal dial" stopwatch,

Mambert used the accounting machine and the audit to promote his conception of efficiency. His influence spread from financial affairs to every aspect of work in TAE, Inc. He could go anywhere and see everything. The only information in TAE, Inc. kept from him was the payroll of executives and departmental heads. He answered only to Edison and Charles Wilson, the vice-president in charge of operations. By 1916 Mambert ran his own empire within the organization, and his power had reached the point where he portrayed himself as the means to translate "Mr. Edison's policies relative to Management, Labor and Sales" into action.[33] His usefulness was probably more as a liaison with the financial community than as an efficiency expert. The banks upon which TAE, Inc. was becoming more dependent wanted the kind of information that Mambert had and were doubtless receptive to his crusade to control costs. Edison thought that Mambert was the most valuable man he had ever employed, revealing that "Mambert's methods and showings have now enabled us to borrow very large sums . . . and the banks have confidence.[34] Edison's instructions to Mambert were to cut the payroll. Mambert had no use for sentimentality in running a business and dutifully obliged. All this was done in the cause of efficiency.

Mambert's methods did not go unchallenged within TAE, Inc. The same informal resistance that undermined the impact of scientific management also hindered the spread of efficiency. Not all managers shared Mambert's love of better financial practices, and some thought them an utter waste of time. The most dissension came when Mambert's methods clashed with the shop culture, which put little value on accurate record keeping and the prompt reporting of information to a central authority.

The men who thought the least of Mambert and his methods were the engineers who had been placed in charge of the manufacturing divisions. Many had begun their careers as muckers in the lab. Some had even worked next to the Old Man, to their delight and edification. As division managers, their concern first and foremost was technical matters. Record keeping came somewhere at the bottom of their priorities. Elisha Hudson, who managed the Primary Battery Division, made a point of ignoring Mambert's memos, telephone calls, and committee assignments. Walter Dinwiddie took resistance a step further and actively opposed Mambert's activities within his division.

A graduate of the University of Virginia, Dinwiddie rose through the lab hierarchy after his successful work in disc record duplication.

In this project he worked closely with Edison, who had an office near to his room on the third floor of the laboratory. Edison's decision to switch Dinwiddie into educational films took him from the complexities of estimating wax shrinkage to finding swarms of bees to be filmed. As director of educational films, Dinwiddie had to display the initiative expected from all muckers. Among his many tasks was the development of a strong cement to bond glass and the search for a powerful lifting magnet (finally located in a scrap yard in Elizabeth, New Jersey), to be filmed in a segment about electromagnetism.[35]

After gaining management experience in the educational film program, he was put back to work on the problems of duplicating discs and finally placed in charge of the division. He had come under the pervading influence of the Old Man in the experimental rooms of Building 5. Dinwiddie demonstrated his absorption of the machine shop culture during his tenure as manager of the Disc Reproduction Division. The high priority given to experimenting, the informal style of management, and the acute scientific curiosity reflected Edison's style perfectly. Unfortunately these qualities were out of step with the new corporate culture of TAE, Inc. Walter Dinwiddie was, at heart, an experimenter, and it infuriated Mambert that he ran his division in a democratic and (to Mambert) unbusinesslike manner. Even worse, Dinwiddie had no respect for the organization (always spelled by Mambert with a capital O) and no interest in the financial reporting that the efficiency expert thought was vital for the successful operation of the business.

Dinwiddie's Disc Reproduction Division was considered by Mambert to be the weakest division within TAE, Inc. and was immediately targeted for financial reform.[36] The issue was the estimation of the costs of manufacturing disc records. As a manufacturing division, its output was transferred to the marketing division at a fixed cost. The price of disc records was decided by the upper management to suit the competitive situation rather than the cost of making them. Consequently, one of the first divisional policies, which was delivered down by Edison himself, was to make intensive studies of costs of operations of this manufacturing division. The disc reproduction operation was expected to be a major profit point in the business. When Dinwiddie took over, it produced 35,000 records a day and was set to increase this output, yet the exact cost to make each disc was not known.[37] The rough estimates of the cost of a disc record ranged from 40¢ to 50¢ per disc, but when a fraction of a cent could alter the balance between profit and loss, an exact knowledge of costs was essential.

Mambert sent his men to the Disc Reproduction Division to check the payroll and keep an exact tabulation of costs of raw materials and the interdivisional transfer of supplies and labor. At this point Dinwiddie did not know how much it cost to make a disc because he was still perfecting the process. Mambert set about changing this situation by financial and organizational reforms, insisting that Dinwiddie provide the information to compute cost. Dinwiddie's main concern was with the quality of raw materials, which was fluctuating during wartime and making it difficult to produce good discs. The sound quality of the duplicated disc had to meet Edison's high standards. The cost could be figured out later when the production process was stabilized.

The clash between the two men became a symbolic struggle between the old and new way of doing things, shop culture versus the Organization with a capital O. Dinwiddie did more than ignore Mambert; he fought back by scheming to remove Mambert's agents from his division, criticizing their effectiveness and loyalty.[38] The duel between Mambert and Dinwiddie was ended when the latter was abruptly called back to the lab to carry out special research. This move was part of Edison's massive effort to improve the sound quality of the disc record. It was probably seen as a victory for Mambert who was then able to institute cost studies of disc record production. Charles Edison worked on this task when he joined the company.

The Financial Executive continually stressed the need to project the costs of each division in advance, and division managers were hounded to provide estimates.[39] It was Mambert's goal to institute a system of standardized costing of manufacturing. This is a modern accounting technique that includes intensive research of the manufacturing process and the creation of estimated costs for each part of the manufacturing process. These standard costs make it possible to calculate potential overruns or reductions during manufacturing, providing a means to control the operation.[40] Although he appreciated that standard costing could only come after standardization of manufacturing, Mambert's speedy introduction of his financial controls meant that the system of costing was established before manufacturing could be instituted. This made standardized costing impossible. As John Constable reflected in 1919, division managers were "responsible for the costs on operations and parts of which they knew nothing . . . There lies one of our greatest curses today."[41]

Although a great deal of publicity was given to business forecasting in the Edison industries, planning did not work out in practice. The divisional policy successfully established a chain of command and financial responsibilities. Its perceived advantage was that it could better employ resources through coordination of effort. Yet the emphasis on reduction of costs undermined this goal because all attention was on the bottom line. The task of running a division became a game of fabricating the right numbers. Efficiency, as defined by Mambert, had its disadvantages. In the laboratory, financial control placed innovation in a straightjacket from which it was unable to escape.

The Reorganization of the Laboratory

The divisional policy had divided TAE, Inc. along product lines, and it was inevitable that the engineering services provided by the laboratory should follow suit. In the years before the divisional policy, the work of the laboratory had become more product specific as the first steps had been taken to organize engineering services around each product. A new direction in labor policy was begun in the new century when employees with specific skills were sought instead of the generalists hired earlier. For example, Alex Werner was hired in 1900 as a photographic experimenter to work on color films.[42] Whereas in the 1880s a man would join the laboratory, report to Edison, and be assigned to an experimental team, after 1900 many employees were hired for specific projects. The conditions of their employment were decided before they came to West Orange.

The Edison enterprise was moving toward hiring professional engineers with specialized knowledge, who were permanently assigned to a product, such as Newman Holland and the business phonograph. He had fifteen years experience in the design of electrical apparatus, including five years at Western Electric, where he had worked in the original design section and rose to head an experimental laboratory in Chicago. His appointment was the result of management's goal of developing a phonograph (the telephonograph) that could take dictation directly from the telephone.[43]

An important precedent in the development of product specific engineering had been established by John Constable in his work on the Amberola phonograph. Hitherto each part of the development process had been carried out by a different worker. From the experimenter who produced the first rough design to the clerk in the Phonograph Works who listed the parts, the innovation passed through many different hands. The engineers working on the Amberola line, in the years before the fire, decided that one person should be made responsible for all stages of a product's development; the man who made up the model should also lay out the method of its manufacture and supervise its routing through the Works. John Constable took responsibility for the Amberola from the lab to the Works. The experience with the Amberola showed that this policy was sound, and it was soon extended to other models, such as the disc phonograph.[44]

The product engineer was the logical outcome of the product specific work carried out in the lab since the formation of TAE, Inc. In the case of the business phonograph, Holland had been acting in this capacity since 1910, and the changes of 1915/16 merely confirmed his role. The product engineers were either recruited from industry or from the manufacturing operations of TAE, Inc. The emphasis on manufacturing and marketing was reflected in the job descriptions that were classified under experiments, development, manufacture, and sales.

The task of the product engineer was to focus all the services of the lab, including toolmaking, drafting, and testing, on his specific interest. Although he was a member of the laboratory staff under the supervision of the chief engineer, he was expected to remain in close contact with the manufacturing and sales departments, which often initiated his work. He was to have a foot in both camps; he carried out experimental and development work in the laboratory, and at the same time, he kept a close watch on problems in manufacture and feedback from the sales force. This dual role reflected a tradition in the Edison lab, in which muckers were expected to carry out both research and engineering. After the formation of TAE, Inc., the emphasis moved to the latter, for it was the engineer's duty to keep his product "one jump ahead of the sales requirements."[45]

With its emphasis on engineering support for sales, the divisional policy shifted laboratory resources from experimenting to supporting manufacture. The ideas of the muckers in the laboratory had once been the moving force of the Edison enterprise. Now the initiative was to be put in the hands of the divisions—technological pull was to be replaced by marketing push in the development of new products. The product engineers found themselves under pressure to work to other people's timetables and plans. Yet the measure of an engineer's support for the work of the divisions depended on his attitude. The shop culture of the laboratory placed a high value on individualism and initiative. The primacy of experimenting was taken for granted. Thus some of the product engineers in the main laboratory did not feel the need to justify their work by producing immediate results for the division manager. Newman Holland, for example, was said to be "a purely experimental man [who] cared little for the manufacturing end." He worked in the laboratory at his own tempo, "totally distinct and oblivious of the factory."[46]

Newman Holland's work as product engineer on the business phonograph illustrates the tension between the traditional experimental outlook of the laboratory staff and the new requirements of the divisions. Holland had to please several interest groups, not the least of which was the office of the division manager, Nelson Durand. He also answered to the various manufacturing committees and to the chief engineer. Finally, Edison had to have his say in any new development. Although Holland was supposed to take note of all this input, he clearly did not have the time or inclination to do so. After failing to satisfy these groups, Holland was assigned an assistant, W. Telfair, who concentrated on the production aspects of the new designs and dealt with the Works. Holland was to confine himself to research. This further division of function reveals how difficult it was to maintain the old flexibility in a large organization producing a complex product. Specialization seemed to be the answer; the same arrangement was made for the amusement phonograph engineer.[47]

The position of product engineer formalized the specialization of the laboratory work force. The purpose of this administrative innovation was a more efficient application of engineering services. The work of the product engineer was framed by the division manager, and the resulting designs were defined by the needs of the market. The chief engineer acted as the interface between the divisions and the product engineers, assigning work to his staff on the instructions of the various committees that initiated development. He wrote the Engineering Notices that authorized changes in product design. Much of his work was promulgating, recording, and enforcing these documents. His role in the organization had evolved from engineering and experimental work to primarily clerical and administrative tasks. Miller Reese Hutchison thought that the job of chief engineer "seems to resolve itself into issuing engineering notices on screws, nuts, bolts, gears, etc."[48]

The creation of separate laboratories near the West Orange lab, under the direct control of the divisions, reinforced the movement to bring engineering and research services under the control of management. The first was a testing laboratory at the Silver Lake manufacturing complex which opened in the 1890s. The expansion of Edison's manufacturing plants soon led to the creation of small laboratories in the many factories built in the early years of the twentieth century. By 1916 much of the product specific experimenting and testing at West Orange was being done in the manufacturing buildings that clustered around the original laboratory.[49] As the primary function of these laboratories was testing and quality control, they had to be close to the point of manufacture.

Testing was especially important in battery manufacture because the end product had to be reliable. One tester argued that "the importance of the work here, as related to the marketing of the product, is to my mind unquestionable."[50] Testing was first carried out at the West Orange laboratory in the battery research department and then in the small cell test department. The increase in work led to the formation of a separate testing department with its own facilities at the factory. The Edison Storage Battery Company test department slowly enlarged its activities; this led to the formation of a research department at Silver Lake in 1913, which began to carry out experiments in addition to testing.

The recording and film studios in New York were equipped to carry out both testing and research. The work of these studios was theoretically under the control of the main laboratory at West Orange where final approval of all Edison records and movies was made by the committee of lab personnel.[51] Yet the management of the studios often asserted their independence. This was especially true of Horace Plimpton at the Bronx studio who fought to keep artistic autonomy over the production of negatives.

The relationship between central laboratory at West Orange and the satellites was not merely one of master and servants. Many projects required coordination of effort between the two, such as the installation of special equipment that was designed at the West Orange laboratory and installed on site by the staff of the Works laboratory. In some cases, processes that had been developed in the Works laboratories were transferred to the central lab at West Orange. The improvement of the phenol resins used in disc manufacture was dependent on good communications between the experimental rooms at West Orange and the vats and ovens of Silver Lake. Harmony was often lacking in their relationship. There were

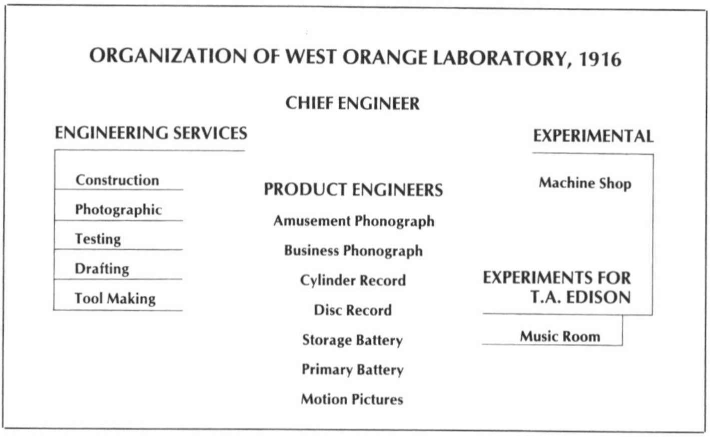

ORGANIZATION OF WEST ORANGE LABORATORY, 1916

CHIEF ENGINEER

ENGINEERING SERVICES
- Construction
- Photographic
- Testing
- Drafting
- Tool Making

PRODUCT ENGINEERS
- Amusement Phonograph
- Business Phonograph
- Cylinder Record
- Disc Record
- Storage Battery
- Primary Battery
- Motion Pictures

EXPERIMENTAL
- Machine Shop

EXPERIMENTS FOR T.A. EDISON
- Music Room

Figure 11.3. Organization of the West Orange laboratory.

doubtless many personality conflicts, but underneath these squabbles was the growing rivalry among laboratories and the natural feeling of superiority of those closest to Edison.[52]

Many of the disputes arose out of the West Orange laboratory's historic role as the final authority on standards. Beginning with phonograph manufacture, the staff of Building 5 had exercised their right to enforce Edison's standards of quality on all his products, often to the dismay and annoyance of the engineers in the divisions. Testing became one of the most important functions of the West Orange laboratory and it grew in step with the expansion of the industrial empire. The importance of neutral testing by disinterested third parties led to the creation of a new department under the control of "Doc" Halprin. This testing facility joined with drafting and toolmaking to form the Engineering Services Department that was one of the three elements of the laboratory in the divisional structure. (See fig. 11.3.)

Draftsmen had been employed in the lab since the 1880s. By the time of the great fire, drafting was done by many different employees in both lab and Works. After the divisional policy a reform of drafting was undertaken. The drafting service of the Works was combined with the laboratory's drafting department, and a lab employee, Werner Olson, was given control of the new organization. Olson had begun his career as a tool designer and was more of a team player than independently minded Otto Weber whom he replaced. As chief draftsman Olson had the power to schedule draft-

Figure 11.4. A retouched photograph showing part of the Engineering Office on the second floor of the main laboratory building, West Orange, a space once occupied by experimental rooms. Starched collars and ties had become the correct attire when working in Edison's laboratory.

ing work in an orderly manner. This replaced the initiative of several men working independently.[53] He answered to the chief engineer and sat on several manufacturing committees.

A new drafting room was laid out on the top floor of the main laboratory building. It was part of a complex of offices that made up the Engineering Services Department. Next to it was the office of the chief engineer and the rooms occupied by the product engineers. The third floor also contained the office of R. W. Kellow, one of Edison's secretaries who also handled all the bookkeeping and work scheduling of the lab. The second floor contained the office of the superintendent of the lab machine shops and the Construction Engineering Department that carried out repairs in the physical plant.[54] This department had its own drafting room, which occupied space that had once been experimental rooms (fig. 11.4). The move to specialization brought about permanent offices for product engineers and administrators to replace the experimental rooms that had been changed to suit whatever campaign was underway at the lab. Flexibility was slowly giving way to specialization at the West Orange laboratory.

Last of the three parts of the laboratory was the Experiment and Development Department. This covered the management of the machine shops, the music room, and Edison's personal staff. The music room on the top floor of the main laboratory building was under the direction of Clarence Hayes, who was responsible for recording, scouting out new musical talent, purchasing music, and making up the lists of records. Edison's desire to control the musical side of the phonograph business was probably the rationale for placing this department under his jurisdiction.[55]

Edison's personal staff assisted him in his three experimental areas in buildings 2, 4, and 5. Building 4 had been refitted to carry out experiments on disc records, and it was where Edison led the struggle to eliminate the surface noise of the records. Despite the changes in the organization and work of the laboratory, Edison's method remained the same. He used the same team approach, expected the same level of commitment from the muckers, and encouraged informal communication among experimenters. The small group of men working with him retained the old traditions and work practices of the nineteenth century lab.

Edison was determined to capture all the "overtones" of a recording which meant that every unwanted crackle and hiss on the sound track was reproduced with alarming clarity. He detected a steamy, hissing sound from the Diamond Disc recordings and set out to find the cause. This quest became an obsession with him. He indulged in an orgy of all-night experimenting sessions throughout 1915 and 1916 to eliminate all surface noise, vowing that "I shall never stop until I get rid of it altogether"[56] (fig. 11.5). Edison did not relinquish his experiments on surface noise until the First World War took him away from the problem in 1917. He spent most of his time "enjoying himself hugely," in the words of his secretary, in the chemical lab (Building 2) which had become his favorite place to work in the West Orange complex.[57]

Elsewhere in the laboratory, the forces of change had made their mark. Hutchison installed a time clock at the entrance of the main laboratory building so that experimenters and product engineers could conform to the same industrial discipline that was imposed in the factories of the Phonograph Works. The most symbolic changes were made to the electrical laboratory in Building 1 because of the slow decline of the experimental work on electricity. Newman Holland had his office in this building before he joined the other product engineers on the third floor of Building 5. His place was taken in Building 1 by Mambert's men. The fact that this building, once the

Figure 11.5. Edison examining the surface of a Diamond Disc record with a microscope. A stereoscopic microscope sits on a test machine on the right.

main attraction of Edison's laboratory, was now occupied by clerks and accountants is evidence of the changing times at West Orange.[58]

The organizational changes in the West Orange laboratory reflected its changing function within the Edison organization. The primary goal in creating divisions was to efficiently manufacture and energetically market existing products. The mission of the West Orange laboratory, as envisaged by Edison in the nineteenth century, was to develop new products. After 1914 the pressure to coordinate engineering support for the manufacturing and marketing branches grew more powerful, leading to the compartmentalization of the West Orange laboratory into service departments. The laboratory created by Edison in 1887 did not fit the needs of a large corporation in 1915, which had become primarily concerned with maintaining market shares in established businesses. By 1916 the laboratory's work was directed toward support of the product line, and according to Edison's secretary Meadowcroft, "There is little of mere research done here."[59]

CHAPTER TWELVE

Business and Technology: The Dictating Machine

he organizational changes in the Edison enterprise brought most benefit to operations that required efficient coordination of engineering and sales. This was certainly the case with the business phonograph, which had languished after its disastrous introduction in the 1880s. This complex product needed a well-organized sales force, supported by a network of specialized dealers, to ease its introduction into the mainstream of American business. The failure to provide this in the 1880s was not because Edison did not appreciate the need, for he told his English agent, Gouraud, "All my experience with marketing novel machines points to the need of experts."[1] It was because he did not have the resources. Not until the divisional policy created a sales organization and gave it the upper hand in product development was the goal of the 1880s realized, and the dictating machine became as commonplace in offices as the typewriter or telephone.

The history of the dictating machine begins again at the turn of the century. It encompasses two separate but interdependent paths: the first covers the development of an appropriate technology, suited to both the needs of the user and the requirements of low-cost manufacture; the second follows the evolution of a business organization equipped with sufficient resources to educate potential customers in the operation of the machine.

When Edison reentered the business machine field in 1905, doubtless encouraged by the booming sales of the musical phonograph, the staff of the West Orange laboratory had improved the phonograph considerably. Over fifteen years of research into sound had produced significant advances in the reproduction of the human voice. Equally important was the reduction of the bulk and complexity of the first phonographs that had been offered to businessmen; the requirements of the amusement market had pushed the phonograph format into a smaller, more compact shape.

The development of the Home model at the turn of the century opened the way to a series of designs in which the length of the mandrel assembly was shortened to reduce the overhang—the distinctive feature of Edison's phonographs in the nineteenth century. The Standard and Gem models had short mandrels with the drive screw assembly parallel and to the rear. This was the general arrangement adopted for the new business phonograph introduced in 1905. It retained the spectacle holder for the two diaphragms (recording and reproducing) and had an elongated mandrel to accommodate the longer cylinders used for transcribing.

Despite the fanfare from West Orange and the obligatory articles planted in the press, the business machine was similar to the standard amusement phonographs in production and to the graphophone, its rival in the office machine field.[2] The graphophone was marketed by the Columbia Company, a powerful force in the talking machine industry that was to give TAE, Inc. and Victor plenty to worry about. Originally formed for the purpose of meeting the transcription needs of offices in the nation's capital, the Columbia Company had a strong interest in the business user.[3] The graphophone was a sturdy belt driven machine with the same drive screw that could be found on most cylinder models (fig. 12.1). It was originally powered by a treadle in the same way as a sewing machine.

Although portrayed as the last word in dictating machines, the new Edison product was no more than the basic phonograph with alterations to the reproducer and mandrel. It came in the same wooden casing and still required a steady hand to adjust it to "talk" audibly. Edison's advice, "a little instruction and practice are needed to operate," was an understatement.[4] Only the patient and light-fingered got results from machines described as "so complicated that the average man would have to keep his eye constantly on the book of directions while working them."[5] The slightest mechanical failure brought this technological revolution in the work place to a halt. It was a fragile machine that was vulnerable to dirt, changes in temperature, and anything less than gentle handling. It was extremely difficult to locate and correct mistakes made in dictating a letter onto the cylinder, and even if the dictation was successfully recorded, there was no assurance that the stenographer would be able to understand it.

Figure 12.1. How the modern businessman made use of the talking machine about 1918, according to the Dictaphone Company.

In 1905 the sales support of the dictating machine was as makeshift as the technology. The creation of the Commercial Department of the National Phonograph Company in 1906 was a small beginning, a department on paper with no specialists and scant resources. Its manager, Nelson Durand, was a longtime Edison employee who had once experimented with the great man on incandescent bulb filaments. A Newark native, Durand had found a job in the Harrison Lamp Works as a young man. He rose in the Edison organization, aided by a successful stint as a mucker in the West Orange laboratory—a path taken by many middle managers in TAE, Inc.[6] Durand started from scratch in the dictating machine business. The sales network at this time were dealers of musical phonographs who were trying to make some money in the off-season by selling dictating machines.[7] The most that Durand could offer them was a few pamphlets explaining the operation of the machine and outlining its selling points.

Sales of dictating machines were low in the first decade of the twentieth century. Customers were found in large offices that produced a great deal of correspondence, such as mail order houses, large manufacturers, and railroads. Many prospective users argued that they did not have enough correspondence to justify mechanizing the process. Many more customers were lost because stenographers claimed that they could not understand the faint, scratchy sounds they strained to hear through the ear tubes. Consequently, Durand took as his first priority the improvement of the technology, and he leaned heavily on the staff of the laboratory whose technical knowledge of the machine and its competitors was invaluable.[8] In 1906 an experimenter was assigned to improve Durand's product.[9]

Durand's ability to sell talking machines to office managers was second only to his skill in persuading the upper levels of TAE, Inc. that his product had a commercial future. Once they were convinced that they were "beginning to succeed" in this business, more engineering resources were devoted to the commercial phonograph.[10] Durand successfully argued the advantages of developing new models that would be the basis of a complete office system. The Commercial Department was given the go-ahead to produce an "office appliance" of unique new design, which would set the standard for dictating machines for years to come.[11] To achieve this goal, an engineer was hired to work exclusively on dictating machines. Newman Holland came to the laboratory in 1910. If he expected to spend his time on developing a new machine he must have been disappointed, for like his colleagues in the lab he found himself carrying out a wide range of duties, of which product development was only one.

The engineering effort was coordinated by Nelson Durand who was responsible for shuttling ideas from one part of the organization to another. Any small improvement that began as a suggestion from a salesman in the field passed through various committees and ended up on Holland's bench at the laboratory. Its development in the lab and its journey to the Phonograph Works were under the supervision of Durand—his office in the administration building provided the central processing point in the flow of paperwork. The dictating machine was improved by small increments: an adjustment here, a control there. Years of effort would produce a small change—such as a redesigned recording-reproducer holder that was harder to break with rough use—that materially improved the machine's performance.[12]

Holland's work covered all aspects of the dictating machine: improved reproducers and control devices, completely new machines aimed at special uses (including a spring-driven portable machine), and the development of some of Edison's pet projects, such as the telescribe—a machine to record telephone messages automatically.[13] Edison had first envisaged this application of the phonograph in 1878, when he claimed that it could provide the means of "transmission of permanent and reliable methods, instead of being the recipient of momentary and fleeting communication."[14]

The telescribe combined two inventions: the phonograph and the carbon microphone. It consisted of a sensitive transmitting device, into which the user placed the telephone receiver, and a modified dictating machine that picked up the sound through a telephone link with the transmitter. These two machines were conveniently controlled by a pair of switches placed next to the user, who could also participate in the conversation by using a speaker attached to the transmitting device. The telescribe could, therefore, record both sides of a telephone conversation. After extensive experimental and development work, the first models were brought out in 1914. They were followed by improved machines in 1918.[15]

The telescribe was an ambitious project typical of the West Orange laboratory. A complicated machine, it taxed the performance of the available technology and the patience of the user. Edison had again attempted to steal a march on the competition by a major innovation, yet in trying to transmit telephone messages to the phonograph with no means of amplifying the sound as it moved from receiver to recorder, he was asking too much of the equipment—it was an impractical idea for the average office. Nevertheless, Holland and other Edison engineers devoted much time to this project. The technological ambitions and contagious optimism of the Old Man overcame a rational distribution of laboratory resources, and while experimental time was squandered on a project that would never achieve any commercial success, the competition advanced with the same old machine and extensive advertising. Under the new name of Dictaphone, Columbia's graphophone became the most popular dictating machine on the market in the twentieth century.[16]

The Dictaphone was a straightforward belt driven machine that sold for considerably less than the Edison product. Originally intended to sell at the same price as the typewriter, the first dictating machines produced at the West Orange factory cost about $170 in the 1880s, and even after years of cost reduction, the cheapest models introduced around 1905 cost $70, still at least $10 more than the price of the graphophone.[17] Although its advertising in

1907 bravely stressed the reduction in price brought about by "economical manufacture," the business phonograph still cost more than the average typewriter and the machines of the competition.[18] By 1911 the wooden case had been replaced by utilitarian sheet metal. The new phonograph came with tall metal legs so it stood by the desk instead of on it. The years of improvements and new features had gradually increased the price of the Edison product to about $80 or $90. But the Dictaphone Company was able to keep the $60 price tag for many years.[19]

Durand's department reflected the ethos of TAE, Inc. by maintaining a policy of constant improvement of the product, even though it inhibited cheap manufacture. It did achieve in 1908 an important advance over the competition when it led to the introduction of a universal electric motor that could work on a wide variety of currents, including alternating current. This important innovation multiplied the potential office market for the dictating machine at one stroke. The 1911 model had an improved motor and a new scale and marker system to make it easier to find passages that needed correction or repeating.[20] The models of 1912 came with a pneumatic control mechanism that was attached to the speaking tube, a distinct improvement over the foot peddle that usually controlled the machines.[21]

The most visible improvement in the dictating machine was the development of electrical controls to replace the mechanical and pneumatic systems that were difficult to operate. Using electricity had the added advantage of ease of positioning the controls. The person dictating could control the machine with a switch on the mouthpiece, and the person transcribing had the control placed by the space bar of the typewriter. Electrical control provided the means to engineer the most important feature of the dictating machine from the user's viewpoint—the correction device.

The major drawback of using dictating machines was the difficulty of locating passages on the cylinder to make corrections or to listen to a part of the dictation again. As mistakes were frequent and users could not always hear the words the first time, much time was wasted in lifting the reproducer from the cylinder and placing it back down where the user judged the passage to be. This difficult engineering problem was solved by a delicate cam mechanism that moved the reproducer assembly back a few turns. It was operated by an electric circuit that activated an electromagnet that moved the cam. A control button next to the typewriter completed the cir-

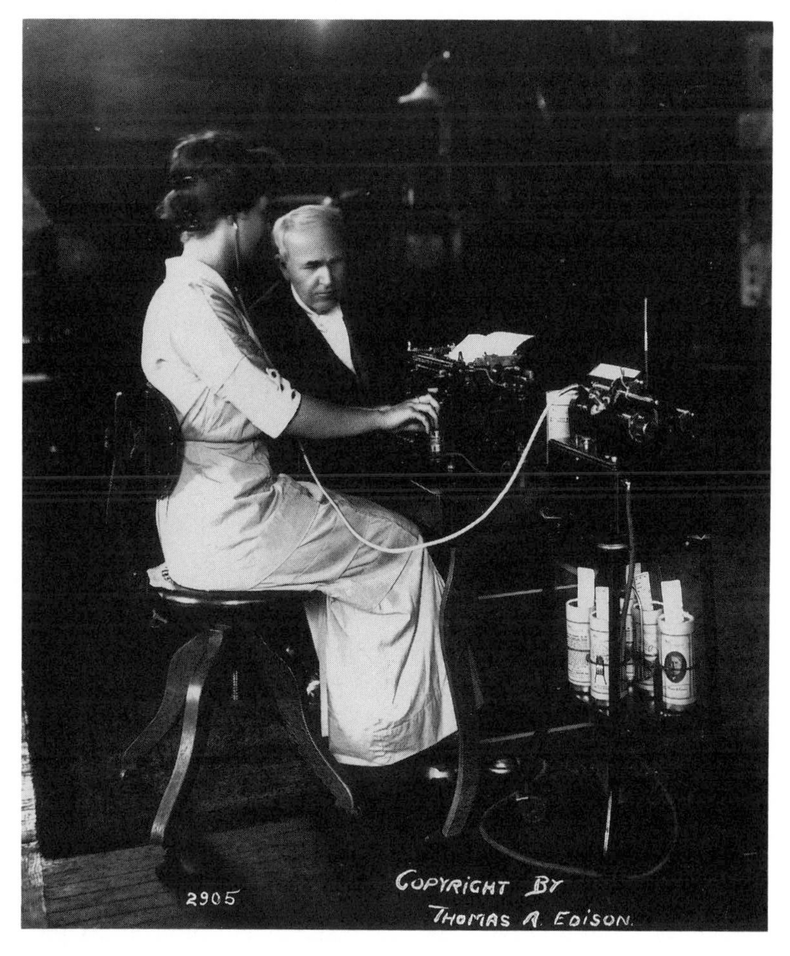

Figure 12.2. The Edison transophone dictating machine in 1914. The typist is using the correction device near her right hand.

cuit.[22] This device, called the transophone, was first introduced in 1914 (fig. 12.2).

Although these design improvements made it easier to use the machine, they were not the whole answer. Success came when the technology was molded to fit the office organization of the user, and the user was educated in a new method of handling dictation and transcription.

Creating an Office System

The first attempts to market a dictating machine showed some of the disadvantages of shop culture values in the marketplace. The view from the laboratory was that this technology could displace a skilled worker in the office: "Your stenographer goes to lunch, goes home, needs a day off occasionally . . . The phonograph don't eat, is on hand at all times, is always ready." The machine was easy enough to use, "Talk into a tube as fast as you can rattle it off—that's all." The economies it offered outweighed the high initial cost and the trouble of maintaining it.[23]

This strategy died an early death at the hands of the workers it was supposed to replace. Stenographers refused to use the machine, claiming that it was unreliable, incomprehensible, and too complicated. Office workers in the early stages of industrialization enjoyed a status that has since declined. They were often from the middle class—as was the case with A. O. Tate in the Edison organization—and they usually were called upon to play managerial roles. Their work in the office has been likened to a craft, such was the importance of their skills and their freedom of action.[24] In the typical American office in the 1880s stenographers were usually male, well paid (up to $20 per week according to the Edison advertising literature), and free to develop their skills of shorthand dictation, transcription, and office management. They stood at the top rank of office workers, above clerks, copyists, bookkeepers, and messenger boys. They were not easily displaced by talking machines.

The introduction of the typewriter and the rapid increase in the amount of administrative work in the late nineteenth century produced a revolution in office work that manufacturers of dictating machines were able to exploit. Women were now entering offices in greater numbers to fill the new position of typist. They did not have the upward mobility and variety of duties enjoyed by male stenographers. By 1900, 75 percent of all stenographers and typists were women. This feminization of the clerical labor force created a layer of low-paid, menial employees.[25] Edison's salesmen now used a different approach to sell dictating machines. The phonograph could be used to replace an expensive stenographer, skilled in shorthand dictation, with a typist who transcribed the correspondence dictated to a phonograph. The slogan of the Commercial Department pointed out the advantage: "You can secure half-priced typists to take the place of your experienced stenographers."[26]

By 1900 the United States was well into the era of big business. The wave of mergers that began in the 1890s had produced a number of large, diversified business organizations, of which the Edison enterprise was a good example. National in operation, with layers of middle management, the National Phonograph Company was typical of the larger organizations that dominated the industries of the Second Industrial Revolution. As the volume of transactions increased, so did the size of the administrative staff and the opportunities to sell dictating machines. In 1910 Durand's department sold over 5,000 machines—its best year ever.[27] The formation of TAE, Inc. in that same year marked a movement that made Durand's job easier. The compartmentalization of work into special functions in which each part of the administration was divided into departments reflected the specialization of work. The division of labor that characterized the laboratory and factory floor was also seen in the administration building, where workers were assigned to one task, such as dictation.[28]

The first salesmen had attempted to sell dictating machines to small general offices, with little success. Now they were given the opportunity to fit the machine into much larger administrative offices where work was becoming more formal and structured and where the division of labor had created a hierarchy of clerical workers. Managers confronted with a greatly increased work load were looking for more efficient ways to handle their operations, and Durand's salesmen had the answer. They were trained to sell the commercial phonograph as a system of dictation, rather than just a machine—an idea that originated with Durand as far back as 1905 when his department was first formed.[29] The concept of introducing new technology as systems rather than independent products is seen as the hallmark of Edison's business strategies. The design of the incandescent lighting system has been routinely described as his great innovation in this industry. The concept of a business system was equally important in the success of the dictating machine; this complicated device had to be marketed as a range of products that complemented changes in the office and made clerical workers more efficient. The idea that one machine could produce revolutionary effects on its own came from the technological optimism of the machine shop culture. It rarely worked in practice.

The systems approach was welcomed by engineers because it separated the functions of dictating and transcribing, permitting them to concentrate on one function in their designs. The dictating machine no longer had to be able to do everything. Now there were

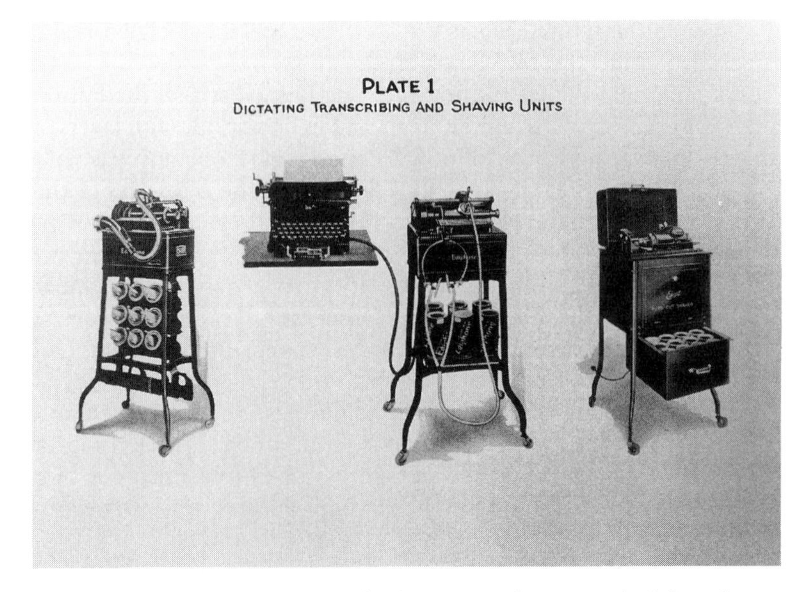

Figure 12.3. The Edison system. The dictating machine is on the left, with its special mouthpiece and recorded cylinders. In the middle is the transcribing machine attached to the typewriter by the correction device control cord. On the right is the special machine that shaved the used cylinder so that more recordings could be made on it.

different models for dictating, transcribing, and shaving. The transcribing machine did not require the complicated recording assembly. In addition, the development of a separate machine to shave cylinders for reuse solved the problem of shavings gumming up the works, which had often been the cause of breakdowns of business phonographs.

Three separate machines made up the Edison dictating system: the executive dictating machine, the secretarial machine, used for transcribing only, and a shaving machine to recycle used cylinders (fig. 12.3). The new dictating system was based on the creation of three levels in the office hierarchy: executives at the top who dictated the correspondence, typists who transcribed, and office boys who shaved the cylinders and maintained the machinery. Those lower on the scale got the dirty work of shaving and cleaning while those at the top leaned back in their chairs and talked into a mouthpiece. The efficiency of the system was found in a central transcribing department that would handle all the correspondence dictated by managers and stenographers (fig. 12.4).

Marketing the business phonograph in this manner was based on

Figure 12.4. The typing pool, key to the Edison system. Mass production techniques were now applied to office work. This office was in the Ferranti Works in Lancashire, England.

a careful assessment of the user's office before centralizing the flow of correspondence into one transcribing department. The Edison salesmen had to be trained to identify the dynamics of office politics before rearranging the offices of their customers. It was not enough to claim increased efficiency, salesmen had to prove it by restructuring office work. In many cases it was unwise to suggest that a stenographer should be replaced by a typist. If the former carried out administrative work in addition to correspondence, the office manager would resist replacing such a valuable employee. The Commercial Department no longer talked about displacing skilled stenographers, for they could be used to manage the correspondence and dictate to the machine. Each office situation had to be treated differently.[30]

Durand's promotional material stressed the uniqueness of the dictating machine system and proudly claimed that it was not sold in stores. The customer was visited by a sales representative who would study the workplace and "adapt the Edison business system" to the existing office arrangement. Customers were told that in buying a machine they were assured of the expertise of the Edison laboratory and the support of the local service organization. Durand emphasized the fact that "people don't buy [the dictating machine] for the motor or the other parts, but for the service."[31]

The divisional policy gave Durand the resources to build up his sales force and maintain a network of dealers. The healthy sales of the Dictating Machine Division justified a large organization to sell and maintain the product. Durand's policy was to keep over 100 salesmen in the field at all times. Selling these machines required technical knowledge and the persistence to overcome resistance to change in offices. Businessmen would rather dictate to a stenographer (who was often an attractive, attentive young woman) than shout into the mouthpiece of a machine. Stenographers often played an important part in composing the letter and were more forgiving of errors than the machine.[32] There were all sorts of reasons put forward to avoid using the dictating machine: fear of making too many mistakes, fear of upsetting the secretarial staff by introducing a strange new technology, and inability to see an advantage in mechanizing correspondence.

Durand's salesmen were trained to overcome these objections, yet many potential customers still clung to the old ways of dictation. The office workers of TAE, Inc. were no exception. Charles Edison was most annoyed to find out that the division managers did not use the dictating machines provided for them.[33] Perhaps they were members of the faction who believed that their correspondence was too technical to be dictated to a machine. Customers needed to be persuaded to give the machines a try. Durand had to maintain the sales pressure, urging his salesmen to pursue prospective customers. His men were trained in the demonstration of dictating and transcribing, an exhibition worked out to the finest detail, even to the appearance of the transcriber.[34] The machine had to perform perfectly every time.

The divisional policy made Durand's job easier by giving him formal access to all the engineering work undertaken in the organization. Improved communication between the various engineering activities of the Edison enterprise offered division managers many opportunities to improve their products. In this way the Dictating Machine Division was the beneficiary of sound experiments made in the new Columbia Street phonograph recording studio (built around the block from the West Orange laboratory), and studies of electric circuits made in the laboratories of the Primary Battery Division.[35]

The reform of engineering at West Orange, especially the creation of the Engineering Department, helped Durand in the vital work of lowering the cost of the dictating machine. Better cooperation between laboratory and Works was to his benefit. Laboratory men like Constable collaborated with engineers in the Works to apply cost reductions developed in phonograph manufacture to business phonographs—such as using parts of the Amberola phonograph in dictating machines.[36]

The reorganization of the laboratory that followed the divisional policy placed more of the business of innovation into the hands of the division manager. Durand assumed the power to direct research and development of his product and usually initiated experimental projects. His insistence on pleasing the customer led to changes in the lab that Holland would never have considered, such as improving the headband holding the earpiece of the transcribing machine, which often caught in the typist's hair.[37] Durand controlled the research effort through the chief engineer who supervised the daily work of the product engineer in the laboratory.

After a collective decision had been made in TAE, Inc. to develop a completely new dictating machine, Newman Holland reviewed all the previous ideas for business phonographs. The critical element of the design was the arrangement of the recorder-reproducer. This assembly determined the quality of recorded dictation and dominated the format of the top plate. It was also the most fragile part of the machine. Many designs had been considered, including one in which all the working parts except the recorder-reproducer had been placed beneath the level of the top plate. The great advantage of this arrangement was that the machine was better protected from dirt and abuse by the user. The main criterion developed in the West Orange laboratory for the new business machine was simplicity of operation and appearance—all working parts should be concealed in the housing. It was to be a simple, sturdy machine that minimized the number of breakdowns in the hands of a careless user.[38]

Figure 12.5. Three generations of Edison dictating machines. On the left, the 1916 Ediphone, with wood-paneled sheet metal case; in the middle, the sleek Ediphone "Hood" design introduced in the late 1920s; and on the right, the electronic Voicewriter of the 1930s and 1940s, which used a microphone in the mouthpiece to replace the old acoustic horn.

Newman Holland worked on this machine for several years and did not complete his design until late 1912.[39] The management of TAE, Inc. decided that as soon as the disc machine tooling was completed in 1914, the Works would turn its attention to manufacturing business machines. But numerous delays held back the start-up until 1916.

The new machine, named the Ediphone, broke away from the format established in the nineteenth century. The first innovation of the Ediphone was the much cleaner design of the top plate (fig. 12.5). Gone is the clutter of the spectacle arrangement of two diaphragms and the many adjustment screws. The Ediphone was designed primarily as a dictating machine. It had one totally enclosed diaphragm and a single lever arrangement to lower it onto the cylinder. The heart of the new machine was an improved universal electric motor, which was smaller and used less current than previous designs. Production of these motors was under the control of the new small motor department which in 1916 was producing about 100 a day.[40] The belt drive and gearing were enclosed in a metal casing. The Ediphone came with a long mandrel for the five-inch cylinders. It had an ejection device to move the cylinder from the mandrel when dictation was over.

The Ediphone incorporated several years of improvements and modifications, such as the new glass mouthpiece (which made it easier to dictate) and a semi-automatic correction device. The rubber tubing of the earlier machines was replaced with a flexible metal one that gave better recordings.[41] Subsequent models incorporated improved correction devices that were based on an electromechanical start, repeat, and stop mechanism. Although the correction feature was the work of Holland, he depended on feedback from the users and the cooperation of Sam Langley, the engineer in charge of the Motor Department of the Phonograph Works.[42] The design of this electrically powered repeat mechanism, operated by a switch next to the keyboard, was based on motion studies of typists.[43]

The development of the dictating machine owed a great deal to the spread of scientific management to American offices in the postwar years. One of its proponents, William Leffingwell, worked out the scientific basis of office work along lines developed by Taylor, applying the same time and motion studies that had been used in industry. Scientific management in the office promised greater productivity, less waste, and higher profits.[44] It also gave managers the means to control office work by removing all the freedom and flexibility enjoyed by previous generations of stenographers and clerks. Durand's salesmen were more than willing to take part in the mechanization of the office. Their product helped specialize work and sustain a strict hierarchy, at the expense of the office workers who now found themselves part of a clerical proletariat. The typing pool turned the production of correspondence into a series of repetitive tasks that could be easily managed. It became commonplace in most large businesses.

The dictating machine business enjoyed steady growth during the years of the World War and by 1918 sales of machines exceeded 10,000 for the year, total sales volume reaching over $1 million.[45] The war marked a dividing line between an attitude that the dictating machine was something of an experiment and the realization that it was a necessity in a modern office.[46] During the 1920s office work became the fastest growing occupation in the American economy as large business units increased the size of their administrations.[47] Durand's department now sold complete office systems, major installations of linked equipment, instead of the single all-purpose machine marketed in the nineteenth century. In 1928 the average sale was about $1,000.[48]

Driven by Durand's expectations of a highly competitive market, the development path of the dictating machine in the 1920s was toward a smaller, cheaper, simpler machine.[49] A portable, spring-driven model was added to the line to allow the busy executive to dictate at home or on the road. Over twenty years of experience with spring motor phonographs was obviously a help in creating this new product. In the 1930s the Edison dictating machine was renamed the Voicewriter, and a new format was adopted (see fig. 12.5). This basic "hood" design achieved great success and was the machine that finally brought substance to Edison's ambitious claims that there would be a dictating machine in every American office.

This machine reflected a new movement in design that was gaining momentum in American industry. It was sleek and streamlined. Its spring-loaded hood automatically closed down over the top works. Although the format and the controls were the same as the Ediphone, they were all enclosed. Only the lever on the recorder assembly was changed to make it easier to reach. This machine is an example of the revolution in industrial design that took place in the 1930s. The first business phonographs reflected a craft style of the nineteenth century that can be appreciated in the design of telegraph equipment and sewing machines. From the polished wood base to the exposed metal parts of the top works, Edison's phonographs followed this style. His first dictating machines were solid and heavy, built so that the many controls could be adjusted.

As the dictating machine moved from an expensive instrument to a mass-produced staple in every office, sheet metal replaced fine wood and heavy castings. Lighter and more utilitarian materials were sought by a new wave of industrial designers, and the new Bakelite plastics were found to be ideal. These plastics were used in Edison products, including the Voicewriter and a telephone number index made under the Bates Manufacturing Company name.[50] The Ediphone and the Voicewriter were based on a new machine aesthetic that was clean, simple, and beautiful. The style that originated in the German Werkbund was based on efficiency of form and not cosmetic decoration. It was a style that consciously reflected modernity and technological progress. The ideal was a streamlined machine that was simple to operate and easy to clean. The smooth shell that enclosed the operating parts of the Voicewriter was perfectly suited to the function. Beauty of form was found in the continuity of line and the contrasts of color. In the dictating machine, form followed function to perfection.

CHAPTER THIRTEEN

The Impending Conflict

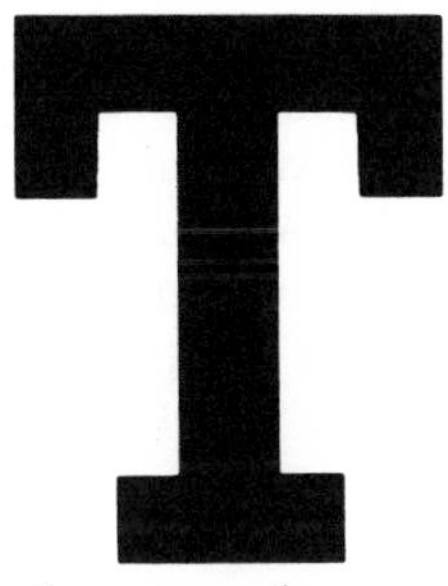

he coming of war to the industrialized West in 1914 presented a great challenge to TAE, Inc. Business had to be carried out in a totally new environment, with some different rules. While threatening the very existence of the enterprise, the war also offered opportunities to make substantial profits and add glory to the name of Edison. It pressed the administration of TAE, Inc. to make further organizational changes and to continue the reform of old practices. For the company the impending struggle would not only be against the Kaiser and his Huns but also against the additional competitors that would emerge after the war.

The West Orange laboratory was built at a time of revolutionary advances in military technology. The benefits of the Second Industrial Revolution were quickly applied to weapons. Bigger guns, better steel, and improved propulsion plants transformed the forces of Napoleon, Bismarck, and Victoria into instruments of imperialism and total war. The world's navies were most affected by new technologies. Radically new warships such as the dreadnoughts and the use of wireless helped transform naval warfare in the twentieth century. The first military customers for electricity were modern navies that used generating plants to light and power their iron warships.

Europe was on the road to war. Many of its leading industrial concerns, such as Krupps and Vickers Armstrong, sustained rapid international growth as a result of their participation in the naval arms race before the war. European electrical manufacturers found new markets for their products, not so much for offensive weapons as Edison had predicted, but for component parts in weapon systems—electricity was used to power, to control, and to communicate.[1]

Although committed to neutrality and isolationism, the United States joined in rearmament. The Spanish-American War had revealed imperialist sentiments in government and the appeal of patriotism to the general public—a movement in which the Edison enterprise had played its part with its bellicose films and recordings. Primarily a naval war, the conflict in the Pacific had also shown the effectiveness of modern sea power, as promoted by Captain Mahan in his book, *The Influence of Sea Power on History*.

The United States Navy wanted to keep up with the modernization of the navies of other industrial nations. Here was an excellent market for the products of Edison's laboratory. He had always seen military applications for his inventions, especially for war at sea. He formed a company with Winfield Scott Sims (the Edison-Sims Electric Torpedo Company) in 1886 to design and manufacture submarines, torpedoes, and other naval weapons. This company is credited with the design of the first remote-controlled weapon, an electrically powered torpedo guided through the water by an elaborate cable system.[2]

The naval arms race of the early twentieth century offered to revive Edison's moribund storage battery business, which had followed the electric automobile into decline. The armed services were much more receptive to stored electricity than the utility industries that had done little to explore the benefits of storage batteries as backup power in their central stations.[3] Military buyers saw many uses for reserves of electric power on ships, in vehicles, and in communications systems. Their requirements put more emphasis on reliability and durability than on low prices, making them receptive to the advantages of the Edison cell.

One revolutionary new weapon promised to be a major market for the storage battery. The idea of an underwater vessel armed with a torpedo had been circulating for many years before Edison built his West Orange laboratory. The availability of more powerful electric motors in the 1880s took the submarine from the realm of fantasy to the threshold of a practical technology. In a short time experimental vessels were designed with electric power plants, including one in England that was powered by Edison-Hopkinson motors and fifty-two cells.[4]

The inventor-entrepreneur who built the prototype of the weapon that transformed naval warfare in the twentieth century worked a few miles from Edison's laboratory. John P. Holland was one of the many inventors exploring the possibilities of submerged vessels, and he devised a submarine to be used against the British navy. Born in Ireland, he had emigrated to the United States and settled in Paterson, New Jersey, a few miles northeast of the Oranges. His basic idea was a streamlined submersible, powered by a dual propulsion system and armed with torpedoes.

His most noteworthy experimental boat, the *Holland VI,* was launched in Elizabeth, New Jersey, in 1897.[5] It was fitted with a gasoline engine for surface propulsion and an electric motor (powered by storage batteries) for underwater running. The forerunner of the modern submarine, this design was rapidly developed by several navies, including the U.S. Navy, as a promising new weapon. The technology evolved rapidly at the beginning of the twentieth century as Holland's basic design was refined. The most important innovation was the substitution of diesel power for the unreliable and dangerous gasoline engines. The electric propulsion system for underwater cruising remained the standard of submarine construction. Powered, illuminated, and eventually controlled by electricity, the submarine was the most complex electrical machine ever built.

Submarine batteries required a steady charge that lasted a long time, and these were the advantages of Edison's battery. The lightness of the alkaline storage battery compared with the lead acid cell was another point in its favor for submarine designers. Its high price was no discouragement for the navies of the world that were involved in a ruinous arms race and had stopped counting the cost. As Edison's name stood behind the cell, its reliability could not be questioned. According to its advertising, it was as durable as a battleship. In Miller Reese Hutchison, Edison had the best salesman he could find; he was a technical man, an inventor-engineer like Edison, who also had social graces and the right connections in Washington. As Edison's personal representative, Hutchison had the prestige to knock on any door. He was given lucrative rights to sell storage batteries, in the spirit of the old shop culture, and was allowed free rein to go after any business he could find. The naval powers were receptive to Hutchison's sales pitches. Representatives of the navies of Europe and South America began to visit the West Orange laboratory to view the experimental rooms and the battery factories nearby. They all left impressed. By 1911 Hutchison was convinced that the storage battery had found its place in the modern navy, claiming that the Edison battery "will be used to the exclusion of all other types in submarines in future."[6]

The main selling point of the new battery was its steady, long-lived charge that gave submarines a longer underwater range. The original role of the submarine in naval strategy was as a defensive weapon, but the improvements in its speed and endurance convinced several naval men that the submarine could be an offensive threat to warships. The trend in submarine design was to larger, better-armed boats capable of ocean-going service. As the Holland boats entering service around 1900 could only cruise at full speed underwater for about an hour, improvements in the battery and electric motor had to be made. This situation perfectly suited the experimental resources of the West Orange laboratory. Hutchison persuaded the U.S. Navy to try Edison batteries in its newest submarine, the *Skipjack* or E2, which was launched in 1911. This was the Navy's latest design. It was the first to employ diesel engines for surface propulsion.[7] At over twice the length of the *Holland VI*, and armed with four torpedoes, the E2 was an ocean-going boat with a powerful armament. It promised to be the prototype of the USN's submarine fleet. The center sections of the boat were filled with lines of storage batteries that had been specially designed in the laboratory to fit into the narrow hull of the E2. Hutchison redesigned the cell to decrease the recharging time and to limit the escaping gases, which might cause an explosion. As several British and American submarines had blown up when accumulating gases in the hull had ignited, much thought was given in the West Orange laboratory to avoiding this type of accident.[8] With such a large market at stake, Edison took a special interest in the development of a submarine battery, even taking part in the sea trials of the vessel[9] (fig. 13.1).

Any doubts that might have existed in Washington about the effectiveness of submarines were soon dispelled in the first month of war when one German submarine, the U-9, dispatched three British cruisers in one action. If the trials of the E2 proved satisfactory, the future of Edison's storage battery seemed assured.

The sinking of the *Lusitania* in 1915 outraged the American public and underlined the lack of defense against the submarine. The ease with which the U-boats destroyed merchant shipping, disrupting the vital lines of communication across the high seas, brought a hurried response from the Allies. This "new and terrible engine of warfare," as Secretary of the Navy Josephus Daniels put it, was visible proof that military power was now based on technological strength. Observers on both sides of the Atlantic had noted Germany's use of its scientists and research facilities in preparing for

Figure 13.1. Edison on board the U.S. Navy's E2 (*Skipjack*) during its trials in December 1915.

war and joined Edison in advocating that Germany's enemies should apply their brains to the conflict or risk losing it.[10] The same efficiency that was applied to industry could easily be transferred to the front, in the factory of death, as Edison described it. His notion that the winning side would have the best machines was appreciated by most of the Great Powers. The First World War marked the beginning of the service of science to the state, in which industrial research was applied to warfare on a grand scale.[11]

A fervent patriot, Edison was more than eager to play his part if the United States entered the conflict. As early as 1895 he publicly announced his intention "to drop everything and serve the country" should there be a war.[12] Like many Americans, he saw the world war as a struggle to preserve democracy against militarism, the American way of life against the forces of absolutism. It was fitting, and perhaps inevitable, that the industrial technology that was the mainspring of the national economy should become its savior in times of crisis. It was also appropriate that the nation's best known inventor should lead the effort when the U.S. government decided to mobilize the industrial and scientific resources of the nation.[13] The secretary of the navy approached Edison in 1915 with the idea of a brain trust of technical experts who would advise the armed forces in matters of military technology. The Naval Consulting Board, as it was later called, rested on the belief that science and technology had an important part to play in the war and that the "natural inventive genius" of Americans could be harnessed. Edison enthusiastically embraced this idea. His longstanding belief in the efficacy of technical solutions now reached a higher plane; in the "impending conflict" science was to come to the service of democracy in its hour of need. Edison was particularly pleased with the message of a cartoon that showed its heroic inventor figure standing tall on the side of the Atlantic that considered itself the home of ingenuity and innovation, and had it mounted on the wall of the laboratory library[14] (fig. 13.2). Secretary Daniels not only wanted Edison to advise but also to employ his "own magnificent facilities" at West Orange in important experimental work that the navy was not equipped to handle.[15]

The Naval Consulting Board (NCB) was officially formed in the fall of 1915. Edison was appointed chairman. Many of America's leading scientists were called to the board, including a number of inventor-entrepreneurs, such as Frank Sprague, Leo Baekeland, and Elmer Sperry. An early meeting of the board resolved that its work should be carried out by a special naval laboratory that would test new machinery and carry out experiments. The facility would ensure that the navy got the best equipment available at a fair price. A secondary task was the development of new technology to fit the navy's requirements. Edison took a leading role in the committee that planned the laboratory, and, not surprisingly, its recommendations described an installation that bore a close resemblance to his West Orange lab. Built around electrical, chemical, and physical laboratories, it was to contain a machine shop, pattern shop, foundry, drafting rooms, testing rooms, and a motion picture developing department.[16] It was to be stocked with materials of every kind and situated close New York City, "the great mart" where supplies could be quickly obtained. Edison saw it as a national "junk heap" of supplies and ideas, envisaging a facility that could work on a broad spectrum of different experimental projects in exactly the same way that his laboratories operated.[17] Supported with facilities to turn ideas into experimental models, its staff would work on a variety of projects. He described it as "not a research laboratory; it is a construction laboratory—more properly a universal machine shop" that could produce "anything from a submarine to a microscope." Speed of innovation was the primary consideration in the design of the laboratory; "the value of the lab is to turn things out quickly."[18]

The experimental directions of the planned naval laboratory were dominated by the new weapons emerging from the Second

Figure 13.2. A message that inspired and delighted Edison. With the diadem of science at her brow but the more practical sword of invention in her hand, Columbia prepares to deal with the nasty Hun.

Industrial Revolution—the airplane and the submarine. Chief among the experimental projects considered by the NCB were an improved aero engine, motive power for torpedoes and submarines (including storage batteries), improvements in submarine equipment and guns, and detection of vessels underwater. The laboratory was to be big enough to build experimental submarines.[19]

Edison's involvement in the development of a modern American submarine came to a sudden end when the E2 blew up, killing several of her crew. The cause of this accident was traced to hydrogen that escaped from the batteries and gradually accumulated in the confined hull. Although the crew did not follow proper procedures to vent gas from the hull, the blame was placed on the batteries. Ironically, at the same time that the navy was proudly announcing the great inventor's connection with them, his batteries were blamed for the tragic accident, ending any hope of developing the submarine market for Edison's alkaline battery.[20]

After this disaster, Edison's attention was focused on methods of locating submarines. His deafness made him a poor choice for a leader of a committee, and he soon asked to be relieved of this duty in order to concentrate on experimental work. Edison gave up all other research activities to devote himself exclusively to war work until the signing of the armistice in 1918.[21] He was joined in this endeavor by experimenters on his payroll and volunteers from universities and industry.

The work of the Naval Consulting Board, and several other research facilities, was directed toward the threat of the German submarine. When the war began, the only means of detecting a submarine was visual sighting. As there was no weapon available to attack it underwater, surface ships had to rely on gunfire or ramming the target to destroy it. Perhaps because the submarine was regarded as little more than a toy by many naval officers, little thought had been given to defending ships from its attacks. The British Admiralty soon set up an antisubmarine department in 1916 while scientists from Great Britain, France, and the United States applied themselves to the problem of locating submarines.

At the beginning of the war it was realized that the best way, if not the only way, to locate submerged vessels was to listen for the noise made by their propeller screws as they cruised underwater. The first attempts along these lines were shore installations that used sets of sensitive microphones. This approach met with some success—submarines could be detected over a distance of several miles—but as a tactical weapon it was useless in the hunt for U-boats at sea. The attachment of listening devices to ships was the next step and it was soon done; the first ships to be equipped with hydrophones entered service in 1916. Yet this method had the great disadvantage of requiring the search vessel to stop engines in order to distinguish the sounds of a submarine from its own screws.[22]

The beginning of unlimited submarine warfare in early 1917 brought the United States into the war and energized the NCB. Although Edison undertook over forty different experiments between March 1917 and November 1918, ranging from plating search light reflectors to airplane detection, he concentrated his efforts on the submarine menace.[23] This problem gave free rein to his ingenuity. He suggested laying a film of oil on the surface to obscure the submarine's periscope and firing nets into the path of torpedoes to stop them from hitting ships.[24] As he had done many times before at West Orange, he approached the problem from all angles, from finding submarines to devising methods of making their prey harder to find. His work on ship camouflage was the beginning of a tactic that had great success in two world wars.

One of the chief difficulties in locating submarines by their noise was to distinguish the sound of their screws from the many other noises that could be heard through the hydrophones. Edison was no stranger to this problem because he had spent many years trying to eliminate the crackle and hiss of surface noise from recorded sound. The sea was a far more difficult subject than the malleable wax of the phonograph cylinder. Its ever changing spectrum of underwater noise confounded the minute adjustments that Edison and his team made to the listening apparatus.

Another critical experimental area in submarine detection, which had already been investigated at the West Orange laboratory, was the problem of amplifying selected sound waves. The listener on board ship needed to increase the volume of the distant submarine noise in order to identify and locate it. Edison went through his repertoire of amplification methods: mechanical diaphragms, telephone amplifiers, audions, resonators, and several devices developed for the "loud-speaking" phonographs. Although he did manage under experimental conditions to pick up and amplify the sound of the submarine's screws as it came through the water, he found that none of these methods could provide a reliable location of a submarine, especially at the distances that could be expected in wartime operations. Once he had fully grasped the difficulties of this approach, he looked for alternative methods to solve the problem, as he had often done in the experimental rooms of West Orange. His next series of experiments returned to familiar technology that had figured in previous inventions. He tried to locate the submarine by measuring changes in magnetic fields caused by its metallic bulk passing through the water. He devised an induction balance but it was not sensitive enough. The idea was good—it is the basis of the modern magnetic anomaly detection system—but the technology was not available in 1917.[25]

After repeated failures, Edison gave up trying to locate submarines at a distance and concentrated on detecting torpedoes as they came close to merchant ships—an easier experimental task but of limited tactical use. Employing technology that had been developed for the phonograph and telephone, Edison constructed simple listening devices (based on a phonograph reproducer) attached to a long speaking tube that was connected to an amplifying device on board ship. Although much of this experimental ground had been covered in years of phonograph experiments, Edison faced the problems of increasing the scale and the sensitivity of the equipment and applying it to operational conditions that were very demanding. The navy gave him the use of a vessel, and during the summer of 1917 he cruised Long Island Sound testing his listening devices. The apparatus was slung out of the vessel and lay in the water about twenty feet from the ship. In perfect conditions it could detect a torpedo at 4,000 yards, but in wartime, conditions are never perfect—the background noise of the search vessel screws, sounds from fish, and water turbulence reduced its effectiveness.[26]

The lessons of Long Island Sound had been discovered by the many other experimenters looking for an answer to the U-boat problem. Professional inventors, engineers, seamen, and cranks had all been attracted to the problem, and some of their ideas were rather eccentric, such as using trained sea lions to find submarines. Yet the threat was so real in 1917 that any proposed solution, however silly, had to be explored. The allies naturally put the best resources at the disposal of the trained scientists and engineers with backgrounds in industrial research. One such experimental team, composed partly of scientists from the research laboratories of GE and Western Electric, was installed at a naval station at Nahant, Massachusetts. Instead of an acoustic amplifier, as found in the phonograph diaphragm, they tried electronic amplification of a narrow range of frequencies received by a microphone produced by Reginald Fessenden. This listening device was developed to detect icebergs after the tragic sinking of the *Titanic*. The one-time mucker in the West Orange laboratory had become a pioneer of radio technology. Fessenden had invented a high-frequency alternator that transmitted continuous (rather than damped) radio waves and used this technology in several systems of wireless communication.[27] He also developed a system of underwater sound telegraphy that transmitted an audible signal underwater. Despite his experience and the scientific expertise of the men from GE and Western Electric, the group at Nahant also would not solve the problems that had confounded Edison. The most successful submarine detection system that came from this research effort was still an acoustical device—a directional hydrophone that resembled an elaborate stethoscope.[28] None of these methods of passive detection played an important part in sinking submarines, the majority of which were sighted on the surface or revealed by the presence of oil or air bubbles. An active sonar system that used reflected sound waves to detect submarines instead of the passive system of listening for them was not developed until after the war.[29]

The most effective answer to the submarine menace was a simple idea that depended on the vast area of the sea to hide merchant ships. The naval authorities were unreceptive to the idea of convoys, but not Edison. After a trip to Washington to study the statistics of ship sinkings—a typical first step in an experimental campaign—he came to the conclusion that merchant ships were making it easy for the U-boats to find them.[30] By grouping ships in convoys, the Allies made it harder for the U-boats to locate a target and also forced them to engage the convoy's armed escorts. This early foray into operations analysis brought immediate results; the introduction of the convoy system in 1917 radically reduced the number of sinkings and held the U-boats at bay until the war ended.

The role of science in this partial victory was not decisive. In fact submarine detection remained unreliable and haphazard until well into the Second World War. Science had not saved democracy as expected. Victory in the Great War had again gone to the side with the most men and materiel. The achievements of the Naval Consulting Board were slight, for very few of the host of experimental projects produced a useful addition to the war effort. The one lesson that emerged from this massive endeavor was that industrial research was often an expensive failure. This was something that Edison, Willis Whitney, and the other managers of innovation already knew. The experience of the war had glorified industrial research, making it more visible to the public and more attractive to business. Yet however well organized and financed, it remained expensive, unpredictable, and often a waste of time.

West Orange During the Great War

The outbreak of the war had a great impact on the Edison enterprise, which was based on the international flow of materials. His phonograph business had made him a major importer of chemicals from Germany, including phenol, a strategic good used in the manufacture of explosives. The Royal Navy's blockade of German ports in the fall of 1914 cut off the supply of chemicals needed to make phonograph records at West Orange, forcing layoffs of men.[31] The sudden elimination of these raw materials might have crippled another company, but TAE, Inc. had the benefit of the West Orange laboratory, a facility especially suited to industrial chemistry. One of Edison's aims in building this laboratory was to find substitutes for expensive materials. As soon as the war began, he quickly put it to work to make his enterprise self-sufficient in strategic chemicals—a task that fitted the resources of the lab and the ingenuity of its leader.

Edison assigned about forty men to devise a process to make synthetic phenol. The experimental and design stage was concentrated at West Orange, as usual, and the pilot plant was erected at the Silver Lake complex under direction of the laboratory, with equipment fabricated in the machine shops of West Orange. Edison and his men took pride in the rapid establishment of phenol plants and in the phenol they produced that was of higher quality than that obtained previously. The result was better sounding records.[32]

The laboratory staff also found a way to produce paraphenylene that was also needed to make disc records. After a manufacturing process was devised for paraphenylene, a small plant was constructed at the laboratory.[33] Edison started an experimental program to find a substitute for potash and eventually found that caustic soda would do the job. Many of the chemicals needed in Edison's factories, such as benzene, phenol, toluene, and aniline, were byproducts of the carbonization of coal (to make coke). These chemicals were produced in plants that Edison constructed next to the furnaces of the Cambria Steel Company of Johnstown, Pennsylvania.[34]

To increase output, Edison's engineers streamlined the manufacture of chemicals. His plants were soon making six tons of phenol a day to meet a daily requirement of one ton. Soon he was able to sell his surplus phenol, and many other chemicals he made, at a profit.

By 1915 the factories of West Orange were fully committed to the policy of preparedness—to ensure that the transition to wartime production would be an easy one. Edison was ready for the coming of war.[35] In Camden, Eldridge Johnson was making the same transition. He too had been forced into using his laboratory to find substitutes for raw materials needed in the war. In his case it was the hard steel for the needle points of his Victrolas. His engineers and chemists found that tungsten could be used in needles, improving

Figure 13.3. Laboratory Home Guard. This group includes mechanics, laborers, experimenters, recording experts, and the chief engineer of the lab. The dapper young man on the right is Edison's son Charles.

sound quality.[36] Like Edison in West Orange, Johnson was preparing his industrial plant to manufacture munitions and war supplies.

The official entry of the United States into the World War in 1917 brought a surge of patriotism that is reflected in the photographs of the West Orange lab: buildings covered with patriotic bunting, victory parades passing outside on Main Street, war bond drives in the factories, and workers in uniform, ready to go to war (fig. 13.3). The labor force decreased rapidly as men joined up. Mambert's department estimated that by 1917 the administration had lost 50 percent of its staff.[37] While Edison chased submarines in the sound, all the resources of TAE, Inc. were devoted to the war effort. There was very little choice. The First World War marked the increasing power of the national government in the economy. Federal authorities placed limits on production, requisitioned key raw materials, and scrutinized accounts for excessive profits.

Edison products were used in many different ways during the war. His batteries powered gun turrets on battleships and provided the weak current for the signal corps. His dictating machines found many uses in the bureaucracy of the armed services, and a special portable model was developed for use in the field. The lab staff designed a special musical phonograph, the Army and Navy model, for the American troops on active duty. Talking machines provided some relief from the horrors of war for the men of both sides, who listened to their favorite songs in trenches and camps. The World War gave a boost to recorded music by bringing more people into contact with it and providing a steady demand for music.[38] Edison's recording studios in New York and West Orange produced patriotic records of marches and sentimental songs for the doughboys and those they left behind.

The Phonograph Works output of talking machines soon declined as it switched to military production under government contracts. It made bombsights, gas masks, shear wire pistols (to cut through barbed wire), and several types of electrical adapters. The war work accounted for most of the capacity of the Works by 1918. The West Orange laboratory was also fully committed to the war effort.[39] The government devised specifications that the Engineering Department made up into blueprints and drawings for the Phonograph Works. F. J. Riker of the lab was appointed product engineer for the government work and given the power to approve all drawings, parts, and raw material orders.[40] Before the war most planning for manufacturing was done in the divisions. The lab took over this function and planned four government jobs, introducing control sheets to keep track of all stages of production.

Although the administrative staff was reduced at the beginning of the war, the management of TAE, Inc. operated smoothly through the years of conflict. The years of disruption and political infighting that followed the reorganization of the Edison enterprise were more or less over. The lines of financial responsibility had been established in the divisions.[41] Most of the grumbling of the division managers had dissipated or at least was quieter than before. Charles Edison ran the business while his father hunted submarines. An early advocate of the divisional structure and a supporter of enlightened management, he established a large personnel office and other service departments in TAE, Inc. to increase productivity and improve industrial relations. The war gave him the opportunity to put his ideas of corporate welfare into practice. Charles was proud of the accomplishments of this style of management. Alfred Sloan, of General Motors, sent his men to study the West Orange operation before carrying out his own divisional policy. Sloan's reorganization of General Motors is taken as the pioneering divisional structure that laid the basis of modern management of large business organizations. The new organization was based on the Edison example, with divisions centered around each product and central service department, such as a research laboratory. Charles Edison felt complimented.[42]

The final tribute to Charles Edison's administration came from the Old Man. On his return from Florida, where he had been carrying out submarine detection experiments at the Key West Naval Station, Edison told the press that he was pleased with his son's work, adding, "I have gradually developed an organization which relieves me from the details of business."[43] Edison was preparing to spend more of his time in Florida at his fully equipped laboratory in Fort Myers. With all his businesses showing profits, the time seemed right for the greatest inventor of his age to take a well-earned rest.

After the war Edison became increasingly isolated from the general work of the laboratory. In 1918 when the leading managers of the company (including the chief engineer) came together in a meeting of the Executive Committee, nobody had any idea what the Old Man was doing on the top floor of the laboratory.[44] He had withdrawn from the day-to-day running of the laboratory. A memo sent by his secretary in 1919 informed company executives that communications about laboratory operations should be sent to J. P. Constable, manager of the laboratory, and not to Edison.[45]

The laboratory was now a division of TAE, Inc. and the chief engineer was a division manager. John Constable had replaced Miller Reese Hutchison after the latter's independent role as free-lance engineer and storage battery salesman had upset so many people that the administration of TAE, Inc. demanded his removal.[46] Edison's failure to protect his favorite is significant in the decline of the machine shop culture at the laboratory. Constable had none of the authority of Hutchison and played no part in the business of innovation. His job centered on issuing Engineering Notices and supervising the product engineers. Edison's secretary, R. W. Kellow, remained in charge of the business affairs of the lab, including billing, scheduling, and accounting. Although theoretically superior to him in the hierarchy of the organization, Constable was ordered to give him full cooperation, leaving no doubt about who was the boss. Edison commented, "If he is a good engineer, he will be a poor biz man as a rule," revealing his opinion of the engineers who now worked in his laboratory.[47]

Like all the other division managers, Constable was expected to show a profit from his operation. Mambert continually pressed him to run the lab "a little more on the business principle."[48] This was a fundamental change in the laboratory's role in the Edison enterprise. It had always run at a loss, the great experimental campaigns supported from Edison's own pocket or the treasury of the Phonograph Company. Long-term development projects had been assigned to the profit and loss account. The changing structure of TAE, Inc. forced the lab to justify its existence and the money spent on research. As it was no longer the sole source of engineering services in the organization, it had to compete with other laboratories for business. The laboratory at Silver Lake had begun with testing raw materials used in record and storage battery manufacture. By 1918 that laboratory was also carrying out chemical analysis for several other divisions, research work on the primary battery and Amberol records, and electrical testing. Its manager hoped to take over some of the electrical testing work done at West Orange.[49] An important part of John Constable's job was to sell the services of the West Orange lab to the divisions of TAE, Inc.

The policy of giving financial responsibility to the division managers often made them reluctant supporters of research and development. For much of the work of the laboratory, finance did not present a problem. Testing, toolmaking, pattern making, and drafting were clearly identified and accounted for. It was general experimenting that was harder to finance. This involved high-risk research, perhaps to develop a totally new product, that might not show a financial benefit on the monthly or yearly reports. Division managers resisted this expenditure because it ruined the financial statements sent to the central administration. Managers, such as Nelson Durand, who were committed to long-term research to develop new products had to argue their case to make the development program an "Edison family affair" supported from the treasury of TAE, Inc., not from the budget of their division.[50]

The years of inflation during the war had aggravated this situation. In a tighter financial environment in which division managers had to watch their costs very closely, it was sometimes hard to justify the expense of experimental work. All costs had risen, especially the wages of skilled men who were in short supply.[51] Changing economic conditions meant that the Edison organization had to get the most out of every dollar spent, especially in experimental work.[52]

The issue between the laboratory and the manufacturing divisions was who was going to pay for experimental and development work. The years after 1915 were marked by interdivisional squabbles over laboratory bills. The divisions often claimed that they had been billed for work done for other people.[53] (This kind of complaint was not new to Edison; he had been answering charges like this since he managed machine shops in Newark.) The administrative reforms had created a more formal and precise system of internal accounting in which every penny spent had to be accounted for, forcing laboratory managers to charge even the smallest job to a purchase order that had to be paid. Installation of special machinery for development work normally led to a heated dispute over who should pay. It was decided that the lab could not order or build specialized equipment without the prior consent of the division affected.[54] Capital expenditure was a critical part of keeping up with other research organizations; their move into electronics, for example, required new testing apparatus.

The equipment of the West Orange laboratory had once been the envy of other research institutions, but now it was the other way around. While the industrial research laboratories of GE and Western Electric spent freely on new scientific equipment, Constable had to continually petition management to replace the aging capital stock of his laboratory. He tried to persuade Edison to buy more modern equipment, arguing that the lab should be "the last word in high class machine shops." His appeals fell on deaf ears. Edison told him to repair the old machines. As a result of his parsimony, visitors to the Edison National Historic Site can still view some of the machine tools installed in the lab in the 1890s. The lab was the last to get any new machinery with the result that in 1918 the equipment in its machine shops was described as "very ancient material."[55]

There was very little need for the latest equipment because the basic function of the laboratory had moved from innovation to engineering services needed by the divisions. It concentrated on manufacture rather than development. Long gone were the days of experiments on extravagant and ambitious projects such as pyromagnetic generators or flying machines. The stream of inventions had dried up, and with it the ideas for new products and new businesses. In 1916 Edison claimed, "I have no dept [sic] in which men are employed to perfect inventions."[56] Instead, formally trained engineers worked under close supervision in routine activities such as testing and the listing of parts. Every part of each Edison product was scrutinized in an effort to reduce manufacturing costs, even down to the hair in the cleaning brushes sold with phonographs.[57]

The research and development function in the lab was carried out by the product engineers for the divisions. Their work was initiated by the division manager, controlled by the various committees on which he sat, and supervised by the chief engineer who had to render frequent progress reports. They were known as the experimental staff although much of their work was routine engineering. The activities of the professional engineers in the laboratory were so closely controlled that they could not carry out research on their own volition—formerly a trademark of the nineteenth century lab. A purchase order was required for any experimental work, and this needed the approval of management. By 1918 the division managers' control of the costs of development had empowered them to decide its direction. In the conviction that "the foundation stone of our success is the engineering division," each division manager argued, "The development (product) engineer must be placed under my jurisdiction and I must govern his research and development activities and control expenditures."[58]

The business of innovation was firmly in the hands of the division managers by 1919 when Edison reviewed the activities of the lab and concluded that it was "mostly routine work," adding that he would personally attend to any experimenting with his present staff.[59] His experimental work either on his old loves, such as phonograph recording, or on personal projects, such as the work done for Henry Ford, was the remaining link to the independent laboratory of the nineteenth century. As John Constable noted in 1919: "The only experimental work I know of now is carried out under the direct supervision of Mr. Edison."[60]

The lab no longer gave employment to experimenters and muckers but to experienced engineers who came to the lab from academia or business.[61] In fact Edison said that it was "undesirable to employ young men with inventive talents."[62] In the twentieth century new employees wanted to know exactly what their duties would be before they arrived at the lab. They were high-grade men, and their time was too valuable to move them from job to job in the fashion of the 1880s. The policy of employing specialists at West Orange paralleled the national tendency toward professionalization of work. By 1920 the lab employed seventeen categories of workers paid on an hourly basis and over thirty on a weekly basis. Included in the latter were librarians, stenographers, and product engineers.[63]

In the old machine shop culture at West Orange the engineer was a law unto himself, and the laboratory was the guiding force of the Edison enterprise. Now the engineer was a corporate employee in a service organization. Instead of the Old Man, he had to answer to administrators and division managers who let him know that the old "aloof and scientifically impersonal viewpoint" was inappropriate for a modern business organization.[64] The laboratory had once enjoyed the highest status in the organization, its work a source of pride and confidence. In the twentieth century it was forced to justify its existence and overcome the prevailing attitude within TAE, Inc. that it was an expensive luxury. The low status of the lab in 1920 is revealed in the following comments made by Edison's secretary: "The laboratory has been quite popularly conceived to be a place where sundry mysterious experiments are carried on by Mr. Edison and charged to other Edison functions arbitrarily without any corresponding advantage to them."[65]

The lab was no longer dominated by the uncouth behavior of Edison's "insomnia squad." After the war many of the employees were clerical staff and the time-honored tradition of working day and night could not be continued. Edison's personal experimenters did maintain some of the work practices of the invention factories, including working on Saturday afternoons after the rest of the staff had left.[66] Although Edison's personal work in 1918 was described by his son as "small stuff, knick knacks," it still received top priority from the draftsmen and skilled machinists in the lab who held him in awe and said, "If the old man comes down and wants something of course everything stops."[67] Edison had ceased to be the driving force in the lab. He became its figurehead, a living legend that tempted young boys to leave their benches and get a glimpse of him before he passed away.

Edison did not change his work habits; he still put in his eighteen- or twenty-hour days as recorded on the time clock in the hallway or in the notes of his devoted secretary, Meadowcroft.[68] He surrounded himself with old retainers like the Otts and continued with machine shop practices. John Ott had been worn out by the demands of the invention factory. But Edison still had a right-hand man in the machinist Sam Moore, with whom Edison worked closely, devising and altering experimental models. He had a telephone and a dictating machine at his fingertips yet he still used boys for his fetch-and-carry work and communications. He had become an anachronism in a modern industrial laboratory.

CHAPTER FOURTEEN

The End of an Era

In physical terms, the lessened status of the laboratory can be seen in the transition from the freestanding structure of 1887 to a small part of a great industrial complex in 1920. Figure 14.1 is an aerial view of West Orange. The laboratory is dwarfed by many factories, its brick a contrast to the modern cement buildings around it.

The community of West Orange had been dramatically changed by the growth of Edison's plant. Once a rural retreat for affluent New Yorkers, it was now an industrial suburb of Newark. The Edison enterprise was the leading employer in the area. Its workers included many immigrants—Italians, Greeks, Poles, and Ukrainians—who spilled into the Oranges from Newark and New York. The labor force had grown from a few hundred men in the nineteenth century to thousands in the twentieth. In 1900 there were about 3,000 men in the West Orange complex but by 1914 there were about 8,000. The steady flow of government contracts provided many new jobs during the war. After the initial loss of workers to the armed forces, there was a rapid increase in the West Orange labor force, and the total number of employees could have reached 10,000 by war's end.[1] A rough and ready community of bars and boarding houses had sprung up around the factories. Workmen often indulged in drunken rowdiness from Friday payday until work began on Monday morning.[2]

The West Orange complex was not isolated from the growing labor strife which marked the prewar years. New Jersey was a hot bed of union activity and strikes. In 1913 the silk workers of Paterson took part in a bitter strike that followed similar conflicts in the textile mills of New England. These strikes were landmarks in labor history because they were supported by a union of skilled and unskilled workers—an ominous development for employers.[3] Edison

Figure 14.1. Aerial view of the West Orange complex in 1930. On the left is the huge storage battery factory. The Phonograph Works takes up much of the right side. The lab buildings are the small, dark buildings between the two great factories.

dealt sternly with any industrial agitation, resisting demands for wage increases and locking out the metal workers in 1913. Nevertheless, his workers became more radical over time as the influence of unions grew. The swelling numbers of the work force and the inflation of the World War period brought more labor problems to West Orange. The Industrial Workers of the World (the Wobblies), a radical group who preached strikes and sabotage, were active in New Jersey and infiltrated the West Orange factories. A strike over working on Washington's birthday closed the Works in

1916. The strikers were quick to point out this flaw in Edison's patriotism.[4]

Edison's already low regard of labor deteriorated through the strife of the war years. He had no sympathy for the workingman who wanted his wages to keep pace with inflation, telling Mambert in 1919: "Wages are certainly up with more loafing than ever . . . Only remedy is automatic machinery and less men." In fact the shift to a more capital-intensive mode of production had been resisted by Edison who did not want to spend the money on automatic machine tools that reduced the input of labor. This was in direct contradiction to his well-publicized support of the "machine made freedom" that would follow the automation of the work place.[5] The decreasing rate of immigration into the United States after the war slowly cut down the number of foreigners working in the Edison industrial complex, leading to shortages of skilled labor. This deficiency had been offset in many factories by the introduction of labor-saving, automatic machinery, but not at West Orange. The keenest loss was the highly skilled machinists and toolmakers who had staffed Edison's machine shops since the 1870s—the men who played a vital, if unrecorded, part in the process of innovation. The personnel office was forced to institute an apprenticeship program to fill the requirements for skilled workers. The low wages paid by the Edison enterprise played a major part in reducing the flow of skilled machinists and engineers into the laboratory. The weekly wage of a mechanical engineer at West Orange was about $35 to $40, the same amount as fixed scale wages for laborers, electricians' helpers, and derrick men.[6] Clerks in administration could earn more than the men who did the engineering work in laboratory and Works.

Even more damaging to the Edison enterprise was the unrest of the white collar staff. Some middle managers put on a good front at West Orange while desperately seeking work elsewhere, while others, such as E. E. Hudson, left their divisions in disgust and joined the competition.[7] The labor problems within TAE, Inc. were masked by the full employment and good business of the war years. The total sales of TAE, Inc. grew during the war, bolstered by government contracts and healthy sales of batteries. Americans had been unable to buy phonographs and other consumer durables during the war because factories were devoted to military production. Demand soared when American industry returned to peacetime production; 1920 was a year of record profits for the Edison enterprise, as table 14.1 indicates.

Underneath the impression of prosperity and expansion were

Table 14.1. Sales and Profits of TAE, Inc., 1915–1921

Sales Year	Sales	Profits
1915	$13,201,000	$ 936,000
1916	18,002,000	1,663,000
1917	20,629,000	1,166,000
1918	22,992,000	291,000
1919	30,252,000	2,930,000
1920	34,371,000	820,000
1921	13,833,000	−1,311,000

Source: Mambert files, ENHS.

some serious problems within TAE, Inc., chief of which was an inability to control rising costs. The sharp inflation of the war years had led to price increases in phonographs, even though past experience had shown that this inevitably led to a decline in sales. Demand for the flagging cylinder line—now reduced to the Amberola—was most sensitive to price changes. In 1918 it was reported to be in critical condition because the price of the Amberola was too high.[8] The Dictating Machine Division was fighting a losing battle to control costs, and consequently its prices rose while its sales declined. Similarly, the primary battery was slipping from its dominant position in the market because, as the division manager wrote, of an "inability to maintain our old position . . . of furnishing practically the best cell for equal or less money than the bulk of our competitors."[9]

The first casualty of rising costs was Edison's motion picture business. The dissolution of the Motion Picture Patents Company (MPPC) had bankrupted many of its members. The survivors struggled on as KESE: Kleine, Essanay, Selig, and Edison. The World War struck a sharp blow to Edison who leaned heavily on foreign distribution of films. Film exhibitors were taxed after the United States entered the war, and the prospect of further taxation on the entertainment of the idle caused pessimism in the industry.[10] Longer movies and the star system were forcing up costs. In 1909 a one-reel feature could be made for about $200. The average length and cost of Edison's movies had increased considerably by 1916, when a typical multi-reeler might cost as much as $20,000.[11] As fewer pictures were made each year, filmmaking was becoming more of a gamble. Although costs were rising dramatically, one blockbuster could easily repay the cost of production and a stable of expensive stars. Success in this business came down to nerve and capital, both of which were in short supply in the Edison enterprise after the difficult introduction of the disc line.

During 1917 the management of TAE, Inc. pondered the question of abandoning the business or putting more resources into it. The war provided the answer. Against the threat of more regulation and taxation were the bright prospects of procuring government contracts. In the first months of 1918 the entire motion picture business was sold to the Lincoln and Parker Film Company.[12] The last sets were struck in October. In the meantime the motion picture equipment was moved out of the Works to make way for government work. Manufacturing anonymous equipment for the war effort was far less glamorous than making movies, but it was more predictable and profitable. For the second time Edison had to leave an industry that he had created. The circumstances of his departure from the electrical industry were somewhat different, but it can hardly have comforted him in his second retreat from a competitive market.

The next part of TAE, Inc. to come under pressure was the core phonograph business. The boom year of 1920 came to an unhappy end as the economy suffered a sharp downturn. The danger signs were apparent in the early part of the year: a collapse in the currency markets, rising interest rates, and some steep declines in share prices.[13] The U.S. government embarked on a policy of sharp deflation, cutting its budget by about $6 billion. The money supply decreased alarmingly. At the end of 1920 the West Orange complex reeled under the shock of a sudden decrease in sales of its products, especially phonographs. The postwar depression started a difficult decade for the phonograph industry, which had always been quick to reflect the ups and downs of the economy. The boom years had begun in 1896, the year traditionally chosen as the end of the Great Depression of the nineteenth century. The decade of 1899–1909 saw a 250 percent increase in phonograph production over the previous decade, despite the downturn of 1907. The next ten years became the golden age for the phonograph industry, which recorded an increase of output over the previous decade of 520 percent. In contrast, the 1920s marked increased competition, the threat of radio, and a 45 percent decline in production compared with the 1909–1919 period.[14]

TAE, Inc. was not alone in watching its market diminish. In 1921 General Motors' sales dropped 21 percent; Ford Motor Company's fell even more; Victor's sales plunged 43 percent; and Columbia's loss of sales led to bankruptcy.[15] Well-established giants of American industry, such as Du Pont and Goodyear, found themselves in financial difficulties. The response of big business to the postwar depression was a massive reorganization of their managements to make them more efficient. This was the beginning of the divisional structure movement that decentralized administration.

Most companies attempted to increase their management capabilities to meet adverse conditions and overcome changes in the market. Others took the path of cost reduction and retrenchment. TAE, Inc. and Ford Motor Company were among the latter. Both were founded by strong-willed individualists who could not bear to hand over control of their companies to professional managers. Henry Ford spent money he did not have to buy out minority stockholders, and when depression came, he sold capital stock (including telephones) and fired half his office force and engineers.[16] In 1921 Edison was so alarmed at the depression in the core business that he returned from retirement to help his son Charles weather the crisis. He noted that "these hard times keep me busy day and night, and I have to work about 18 hours a day to keep the ship afloat."[17] Cost reduction through deep cuts in the payroll became an obsession with Edison senior. In 1921 he took the drastic step of going on what was described as a "rampage" through the Works, firing men right and left, including John Constable, chief engineer of the lab.[18] After Edison's purge, it was estimated that less than 2,000 workers remained in the factories at West Orange.[19] To economize, administrative positions were abolished or merged as departments were amalgamated. Mambert survived the crisis but could not maintain his powerful position much longer, and in 1924 Edison personally removed him from office.[20]

"Merrily the axe swings," noted Charles Edison, who could not bring himself to dismantle the administration he had so carefully built up.[21] Corporate welfare at West Orange had been swept aside by a strong dose of nineteenth century capitalism. It was a much reduced work force that survived the phonograph depression of 1921. The great purges left a deep scar in the organization, adding to Edison's reputation for firing men whenever sales dropped and creating an atmosphere of gloom and foreboding throughout the 1920s. Many of the surviving professional staff looked elsewhere for employment. Several senior executives left of their own accord to find better, more secure positions in other firms. The Edison enterprise could not afford to lose such experienced men before the impending struggle to remain in the phonograph business.[22]

The departure of so many sales managers made it harder for TAE, Inc. to adjust to the demands of the postwar market. It was ill-equipped to compete in a business environment dominated by falling prices. The consequences of the failure to engineer an economy disc phonograph were now apparent. Customers wanted value for money and were buying more modest machines, such as Victor's Victrola XI—the most popular talking machine of the early 1920s.[23] This shift in demand had been anticipated by several managers within TAE, Inc., including Nelson Durand, but the organization as a whole struggled to move from one end of the market to the other. The transition was especially difficult for a sales force that had grown accustomed to selling the quality of the Edison phonograph, depicting it as "the superior instrument." Now they had to change their tune to keep up with Victor and a reorganized Columbia, which had responded faster to the changing times.[24] TAE, Inc. introduced a new line of disc machines in 1922. The London series was the first to include an upright model selling for $100, yet this was more than the ideal price range of $75 to $80 needed to compete with Victor.

The postwar depression hurt Edison's cylinder business the most. Sales of Amberolas declined steadily as dealers deserted the cylinder format. The introduction of the Diamond Disc line before the war should have been the death knell of the cylinder machine, and both dealers and the competition expected TAE, Inc. to withdraw from a losing proposition. But Edison ordered the lab to continue research and development in the cylinder format. The steady improvement of cylinder technology made little impression on a declining trade. Despite tumbling sales, TAE, Inc. stayed true to Edison's wishes and supported the cylinder users to the very end, incurring heavy loses in the 1920s[25] (table 14.2).

Small improvements in the technology of acoustic phonographs did not require the services of trained scientists, and consequently few were employed at the West Orange lab. TAE, Inc. occasionally hired professional chemists, and at least one product engineer had formal scientific training, but on the whole the flow of first-class men dried up in the 1920s. There were even problems finding formally trained engineers with commercial experience (to be product engineers) because Edison's prestige could no longer offset the low wages and job insecurity in West Orange. In fact there was little

Table 14.2. Phonographs Sold to Jobbers, 1911–1924

Sales Year	Discs	Disc Records	Amberolas*
1911			1,081
1912	3,833	13,859	1,945
1913	16,251	437,579	30,188
1914	29,079	1,015,506	20,650
1915	52,632	2,417,911	28,759
1916	73,346	2,621,113	40,845
1917	76,256	3,351,098	54,293
1918	67,405	3,260,534	51,971
1919	121,537	7,163,028	54,178
1920	140,149	7,596,856	44,310
1921	32,343	4,348,471	3,090
1922	59,608	3,626,481	22
1923	65,925	3,877,450	2,001**
1924	31,900	2,573,508	—

Source: 1925 Phonograph folder, ENHS.

Note: This table does not include several sales categories, including export; and several models including the Army and Navy model.

* Many types of cylinders with horns are not included.

**This high figure reflects a discount sale to clear inventory.

need for them as the work of the laboratory became more routine. As an economy measure the lab had assumed many small engineering and machining jobs that had previously been sent out to contractors.[26] Its machine shops were becoming more important as manufacturing resources than as facilities to make experimental models.

The work force of the West Orange laboratory in the 1920s was relatively underqualified when compared with other industrial research laboratories. GE's laboratory at Schenectady had a staff of about 300 in 1920, including about 50 scientists and engineers. Western Electric's research labs had 413 employees in 1921, many of whom had Ph.D. degrees and were professional engineers.[27] In the early 1920s the West Orange laboratory employed about 200 men. The majority of them were machinists and laborers who worked in the machine shops on the first and second floors of Building 5. The number of professional engineers and formally trained experimenters was about fifteen. The cutbacks of the early 1920s probably reduced this number considerably.[28]

After 1900 as organized industrial research became more widespread in American industry, the West Orange laboratory no longer dominated the field as it had done in the nineteenth century. Many small laboratories that had begun as testing labs or Works engineering departments in the nineteenth century were enlarged and given more important tasks in the twentieth. Eastman Kodak, for example, established a small lab in 1893 for testing. The company increased its duties and hired first-class men with scientific credentials. By 1919 the company enlarged the scope of the laboratory's work to cover any area that had relevance to its product line. The number of staff in its laboratory reached twenty in 1913 and had climbed to eighty-eight in 1920 when expenditure on research amounted to $338,680.[29] The West Orange laboratory now stood in this second rank of industrial research.

Once established, corporate research laboratories experienced rapid growth. The management of Du Pont decided to form a laboratory in 1902. In 1911 a larger department was created, which brought the work of several labs under one organization. The salaried staff of this Chemical Department stood at 111 with a budget of $300,000 in 1912.[30] Expenditures on research at Du Pont exceeded $2 million by 1926. In contrast, experimental activity at TAE, Inc. declined during the 1920s. In 1920 its experimental expenditure was $669,000, which was more than Eastman Kodak but considerably less than GE and AT&T that each spent over a million dollars on research in 1920. During the 1920s, research expenditures at Du Pont, GE, and AT&T rose steadily while those of Edison's laboratory dropped to about $150,000 in 1926—one-tenth of GE's outlay.[31]

There was also a significant difference in the type of research undertaken. The corporate laboratories had proved their worth by 1918. After the war they were allowed to pursue ambitious research and development projects with long-term goals. AT&T, for example, appropriated $500,000 in 1919 for research into radio.[32] In contrast, TAE, Inc. allotted $15,000 to design a cheaper dictating machine and $20,000 to find a better method of affixing diamonds to the stylus of the Diamond Disc.[33] In terms of quality and quantity of staff and the type of technology it hoped to develop, Edison's laboratory could not compare with the front rank of American industrial research. The "largest lab extant" was now a shadow of its former self as the leading research institution in the United States.

Innovation in a Declining Industry

The decline of innovation at the West Orange laboratory was most evident in the phonograph business where the Edison enterprise lagged behind in key innovations, such as a universal reproducer (that could play any record). Although Edison claimed in 1920, "We are in front and we intend to stay in front," the next decade saw smaller companies moving ahead by introducing new features for the talking machine.[34] The boom years that preceded the First World War had attracted several new companies to the talking machine business. The Brunswick, Aeolian, Sonora, and Cheney companies also benefited from the expiration of some of the key patents that had helped sustain the Big Three of Edison, Victor, and Columbia. Brunswick and Aeolian were perceived as the most dangerous new competitors.[35]

The technological lag in TAE, Inc.'s phonograph business was not the result of inability to identify the problems in the product, for the drawbacks of short playing time of records, lack of sufficient play back volume, and inability to play the competition's records had been evident for some time. It was the consequence of slow progress in producing the technological answers to these problems. Speedy innovation had been one of the cornerstones of Edison's conception of a research laboratory and a major part of his planning for the West Orange facility in the 1880s. Although he proved that his teams of muckers could still move rapidly and successfully into new areas, such as the chemical plants of the First World War, the corporate part of his laboratory—the product engineers and the committees that governed their work—worked at a much slower pace.

Innovations, such as a universal reproducer and an automatic (record) changer, were included in the lists of experimental work made by Edison product engineers, but the credit of introducing them went to other companies. The Brunswick company was the first to bring out a universal reproducer and quickly gained a large share of the market.[36] Important development projects, such as an automatic changer, spent years in the laboratory when their rapid introduction might have turned the tide in favor of TAE, Inc. It was left to Victor to bring out the changer.[37] Other projects were begun and never finished. Work started on an automatic stop mechanism (that stopped the machine at the end of the record) in the first decade of the twentieth century. This was a comparatively simple mechanical device based on the stop developed by Victor. Although many experiments were carried out at the West Orange laboratory, no automatic stop was ever introduced.[38]

While the more formal structure of engineering had produced an ossification of the innovative process in the laboratory, this was only half the problem. Behind many of the delays stood Edison's continued interference in the work of the laboratory and his input into the business of innovation. His opposition to lateral-cut records impeded the introduction of a universal reproducer and prevented this system from replacing the hill-and-dale cut pioneered in the lab.[39] Edison was no longer the commercial inventor he had been in his thirties. In his seventies he was a technological conservative.

This resistance to change did most damage in the critical area of record selection. The talking machine was no longer a novelty but an item of mass consumption in the postwar period. Success in this mature industry would come to those companies that met the public's needs for recorded music, as the salesmen in TAE, Inc. duly informed upper management. The task was to accurately predict the public's musical taste and meet it before the competition. The manager of the Musical Phonograph Division concluded that "the company who correctly solves these problems will dominate the trade."[40]

The division managers had control of the mechanical development of the machine through their product engineers, but record selection was still inexplicably left to Edison—who had his own unbending views on the subject. The divisional structure should have established a sales organization that was willing and able to pander to its customers' changing tastes. Instead, Edison still dominated the proceedings with his desire to perfectly reproduce classical and operatic music. He preferred to educate the user about the joys of "good music" (as he defined it), just as he had educated customers before about new technologies. He claimed, "You can like a jazz tune but you love good music."[41] His professional managers knew better.

The era of the First World War marked a period of social upheaval in the United States. Tastes were changing; values were changing; and even Prohibition could not quench the enthusiasm of the younger generation. Postwar America was a time of new ideas and new fashions as hemlines rose and hair was bobbed. Musical tastes underwent a dramatic change. Sentimental favorites, such as "Sweet Adeline" and "By the Light of the Silvery Moon," gave way to "Alexander's Ragtime Band."[42] The wartime movement of southern blacks to northern cities had exposed many to the swinging music of New Orleans. It was soon given a name, "jazz," and it quickly gained an audience in the urban centers of the northeast. The jazz craze was just one indication of new American tastes.

TAE, Inc. did not follow its competitors in exploiting the popularity of jazz because the chief executive disliked it, believing erroneously that people who bought phonographs did so to listen to good music. Edison was not in tune with the predominately urban market for recorded music; his favorite song was "I'll Take You Home Again Kathleen," and he claimed that this was also the most popular Edison recording.[43] His emphasis was still on the machine rather than the media. The word from the customer was, "The Edison is a good machine, but has a very limited selection of records."[44]

Victor's recording engineers were free to follow public taste and encouraged to find new music that might have an audience.[45] Victor was the first to record jazz in 1917 with its Original Dixieland Jazz Band.[46] The company's domination of the disc market rested on its broad recording catalog, which bridged the gap between grand opera and sentimental songs with popular singers, such as John McCormack. Furthermore, the owner of a Victor machine could also listen to records from the Brunswick and Columbia companies that also used lateral-cut recordings. But TAE, Inc. maintained its position of splendid isolation with its hill-and-dale records. The decision to keep a unique format for records was an awful handicap in the 1920s. The rigid adherence to an outdated standard cost TAE, Inc. many customers, yet it perfectly suited the disposition of its founder. The company, however, paid the price, for by 1927 its records accounted for only 2 percent of total record sales.[47]

The years of sound research at West Orange constituted a valuable investment that Edison could not bear to give up. The goal of the most realistic recreation of the sound of the soloist and accompaniment was achieved after the war, and the tone tests proved it—in demonstrations sponsored by TAE, Inc., thousands of people were unable to distinguish the sound of a Diamond Disc machine from a live performance.[48] Edison and his laboratory staff had advanced the science of acoustical recording almost to perfection, especially in reproducing the articulation of the human voice. The diaphragms were so sensitive on the dictating machine that they required adjustment when women began to dictate more letters during the war.[49] Although the extensive postwar research on disc reproduction had produced a highly advanced technology, the public were rarely as discerning as Edison would have liked. Customers who had no intention of listening to the prestigious Victor Red Label records were still impressed and took them as proof of the quality of Victor's recording technology. Victor had brought talking machines to the status of musical instruments by astute marketing as well as research and development.

The vaunted Edison phonograph technology was out of step with the market. A younger record-buying public wanted to dance to its talking machines rather than listen intently to its lucid sound, as depicted in TAE, Inc.'s advertising. Market research showed that the customer wanted loudness first and a good tone second.[50] Unfortunately, the Edison machine was designed for accurate reproduction rather than loudness. Increases in volume soon overburdened the reproducer that blasted the sound. Edison was prepared to make sacrifices for fidelity of recorded music. Diamond Disc records had to be twice the thickness of the competition's discs in order to provide an absolutely flat surface to take the hill-and-dale cut. Edison felt that a heavier record was a small price to pay for good reproduction, but customers, who were more concerned with ease of storage, saw it as a nuisance.[51]

The introduction of radio led to radical changes in musical tastes. Its high volume and extenuated bass did more than suit the dance music of the 1920s; it redefined what customers expected of recorded music. An even louder play back was now required. Electronic amplification had a special quality that Americans took to heart and soon wanted to hear from their phonographs. Special attachments were soon available to add onto acoustic phonographs to give them the amplified radio sound.[52]

Thomas Edison is often placed at the beginning of the radio story with his discovery in 1883 of the "Edison effect," an electrical force moving between two electrodes in an incandescent bulb.[53] A scientist employed by the English Edison Electric Lighting Company found that Edison effect bulbs (with two electrodes) could be adapted to detect high-frequency radio waves. Ambrose Fleming's valve, called a diode, was an important step in the development of vacuum tubes. The critical innovation was the work of the inventor-entrepreneur Lee De Forest, who added a third element to the Edison-Fleming tube and produced one of the great inventions of the twentieth century.[54] De Forest's audion (or triode) could be used not only to detect radio waves but also as an oscillator to generate them. His use of a (third) control electrode opened up a vast new horizon of opportunity for those laboratories equipped to develop the technology of vacuum tubes.

Edison, the "father of electronics," followed the development of wireless technology with some interest; he had been one of the first to see it had a commercial value and had carried out many experiments in the general area. Several phonograph engineers at West Orange took an interest in radio, buying books on the subject and making up experimental apparatus.[55] The laboratory staff were also well aware of the advances of vacuum tubes. De Forest visited the lab, corresponded with Edison, and worked with some of the muckers on experiments in broadcasting. When Edison asked De Forest what parts of his invention were free and open for development, Edison made it clear that his interest was in amplification rather than radio.[56]

The development of electronic amplification obviously had important applications in both business and amusement phonographs, primarily in providing undistorted increases in volume. The telephone companies also had a use for amplification of feeble electrical currents. AT&T's purchase of the audion patents not only provided the means to dominate long distance telephone communication but also gave them a foothold in the attractive new field of radio telephony.[57]

The First World War brought the U.S. government into the field of wireless telegraphy and provided a respite from the continuous litigation over patents. Radio technology advanced rapidly during the war years as cooperation replaced conflict. The staff of the West Orange laboratory got their chance to get fully involved in the exciting new field. Several experimental projects were undertaken in wireless technology as part of Edison's work for the Naval Consulting Board. Holland and Constable soon obtained a spare audion which they rigged up to a disc machine. The audion acted as an amplifier and was connected to a loudspeaker. This apparatus remained an experimental proposition and was not developed any further.[58]

At the end of the war, the government acted to keep the future of radio in American hands. The formation of the Radio Corporation of America (RCA) in 1919 was the first step in the pooling of patents that opened the way for the further development of the technology. GE, AT&T (and its manufacturing subsidiary Western

Electric), and Westinghouse could now use their considerable experimental facilities for making radio telephony a commercial proposition. Up to this point attention had been focused on communications. In 1916 a young employee of the American Marconi Company had anticipated the use of wireless communication to bring music into the home, making radio "a household utility in the same sense as the piano or phonograph." David Sarnoff's idea of a "radio music box" lay dormant until 1920 when Sarnoff, now commercial manager of RCA, persuaded the company to develop it as a commercial product.[59]

When radio began to achieve universal popularity, Edison's secretary, Meadowcroft, ruefully pointed out to his boss that here was "another one of your stunts gotten away from you."[60] This was hardly a fair criticism of the lab for very little effort had been put into wireless experiments. As usual Edison's time had been taken up by more pressing matters. Yet he resisted every effort to push TAE Inc. into the new business. This opposition was not the result of his lack of understanding of wireless technology, nor was it because he failed to see the commercial potential of radio. The number of other inventors in the field discouraged him. Entering into radio research would be "going against the boys on wireless," some of whom had connections with the Edison enterprise or the West Orange laboratory.[61] In terms of the business of innovation, Edison was reluctant to enter a field where the patent situation was complex, and it was unlikely that one company would be able to control all the parts of a radio system. In this judgment he was not alone. Eldridge Johnson of Victor had come to the same conclusion: "The radio industry is going to be fighting over patents for years. We are not going to put out a radio."[62]

Edison's stubborn opposition to radio was the stance taken by the rest of the phonograph industry that would not let their recording stars perform on the air, depriving the broadcasting stations of a pool of popular talent. Edison doubted that radio had a commercial future. He felt that it could never be more than a novelty, forgetting the enormous novelty appeal of his own phonograph, which kept it going during its first difficult years.[63] Just as people in the 1880s were impressed that sound could be reproduced, Americans in the 1920s bought radios because of the wonder of wireless—there was the same excitement in listening to recorded music on the air even though the sound quality was very poor. Edison had trained his ear to hear the near perfect articulation of the phonograph and could not bear the rasping, rumbling roar of the first loudspeakers. Fully convinced that the laws of nature prevented the radio from ever sounding like a musical instrument, he condemned radio by arguing that amplification distorted music. The broadcasting industry shot back that Mr. Edison might be the world's greatest inventor, but he was also deaf.[64]

Edison's opposition to radio is easy to understand. His failure to exploit the potential of electronic amplification in talking machines almost defies explanation. It was clearly destined to play a major role in the future of recorded sound in all forms. It was an accessible technology that was well within the competence of his laboratory staff. All the signals from the marketplace pointed to the popularity of its sound, yet electronic amplification took a back seat to his lifelong quest to perfect acoustic recording. In 1927 he listed his research goals for his son. Electric recording was second on this list; first was, "I want to perfect Super phono."[65]

Radio had a strong supporter in Charles Edison who summoned all the reasons why TAE, Inc. should enter the business: the lab had the expertise to successfully develop a product, the Works had the facilities to mass-produce radio sets, the marketing organization was accustomed to the leisure market, and, above all, the magic name of Edison on a radio would ensure its acceptance by the public. Charles failed to convince his father who had already decided: "I do not intend making radio sets."[66]

The first commercial radio station, KDKA in Pittsburgh, began transmission in late 1920. Its programs of recorded music and advertising announcements were broadcast from the roof of the Westinghouse factory and heard by amateur radio operators on their crystal sets. Westinghouse and RCA worked quickly to put simple single-circuit receivers onto the market.[67] Loudspeakers soon replaced earphones, and more complicated sets with better reception and dignified wooden cabinets were offered for sale. By 1921 there were over 250,000 radio sets in the United States. Yearly sales increased to 400,000 in 1922 and the number of broadcasting stations exceeded 200.[68] The first radio boom had begun. There was a noticeable decrease in the sales of pianos, sheet music, and phonographs.

Edison labeled this social and economic revolution a passing craze and would have nothing to do with it. Worried about the number of companies in the new industry with little technical know-how, he told Charles to stay away from the "radio gang of crooks" who were going to sell inferior equipment.[69] His predictions for radio are worth repeating: "In three years, it'll be such a cut-throat business that no one will make any money. Those large cabinets they are selling now at high prices will be reduced to a little box."[70]

In 1923 the radio boom appeared to be running down. In sound quality there was no comparison with a disc record, and Victor felt confident enough to parody in a popular recording the awful sounds coming out of radio loudspeakers. The talking machine industry hoped that the novelty of radio would soon wear off. But in 1924 improved radios dominated the important Christmas sales season. "Almost everybody immediately bought a radio set," remembered one Victor employee, and all phonograph manufacturers suffered as consumers rushed to purchase radio sets.[71] The main profit maker of TAE, Inc. went into "a terrible slide into the valley of depression," from which it never recovered. The amusement phonograph business lost $1 million in that year and the level of loss was expected to continue.[72]

The manufacturers of talking machines were not going to give up without a fight, and a wave of innovation swept through the declining industry. The desperate situation of the Edison phonograph business required immediate action from the laboratory. Walter Miller reiterated a well-established policy when he told Edison, "Under the present conditions, I feel that we have got to come out with some improvement."[73] In previous crises the lab had improved the recording medium and brought out superior reproducers. This time the response was the development of a louder dance reproducer and a twelve-inch long-playing record, a double-sided disc that was cut with 450 grooves to the inch instead of the normal 150. This problem of crowding more microscopic grooves onto the record had been under investigation at the lab since the 1890s when the first efforts were made to cut 400 grooves to the inch. Soft wax could not maintain the shape of the groove, but Edison had put his faith in long-playing records and continued the experiments. The harder condensite recording medium finally made it possible to keep the stylus from skipping from one microscopic groove to another. The long-playing record was introduced in 1926. Touted as a "revolutionizing invention," the new record would permit the reproduction of entire symphonies in the forty-minute playing time of both sides. In fact the innovation was a complete failure because the long-playing record could not stand up to repeated play and its sound was too faint to please the customer. There was hardly any point in introducing it as the library of compatible recordings did not include complete symphonies or operas that could be played on the new machines.[74]

The Victor Company faced the same decrease in sales as TAE, Inc. and likewise initiated a research and development program to improve their product—with different results. Victor's crisis in 1924 was not merely the result of losing customers to radio, for Eldridge Johnson had made a major marketing error that left his company with a whole line of unwanted machines. Since the introduction of the Victrola in 1906, it had been clear that style sold talking machines as well as sound. The development of different styles of cabinets assumed great importance in the industry. After elaborate period styles reached a high point during the war, consumer tastes turned to horizontal designs with flat tops that made phonographs look like end tables or cabinets. Still married to the vertical format of the Victrola, Victor completely missed the "console craze" and paid a heavy price. Sales declined by 60 percent in 1924.[75] In January 1925 the Victor manufacturing plant at Camden, New Jersey, was closed down as the specter of bankruptcy loomed.[76]

The Victor Company started to cater to radio listeners by introducing machines with space to accept radio receivers made by RCA. It allowed its recording stars to broadcast radio programs. Meanwhile it explored systems of electrical recording that were under development at several laboratories, including RCA, Western Electric, and Westinghouse.

The development of electrical recording—the technology that took the talking machine from the acoustic phonograph of the nineteenth century to the high fidelity of the twentieth—can be traced back as far as 1877, when Edison noted that a telephone repeater could be used to record and reproduce sound. In the 1920s he had continued experimenting in this vein with little success.[77] Several of Victor's engineers had also examined telephone and microphone technology as the basis of a better recording system.[78] In 1924 a team of researchers at Western Electric's laboratory, under the leadership of Joseph Maxfield and Henry Harrison, began to systematically apply their knowledge of telephone amplification to electrical recording. Their work yielded a recording system in which the sound was picked up by a microphone, amplified by a vacuum tube, and then inscribed on the disc with an electromagnetic recorder. The diaphragm in the microphone required only the smallest movement to record the sound that was then amplified to power an electromechanical recording cutter. The result was a diaphragm of uniform response and none of the distortion of the acoustic method.[79]

The breakthrough at Western Electric's lab did not take long to reach the talking machine industry. Representatives from AT&T and RCA had demonstrated the new recording system to the executives of Victor, who remained cool to the new technology until it was combined with an improved system of acoustic reproduction (also designed by Maxfield and Harrison). The dramatic increase in volume, the clear sibilants, and most of all, the amazing reproduction of bass notes of the new Orthophonic machine could not be ignored. Its reproducer was an entirely new design that incorporated a metal spider in the air chamber to accommodate the wide range of frequencies picked up by the microphone. Maxfield and Harrison had taken the development of the exponential horn one step further by devising the reentrant (or folded) horn that increased in size exponentially from the sound box to the outer rim of the horn. The longer passage along the airtight confines of an enlarged sound box and coiled horn (up to seventy-two inches in some models) helped produce a level of acoustic reproduction that was equal to the extended frequency range of electrical recording.[80]

After acquiring the rights to the new technology, Victor acted with a speed that should have made Edison jealous. The first electrical recordings were made at the Victor plant in Camden in February 1925, with the assistance of Maxfield himself. The company introduced nineteen Orthophonic models in the same year, followed by twenty-four more models in 1926. The line included one model that contained both acoustic and electric (electromagnetic pickup and vacuum tube amplifier) reproduction systems, and radio-phonograph combinations. The Orthophonic machines maintained Victor's preeminent position in the phonograph industry and set the standard in acoustical reproduction. They gave the company one of the most profitable years in its history; in 1926 Victor made about $7 million profit, which eradicated the $6,500,000 loss of the previous year.[81]

Victor was soon joined in the new electrical field by the Columbia Company that acquired rights to the Western Electric process. Brunswick, which had been the first of the large talking-machine companies to move into radio, introduced a system of electric recording and reproduction developed by GE and Westinghouse. Of all the leading phonograph companies only TAE, Inc. stayed with the acoustic system. This technical isolationism did not dismay Edison, who was pleased that his was now the only "straight phono" company in the running.[82] While the competition charted new waters, the Edison line remained unchanged; its louder dance reproducer of 1926 was a redesign of the standard floating head format adopted well before the war. Its advertising still referred to the "volume fad"—anything that diverged from Edison's view of the phonograph was still called a fad, even though the managers of TAE, Inc. realized that the only machines that the public would accept were those that had the radio timbre of electric recording.[83] "There must be something to electrical recording," wrote one salesman, doubtless frustrated by the technical conservatism coming from the lab. It was clear that Edison recordings could not match the sound of the competition.[84] Victor and Columbia continued to enjoy high sales and profits in 1927, an indication that the radio craze was not eliminating the demand for phonographs but rather forcing manufacturers to adapt to the new order in recorded music.[85] The phonograph sound was out, and so was the name. Charles Edison had to gently explain to his father that the competition had discredited the name given by Edison to his favorite invention. Victor had successfully promoted the idea that the Orthophonic was a totally new technology and that the phonograph was old fashioned. The new Edison phonograph had to have a new name, preferably ending in -phonic, to attract customers.[86]

Even though the losses of its phonograph business slowly assumed disastrous proportions, TAE, Inc. moved very slowly toward the new electrical technology. Walter Miller and Newman Holland had covered much of the experimental ground (probably in secret or in unofficial trials) by 1927 when the first electrical recordings were made at the West Orange laboratory. The introduction of the Edisonic line with electrical pick up and radio-phonograph combinations did not come until 1928, and by that time it was too late to save the business. TAE, Inc. threw in the towel in October 1929. Manufacture of phonographs and records ceased. All recording was stopped. The inventor of the talking machine had bowed out of another industry he had created. The Great Depression had begun at West Orange.

CHAPTER FIFTEEN

The Last Years

By 1928 Edison's stubborn opposition to radio had been overcome. The board of directors allotted the money to develop a radio and the Splitdorf Radio Company was purchased to give TAE, Inc. access to the RCA pool of patents.[1] Theodore Edison, the youngest son of Thomas and Mina, was put in charge of the development project. A graduate of MIT, he was well qualified for the task. He also had some of his father's energy, for his brother recalls that he worked day and night and produced the Edison radio in a remarkably short time.[2] Theodore had the benefits of a formal education and few of the technological attachments of his father. He devised a system of electrical reproduction for the phonograph, inventing a pickup that could play both hill-and-dale and lateral-cut records—something his father would never have done. The introduction of Edison phonographs that could play the records of the competition was evidence that TAE, Inc. was coming to terms with its lowly position in the talking-machine industry.

The new line of phonographs, the Edisonics, came complete with electrical amplification and radio-phonograph combinations. Advertised as the "Steinway of the radio business," the Edisonic was the vanguard of TAE, Inc.'s march into a new market. Striking wooden cabinets and impressive bulk added to the cost, and the top of the line sold for more than $1,000. "The new voice of the skies" promised the sound quality that had become synonymous with the Edison trademark. Its advertising emphasized Edison's pioneering role in the development of radio and immodestly claimed that it was his research that made it all possible.[3]

TAE, Inc.'s foray into the radio field proved to be disastrous. The technology was adequate, if not exceptional; the problem lay in the design criteria of the new product and the timing of its introduction. The Splitdorf Company had warned its investors that the radio business was a hazardous one, and Edison's pessimism about the profits in radio proved to be correct. Competition was tough in the volatile new industry.[4] By 1925 the radio boom had turned into a bust; even RCA began to experience losses and many of the smaller companies went out of business. The expiration of the audion patents in that year enticed many more into manufacturing, which exerted a downward pressure on prices.[5] By the time that TAE, Inc. entered the market a price war had begun. The trade was now in "little boxes" as Edison had predicted. With no truly outstanding features to differentiate its product at the high end of the market, and unable to compete at the low end where most of the sales were, TAE, Inc. lost about $2 million dollars before it withdrew from the industry.[6]

After the radio debacle the management of the Edison enterprise had to find another product to replace the earnings lost in the declining phonograph business. A committee made the decision to manufacture electrical appliances for the kitchen. A fixed amount was allotted for development, and Theodore Edison was put in charge of designing electric coffee makers, waffle irons, and toasters. In 1926 the board of directors appropriated $20,000 to investigate electrical refrigeration, a paltry sum compared with the money spent by General Motors on this technology.[7] The project began at GM's lab in 1920 and continued through the decade under the direction of Charles Kettering. GM and Du Pont finally pooled their technical talent into a joint operation to produce a new system of refrigeration based on freon.[8]

TAE, Inc. had neither the time nor the money for long term experimental projects; it needed an inexpensive item that could be easily manufactured in the Phonograph Works. It looked for a product in the less challenging area of small appliances.

Thomas A. Edison, Jr. returned to the West Orange lab to work on developing the line of consumer durables. The son of Edison and his first wife, Mary Stillwell, Tom, Jr. had achieved little in a troubled life that was littered with failed businesses and health problems. Edison allowed him to come back to the laboratory in the 1920s. Tom, Jr.'s careful drawings of toaster filaments and heater elements bear a signature quite similar to that of his famous father. These drawings provide a symbolic commentary on the new role of the West Orange laboratory. Tom, Jr. carried out endurance tests of the electrical components of the toaster and continuously monitored repeated toasting of all kinds of bread, hardly the stuff of a heroic invention.[9]

It was a long road from the incandescent bulb to the electric toaster. The only innovative part of the Edicraft appliances was the design of the outer casing. Appearance had become a major consideration in phonographs and it took on an even greater importance in the new consumer durables. Again the marketing strategy was to sell impressive technology in stylish designs to the most affluent households. Subsequently, a free-lance designer was hired to give these products a distinctive look. The toaster was described as resembling "a beautiful metal book" and even the advertising material was a striking example of tasteful art deco.[10] Although these designs were simple and elegant, they were in no way revolutionary. They do not have the flair of the work of Peter Behrens for the German electrical manufacturer, AEG, or the dramatic clarity of Gerhard Marck's designs for coffee makers. The Edicraft line was a conservative rendition of a new style in industrial design.

Advertised as "the final achievement of years of research and experiment," these products took two years to develop, despite the urgent need for a speedy introduction.[11] Their technical features were unimpressive. The selling point of the toaster was that it could simultaneously brown both sides of two pieces of bread.[12] Although they bore Edison's image, his famous signature, and his personal assurance that these were the only toasters and coffee makers developed in his laboratories, the consumer durables did not sell. Nineteen twenty-nine was not a good year to introduce a line of premium-priced kitchen equipment; prices were falling, demand was sluggish, and unemployment was growing.

The management of TAE, Inc. had followed a path of cost-effective, low-risk innovation to failure. The research had been placed within predetermined limits; all expenses had been accounted for; and the result was mundane products with little to justify their premium price. The lack of money had encouraged a policy of safety first in the business of innovation, yet Edison's own career proved that revolutionary products could not be created without some risk.

After leading the way in the nineteenth century, the Edison enterprise was unable to develop the next generation of commercial technologies. Its hopes for the future, such as the talking picture, had all fallen by the wayside, victims of technical problems and conservative financial policies. The laboratory no longer had the resources to bridge two areas of research, as it had done in the 1880s and 1890s. The rapid technical advance in electronics and acoustics had occurred with hardly any change in the physical facilities at West Orange. It was a sign of the times that the recording studios of the complex were no longer the state of the art and that the fruitful research collaborations were between other companies: Victor and Western Electric in electrical recording, or GM and Du Pont in refrigerants.[13] Edison's once great laboratory was not a participant in these efforts. The facilities in West Orange had been run down. When RCA cast around for a manufacturing facility, it was immediately drawn to the large, modern plant Johnson had built in Camden, New Jersey.[14]

What remained of the West Orange laboratory was devoted to engineering support of TAE, Inc.'s manufacturing operation. The facility that had been built to generate new products and to create new businesses was now a thing of the past. In the 1920s there were no new ideas coming from the laboratory. In the nineteenth century Edison's departure from one industry had always been marked by the rapid introduction of a new product and the development of a new business to make up for the losses caused by competition and obsolescence. Ore milling and motion pictures had filled the gap left by the retreat from the electrical industry.

In the twentieth century, retreat was followed by further retrenchment. The replacement products developed in the laboratory were in no way revolutionary. TAE, Inc. now chose to follow the pack rather than strike out into new territory where the risks and the profits were greater.

The West Orange laboratory had become a research laboratory in form rather than substance, a potent symbol of innovation long after the glory days of heroic inventions were over. The image of the great inventor and his famous laboratory had served the Edison companies well. The products of TAE, Inc. were always labeled as products of the Edison Laboratory (or Laboratories) rather than "Made in the Edison Phonograph Works, West Orange." "Every American citizen knows Edison and his work," said one advertisement in 1907, "and therefore feels convinced that the phonograph . . . must be better . . . because Edison made it."[15] The standards of the laboratory were the highest, and the Official Laboratory Model phonograph was advertised as the ultimate in reproduction. Notwithstanding Edison's deafness, dealers told customers that they could hear the same sound quality that the great man had heard and approved in his laboratory. The actual process of innovation, the years of experiments, became part of the promotion of the Diamond Disc—the "supreme effort" of the inventor and his lab staff.[16]

The Edison laboratory existed in the minds of the public as the home of the Wizard, and he never shook this image, still being known as the Wizard of Menlo Park until his death. It irked him when letters arrived at West Orange addressed to the Edison laboratory, Menlo Park. The fact that he moved from Menlo Park and became a great industrialist was lost on many Americans.[17]

Edison's image as a great inventor grew while his laboratory declined. The myth of the heroic inventor who went from rags to riches had been taken to heart by Americans who saw in his life a realization of the American dream. He was regarded as the man who could turn visions into reality. His career was described in a growing number of books and articles. It read like a Horatio Alger story. Edison's long life bridged two Americas: the nineteenth century world of his midwestern upbringing and meteoric rise to fame and fortune, and the modern industrial society of the twentieth century that his inventions had helped create. The appeal of the Edison legend was that it combined the values of rural America with the trappings of an industrial state. By the time of his death he had become "a symbol of our Machine Age."[18]

By the 1920s Edison had taken his place among America's great men. His face was as well known as that of Washington or Lincoln. His annual camping trips with fellow millionaire industrialists Harvey Firestone and Henry Ford captured the public's imagination. Here were the heroes of capitalism enjoying the simple life of camping and hiking, and newsreel audiences were led to believe that these great men acted no differently than the average American. "Genius to Sleep under the Stars" was shown in almost every movie theatre in the country.[19]

The climax of Edison's popularity came with Light's Golden Jubilee of 1929, a celebration marking the opening of Henry Ford's Greenfield Village and a milestone in corporate public relations. What had begun as the electrical industry's commemoration of the fiftieth anniversary of Edison's invention of the incandescent lamp ended as a national tribute to him and his achievements. The famous invention was reenacted in a reconstruction of the Menlo Park Lab that Ford had installed in his village. Ford had taken pains to ensure the authenticity of the lab buildings, even down to shipping the actual soil on which the lab stood from New Jersey to Michigan. The most important dirt, however, the muck and grime that covered the inside of the lab, was left out of the replica. So was the mass of junk that Edison considered vital in the work of innovation.[20] In glorifying the act of invention, Ford had added more myth to the legend of the Wizard of Menlo Park.

In addition to honoring a great man and a great invention, Light's Golden Jubilee was a carefully choreographed public relations exercise by the electrical industry. General Electric and the utilities had come to appreciate the value of a good image and the power of Edison's name, rarely missing an opportunity to promote the myth of the great inventor whose dream of "electric light for everybody" they had fulfilled.[21]

Henry Ford's themes for Greenfield Village in Dearborn stretched back to an idealized America in which ingenuity and enterprise played a significant role. As befitting his mastery of both these distinctly American characteristics, Edison was greeted at the village site by the president of the United States. An engineer and a businessman, Herbert Hoover was a man after Edison's own heart. It was a tragic irony for both men that while they paraded through Detroit, a falling stock market heralded the beginning of the Great Depression. It was doubly ironic for Edison. He had left the electrical industry more than twenty years before Light's Golden Jubilee, and while he was being lionized during the celebrations of an America gone by, his customers were deserting the products based on his traditional values and aesthetics. The competition had successfully labeled his products old fashioned.

The failure of his motion picture and phonograph businesses shows that he was out of step with the consuming public of the twentieth century. His influence on the choice of film subjects tended toward themes that would not provoke censors. The Conquest series of films were advertised as clean and wholesome fare that would entertain and educate, but they would not provide the titillation and violence that sold seats in movie houses.[22] This series was the last great effort of the Motion Picture Division of TAE, Inc. to recapture the audience lost to longer narrative films with great stars. The Conquest films flopped, of course, and it was only the unwholesome melodramas, such as *The Cossack Whip,* that made a profit for TAE, Inc.[23] The same problems were evident in the catalog of Edison phonograph recordings. The closing down of the phonograph operation—a humiliating step for the inventor of the talking machine—while he was being honored as the great innovator is an indication of the gulf that had arisen between the Edison image and reality.

Ironies abounded in the Edison myth. As the originator of industrial research, he had pioneered new business organizations to administer the diverse products coming from his laboratory. TAE, Inc.'s divisional structure was years ahead of its time. Yet he stayed true to the machine shop ethic of personal leadership and was reluctant to hand over control of his business to professional managers. His frequent retirements were followed by sudden returns to West Orange and sharp changes of course from twentieth century enlightened management to nineteenth century capitalism. He hired professional managers and then overruled their decisions, making several critical errors of judgement that were to cost TAE, Inc. dearly. Edison knew how to delegate responsibilities—this was part of his method of innovation—but he found it impossible to let the subordinate do the job without regular oversight and interference. This trait was more pronounced in the administration of his business than in the operation of his laboratory.

Eldridge Johnson's attitude toward professional management was as different from Edison's as the record of Victor's successes in the phonograph business were from TAE, Inc.'s failures. Johnson organized Victor around functional departments: manufacturing, selling, legal, recording, engineering, experimental, and advertising.[24] He put a man in charge of each and let him get on with the job. His motto was, "You can't furnish all the brains yourself, nor can you draw all the pay."[25] Edison's was, "I am the organization." Instead of being the dynamic force at the center of the enterprise, Johnson spent a great deal of his time smoothing relations among the various parts of the administration and the committees where policy was decided. He put a high value on his subordinates in the administration and his collaborators in the laboratory. He paid high salaries to key personnel, such as Leon Douglas of the selling department. Douglas was one of several managers who became vice-presidents and members of the board of the Victor Company.[26] This process was not evident in the Edison enterprise where department heads were low paid, without executive powers, and came and went with alarming rapidity. Where Johnson made his collaborators equals, Edison found it difficult to support another powerful influence in either his laboratories or his companies, preferring to keep to his retinue of the boys who served him without challenging his absolute authority.

It is significant that all the independently minded inventors and engineers employed at the West Orange laboratory soon left. Although open minded in the lab, Edison jealously guarded his power to decide on the business side of innovation. As the machine shop culture gradually diminished at West Orange and the numbers of the boys declined, he became increasingly isolated in the organization. He became more autocratic and his "hire them and fire them" labor policy became more pronounced. Only his son Charles could feel secure in his job. Administrators like Stephen Mambert enjoyed a rapid rise in the organization and suffered an even faster fall. In the 1920s TAE, Inc.'s inability to hold onto valuable managers became a severe handicap. In the 1890s the key defections from the Edison enterprise had been machinists and experimenters. In the twentieth century they were white collar staff, sales managers, and factory superintendents.

Both Johnson and Edison came from the same machine shop background. Johnson's love of experimenting, his many patents, and his rise from machine shop operator to manufacturer was much the same as Edison's. Yet Johnson, unlike Edison, put industrial research into its proper place in the organization. He had Edison's commitment to continually improve the product, and he set up a research department before he incorporated the company, but he never allowed the laboratory to lead the business enterprise.[27] Johnson thought that the most important activity in the phonograph business was selling, and he made the sales departments the strongest in the company, giving them the lion's share of resources and the best personnel.[28]

With Johnson's wholehearted approval, Victor became one of the largest advertisers in American business. His Master's Voice became one of the most memorable trademarks ever devised, and Nipper, the dog listening to the talking machine, one of the most effective corporate promoters.[29] Edison preferred to spend his money in the laboratory and had to be convinced that advertising was not a waste of money. He took promotion seriously for he saw that it held the key to obtaining financial support for his laboratory. Manipulation of the press and the resultant free advertising was one of Edison's great skills. His personal involvement in advertising meant that it became self-promotion or promotion of the work of the laboratory, and this remained the basis of TAE, Inc.'s advertising in the twentieth century. Edison had his own strong views on marketing. He tended to favor technical demonstrations over national advertising and often imposed his own schemes on the advertising department.[30] Throughout the 1920s, when advertising became a major part of American business activities, TAE, Inc. spent only thousands of dollars a year while Victor spent millions.[31]

When confronted with a decline in sales, the first response of TAE, Inc. was to initiate a development program to bring out improvements. Victor often resorted to more advertising. In contrast to Edison's continual dependence on technological answers, Johnson recognised that incremental improvements in the talking machine were not going to prevent the loss of sales to radio. He did not believe that increasing the playing time of recordings would end declining sales, and he was right.[32]

Within TAE, Inc. technological push still maintained a balance with marketing pull in the development of new products, despite the reform of engineering in the laboratory. This is evident in new products such as the Ediscope, a fusion of two partly developed technologies (long-playing record and still projector), that flashed pictures on the front screen of the machine. As this was an attempt to salvage something from two experimental projects that had gone wrong, no thought had gone into its potential market, and the best that TAE, Inc. could do was claim that it would "entertain and educate children." Not surprisingly, few people bought it.[33]

Although the machine shop culture gradually died out at the laboratory, it remained at the center of Edison's work and his style of management. Experimenting was a way of life for him. Unlike the other great industrialists of his day, Edison had no life outside the laboratory. He did not become a philanthropist like Carnegie or found colleges and museums like Rockefeller. He never took a vacation from experimenting because he did not consider it work. His one love in life was messing around in the lab. Technological perfectionism proved hard to eradicate. It played a time-consuming and ultimately destructive role in the talking-picture and sound recording projects.

Edison built the West Orange complex around his laboratory in his desire to make innovation the keystone of his business enterprise. In contrast, Johnson kept his research and development organization small, while allowing the recording and manufacturing parts to build up their own research facilities.[34] In this way he divided the functions of research and production engineering and put the latter under management control well before 1910 when Edison was forced to do so in his reform of engineering. Instead of allowing one central lab to dominate the enterprise, Johnson chose to set up small labs focused on one problem when the need arose. After dividing up a large job over several experimental teams, in the Edison manner, Johnson maintained tight control over them, making sure that they stayed on track.[35] Unlike Edison he had to be sold on the value of an experimental campaign needing capital investment; he resisted the temptation to set up new labs. He argued that the disadvantage of setting up a new laboratory was that its output was dependent on the fertility of one man's brain, and this usually declined over time.[36]

This statement condensed forty years of history of the West Orange laboratory. It had been built in the great inventor's image. Speedy, versatile, and constantly changing its center of attention, the laboratory reflected Edison's most pronounced character traits. Over the years his capriciousness, short attention span, and tendency to spread himself too thinly over too many projects were reflected in the confused work schedule of a facility trying unsuccessfully to do several jobs at once. The laboratory embodied his best and worst characteristics. Edison's ability lay in starting things—experimental projects, new ideas for products, new industries—not in finishing them. The history of the lab was littered with half-finished projects that had not held his attention long enough. Some of these stunts got away from him, as his secretary noted; radio, electric traction, and standard electrical measuring instruments all came to successful fruition in other inventors' laboratories.

The Edison enterprise was regarded in the phonograph industry as technically proficient but not progressive in marketing and promotion.[37] This view was substantiated in the twentieth century when it produced many new ideas but was unable to make them pay in the marketplace. Edison's fate was to open the doors to new industries and show others the way. In this he proved conclusively that pioneering did not pay.

After setting the pace in the early growth stages, Edison's companies were soon overtaken and never regained the place of front runner. This was most evident in the phonograph business where Edison started as the leader, began the twentieth century as one of the Big Three companies (with Victor and Columbia), and gradually slipped back. Even when TAE, Inc. was selling record numbers of machines in the prewar period, Victor did not regard it as a threat.[38] In most of his businesses Edison had to settle for second place. This was the case in the battery and dictating machine businesses. In some industries Edison dropped out completely. The final evaluation of his career as a businessman must come down to the fact that he created more opportunities for others than he did for himself.

Yet it should be remembered that Edison's purpose in setting up businesses was to provide the money for inventing. As long as he had a lab and money for more experiments, he felt he was successful. Perhaps the greatest difference between Johnson and Edison was that one saw experimenting as a means to an end and the other saw it as the end itself. Edison's perception of his business organization followed this priority. He could never see the enterprise as something more than an extension of his laboratory and the laboratory as the extension of his personality. He dominated the proceedings for so long that when old age and ill health forced him to step down, there was nobody in the organization who could take his place.

Retirement

Edison's advancing years slowly loosened his grip on the management of TAE, Inc. As he approached his eightieth birthday in 1927 his health deteriorated alarmingly. He was now totally deaf and increasingly troubled with stomach pains. When he came to the West Orange laboratory, he stayed on the ground floor, working at his desk in the library. In 1926 he relinquished control of the company to Charles Edison, who became president. In the same year Edison sold his remaining patent rights to TAE, Inc. for $78,200.[39] Thereafter, Edison spent more of his time in Florida at his Fort Myers home. He had acquired the property for his winter home in 1885 and gradually built it up over the years, erecting houses, providing it with a pier, and constructing a laboratory. Edison and his wife customarily spent the winter months at Fort Myers. These vacations did not signal a break from experimenting; he equipped his laboratory with everything he needed to continue work and often took assistants to Florida with him when he was in the middle of an experimental campaign. The experimental facilities are exactly what one might expect: machine tools, experimental benches, chemical supplies, a small library, and a storeroom.[40]

With his son Charles running the business affairs of the company, he could spend more of his time experimenting, concentrating his efforts on the recording of sound and the design of an electrical system for automobiles based on his alkaline storage battery. The latter was commissioned by Henry Ford, a friend and admirer who first visited the West Orange laboratory in 1912. Edison attempted to adapt his battery to suit Ford's Model T, and he also worked on a generator and starter motor. Very little progress was made in this experimental campaign because the alkaline battery did not have the power to turn over the engine.

This failure in no way affected Ford's adulation of Edison, and

Ford soon came up with another project to keep the eighty-year-old inventor busy. During the First World War, the United States' dependency on foreign raw materials was underlined to industrialists. Edison was not the only one who had to scramble to find substitutes for materials that were no longer available. Several vital strategic materials were concentrated in a few places, subject to being cut off in wartime. A case in point was rubber, which was largely in British and Dutch hands. In the early 1920s the United States consumed about 70 percent of the world's supply of rubber, which went mainly into the automobile industry. Suspicions that a monopoly of rubber production was in the making led to several unsuccessful attempts to produce synthetic rubber. An alternative solution to this problem was to find another source of organic rubber that could be grown in the United States. Henry Ford, along with Harvey Firestone of the Firestone Tire and Rubber Company, persuaded Edison to look for a domestic source of rubber. The idea was not to substitute American for Malayan and South American production but to ensure a supply of domestic rubber in an emergency. The Edison Botanic Research Company was formed in 1927. Edison, Ford, and Firestone provided the capital.[41]

Edison's interest in rubber had begun when the laboratory first opened in the late 1880s. Several of his competitors in the phonograph business (especially Berliner) were using records made of hard rubber, which offered a durable recording medium. In 1888 Jonas Aylsworth experimented with hard rubber records in the West Orange laboratory.[42] The experiments in rubber continued into the 1900s when rubber became an important part of the storage battery project. Rubber seals were used to contain the chemicals used in batteries. By 1920 the laboratory staff were well acquainted with this material and Edison had a knowledge of the chemistry of rubber that impressed Firestone.[43]

The rubber project began with the customary search of the literature. Edison made several trips to obtain information and examine botanical operations. Many studies had been carried out on native plants that might contain rubber, but Edison wanted to go further and mount a systematic search that would cover thousands of plants, ever hopeful that kindly nature might have provided a substitute for the latex of the Hevea tree. The search for a domestic source of rubber suited Edison's experimental style. It was clearly a line of research that needed an empirical approach, and it combined the adventure of the hunt for bamboo filaments with an opportunity to study botany—a new interest that his friend John Burroughs had stimulated. As it was privately financed, he was able to run the program outside the bureaucracy of TAE, Inc. Finally, the project could be carried out in the balmy weather of Florida. Edison began the search for a new source of rubber with his usual gusto, gathering a library of books on rubber at Fort Myers and planting many different species in the gardens around his home. Age and ill health had not dimmed his intellectual curiosity. He built a new chemical laboratory across the street from his home and soon filled it with supplies and equipment.[44]

With characteristic energy, Edison found time to develop Fort Myers as a garden city, planting a variety of trees and shrubs in the community and lining the road to his estate with tall palm trees from Cuba. He enthusiastically turned the land around his house into a large experimental farm and proudly showed off his results. It was reported that he had 4,800 varieties of shrubs and trees under cultivation. He also had a small army of amateur and professional botanists in the field collecting samples. In a short time his assistants had gathered over 3,000 examples of wild plants and tested them for rubber content. Meanwhile, the flow of specimens collected by well-wishers increased and the lab was inundated with plants.[45]

Much of the testing was done at the laboratory in West Orange. Building 4 was converted to receive the incoming flow of plants gathered from all parts of North America. (Hundreds of the specimens are still stored with the Edison archives.) In the event of finding a source of rubber, he still had to devise an economical production process. Edison was again in his element; he had considerable experience in chemical engineering. He designed a machine to extract the sap from his specimens. The rubber project shows that time had not altered Edison's method. He worked with a small team of experimenters and brought in chemists and botanists when needed. His deafness did not interfere with the management of the experimental team, for written communication proved adequate (fig. 15.1). His experimenters were directed by a steady flow of written notes. Even the men in the field were instructed to keep in constant contact by mail with the West Orange laboratory.[46]

After exhaustive testing, the goldenrod plant was found to offer the most rubber content. Two varieties, *Solidago rugosa* and *Solidago leavenworthii,* were singled out for intensive cultivation and crossbreeding. Edison spent all his time on this project, studying botany in the house at Fort Myers, examining specimens in the special lab built across the street, directing a small staff of chemists

Figure 15.1. The Old Man at work in the chem lab about 1930.

and botanists in the crossbreeding program, and examining the progress of the goldenrod plants in the gardens. He was totally absorbed, some might even say, obsessed, with the work. In 1929 he acknowledged, "I can't get my mind off rubber just now."[47] He was able to produce a strain with high rubber content, but the results did not justify production. Yet the rubber project kept the aged inventor busy and content. The staff at West Orange were instructed not to send him any bad news. He took pride in telling the press that he was busy "getting rubber from weeds," and what could be a more appropriate task for a patriot and a practical man?[48]

The newsreels taken at Light's Golden Jubilee show that age and illness had begun to take their toll. Yet Edison was still determined to keep on experimenting. He had lost none of his enthusiasm for the work and when he came to the lab he still maintained an interest in whatever the men were doing. By 1930 he was confined to a wheelchair but still continued working on the rubber experiments. He normally returned to West Orange from Florida in March. But in 1931 his health was so poor that his return was delayed until June. When his staff asked him what was to be shipped back to New Jersey, a one word answer indicated that the end was near: "Everything."[49] Edison did return to the laboratory in the summer of 1931 and reported for work several times. Some progress was being made in producing rubber from goldenrod and he was well pleased.[50]

In September Edison was confined to his bed and the family began to prepare the funeral arrangements. He died on 18 October 1931. A national outpouring of grief ensued. The great inventor's last days had been followed by the world's press. President Hoover had asked for daily bulletins. There was no question but that Edison should lie in state to receive one last round of homage. His body would lie in the laboratory and not Glenmont, his West Orange home (fig. 15.2). Thousands of people came to pay their last respects. The five brick buildings on Main Street had already begun the transition from laboratory to museum.

The Great Depression was just beginning when Edison died. The first shocks to the stock market had struck while Edison, Ford, and Hoover were parading through the streets of Detroit during Light's Golden Jubilee. Charles Edison had a difficult task ahead of him as chief executive of TAE, Inc. Yet the organization did survive this Great Depression, just as it had survived the Great Depression of the nineteenth century and the depressions of 1907–8 and 1920–21.

Edison senior had turned over a resilient business to the next generation. In 1887 he had looked into the future and envisioned a great industrial complex, telling a financial backer, "I honestly believe that I can build up a works in 15 to 20 years that will employ 10 to 15,000 men and yield 500 per cent to stockholders."[51] Although he did not produce a business enterprise to rival that of Standard Oil or U.S. Steel, he did go a long way in achieving this

Figure 15.2. Edison's Funeral. Lines of mourners wait on Lakeside Avenue to enter the library (in Building 5 on the left) to view the body. The administration building is in the background, left.

goal. At the time of his death, TAE, Inc. was returning a dividend and had an average yearly surplus of $7 million, which was roughly the same as the profit of the Victor Company in the late 1920s. TAE, Inc.'s indebtedness of over $2 million, acquired during the war years and postwar depression, was fully discharged.[52]

The divisional structure of TAE, Inc. had helped it weather the storm of the depression and the decline of its core phonograph business. Edison's policy of diversity, established in the 1890s, proved to be the salvation of the company, for as one product went into decline, another took its place. The storage battery and dictating machine kept TAE, Inc. solvent in the 1930s and made up for the loss of the phonograph business. The Instrument Division matured in the 1940s and helped arm the United States during the Second World War, providing a basis for the broad range of electronic and precision products made in the 1950s at West Orange.

Edison's contributions to the quality of life in the industrial West have been acknowledged. While he was alive, the U.S. Congress passed a resolution expressing the nation's thanks to the great inventor "whose conquests in the realm of science have enriched all human life."[53] After his death a host of tributes underlined his pioneering role in many of the wonders of modern industrial society, making Edison the "father" of everything from electronics to charcoal barbecuing.[54]

His contributions to American business have received much less attention. To Edison must go the honor of inventing the idea of industrial research—a permanent research establishment devoted to improving old products and finding new ones. By the 1920s TAE, Inc. could say that "invention is an organized service from the Edison laboratories."

Edison originated a method of managing industrial research and organizing a business enterprise around the flow of new ideas and products coming from the lab. He was the pioneer of a diversified business based on industrial research. By the time that TAE, Inc. was formed, he had established an enterprise based on five product lines: musical phonographs, dictating machines, primary batteries, storage batteries, and cement. "The strength of our situation," said Charles Edison, "was that such a diversified business could ride out depressions and competition."[55] This was the strategy adopted by companies such as Du Pont, General Electric, and RCA. According to RCA, "Research and diversification have always been RCA Victor's planned investment for the future."[56]

As a manager and organization builder, Edison was not without his failings; the diversification policy he pioneered was both the making and the undoing of his business enterprise. His financial and laboratory resources were always stretched to the limit. This prevented him from completely dominating one business. Yet up to the 1920s, Edison always managed to recover from business disasters by moving into a new field.

In one achievement his idea and its execution were perfect. He is unquestionably the man who sold the idea of industrial research to both businessmen and customers. The West Orange laboratory was the place where experimenting was elevated to the high plane of improving the quality of life while providing unlimited opportunity to entrepreneurs and manufacturers. It took some years before its output was dignified with the name of technology—a word absent from the vocabulary of Edison's laboratory and machine shop. At the time of his death, the landscape of American business was dotted with hundreds of research laboratories. In 1931 there were

1,600 industrial research laboratories with a combined staff of 32,000. There were famous research institutions, such as Bell Labs, and smaller affairs, such as the laboratories of Kellogg's cereals, Hershey's chocolate, and Jell-O.[57] The message of the sponsors of industrial research was a simple one: technology was good, technology was controllable, and most of all, industrial research was profitable. Articles appearing on the subject had titles such as "Research—the Doctor of Business," "Industrial Science—A Gilt Edge Security," "Industrial Progress Made through Research," and "Research the Beacon of Progress."[58]

This idea was Edison's greatest contribution to the electric utility industry, along with the magic of his name that embodied quality, safety, and reliability. "The use of the name EDISON in our business relations and corporate title," said an executive of the New York Consolidated Edison Company, "is of inestimable value. Everyone is familiar with the name, his accomplishments, and the standard of apparatus, equipment, of service, and performance implied."[59] In the days of introducing the electric light, he learned that the public would buy a radically new product if a familiar face and an international reputation stood behind it. He had a craftsman's pride in his work, especially as it bore his name. He took a personal responsibility for his companies and their products. In his later years, his white hair and craggy, kindly features put a benign face on new technology that often came with some risk, giving substance to his company's proud claim that "Edison has invented no machine that will do you harm. All of his inventions are for the benefit of humanity." TAE, Inc. spoke for America when it proclaimed "Trust Edison for Progress."[60]

Epilogue

The West Orange laboratory continued to carry out research for TAE, Inc. during the 1930s and 1940s, although the laboratories in the divisions did most of the work. The Annual Report of 1948 stressed that "research assures progress and quality."[1] In 1952 a new central laboratory for the company was planned, and a modern facility was built on nearby Watchung Avenue. The new laboratory was an unexceptional, one-story building in the international architectural style. It contained nine experimental rooms and began with only twenty-eight employees—serious young men with white coats and clipboards. In the publicity photographs published by the company, the experimental rooms appear clean, quiet, and uncrowded. Industrial research had come a long way in seventy years.

Much of the original laboratory on Main Street had been left unchanged as a tribute to Edison. In 1956 TAE, Inc. gave it to the U.S. government and a few months later the company was merged with the McGraw Electric Company.[2] McGraw Edison, as it was then called, slowly dismantled the West Orange complex. Today all the factory buildings are gone except the storage battery plant across the road from Building 5. This has been revived as a conglomeration of small businesses, now called the West Orange Edison Industrial Center. Some of the factory space is still used for manufacturing, concentrating on textiles, transformers, and bicycle parts, but most of it is devoted to wholesale outlets that sell things like raffia furniture and wallpaper. Immigrant workers, this time from the Caribbean and Asia, still wait for public transit outside the Works. Around the block the modern laboratory is deserted and unsold—a sad commentary on the industrial decline of the Northeast. The old Edison laboratory stands alone, still giving the mistaken impression that it was a free-standing building and not part of a great industrial complex. The energy and activity that once permeated the site are long gone, and instead a feeling of depression, both spiritual and economic, pervades the area. Yet to those who have visited the Edison National Historic Site, the spirit of the Old Man lives on; if not in the hallways and machine shops, then in the wide eyes of the children as they are shepherded through the laboratory.

Notes

Chapter One. The Largest Laboratory Extant

1. The story of Edison's work at Menlo Park is told in Robert Friedel and Paul Israel, *Edison's Electric Light: Biography of an Invention* (New Brunswick, N.J.: Rutgers University Press, 1986).

2. The Paris Exhibition of 1881 began the diffusion of the Edison system in Europe, and the equipment of the Edison display at Paris became the basis of the French and English Edison companies.

3. Leonard Reich, *The Making of American Industrial Research: Science and Business at GE and Bell, 1876–1926* (New York: Cambridge University Press, 1985), p. 45.

4. Edison chose the "boys" for positions in the new electrical companies and gave them their orders without much ado: his list for the Edison Machine Works covers only one page and includes experimental jobs to be carried out at Schenectady and West Orange. Undated document in 1886 document file, Edison National Historic Site (henceforth cited as ENHS).

5. George Prescott (*Dynamo Electricity* [New York: Appleton, 1888], p. 228) estimates that there were 164 Edison installations in the United States in 1886 and 142 in Europe. Harold Passer (*The Electrical Manufacturers, 1875–1900* [Cambridge: Harvard University Press, 1953], p. 121) estimates 121 central stations in October 1887. The European installations were in Western Europe, Italy, and Russia. The figure for capitalization is from Matthew Josephson, *Edison: A Biography* (New York: McGraw Hill, 1959), pp. 300–301.

6. Robert Stern, *Pride of Place: Building the American Dream* (New York: Houghton Mifflin, 1986), p. 129.

7. Edison's biographers include Matthew Josephson; Robert Conot, *A Streak of Luck* (New York: Seaview, 1979); and Frank L. Dyer and Thomas C. Martin, *Edison: His Life and Inventions* (New York: Harper and Row, 1910).

8. This is the view of Josephson (*Edison: A Biography*, p. 133). David Noble quotes Norbert Wiener of MIT, "Edison's greatest invention was that of the industrial research laboratory." David Noble, *America by Design:*

Science, Technology, and the Rise of Corporate Capitalism (New York: Alfred A. Knopf, 1977), p. 113.

9. Thomas A. Edison (TAE) to Hood Wright, Nov. 1887. Laboratory Notebook N 871115 (henceforth these books cited as N plus number).

10. *Measuring Invisibles* (Newark: Weston Electrical Instrument Co., 1938), pp. 14–17, Edward Weston Papers, box 353, Special Collections, New Jersey Institute of Technology (henceforth this collection cited as NJIT).

11. The *Daily Herald* (28 Oct. 1887) called Weston "the new competitor of the great Edison," and claimed that Weston was a poor boy who rose to fame as a genius and that his lab was the most extensive of its kind in the United States. *The Manufacturer and Builder* (Dec. 1887) saw Weston as a true inventor, "not a haphazard inventor, but a man of science who applies the careful, thorough, laborious methods of science." NJIT.

12. *Newark Call* (7 Nov. 1886) called it "the most complete private lab in the United States." Douglas Eldridge, "Landmark Disappears," *Newark Sunday News*, 3 Jan. 1965, NJIT.

13. *The New Yorker*, 18 Dec. 1887, NJIT. In this interview Weston pointedly announced that he would have nothing to do with stock dealings. When asked if he was going to invent a certain type of exotic dynamo under experiment in Edison's lab, he responded: "No. I leave all those extraordinary things to Mr. Edison."

14. In a personal recollection, an associate noted that Weston was a taciturn man, but "the key to unlocking his tongue . . . I found to be the mention of Edison." The bad feeling seems to have come from Weston's annoyance at Edison's getting credit for many of his inventions. Letter of William H. Onken, 25 April 1938, box 53, NJIT.

15. Louis M. Hacker, *The World of Andrew Carnegie* (New York: J. P. Lippincott, 1968), p. 349.

16. J. P. Constable to product engineers, 11 Sept. 1917, Laboratory Blue Book, ENHS. (These are collections of related documents, probably pulled for the personal use of Charles Edison, and bound together in blue books.) An important analysis of Edison's method is Thomas P. Hughes, "Edison's Method," in William B. Pickett, ed., *Technology at the Turning Point* (San Francisco: San Francisco Press, 1977), pp. 5–22.

17. The first important innovation in armature design was the work of the Italian scientist, Antonio Pacinotti. His slotted ring armature of 1869 was a breakthrough, but the scientific paper describing it languished in the pages of an obscure scientific journal for ten years, unnoticed by the engineers who were trying to make a better armature. Malcolm Maclaren, *The Rise of the Electrical Industry during the Nineteenth Century* (Princeton: Princeton University Press, 1943), p. 114. The lab had this journal and loaned it out. Seely to TAE, 30 Dec 1892, ENHS. The first commercial dynamo that owed something to Pacinotti was the work of the Belgian, Theophile Gramme. His dynamo was obtained by Edison and examined in the laboratory.

18. Edwin Layton, "Mirror-Image Twins: The Communities of Science and Technology in 19th Century America," *Technology and Culture* 12 (Oct. 1971): 567, notes the vast increase in written technical information available in the late nineteenth century. The *Electrician* (18 [July 1888]: 341) described the library. A lab notebook (N 880130) was used to record loans of books to lab staff.

19. The quote is from "The Adventure of the Abbey Grange," *The Strand Magazine* 38 (Sept. 1904): 497. At the same time that Edison started work at West Orange, Conan Doyle was publishing his first Sherlock Holmes story, which appeared in *Beeton's Christmas Annual* (1887).

20. Reginald Fessenden, "The Inventions of R. A. Fessenden," *Radio News* 7 (Aug. 1925): 156.

21. *Electrical Review* 21 (Oct. 1887): 391.

22. Edison accounts, vouchers #1–601, voucher series, 1886–87, ENHS.

23. Maurice Holland, "Edison: Organizer or Genius," National Research Council, April 1927, Edison Papers, box 12, folio 7, Archives of Edison Institute, Dearborn, Michigan.

24. Lab notebooks for 1887, reserve microfilm reel 279. These reels of microfilm were copies made of lab notebooks and minutes of meetings (henceforth cited as RR).

25. TAE to Hood Wright, Nov. 1887, N871115, ENHS.

26. Dyer and Martin, *Edison: His Life,* pp. 648–50.

27. TAE to Hood Wright, Nov. 1887, N871115, ENHS.

28. John Rae, "The Application of Science to Industry," in A. Oleson and J. Voss, eds., *The Organization of Research in Modern America* (Baltimore: Johns Hopkins University Press, 1979), pp. 249–63; Robert V. Bruce, *Alexander Graham Bell and the Conquest of Solitude* (Boston: Little, Brown and Co., 1973), pp. 340–45; Conot, *A Streak of Luck,* p. 246.

29. *Scientific American* 57 (5 Nov. 1887): 287–90.

30. Darwin H. Stapleton, "Early Industrial Research in Cleveland," manuscript, April 1987.

31. W. Bernard Carlson, "Invention, Science, and Business: The Professional Career of Elihu Thomson" (Ph.D. diss., University of Pennsylvania, 1984); Noble, *America by Design,* p. 113.

32. George Gouraud, "The Phonograph," *Journal of the Society of Arts* 36 (Nov. 1888): 33.

33. List of experimental projects, Exhibit A, 1888 West Orange laboratory document folder, ENHS (henceforth cited as lab folder, ENHS).

34. Draft laboratory agreement with Henry Villard, 1888 lab folder, ENHS.

35. Wyn Wachhorst (*Thomas Alva Edison: An American Myth* [Cambridge: MIT Press, 1981], pp. 34–35) describes the role of science in Edison's image. Reich uses a science-based definition of industrial research in *The Making of American Industrial Research,* p. 3.

36. This is Hugh Aitken's conclusion. *The Continuous Wave: Technology*

and American Radio, 1900–1932 (Princeton: Princeton University Press, 1985), p. 533. In deciding if Reginald Fessenden, one of Edison's experimenters, was a scientist or an engineer, Aitken argues that Fessenden had a foot in each camp. See also David Hounshell, "Edison and the Pure Science Ideal in America," *Science* 207 (Feb. 1980): 612–617.

37. Quoted in Bob Coleman, "Science Writing," *New York Times Book Review* (27 Sept. 1987); see also David S. Landes, *The Unbound Prometheus* (New York: Cambridge University Press, 1969), p. 235.

38. Silvanus Thompson, *Dynamo-Electric Machinery* (New York: F. N. Spoon, 1886), p. 20.

39. "The master cut-and-try genius of his day," is John Rae's comment on Edison in Oleson and Voss, *Organization of Research,* p.259.

40. George Wise comments on Edison: "His use of trained scientists was as limited and selective as his use of science." *Willis R. Whitney, General Electric, and the Origins of U.S. Industrial Research* (New York: Columbia University Press, 1985), p. 68.

41. Layton, "Mirror-Image Twins," pp. 562–580.

42. When Fessenden asked Edison if he could publish some of the theoretical work done on electrical insulation and conductors, Edison refused, saying that he had not finished experimenting on it yet and so did not "want anything said at present." Fessenden to TAE, 16 April 1893, and annotation, ENHS.

43. *Electrical World* 12 (Nov. 1888): 238.

44. Arthur Kennelly to Dr. Muirhead, 28 Dec 1888, Kennelly Reports (a series of Kennelly's letterbooks and notebooks), 060788, p. 229.

Chapter Two. The Machine Shop Culture

1. Charles Edgar ("An Appreciation of Mr. Edison," *Science* 75 [15 Jan. 1932]: 60) noted that many worked in the Edison organization for "glory and not for pay."

2. David Trumbull Marshall, *Recollections of Edison* (Boston: Christopher, n.d.),p. 60.

3. George Creel, "Memoir," 24 Aug. 1916, Naval Consulting Board (NCB) records, box 4, ENHS.

4. This was John Ott's impression, cited in Josephson, *Edison: A Biography*, p. 88.

5. David Montgomery, *The Fall of the House of Labor* (New York: Cambridge University Press, 1987), p. 192. An editorial in *American Machinist* (14 [21 May 1891]: 8) agreed with many machinists that shop rules contributed nothing to the success of the operation.

6. M. A. Rosanoff, "Edison in His Laboratory," *Harper's Monthly* 165 (Sept. 1932): 402.

7. My understanding of the machine shop culture comes from Monte A. Calvert, *The Mechanical Engineer in America, 1830–1910* (Baltimore: Johns

Hopkins Press, 1967). Craft practices in New Jersey are described in David Bensman, *The Practice of Solidarity: American Hat Finishers in the 19th Century* (Urbana: University of Illinois Press, 1985), and in Herbert Gutman, *Work, Culture and Society in Industrializing America* (New York: Vintage Books, 1976).

8. Edwin Gabler (*The American Telegrapher: A Social History, 1860–1900* [New Brunswick, N.J.: Rutgers University Press, 1988], pp. 79–85) describes "the unique craft culture of the Knights of the Keys," including their sense of community, special vocabulary, social organizations, and "hazing freshmen."

9. Montgomery, *House of Labor*, pp. 16–17, 25; *American Machinist* (11 [11 Feb. 1888]: 4) argued that "men in mechanical life should have the same dignity in their calling as men in professional life."

10. Quote is from Lillian Hoddeson, "The Emergence of Basic Research in the Bell Telephone System 1875–1915," *Technology and Culture* 22 (July 1981): 517. For a detailed study of innovation in telegraph machine shops, see Paul B. Israel, "From the Machine Shop to the Industrial Laboratory: The Changing Context of American Invention, 1830–1920" (Ph.D. diss., Rutgers University, 1989).

11. Eldridge Reeves Johnson, "The History of the Victor Talking Machine Company," Feb. 1913, Eldridge Johnson Papers, box 8, Experiments and History, American Heritage Center, University of Wyoming, Laramie (henceforth cited as AHC).

12. Richard N. Current, *The Typewriter and the Men Who Made It* (Urbana, Ill.: University of Illinois Press, 1954), p. 10.

13. Josephson, *Edison: A Biography*, p. 85.

14. *New Jersey's Leading Manufacturing Centers* (Newark: International Publishing, 1887), p.47. Susan Hirsch (*Roots of the American Working Class: The Industrialization of Crafts in Newark, 1800–1860* [Philadelphia: University of Pennsylvania Press, 1978], p. xix) claims that Newark was the "leading industrial city in the nation" in 1860 but does not indicate by what measure.

15. Report and catalog of *Newark Industrial Exhibit* (Newark: Holbrooks, 1872), p. 27, New Jersey Historical Society Archives, Newark (henceforth cited as NJHS).

16. Hirsch, *Crafts in Newark*, p. 103. The population of Newark rose from 6,507 in 1820 to 71,941 in 1860. Breakdown of national origin of mechanics is given in New Jersey, *Annual Reports of the Bureau of Labor Statistics*: 8th (1885), p. 166; 10th (1887), p. 346. Germany, England, and Ireland are mentioned most frequently.

17. The Newark Public Library has detailed maps of the city in the 1870s and 1880s.

18. Rowland T. Berthoff, *British Immigrants in Industrial America, 1790–1950* (Cambridge: Harvard University Press, 1953), p. 44. Among many new industries in New Jersey, British immigrants founded the pottery concerns of Trenton, transferring techniques and work practices, pp. 75–76.

19. Details of the activities of the Newark shops are in *Commerce, Manufactures, and Resources of Newark, New Jersey* (Newark: National Publishing Co, 1881); *Holbrook's Newark City Directory*, 1872–75, Newark Public Library.

20. Comment of a machinist recorded in the New Jersey, 6th *Annual Report of the Bureau of Labor Statistics* (1883), p. 3. Gutman (*Work, Culture and Society*, pp. 231–32) provides several examples of upward mobility. "Is there a mechanic worthy of his name who has not the hope of someday owning a shop of his own?" argued a mechanic in *Northwestern Mechanic*, cited in Calvert, *Mechanical Engineer*, p. 192. Susan Hirsch (*Crafts in Newark*, p. 10) points out that artisans were the pillars of the community in Newark. As they expected to be masters themselves one day, they acted accordingly.

21. *New Jersey's Leading Manufacturing Centers* (Newark: International Publishing, 1887), pp. 39, 49, NJHS.

22. Statistics for Newark Industrial Exhibit, 1872; *Holbrook's City Directory*, 1872–73, Newark Public Library.

23. See David Montgomery, *Worker's Control in America* (New York: Cambridge University Press, 1974), p.11.

24. Merritt Roe Smith, *Harper's Ferry Armory and the New Technology* (Ithaca: Cornell University Press, 1977), pp. 239–40.

25. Johnson, "History of Victor," p. 4.

26. Ibid., p. 2.

27. E. P. Thompson (*The Making of the English Working Class* [New York: Vantage, 1963], p. 241) called the tramp "the artisan's equivalent of the Grand Tour." W. J. Rorabaugh (*The Craft Apprentice* [New York: Oxford University Press, 1986], p.6) describes the tramp's educational role; *American Machinist* (11 [6 Oct. 1888]: 7) took the view that travel should be encouraged among young mechanics, "as part of a system of industrial education." It noted that it was "not so easy to walk in and out of shops as a visitor as it used to be."

28. Roe Smith, *Harper's Ferry Armory*, pp. 245–47. On the diffusion of technical knowledge in machine shops, see Nathan Rosenberg, "Technological Change in the Machine Tool Industry," *Journal of Economic History* 23 (1963): 414–46.

29. Hoddeson, "Basic Research in Bell System," p. 517, describing Williams' shop.

30. Here is one example. Charles F. Beers apprenticed as a machinist at the Allen Manufacturing Co. in Norwich, Conn. He came to Newark and worked for a concern in its large jewelry industry as a machinist and toolmaker. He moved to a shop manufacturing fare registers for horse cars. He was making patent models for sewing machines (presumably in another shop) when he was personally recommended to Weston. Although Beers had no electrical experience, he quickly added these new skills to his repertoire, making lamps and other equipment in Weston's factory. Defen-

dant's Record, C. *Brush v. J. Owen*, pp. 7–12, in Weston Papers, box 18, NJIT.

31. For an example of how printers adapted craft culture, see William S. Pretzer, "The Printers of Washington, D.C., 1800–1880: Work Culture, Technology, and Trade Unionism" (Ph.D. diss., Northern Illinois University, 1986).

32. Francis Jehl, *Menlo Park Reminiscences* (Dearborn, Mich.: Edison Institute, 1936), 2: 858.

33. Describing the fraternization of German and English soldiers in No Man's Land during World War 1, one soldier remembered: "We mucked in all day with one another." Quoted in Peter Mason, *Blood and Iron* (New York: Penguin, 1984), p. 84; Montgomery, *House of Labor*, p. 25; David Montgomery, *Workers' Control in America* (New York: Cambridge University Press, 1974), pp. 13–14.

34. *American Machinist* 12 (5 Dec. 1889): 6.

35. In this case the table was his workbench and the food brought down from Glenmont or from his house in Menlo Park. The master gave food and board to his apprentices in the European tradition. With all his apprentices around him in front of the fire, he led "a sort of debating club," as W. J. Rorabaugh puts it (*The Craft Apprentice*, p. 102). Edison did the same thing after his late night feasts in the invention factory.

36. Jehl, *Menlo Park Reminiscences*, 2: 858.

37. The classic study of this issue is E. P. Thompson, "Time, Work-Discipline and Industrial Capitalism," *Past and Present* 28 (Dec. 1967), 56–97.

38. Roe Smith, *Harper's Ferry Armory*, pp. 272–75 on the 1842 "Clock Strike."

39. Quote is from Marshall, *Recollections of Edison*, p. 89. Comments on the all-night work at the lab can be found in the Columbia Oral History Project of Edison's lab workers on file at ENHS; and Mary Childs Nerney's notes for a book about Edison, which were based on interviews with laboratory staff and are in N 281101. Edison was one of those lucky people who can sleep at any time. He catnapped through the day, and this probably enabled him to stay up all night.

40. Arthur Kennelly noted, "Edison is very courteous to visitors and lets them, as a rule, experiment upon everything." He reported that there were several instances of this happening and Edison letting costs run up. Hearing on the electrocution of William Kemmler, New York State Court of Appeals, 1847–1911, 893 (Buffalo, New York, 1890) 2: 751.

41. Marshall, *Recollections of Edison*, p. 59.

42. Edgar, "Appreciation of Mr. Edison," p. 61

43. Montgomery, *House of Labor*, pp. 54–57. James Green (*The World of the Worker: Labor in Twentieth Century America* [New York: Hill and Wang, 1980], p.9) points to "a kind of defiant egalitarianism which put the producer craftsmen on the same plane as the capitalist manufacturer."

44. John T. Cunningham, *Newark* (Newark: New Jersey Historical Society, 1966), p. 178.

45. Edison's testimony in the complainant's record, *Edison v. American Mutascope Co.*, 1910, p. 119, legal box 173, ENHS.

46. This was written inside a notebook for 1871, cited in Josephson, *Edison: A Biography*, p. 90.

47. Rosanoff, "Edison in his Laboratory," p. 406: "Nothing that's any good works by itself, just to please you, you got to make the damn thing work."

48. Jehl, *Menlo Park Reminiscences*, 2: 336 on Batchelor and Edison together. Edison's comment on his assistants: "Without their aid I would have been tormented and hampered and delayed by details," cited by John Winthrop Hammond, "The Edison Pioneers," *The Mentor* 16 (June 1928): 10. The construction and altering of models in the electric light project are described in Friedel and Israel, *Edison's Electric Light.*

49. TAE to Bergmann, 14 Dec. 1910, LB 101208, ENHS.

50. Marshall, *Recollections of Edison*, p 61.

51. F. R. Welles to Dr. Jewett, 17 Feb. 1932, ENHS.; (The Edison) *Phonogram* III (June 1893).

52. Rosanoff, "Edison in his Laboratory," p. 411.

53. TAE to Bergmann, 28 Dec. 1904, Letter Book LB 040727, p. 220, ENHS (henceforth letterbooks cited as LB number then page number).

54. In considering whether to promote an experimenter, A. Wangemann, to a management position, Edison and Tate initially decided against him because "he experiments too much . . . He does not seem to get results." Tate to TAE, 9 Jan. 1893. Wangemann had some formal education in Germany and was known as "professor" by the muckers.

55. TAE annotation on letter from the University of Illinois, 12 Feb. 1920, ENHS.

56. Jehl, *Menlo Park Reminiscences*, 2: 858.

57. Bergmann to TAE, 3 April 1888, ENHS.: "You asked me to send a good man if one comes around."

58. Complainant's record, *C. Brush v. J. Owen, P. Owen, J. T. Brush*, 1885, p. 274, Weston Papers, box 17, NJIT.

59. Reginald Fessenden, "The Inventions of Reginald A. Fessenden," *Radio News* 6 (June 1925): 2218.

60. Ibid.

61. Edison to Tate, 21 March 1889, ENHS.

62. Holland, "Edison: Organizer or Genius," p. 7.

63. Montgomery, *House of Labor*, p. 189.

64. Marshall, *Recollections of Edison*, p. 57.

65. Frank to Meadowcroft, 10 Aug. 1922, ENHS.

66. TAE to Tate, 1888; Mary Childs Nerny's notes in N 281101; Recollections of Richard G. Berger, *Beacon Herald*, 30 Aug. 1937, ENHS.

67. Marshall, *Recollections of Edison*, p. 59.

68. TAE to F. Applegate, 26 Dec. 1916, LB 160913, p. 606, ENHS.

69. Some of the few surviving wage records for the laboratory (not the machine shop) are in the 1890 laboratory payroll folder. Edison was notorious for paying his men the lowest possible wages. There can be no doubt that he capitalized on his fame and exploited his prestige by paying low wages to those who wanted the honor of working with him. His practice of using foreign workers has been interpreted as part of this policy to pay low wages. The practice in American industry was to pay foreigners less than Americans. George Wise reports that GE's lab paid 50 percent more to Americans. The complex wage structure at West Orange makes it impossible to determine if Edison discriminated against foreign engineers. Edison was concerned with keeping labor costs as low as possible and getting the most out of his laboratory employees. He wrote in 1917: "The men I use receive only ordinary workmen's salaries." But he expected them to work harder than the ordinary worker. TAE to N. Baker, 30 Aug. 1917, NCB box 21, experiments, ENHS. The annual reports of the New Jersey Bureau of Labor Statistics provide some means of comparison: In 1886 yearly wages for a skilled machinist ran from $420 to $768 a year ($10 to $16 per week); in 1897 from $7.50 to $21 per week; in the late 1880s and early 1890s around $12 per week. (20th Annual Report [1897], p. 78).

70. TAE to Villard, "Mr. Edison's Reply to Thomson-Houston." 23 March 1889, Villard Papers, Harvard Business School, ms. 78, box 63.

71. Mary Childs Nerney's notes in N 281101, ENHS. See also Mary C. Nerney, *Thomas Alva Edison: A Modern Olympian* (New York: H. Smith and R. Haas, 1934).

72. Francis Upton, quoted in Friedel and Israel, *Edison's Electric Light*, p. 138. The lab acted as a school for instruction in new technology, Tate to C. Cheever, 31 Aug. 1888, ENHS.; A. O. Tate, *Edison's Open Door* (New York: E. P. Dutton, 1938), p. 175: "We had a number of young men under instruction at the laboratory for graduation as experts in the operation of the phonograph."

73. Interview with A. Kennelly by C. Withington, Edison pioneers file, ENHS.

74. Fessenden to TAE, 15 June 1915, ENHS.: "I attribute what luck I have had in the inventing field and scientific line to the fact that you taught me the right way to experiment." A. O. Tate agreed with this sentiment: Tate to TAE, 9 July 1920, "Above all you taught me how to think . . . you taught me not to be afraid of failure, that scars are sometimes as honorable as medals."

Chapter Three. The Business of Innovation

1. TAE to M. Platt, 24 Feb. 1911, LB 101218, p. 612, ENHS.

2. TAE to A. Sheetz, 10 July 1908, LB 080601, p. 280. In a note to M. Van Vechter Edison said, "One of the most difficult things in the world is to sell a patent without being cheated." 21 Oct. 1908, ENHS.

3. Article in *Munsey's Magazine*, 20 July 1901; LB 100521, p. 435, ENHS.

4. Insull's notes in Meadowcroft box (MB) 68, ENHS. (These boxes contain material Meadowcroft collected for use in his biography of Edison.)

5. Vincent P. Caruso, *The Morgans: Private International Bankers, 1854–1913* (Cambridge: Harvard University Press, 1987), pp. 270–73.

6. TAE to J. Hood Wright, Nov 1887, N871115. Hood Wright, a Scotsman, moved from Drexel and Co. in Philadelphia to New York in 1874. He became a trustee of the Electric Light Company and a board member of Edison General Electric. Caruso, *The Morgans,* p. 168.

7. Meadowcroft manuscript, box MB 68, p. 53, ENHS.

8. Dietrich G. Buss, *Henry Villard: A Study of Transatlantic Investments and Interests* (New York: Arno, 1979), p. 198.

9. TAE to Villard, 19 Jan. 1888, ENHS.

10. Marshall, *Recollections of Edison,* p. 58; Fessenden, "Inventions of R. A. Fessenden," *Radio News* 7 (Aug. 1925): 156. ENHS.

11. Among the investors in the phonograph company were C. A. Cheever, Uriah C. Painter, and Gardner G. Hubbard, all prominent in the telephone industry. Josephson, *Edison: A Biography,* p. 172; Bruce, *Alexander Graham Bell,* p. 253.

12. The ore-milling agreement is in S. Eaton to Tate, 27 June 1889, H. F. Miller file 121, ENHS.

13. Draft Agreement with Villard, 1888 laboratory folder, ENHS.

14. This interpretation is not shared by several management writers, including Peter Drucker, who see Edison as an inventor who wanted to become a businessman. Edison's version is much different. He claimed, "It was never my intention to act as a capitalist, except as to pioneering my inventions experimentally, which requires a great amount of money." TAE to Ore Milling Syndicate, 19 Feb. 1901, LB 000608. When asked, "Did you ever invent a thing because there was money in it or for the sake of creating new things for mankind?" he answered, "I always invented to obtain money to go on inventing." Handwritten answer to list of questions submitted by Dudley Nichols of *New York World,* 8 May 1929, ENHS.

15. *Brooklyn Eagle*, 2 Aug. 1891, ENHS.

16. It was, in fact, one of Edison's laboratory staff, Reginald Fessenden, who saw the news about Hertz's experiments and asked the boss if he could carry out the experiments. Hugh Aitken, *Continuous Wave,* p. 29. Edison read Hertz's book on electrical oscillations in 1893 and "realized the fact that the fundamental principles of aerial telegraphy had been in his grasp." Meadowcroft to G.A. Walker, 25 Feb. 1919, ENHS.

17. Exhibit A, an 1888 list of experimental projects, 1888 laboratory folder, ENHS.

18. Quoted in Josephson, *Edison: A Biography,* p. 428. Ford's judgment has been picked up by other writers who see Edison as an unsuccessful businessman. See for example John S. Morgan, *Managing Change* (New York: McGraw Hill, 1972), pp. 50–51.

19. Peter F. Drucker, *Innovation and Entrepreneurship* (New York: Harper and Row, 1986), pp. 12–13.

20. Josephson, *Edison: A Biography*, p. 89.

21. William H. Ford, *The Industrial Interests of Newark* (New York: Van Arsdale, 1874), p. 233.

22. TAE to Hood Wright, Nov. 1887, N871115, ENHS.

23. *The Electrician* 21 (July 1888): 330.

24. Sprague's story is told by his wife, Harriet Sprague, in *Frank J. Sprague and the Edison Myth* (New York: William Frederick, 1947). As the title suggests, Mrs. Sprague felt that Edison had deliberately taken credit for her husband's work. Frank Hedley (*Frank J. Sprague: 75th Anniversary* [New York: Engineering Societies, 1932], p. 31) ascribes the breakup to Edison's desire to concentrate on electric lighting and Sprague's determination to perfect his electric motors. Batchelor's diary, 1336, June to December 1886, ENHS, describes the development of dynamos, Sprague's suggestions, and Edison's impatience to get everything done.

25. E. Gilliland to TAE, 10 Dec. 1887, ENHS. Gilliland's proximity to Edison in 1886 and 1887 and his involvement in several key planning sessions (noted in Batchelor's diary) indicate that he was working very closely with Edison at this time.

26. TAE to S. Bergmann, 29 Nov. 1904, LB 040727, p. 169, ENHS.

27. Tate to Fessenden, 2 April 1889, LB 890130, p. 932, ENHS.

28. TAE to A. Sheetz, 10 July 1908, LB 080601, p. 280: "The great trouble with inventions is to get backing before they have been perfected."

29. Tate, *Edison's Open Door*, p. 142.

30. Ibid., p. 52; F. Maguire to Tate, 19 Nov. 1888, ENHS.

31. Tate to Insull, 30 Aug. 1887, LB 870621, p. 73, ENHS.

32. Edison wrote in 1891 that "this raising of several hundred dollars in a panic has been tough on me." LB 900412, p. 190, ENHS. The various interpretations and scholarly debates about the character and extent of the Great Depression can be found in S. B. Saul, *The Myth of the Great Depression 1873–1896* (London: Macmillan, 1969).

33. One of his first estimates of the yearly cost of running the West Orange laboratory was: Interest on investment (opportunity costs of an investment of $184,000 @ 5%) $9,200; insurance $1,100; depreciation $9,200; supplies $7,000; labor costs $62,000. This makes total yearly operating costs $88,900. TAE memo, 1888 laboratory folder, ENHS. The data available for 1890 show that the year's experimental work was billed at about $80,000, so Edison's estimate of operating costs was fairly accurate.

34. Contract research agreements in Accounts, billbook #2; Exhibit A, 1888 laboratory folder, ENHS; Marshall, *Recollections of Edison*, pp. 62–65.

35. Annotation on letter from S. Ritchie, 21 Nov. 1889, ENHS.

36. See accounts ledgers for payments, N871124 for shop order to carry out manufacture.

37. Experiment list, 30 Dec. 1887, ENHS.

38. Edison patented a method of railroad signaling with static induction in 1885. U.S. patent 486,634.

39. Experiments in N880522; claims to have transmitted over 500 feet with grasshopper in Josephson, *Edison: A Biography*, p. 280.

40. TAE to J. Hood Wright, Nov. 1887, N87115, ENHS.

41. Alfred Chandler, "The Beginnings of Big Business in American Industry," in James P. Baughman, ed., *The History of American Management* (New York: Prentice-Hall, 1969), p. 5.

42. Exhibit A, 1888 lab folder, ENHS.

43. TAE to *Pittsburgh Chronical Telegraph*, ENHS: "Promoters came to me and asked me to redesign machine and construct it of steel so it could be made on a commercial basis." Current says that engineers of the Remington Company made the important modifications, (*Typewriter and Men Who Made It*, p. 43).

44. Agreement, 1887 Edison Industrial Co., ENHS.

45. Article for *Munsey's Magazine*, 20 July 1901, LB 100521, p. 435, ENHS.

46. TAE to J. Hood Wright, Nov. 1887, N871115, ENHS.

47. Batchelor's diary, 1336, Sept. 1887, pp. 277–78, ENHS.

48. The original cylinders were called phonograms as they were intended to be sent through the mail. The collapsible cylinder was later named the mailing phonogram and experiments continued into 1889. Experimental lists, 1889, ENHS.

49. A. B. Dick to TAE, 23 July 1887, ENHS; *Scientific American* 57 (29 Oct 1887): 273.

50. *Orange Herald*, 24 Nov. 1888, ENHS.

51. "Prospectus and Plan of Organization" for Edison Industrial Co, 1887, ENHS.

Chapter Four. The Phonograph: A Case Study in Research and Development

1. *Electrical World*, 15 (Feb. 1890): 69; *Scientific American* 69 (8 July 1893): 25.

2. An excellent account of the invention is given by Paul Israel in "Telegraphy and Edison's Invention Factory," in W. S. Pretzer, ed., *Working at Inventing: Thomas Edison and the Menlo Park Experience* (Dearborn, Mich.: Henry Ford Museum and Greenfield Village, 1989).

3. "The New Edison Phonograph," *Scientific American* 57 (31 Dec. 1887): 415.

4. TAE to E. H. Johnson, undated, in 1888 Phonograph folder, ENHS.

5. Bruce, *Alexander Graham Bell*, p. 343.

6. Henry Edwards, "The Graphophone," *Journal of the Society of Arts* 36 (Dec. 1888): 41.

7. Ibid.; Bruce, *Alexander Graham Bell*, pp. 350–54.

8. TAE to E. H. Johnson, undated, in 1888 Phonograph folder; Tate, *Edison's Open Door,* pp. 135–37.

9. TAE to Gouraud, 22 Nov. 1887, ENHS.

10. TAE to Gouraud, 1 Oct. 1888, ENHS.

11. TAE to Gouraud, 21 July 1887, ENHS.; Batchelor diary, 1336, pp. 201–21. In the summer of 1887 Edison promised to bring out an improved phonograph "in a few months." TAE to G. H. Gould, 21 July 1887, LB 870621, p. 24, ENHS.

12. *Scientific American* 57 (31 Dec. 1887): 422; patent analysis by Mark E. Ham, "Edison and the Phonograph," working paper, 19 May 1986.

13. Gilliland to TAE, 10 Dec. 1887, ENHS.: "The present form of machine . . . would not compare favorably with the graphophone, and I have never felt that you would put it upon the market in that condition." The final break came when Gilliland cheated Edison in the sale of Edison's phonograph patents. See note 47.

14. Batchelor Diary, 1337, 11 Nov. 1887, p. 27, ENHS.

15. Gilliland's notes, 5 Oct. 1886, N 860825, ENHS.

16. *New York Journal,* 13 May 1888, ENHS.; *Scientific American* 57 (29 Oct. 1887): 273.

17. Lab notebook N 880601.1, ENHS.

18. John Jewkes, David Sawers, and Richard Stillerman, *The Sources of Invention* (London: Macmillan, 1960), p. 69. The recent development of superconductors reveals a mix of intuition, experiment, and trial and error. See James Gleick, "In the Trenches of Science," *New York Times Magazine,* 16 Aug. 1987.

19. Byron Vanderbilt, *Edison the Chemist* (Washington, D.C.: American Chemical Society, 1971), pp. 121–24; Aylsworth's notebook, N 880823, ENHS.

20. Tate to F. McGowan, 8 July 1888, ENHS.

21. The operation of the new phonograph is explained by G. Gouraud, "The Phonograph," *Journal of the Society of Arts,* 36 (Nov. 1888): 23–33.

22. Tate to F. McGowan, 8 July 1888, ENHS.

23. Edison accounts, billbook #3, statements of North American Phonograph Company, ENHS.

24. L. Glass to TAE, 6 July 1910; "New Model" list, 1888 Phonograph folder, ENHS.

25. TAE to Gouraud, 1 Oct. 1888, ENHS. The new machine was able to work with a wobble of one-sixth inch out of true.

26. TAE to Gouraud, 1 Oct. 1888, ENHS.

27. C. King Woodbridge, *Dictaphone* (New York: Newcomen Society, 1952), p. 13; comment in *Telegraph Journal and Electrical Review* 23 (3 Aug. 1888): 111.

28. Tate to A. Lawson, 22 May 1888; Batchelor diary 1336, 314, ENHS.

29. TAE to G. S. Nottage on 23 March 1878, cited in David A. Hounshell, *From the American System to Mass Production, 1800–1932: The*

Development of Manufacturing Technology in the United States (Baltimore: Johns Hopkins University Press, 1984), p. 333.

30. Robert S. Woodbury, *Studies in the History of Machine Tools* (Cambridge: MIT Press, 1972), 4: 59–60, 76–68, 183; *American Machinist* (11 [8 Jan. 1888]: 2–3) noted good business for machine tool builders and production runs of 100 machines at a time for Brown and Sharpe.

31. *American Machinist* 12 (18 April 1889): 8.

32. Account sheet, 1 Jan. 1889, ENHS. Samuel Gompers said in 1887, "The displacement of labor by machinery in the past few years has exceeded that of any like period in our history." Philip S. Foner, *History of the Labor Movement in the United States* (New York: International Press, 1955), 2: 14.

33. London *Daily Telegraph*, 9 Nov. 1888, ENHS.

34. "Tools for Phonograph," 1887 Phonograph folder, ENHS.

35. Tate to Lippincott, 2 Nov. 1888, LB 870621, p. 394, ENHS.

36. *American Machinist* 12 (9 May 1889): 8.

37. Tate to A. Wright, 16 Feb. 1888, ENHS.

38. TAE to G. Gouraud, 30 May 1889, LB 890521, p. 170, ENHS.

39. F. W. Toppan to Lippincott, 24 Oct. 1888, ENHS.

40. Batchelor diary, 1337, p. 582. Target output was 200 machines a day made by 500 employees. TAE to A. Lawson, 22 May 1888, ENHS.

41. Annotation, G. Morrison to TAE, 1 July 1895, ENHS. *Scientific American* (62 [26 April 1890]: 263) describes the American system in the Edison Phonograph Works.

42. *American Machinist* 12 (18 April 1889): 8; Tate to Lippincott, 17 Nov. 1888, LB 870621, p. 394, ENHS.

43. Tate (*Edison's Open Door,* p. 140) notes, "He had a habit of strolling in there to see how things were progressing. He might have a bright idea and formulate it on the spot without going through the formality of informing anyone." Quote in TAE to Villard, 19 Jan 1888, Villard Papers, Harvard Business School, ms. 78, box 63.

44. This was the view of Frank Sprague, to E. H. Johnson, 20 March 1887, ENHS.

45. Edison loaned $20,000 to the Works initially. Tate to Insull, 23 Nov. 1888, LB 881112, p. 140, ENHS. He followed this with at least $45,000 in 1889. TAE account with Drexel, Morgan, Morgan Archives, New York, E 2530, p. 522. Many notes (of $5,000 each) were purchased by a Mr. Chapman of Newark. billbook #3, p. 296. Insull told Edison that as he could not find Lippincott to borrow money, Edison would have to loan the Works $10,000 and deposit it immediately. Insull to TAE, 3 May 1889, ENHS.

46. Tate, *Edison's Open Door,* pp. 152–54. A model was later demonstrated at Seligman's house with more success.

47. Agreements in 1888 Phonograph folder, ENHS. Edison had attempted to revive the old phonograph company by offering to exchange one-third of the stock of a new company for the interest of the old share-

holders. They refused and Edison formed the new Edison Phonograph Company without them in December 1887. The right to manufacture phonographs was given to the Edison Phonograph Works. Gilliland extracted a large commission out of Lippincott on the understanding that Gilliland would sell Edison's interest to Lippincott for half its worth. Once he got the money from Lippincott, Gilliland fled the country. This story is told in Tate, *Edison's Open Door;* Conot, *Streak of Luck;* and Josephson, *Edison: A Biography.* Edison's faith in the boys was shaken. See TAE to E. H. Johnson undated memo, 1888 Phonograph folder, ENHS.

48. The agreement between Edison and the North American Phonograph Company stipulated an experimental fund that Edison could draw upon: up to $15,000 in the first year, $10,000 in the second, $7,500 in the third, and $5,000 each year thereafter.

49. Gilliland's notes, 5 Oct. 1886, N 860825, ENHS.

50. Tate, *Edison's Open Door,* pp. 161–62.

51. T. C. Hepworth, "The Phonograph," *The Technical Educator* (1890): 154–56.

52. *Atlantic Monthly* 63 (Feb. 1889): 258.

53. Caveat for improvements in phonographs, 21 July 1890, 1890 Phonograph folder; J. A. Beecher to TAE, 1 Nov. 1889, ENHS.; *Electrical World* 14 (Sept. 1889).

54. TAE to Batchelor, 7 May 1889, LB 890408, p. 325, ENHS.

55. Alfred Clark, "His Master's Voice," manuscript about his career in the Edison and Victor companies, in AHC.

56. *Phonogram* 2 (Dec. 1892): 274; List of experiments, 1888 Phonograph folder; Drawing of "plate phono," 1 Oct. 1888, N880001, ENHS.

57. TAE to Batchelor, 7 May 1889, LB 890408, p. 325, ENHS.

58. List of experiments, 30 Dec. 1887, Phonograph folder, ENHS.

59. TAE to F. S. Dyer, 17 May 1889, LB 890408, pp. 434–35, ENHS.

60. "A Boom That Collapsed," *Philadelphia Times*, 2 Jan. 1891, ENHS.

61. *Atlantic Monthly* 63 (Feb 1889): 258.

62. Flyer in Primary Printed Collection, Circa 1890, ENHS (This collection groups together all printed primary source material related to Edison).

63. Ham, "Edison and the Phonograph," p. 17.

64. Edison's 1890 handwritten notes on phonograph development, N 880103, 1890 Phonograph folder, ENHS.

65. TAE to Batchelor, 7 May 1889, LB 890408, p. 325, on "Music Room," ENHS. Clark, "His Master's Voice," p. 181. Clark began work in the West Orange laboratory in 1889; he was sent to Europe in 1899 on phonograph business.

66. TAE to S. Eaton, 11 Dec. 1890, ENHS. Edison's plans for the phonograph business were based on "a monopoly of records," annotation on letter from S. Insull, 28 Jan. 1892, ENHS.

67. Loose drawings in 1890 lab folder, ENHS.

68. TAE to Edison Phonograph Works, 19 Nov. 1890, ENHS.

69. O. Read and W. Welch, *From Tinfoil to Stereo* (Indianapolis: Howard Sams, 1977), p. 50; *Phonogram* 2 (July 1892): 160.

70. TAE memo, 1891 Phonograph folder, ENHS.

Chapter Five. Edison's Laboratory and the Electrical Industry

1. Payson Jones, *A Power History of the Consolidated Edison Company 1878–1900* (New York: Consolidated Edison Co. of N.Y., 1940), p. 209. The cost overruns on building the Pearl Street station, estimated to be 50 percent, played their part in the low return on investment and led to caution. Passer, *The Electrical Manufacturers,* p. 119.

2. Lamp Co. memo, 1889. The *Annual Report of the Edison Electric Light Co.* noted that 1888 had marked "the keenest competition from infringing companies." 1888 Electricity folder, ENHS.

3. TAE to Insull, 10 May 1895, ENHS.

4. Josephson, *Edison: A Biography,* p. 300; Memo on selling price of lamps, 1889 Electricity folder; Edison's notes in N 900104.3. Edison's computations on how lowering the price of lamps would increase demand and profits are in lab notebook N 880001, ENHS.

5. Lab notebook N 880315.2 has notes on cellulose machine; on saving reject lamps, TAE to Lamp Works, 18 June 1890, ENHS. Arthur Bright, *The Electric Lamp Industry: Technological Change and Economic Development from 1800 to 1947* (New York: Arno, 1972), p. 124.

6. Edison moved the Machine Works to Schenectady after labor trouble in the Goerck St. facility. Henry George ran for mayor of New York in 1886 on a radical ticket, which included the eight-hour day.

7. Constant appeals were made to Edison to improve the performance of the lamp. F. Upton to TAE, 20 April 1889, ENHS. In 1886 one improvement in the resistance of the lamp was instrumental in lowering need of copper content of conductor and current requirements to illuminate. It brought increased profits—estimated increase on return to capital 18.6 percent to 34.6 percent, J. H. Vail to Edison, 28 Dec. 1886, ENHS. Edison reviewed all lamp patents and advised the Electric Light Company if they should acquire them. E. H. Johnson said to TAE (10 June 1889) "The board considers you the umpire in all such matters."

8. N 880202, ENHS.

9. TAE to Insull, 2 May 1898, LB 980131, p. 145, ENHS. But there were competent mathematicians at the Menlo Park laboratory, such as Francis Upton who had graduate training in a German university.

10. Percy Dunsheath, *A History of Electrical Power Engineering* (Cambridge: MIT Press, 1962), pp. 114–18. Hopkinson was trained as a mathematician at Cambridge University and was employed in 1882 as a consultant to the Electrical Construction Co., an Edison subsidiary. Maxwell

published a paper in 1865 in which he outlined a mathematical model supplying the foundations of classical electromagnetic field theory. In 1886 Hopkinson and his brother Edward supplied a basis for applying Maxwell's theories in a famous paper delivered at the Royal Society. Hopkinson's improved dynamo was manufactured in great numbers as the Manchester model by the engineering firm of Mather and Platt; James Grieg, *John Hopkinson: Electrical Engineer* (London: HMSO, 1970), p. 28.

11. Explanation of experiments, 14 May 1890, lab folder, ENHS.

12. Edison memo to Villard, 23 March 1889, Villard Papers, Harvard Business School, ms 78, box 63.

13. Arthur Wright of the Brighton supply undertaking in England is acknowledged as the great pioneer in rate making and load balancing. His idea was to restructure rates to reflect the costs of generation. His system and maximum demand meter were widely diffused and used extensively in Germany and the United States. "Technical Aspects of the Period," in United States Bureau of the Census, *Central Electric Light and Power Stations* (Washington, D.C.: GPO, 1910), pp. 120–22.

14. *Measuring Invisibles* (Newark, N.J: Weston Electrical Instrument Co., 1938), p. 19, NJIT. Weston's lab had worked in several areas, including electric light, telegraphy, and telephony, before Weston switched into instrument research. He devised a moving coil that operated inside the curved poles of a magnet. This major innovation required an improved permanent magnet and some new alloys for the spring. His direct reading portable ammeter was a highly successful product and the basis for the world's largest company devoted to electrical measuring instruments. *Weston Engineering Notes* 12 (Sept. 1957), pp. 8–10, NJIT; Dunsheath, *Electrical Power Engineering,* p. 308.

15. The chemical meter used a known proportion of the current to record ampere hours, not watt hours. It had to be calibrated for a given voltage and was incorrect for other voltages. Its accuracy was poor for light and full loads and was affected by temperature. "Metering Electrical Energy," *Electrical Engineering,* 60 (Sept. 1941): 421–22; H. W. Richardson, "The Electric Meter," *General Electric Review* (Aug. 1911): 367. The process of weighing and calculating was done at the central station, and was naturally very expensive in terms of labor. Boston Edison paid $2,500 annually to service 800 meters. Vanderbilt, *Edison the Chemist,* p. 87. The customer could not read the meter and was at the mercy of the utility company when it came to billing. The Pearl Street operation provides us with the first examples of disputes between a utility company and its customers. Jehl, *Menlo Park Reminiscences,* 3: 1085–86.

16. EGE to TAE, 11 Dec. 1890, ENHS.; David Woodberry, *Beloved Scientist: Elihu Thomson* (New York: MacGraw-Hill, 1944), pp. 198–99. The Thomson recording wattmeter (TRW) was the first true energy-measuring device. It was essentially a small shunt motor with its torque proportional to the power. Many were manufactured beginning in 1889.

Edison continued to work on meters. In 1900 he received patents on several meter designs, including balance-beam, electromagnetic, and electrochemical.

17. W. J. Jencks to TAE, 12 May 1888, ENHS. The primitive state of the first central stations is noted in T. Cromerford Martin and S. Leidy Cross, *The Story of Electricity* (New York: Story of Electricity Co., 1919), p. 77. Fire claimed many stations, including the one on Pearl Street. The rising number of electrical fires prompted the search for an insulating material that would not burn.

18. Kennelly was chairman of the American Institute of Electrical Engineers' (AIEE) first committee on standards formed in 1891 and this began his lifelong work on standardization. He served as president of the AIEE from 1898–1900. In 1933 he received the Edison medal for "meritorious achievements in electrical science . . . and his contributions to the theory of electrical transmissions and to the development of international electrical standards." Harlow Freitag, ed., *Electrical Engineering: The Second Century Begins* (New York: IEEE Press, 1986), p. 14.

19. W. J. Jenks to Kennelly, 20 April 1889; Jenks to TAE, 3 May 1889, ENHS, on conductivity tests. Insurance companies published Kennelly's curves of safe carrying capacity of wires. Edison wanted to know more about the physiological effects of electricity to safeguard employees of electrical companies. TAE to H. Bergh, 13 July 1888, LB 870929, p. 273, ENHS.

20. Estimates of copper required in conductors in Kennelly Reports, and in LB 890727, ENHS.

21. Edison Electric Illuminating Company of New York, *Annual Report 1886*, ENHS.

22. A French syndicate had cornered the market on copper. Its price per pound went from 10¢ in 1886 to 20¢ in 1888. Terry Reynolds and Theodore Bernstein, "The Damnable Alternating Current," *Proceedings of the IEEE* 64 (Sept. 1976): 1340. John Hopkinson also patented a three wire system in 1883. See Thomas Parke Hughes, *Networks of Power: Electrification in Western Society, 1880–1930* (Baltimore: Johns Hopkins University Press, 1983), p. 84. Passer notes that the three wire system saved 63 percent of the copper used in the two wire system. The copper cost per lamp went from $1.50 to $1.25.

23. The work on the municipal lamp had begun in 1886, when Edison had his laboratory at the Lamp Factory. J. H. Vail to TAE, 22 May 1886, ENHS.

24. N 871124, ENHS, lists of experiments. Alternating current experiments were underway by January 1888.

25. Patent #369,439 granted 6 Sept. 1887.

26. The development of this system is described in Hughes, *Networks of Power*, pp. 86–95.

27. Siemens & Halske Report of Nov. 1886, Electricity folder, ENHS.

28. Thomas A. Edison, "The Dangers of Electric Lighting," *North American Review* 149 (Nov. 1889): 625–34.

29. TAE to E. H. Johnson, undated, in 1888 Phonograph folder, ENHS.

30. TAE to Villard, 24 Feb. 1891, Villard Papers, HBS, fol. 475, box 63.

31. J. H. Vail to TAE, 10 Nov. 1888, ENHS.

32. TAE to E. H. Johnson, Nov. 1887, ENHS.

33. Insull to TAE, 6 April 1888, laboratory bills. In 1888 Edison proposed to the Electric Light Company management that they pay him a weekly stipend to cover experiments. They put so many conditions on the weekly payments that Edison gave up in disgust and swore that he would never ask the company for any more money. Annotation in E. H. Johnson to TAE, 21 Feb. 1889, ENHS.

34. Linseed oil and beeswax were added to the mixture, which was heated in cauldrons before being poured into the pipe. Vanderbilt, *Edison the Chemist*, pp. 78–81. These wires did excellent service. I am informed by David Sicilia, historian of the Boston Edison Company, that they still lie beneath the city center.

35. J. Kruesi to TAE, with annotations, 24 Aug. 1887, Kruesi collection, ENHS.

36. Insull to Tate, 5 July 1888, ENHS.

37. Insull to Tate, 28 June 1888; Insull to TAE, 24 May 1888, ENHS.

38. Upton to TAE, 1 Aug. 1889, ENHS.

39. TAE to Villard, 23 March 1889, Villard Papers, HBS, ms. 78, box 63.

40. TAE to Villard, 8 Feb. 1890, Villard Papers, HBS, fol. 473.

41. E. H. Johnson to TAE, 1 June 1889, ENHS.

42. W. J. Jencks to lab, 10 July 1889, ENHS.

43. Westinghouse made his money in the air brake and signaling business (for railroads). He formed the Westinghouse Electric Company in 1885; it was capitalized at $1 million. His biographer notes that he showed no great enthusiasm for electric lighting until he saw the potential of alternating current. His business and technical associates did not share his enthusiasm for the new technology. Henry G. Prout, *A Life of George Westinghouse* (New York: Arno, 1972), pp. 95, 103; Passer, *Electrical Manufacturers*, p. 132. Westinghouse engineers developed Stanley's designs and added several other important parts of the commercial system that could be mass produced. F. L. Snyder, "The Transformer and How It Grew," *Westinghouse Engineer* 10 (Jan. 1950): 50–51.

44. Passer, *Electrical Manufacturers*, p. 174.

45. Called the "cut-out" or "safety-catch," the fuse was enclosed in a safety plug with a screw thread, like the modern fuse. It was claimed that this device made fires impossible. "The Edison Light" promotional material, early 1880s, ENHS. Edison repeatedly stressed that research and experiments had been aimed at safe use of the system even by "the inexperienced domestic servant; nor can the most careless person do injury to

himself." "The Success of the Electric Light," *North American Review* 131 (Oct. 1880): 298.

46. Advertising material in 1888 Electricity folder, ENHS.

47. Edison wrote in 1889: "With the increase in electric lighting (which to-day is used only to a very limited extent as compared with its inevitable future use) . . . dangers that exist now in a thousand different parts of the city will be manifolded many times." "The Dangers of Electric Lighting," *North American Review* 149 (Nov. 1889): 625.

48. Cited in Josephson, *Edison: A Biography,* p. 346.

49. The Electric Lighting Act had placed all public supply in England under the Board of Trade. Each local authority had veto power over lighting projects, and the companies were subjected to vigorous regulation of every aspect of their operations. In comparison with the great increase in electricity supply in the United States, England was lagging behind. Many blamed this lag on regulation. See A. J. Millard, *A Technological Lag: Diffusion of Electrical Technology in England* (New York: Garland, 1987).

50. David Sicilia, "History of the Boston Edison Company," produced by the Winthrop Group, 1986, p. 12; Letter from New Orleans Edison Company, 12 May 1887, ENHS.

51. "Confidential Sheet for Agents," 1888 Electricity folder, ENHS; *Electrical World* 11 (Feb. 1888): xviii. Advertising materials, 1888 Electricity folder.

52. Edison called Westinghouse a "shyster" in an interview with the press and Westinghouse was most annoyed. S. Eaton to TAE, 7 Oct. 1889, ENHS.

53. Kennelly's record of the battle is found in the Kennelly Reports, in letterbook LB 870929, and his evidence to the hearing on the electrocution of William Kemmler, New York State Court of Appeals, 1847–1911, 893 (Buffalo, New York, 1890), Vol. 2. The unfortunate Kemmler was the first to be electrocuted as capital punishment.

54. Circular letter of H. P. Brown, 1888 Electricity folder, ENHS. The laboratory also supplied the equipment for Brown's demonstrations at Columbia College in July 1888. Kennelly to F. S. Hastings, 25 Feb. 1889, Kennelly Reports, 060788, p. 350. Brown's work is described in Thomas P. Hughes, "Harold P. Brown and the Executioner's Current: An Incident in the AC-DC Controversy," *Business History Review* 32 (1958): 143–65.

55. Hastings to Kennelly, 20 Nov. 1888, ENHS. Kennelly notebooks for 1888.

56. Record book of Galvanometer Building, N 880524; Record book # 3 of Galvanometer Building, N 880606, ENHS.

57. 25 April 1890, LB 900412, p. 190; costs of "physiological experiments" were billed through 1890, billbooks #3, #6, ENHS.

58. Quote is from *Electrical Engineer* 8 (Dec. 1889): 520.

59. Forest McDonald, *Insull* (Chicago: University of Chicago Press, 1962), p. 45.

60. Letter of 22 April 1889, ENHS.

61. Insull to Tate, 17 June 1889, ENHS.

62. E. H. Johnson to TAE, 22 Aug. 1888, ENHS.

63. Caruso, *The Morgans,* p. 272; Buss, *Henry Villard,* pp. 206–10.

64. Prospectus of Edison General Electric, 1889 Electricity folder, ENHS.

65. C. E. Chinnock to TAE, 17 April 1888, ENHS.

66. Agreement in 1889 Electricity folder, ENHS.

67. "Explanation of Experiments," 1890 lab folder, ENHS.

68. Report of annual meeting of Edison Illuminating Companies, *Electrical World* 14 (Aug. 1889): 136.

69. TAE to F. Gorton, 24 Sept. 1891, ENHS.

70. "Mr. Edison's reply to Thomson-Houston Memo," 1 April 1889, Villard Papers, HBS, ms. 78, box 63, 472.

71. Insull to TAE, 17 Oct. 1890, ENHS.

72. Tate to R. Grimshaw, "Memorandum of Mr. Edison's Work in 1891," 6 Jan. 1892, lab folder, ENHS.

Chapter Six. Diversification in the 1890s

1. TAE to Edison Manufacturing Co., 11 Jan. 1889, ENHS.

2. TAE to Batchelor, 7 May 1889, LB 890408, p. 325, ENHS.

3. EGE to TAE, 10 Oct. 1890, ENHS. See also annotation and answer of 14 Oct. 1890.

4. Tate to F. S. Dyer, 7 Sept. 1888, ENHS.

5. TAE to Batchelor, 6 April 1887, ENHS.

6. F. Maguire to Tate, 30 July 1889, ENHS.

7. Tate, *Edison's Open Door,* p. 140.

8. Dead experiments, conducted in 1888 and transferred, Edison accounts, journal # 5, p. 147, ENHS.

9. *Albany Telegraph,* 3 Nov. 1895.

10. N 900104.3, ENHS.

11. Frank Sprague, "Some Personal Notes on Electric Railways," *Electrical World* 40 (Feb. 1902): 226; Harriet Sprague, *Sprague and Edison Myth,* p. 10.

12. See TAE to E. H. Johnson, handwritten memo, undated, 1888 Phonograph folder, ENHS. Johnson had hired Sprague while in London. Sprague received his engineering education at the Naval Academy at Annapolis. Hedley, *Frank J. Sprague,* pp. 29, 34.

13. Dunsheath, *Power Engineering,* p. 184.

14. Harriet Sprague, *Sprague and Edison Myth,* pp. 10–11.

15. O. E. Carson, *The Trolley Titans* (Glendale, Calif.: Interurban Press, 1981), p. x; mileage figures from *Electrical Trades Directory, Electrical World.*

16. *Electrical World* 20 (Sept. 1892): 151; Bensman, *Practice of Solidarity,* p. 137.

17. N 890815; "Edison's New System of Street Car Propulsion," *Electrical World* 19 (Jan. 1892): 2.

18. Vociferous complaints against the "death wires" occurred in most cities that had extensive overhead wiring. The outcry was greatest in New York and Chicago. When Jay Gould used legal means to hinder the prohibition of Western Union's wires in New York City, *American Machinist* (12 [31 Jan. and 11 April 1889]) saw this as "money vs people," an example of how great corporations ran cities.

19. *Evening Post*, 5 Oct. 1891; *Electrical World* 19 (Jan. 1892): 1.

20. The American civil, mining, and mechanical engineers chartered ships to take them to Paris, where the engineers of the Old World received them. Many engineers from the New World were in Paris to pick up ideas for the celebrations planned for 1892 in Chicago, including the idea to build an iron tower for "our show." *American Machinist* 12 (19 Dec. 1889): 3.

21. Maguire to Tate, 30 July 1889, ENHS.

22. Batchelor to TAE, 19 Aug. 1889, ENHS.

23. Antonia and W. K. L. Dickson, "Edison's Invention of the Kineto-Phonograph," *Century Magazine* 48 (June 1894): 212.

24. *Electrical World* 14 (Sept. 1889): 165–67; Josephson, *Edison: A Biography*, p. 334.

25. Gordon Hendricks, *The Edison Motion Picture Myth* (Berkeley and Los Angeles: University of California Press, 1961), pp.82–83.

26. *Electrical World* 19 (Jan. 1892): 2.

27. Edison kept in contact with the work of Siemens' lab. In 1911 he obtained one of their new tungsten filament lamps. R. Burrows to TAE, 3 May 1911, ENHS.

28. Conot, *Streak of Luck*, pp. 284–85.

29. Inquiry of 27 Dec. 1889, 1889 Electricity folder, ENHS.

30. TAE to Dyer, 17 May 1889, LB 890408, pp. 434–35, ENHS.

31. Charles A. Stansfield, *New Jersey* (Boulder Colo.: Westview, 1983), p. 36; New Jersey, *Annual Report of the Bureau of Labor Statistics*, 1895, pp. 9–13, NJSL.

32. Annotation on J. Birkenbine to TAE, 13 July 1889, Ore Milling folder, ENHS.

33. *Scientific American* 66 (2 April 1892): 216.

34. Quotation cited in Vanderbilt, *Edison the Chemist*, p. 154; expert opinion of S. B. Patterson, superintendent of Andover Iron Company in New Jersey, *Annual Report of the Bureau of Labor Statistics*, 1895, p. 10.

35. This was Passer's conclusion to the battle of the systems, *Electrical Manufacturers*, p.174; quotation is from Tate, *Edison's Open Door*, p.280; Tate to F. Van Dyck, 21 Feb. 1894, LB 930808, p.668, ENHS.

36. Telephone, shop order #540, 1891; description of ore-milling equipment in *Iron Age* 59 (28 Oct. 1897): 1.

37. See W. Bernard Carlson, "Edison in the Mountains: The Magnetic Ore Separation Venture," *History of Technology* 8 (1983): 37–59; "The Edison Concentrating Works," *Iron Age* 59 (Oct. 1897): 1–5.

38. H. Ford and S. G. Crowther, "Edison as I Knew Him," quoted in

Josephson, *Edison: A Biography*, p. 374. Ford's great admiration for Edison probably led him to stretch this point. In addition to the ore-milling works he also had the example of the meat packers of Chicago and other automobile manufacturers, especially the Olds company.

39. T. A. Robbins reminiscence, quoted in Josephson, *Edison: A Biography*, p. 368.

40. TAE to J. Kruesi, nd, Kruesi collection, ENHS.

41. *Scientific American* 66 (2 April 1892): 216.

42. Recollections of Dan Smith and Benny Odel, in Mary Childs Nerney's notes, N 281101, ENHS.

43. Batchelor diary, 1337, 21 Jan. 1891, p. 154, ENHS.

44. *Electrical World* 45 (Jan. 1892): 2.

45. Billed to EGE $17,045; Edison and Villard, $19,225; phonograph companies, $20,984; ore-milling companies, $21,465. My calculations were based on the expenses billed in a series of billbooks, numbers 4 to 8, in ENHS. Because of cross-billing and the lack of any measure of Edison's personal time—an important element in this research—these numbers are only to be taken as a rough estimate.

46. 2nd annual report of EGE, 31 Oct. 1890, quoted in Payson Jones, p. 160; agreements are in box D3, ENHS.

47. TAE to F. S Gorton, 10 Nov. 1891, ENHS.

48. Billing information, LB 041289, 25 April 1890, p. 193, ENHS.

49. B. E. Morrow to C. Wilson, 13 March 1918, ENHS.

50. Billbook #9, p. 489, ENHS.

51. Tate, *Edison's Open Door*, pp. 260–65; Batchelor's diary, 1337, p. 132. The recent television program, "An Ocean Apart," followed this interpretation. One popular version of these events relates: "Edison then withdrew from the industry to his Menlo Park laboratory, his goal of creating an incandescent lighting system achieved." Keith L. Bryant and Henry C. Dethloff, *A History of American Business* (Englewood Cliffs, N.J: Prentice-Hall, 1986), p. 155. The *New York Herald* reported after the consolidation that Edison was frozen out of the electrical trust. The *New York Morning Post* described the inventor as "worn out, shoulders drooped, hair grey" and spending more time at home! Cuttings in 1892 Electricity folder, ENHS.

52. In an interview published in the *Albany Telegraph*, 3 Nov. 1895 Edison described the electrical industry as a "commissary of wolves" and justified his departure from it by saying, "There are too many in it, it offers no inducements." Although Westinghouse purchased the rights to Tesla's induction motor in 1888, it took several years to develop it as a commercial motor.

53. Prout, *George Westinghouse*, pp. 122–25.

54. New Jersey, *Annual Reports of the Inspectors of Workshops and Factories*, 1890–91, NJSL.

55. TAE to C. Coffin, 9 May 1892, LB 920309, p. 440. Edison also had the job of disposing of $20,000 worth of parts for toy dolls, which were in

the Works; Tate to TAE, 31 Oct. 1892; *Phonogram* 2 (Dec. 1892): 263–74, ENHS.

56. *Electrical World* 19 (March 1892): 177.

57. John Winthrop Hammond, *Men and Volts: The Story of General Electric* (New York: J. B. Lippincott, 1941), p. 89.

58. Hammond, *Men and Volts*, pp. 222–23. There were rumors on Wall Street that General Electric would not survive the depression. Prout, *George Westinghouse*, p. 72.

59. Tate to F. Jehl, 11 Sept. 1893, saying industry is "perfectly dead," LB 930805, p. 91, ENHS.

60. *Scientific American* 66 (2 April 1892): 216.

61. New Jersey, *Annual Reports of the Inspector of Workshops and Factories*, 1895–98, NJSL. The cuts were so brutal that the superintendent of the Works, Ballou, resigned in protest. Gordon Hendricks, *Origins of the American Film* (New York: Arno, 1972), p. 43. The amount of work done in the lab declined significantly during the 1893/94 depression. During one month in 1894 (May), the labor cost for work done on the phonograph was less than $100, Edison's personal experiments cost $460, and ore-milling labor was over $1,000. Total cost of the month's work was $2,883, of which GE paid $911. Billbook #12, p. 469. Research expenditure for 1893 probably did not exceed $25,000—much less than the $75,000–$80,000 annual range at the beginning of the decade. Edison cleaned up his affairs in 1894, withdrawing from many honorary posts and cutting back on his personal expenditures. TAE to A. Elliot, 24 March 1894, LB 931030, p. 109; TAE to E. H. Lewis, 15 March 1894, LB 931030, p. 98, ENHS.

62. Charles Hoffman, *The Depression of the Nineties: An Economic History* (Westport, Conn.: Greenwood, 1970), pp. 151–52; Bensman, *Practice of Solidarity*, p. 171.

63. Walter Welch, *Charles Batchelor: Edison's Chief Partner* (Syracuse, N.Y.: Syracuse University Press, 1972), p. 87. In 1891 the price of Lake Superior ore was $6.82; Cuban ore was $5.82.

64. Experiment quote is from Patterson report, New Jersey, *Annual Report of the Bureau of Labor Statistics*, 1895, NJSL.

65. John D. Venable, "Big Rocks and Rocky Years," *New Jersey History* 99 (Spring 1981): 97.

66. *Scientific American* 69 (8 July 1893): 25.

Chapter Seven. Moving Pictures

1. See N 900104.3, ENHS.

2. *Electrical World* 7 (Jan. 1886): 26; experiments are in N 871210.2, ENHS.

3. Edison caveats on motion pictures are in the appendix of Hendricks,

Edison Motion Picture Myth; Dickson and Dickson, "Kineto-Phonograph," pp. 206–14.

4. A. R. Fulton, "The Machine," in Tino Balio, ed., *The American Film Industry* (Madison, Wisc.: University of Wisconsin Press, 1985), pp. 30–32.

5. W. K. L. Dickson, "Some Facts Relating to Moving Photography," manuscript, ENHS.

6. Letter of 17 Feb. 1879, Hendricks, *Edison Motion Picture Myth*, p. 144.

7. Although Hendricks is skeptical of many of Dickson's claims, especially concerning chronology, this one has the ring of truth; newly discovered pay vouchers show that Dickson was with Edison at this time. Edison accounts, vouchers, 1887, #224 (July), #432 (Oct.), and #485 (Nov.), ENHS.

8. See William J. Broad ("Subtle Analogies Found at the Core of Edison's Genius," *New York Times*, Science Times, 12 March 1985) quoting Paul Israel and Reese Jenkins of the Edison Papers Project. On this craft tradition of emulation see Brooke Hindle, *Emulation and Invention* (New York: New York University Press, 1981).

9. W. K. L. Dickson, "A Brief Epiphany of All Facts Relating to Mr. Edison's Invention of the Kinetoscope" 21 April 1928, manuscript, ENHS.; Edison caveat, in Hendricks, *Edison Motion Picture Myth.*

10. Brown's testimony, complainant's record, *Edison v. American Mutascope Co.*, 1910, p. 143, legal box 173, ENHS.

11. The history of the celluloid filmstrip is told in Reese V. Jenkins, *Images and Enterprise: Technology and the American Photographic Industry, 1839–1925* (Baltimore: Johns Hopkins University Press, 1975), ch. 6. Gordon Hendricks argues that Carbutt was the inventor of the filmstrip; Eastman's contribution was in manufacturing methods. *Edison's Motion Picture Myth*, p. 40. Eastman set up a laboratory in 1890 for testing. Jenkins, *Images and Enterprise*, p. 147.

12. This recollection is from Albert Smith, *2 Reels and a Crank* (New York: Doubleday, 1952), cited in Hendrick, *Edison's Motion Picture Myth*, p. 171.

13. TAE annotation on letter of 20 March 1914, commenting that the early film stock was "damned uncertain as Eastman and all of us found," ENHS.

14. Patent #589,168, filed 24 Aug. 1891. Jenkins (*Images and Enterprise*, p. 274) stresses the importance of the technological system of cinematography.

15. Description of kinetograph in Dickson, "Kineto-Phonograph," and Dickson to Meadowcroft, 1 May 1921, ENHS.

16. *Electrical Engineer* 18 (Nov. 1894): 377; Jenkins, *Images and Enterprise*, pp. 268–269.

17. Promotional and advertising material from Primary Printed, ENHS.

The advantage of grouping these machines together was that it made the frequent maintenance easier.

18. Gordon Hendricks, "The Kinetoscope," *The Origins of American Film* (New York: Arno, 1972), pp. 35, 43.

19. The Edison Tower of Light was at the center of the Electricity Building at Chicago. Put up by General Electric, it stood eighty-two feet high and was studded with 5,000 incandescent bulbs. The development and significance of spectacles such as this are described in Carolyn Marvin, *When Old Technologies Were New* (New York: Oxford University Press, 1988), pp. 171–72.

20. Tate, *Edison's Open Door*, pp. 285–87.

21. Robert C. Allen, "The Movies in Vaudeville," in Balio, *American Film Industry*, p.60; Gordon Hendricks, "The Kinetoscope: Fall Motion Picture Production," in John Fell, ed., *Film Before Griffith* (Berkeley and Los Angeles: University of California Press, 1983), p. 20. Dickson's version is in his "Kineto-Phonograph."

22. Shipments list, 1894 Motion Picture folder, ENHS.

23. Shop order #654, N871124; N. Speiden, Historic Site Report on the Black Maria, National Park Service, ENHS.

24. Terry Ramsaye, *A Million and One Nights: A History of the Motion Picture through 1925* (New York: Simon and Schuster, 1926), p. 84.

25. TAE to Hood Wright, Nov. 1887, N 871115, ENHS.

26. N 930724, experiments of 24 July 1893, ENHS.

27. *Electrical Review* 27 (Aug. 1890): 213.

28. TAE to A. Brady, 21 Dec. 1898, LB 980131, p. 305, ENHS.

29. TAE to W. Bowen, 16 April 1898, ENHS.

30. Annotation on letter from O. Glasser to Meadowcroft, 9 Dec. 1929; lab account books, subledger #6, p. 240, ENHS.

31. *Albany Telegraph*, 3 Nov. 1895, ENHS.

32. Annotation on F. P. Fish letter, 11 Oct. 1895, ENHS.

33. Subledger #6, p. 446, ENHS.

34. TAE to F. P. Fish, 15 Dec. 1896, LB 931030, p. 241, ENHS.

35. TAE to F. P. Fish, 10 Feb. 1897, LB 931030, p. 318. Ledger #7, ENHS, provides details of GE's payments; fluorescent lamp experiments cost over $11,000 from 1896 to 1899, and filament experiments cost over $8,000 from 1897 to 1899.

36. Letters in Stanley's correspondence reveal that he felt cheated and deceived by Westinghouse. B. A. Drew and G. Chapman, "William Stanley," *Berkshire History* 6 (Fall 1985): 16.

37. Bright, *Electric Lamp Industry*, pp. 170–73.

38. See George Wise, *Willis Whitney*, ch. 5.

39. Ibid., p. 76.

40. TAE annotation on letter from B. Liedesman, 3 May 1921, ENHS. "One of my assistants was killed and several injured by x-ray . . . I myself under any circumstances wouldn't use x-rays."

41. Kendall Birr, *Pioneering in Industrial Research* (Washington, D.C.: Public Affairs Press, 1957), pp. 48–51, 65.

42. Wise, *Willis Whitney*, p. 97; Reich, *American Industrial Research*, p. 2.

43. Wise, *Willis Whitney*, p. 134.

44. Ibid., p. 79; Reich, *American Industrial Research*, p. 149.

45. Noble, *America by Design*, p. 115, quoting J. J. Carty.

46. Harry Braverman, *Labor and Monopoly Capital: The Degradation of Work in the Twentieth Century* (New York: Monthly Review Press, 1974), p. 164.

47. Leonard Reich, "Industrial Research and the Pursuit of Corporate Security," *Business History Review* 54 (1980): 504–5.

Chapter Eight. An Industrial Empire

1. See Raymond R. Wile, ed., *Proceedings of 1890 Convention of Local Phonograph Companies* (Nashville, Tenn.: Country Music Foundation Press, 1974). This is a facsimile of the original document, with an excellent introductory essay by Wile.

2. Proposed letter to New England Phonograph Co., 1890s; Tate to TAE, 13 Jan. 1892, ENHS.

3. Tate to TAE, 19 Jan. 1892, ENHS. A widely used technique at this time was the performer singing into banks of recording phonographs and making several recordings of the same song.

4. George L. Frow and Albert F. Selfi, *Edison Cylinder Phonographs* (privately published by Mr. Frow in Tunbridge Wells, United Kingdom, 1978), p. 134.

5. Tate to New York Phonograph Co., 1 June 1892, LB 920309, p. 579, ENHS.

6. C. W. Noyes, *The C. W. N. Handbook: Use and Care of the Edison Phonograph* (Cincinnati: E. Ilsen, 1901), p. 22, NJHS.

7. Frow and Selfi, *Edison Cylinder Phonographs*, p. 19.

8. TAE to N. Block, nd. in 1894 Phonograph folder, ENHS.

9. Read & Welch, *Tinfoil to Stereo*, p. 63. In a letter to Steven Moriarty, 16 June 1893, Edison summed up the knowledge of the consumers' use of talking machines, claiming that "the greatest of fields" was in reproduction in households only, adding, "I have never got any one to believe in it until lately." This contradicts Tate's claims that Edison thought that amusement use was beneath his invention.

10. Tate, *Edison's Open Door*, pp. 291–93.

11. Edison claims to have tens of thousands of dollars invested in phonograph R and D in the 1890s were called excessive in the various court cases fought over his phonograph business. In 1895 he claimed to have spent all the money he received from the phonograph companies on experi-

ments and to have thrown $45,000 of his own money into the pot. Evidence in *American Graphophone v. Edison Phonograph Works,* 25 April 1895, ENHS.

12. TAE to N. Block, nd. in 1894 Phonograph folder, ENHS.

13. Allen Koenigsberg, *Edison Cylinder Records, 1889–1912* (New York: Stellar, 1974), p. xx.

14. *The Phonograph and How to Use it* (New York: National Phonograph Company, 1900), pp. 34–37, NJHS.

15. TAE to Edison United Phonograph Co., 19 April 1897, LB 931031, p. 406, ENHS.

16. TAE note to Edison United Phonograph Co., 16 June 1893, ENHS.

17. The history of the duplication project is given in the record of *Thomas B. Lambert v. T. A. Edison,* legal box 173; Brief for Edison, legal box 169; see also Vanderbilt, *Edison the Chemist,* pp. 123–27; German Company to Edison United Phonograph Co., 18 July 1896; Minutes of National Phonograph Co. meeting, 9 July 1896, ENHS.

18. *American Machinist* 26 (9 July 1903): 977.

19. Koenigsberg, *Edison Cylinder Records,* p. xxii.

20. The "high speed" refers to the 160 rpm of the cylinder which was an increase over previous speeds. It should be noted that the user set the speed of the nineteenth century machines; the National Phonograph Company suggested using a watch. The 160 rpm speed of the new play-back machines became the industry standard until 1929.

21. Shop order #1114, ENHS.

22. Batchelor diary, 1338, p. 51, quoting Aylsworth, 1 Sept. 1903, ENHS.

23. Ibid.

24. Clark, "His Master's Voice," pp. 202–8. Another problem in recording classical music was the difficulty of capturing the sound of stringed instruments.

25. Koenigsberg, *Edison Cylinder Records,* has an exhaustive list.

26. History of Will Hayes, an Edison recording engineer, in 1895 Phonograph folder, ENHS.

27. National Phonograph Co. record catalogs; John Fell, "Cellulose Nitrate Roots," in *Before Hollywood: Turn of the Century Film* (New York: Hudson Hills, 1987), p. 43

28. This quote was taken from a Victor catalog (1909) in facsimile, published privately by A. Koenigsberg, NJSL.

29. "Sales of Edison Phonograph Merchandise," 1899 to 1925, in 1925 Phonograph folder; A. Clark to E. R. Johnson (ERJ), 31 Aug. 1908, reporting on sales figures shown to him by Gilmore, box 4, AHC.

30. *Edison Phonograph Monthly*, Feb. 1907.

31. Text of agreement between Edison Manufacturing Co. and Kinetoscope Exhibiting Co., 10 Sept. 1895, ENHS.

32. Charles Musser, *Guide to Motion Picture Catalogues by American*

Producers and Distributors, 1894–1908 (Frederick, Md.: University Publications of America, 1985), p.5.

33. *Electrical World* 23 (June 1894): 799.

34. F. Chrisman interview, *St Louis Republican*, 3 July 1899, ENHS.

35. Edison said that he was not confident in the patent system and thought he would try and keep his motion picture technology a "trade secret." Edison's statement, complainants record, *Edison v. American Mutascope Co.*, 1910, legal box 173, ENHS. He was also short of money at the time and did not want to pay the trifling amount to apply for foreign patents.

36. Ramsaye, *Million and One Nights*, ch. 26.

37. F. R. Fulton, "The Machine," in Balio, *American Film Industry*, p. 39.

38. Ramsaye, *Million and One Nights*, ch. 16.

39. Gordon Hendricks, "Beginnings of the Biograph," in *Origins of the American Film* (New York: Arno, 1972), p. 65.

40. Robert C. Allen, "Movies in Vaudeville," p.60.

41. Annotation on T. Armat to TAE, 11 May 1922, ENHS.

42. Musser, *Motion Picture Catalogues*, p. 11.

43. C. Musser, "American Vitagraph," in Fell, *Film Before Griffith*, pp. 33–34.

44. Brochure for Edison projecting Kinetoscope in Primary Printed, ENHS.

45. Ramsaye, *Million and One Nights*, p. 328.

46. An account of the rise of one dealership is in Erik Barnouw, *A Tower In Babel: A History of Broadcasting in the United States* (New York: Oxford University Press, 1969), p. 130; Clark, "His Master's Voice," p. 176, AHC.

47. TAE to R. Miller, 2 Dec. 1897, LB 931031, p. 612, ENHS.

48. J. Randolph to J. Haines, 29 Dec. 1897, LB 931031, p. 626, ENHS.

49. TAE to W. Cutting, 26 Oct. 1897, LB 931031, p. 567; TAE to H. Dick, 12 Jan. 1899, LB 980131, p. 319, ENHS.

50. Batchelor diary; J. Randolph to S. G. Burn, 15 Nov. 1900, LB 000608, p. 229, ENHS.

51. "Sales of Edison Phonograph Merchandise," 1899 to 1925, in 1925 Phonograph folder, ENHS.; A. Clark to ERJ, 31 Aug. 1908, AHC.

52. *Edison Phonograph Monthly* (July 1906).

53. Roundtable letter, a short informal history of the organization written by Charles Edison, 26 April 1920, in Engineering Department files (henceforth cited as Eng. Dept. files), ENHS.

54. Dyer memo, 10 June 1908, ENHS.

55. Claims were made that Dickson took his notes with him when he left during the *Edison v. American Mutascope* trial, complainants record, legal box 173. Edison said that Dickson was guilty of "double crossing me and selling me out to Latham for his own benefit." Annotation on letter

from T. Armat, 11 May 1922, ENHS. The activities of the pioneer filmmakers are described in Jenkins, *Images and Enterprise,* pp. 270–73.

56. Musser, *Motion Picture Catalogues,* p. 7. Columbia was formed by Edward D. Easton, a court reporter who thought the talking machine would replace stenography. Easton was known as one of the shrewdest businessmen in the industry. Victor also had to re-engineer its product to meet the low-priced competition from Columbia. B. J. Royal to ERJ, 4 June 1902, 29 May 1903, box 26, AHC.

57. Edison detailed John Ott to build a correcting device to print images from the negative so that the positive would have all images an equal distance apart; this was to stop the image from moving on the screen due to the camera being different distances from the subject. Edison's instructions began, "the Biograph people do this." 1909 Motion Picture folder, ENHS. Edison spies were in the Dictaphone and Victor companies, M. Jones to C. Edison, 30 July 1917, ENHS.

58. Bergmann to TAE, 3 April 1888, ENHS. The man in question was a Mr. Umbach, an instrument maker, "a very good hand on experiments," who worked with Bell and Tainter in Washington.

59. Tate to Dyer, 7 Sept. 1888, ENHS.

60. Wile, *Proceedings of 1890 Convention,* p. xliv.

61. Columbia announcement, 1907, reproduced in *Dusting of a Little History: Spring Type Phonographs* (Yorba Linda, Calif.: Phonograph Collectibles, 1981), p.1.

62. *American Machinist* 26 (9 July 1903): 978.

63. *Phonograph and How to Use It,* pp. 152–56.

64. Account of the record duplication project by Will Hayes of Wurth's team (post-1914 but in 1895 phonograph folder, ENHS). Aylsworth had a standard sample of the wax compound that he checked with all other versions. Even the smallest change from advised practice, such as a single rather than double pressing of stearic acid, could ruin the compound. Aylsworth to Gilmore, 10 Jan. 1907, ENHS.

65. A gatehouse was constructed in 1890 to help keep unauthorized people out of the lab, memo of 18 Sept. 1890, ENHS. Edison had made a habit of working in secret rooms since his days in Newark. In 1887 he instructed Batchelor to provide a private room in the new lab for "special things I want sub rosa." TAE to Batchelor, 6 April 1887, ENHS.

66. Kennelly's testimony, "Edison is very courteous to visitors and lets them, as a rule, experiment upon everything." He reported that there were several instances of this happening and Edison letting costs run up. Hearing on the electrocution of William Kemmler, New York State Court of Appeals, 1847–1911, 893 (Buffalo, N.Y., 1890) 2: 751. Billbook #13 records Harold Brown's use of machinists, Dec. 1893. Shop order #841 for Dr. Klein's experiments on X rays, ENHS.

67. Fulton, "The Machine," p. 35.

68. Shop order #777, Sept. 1894, contract with C. Hopflinger to make

model kinetoscope motor; #857, April 1897, contract with Mr. Thronicke.

69. Contract between TAE and J. Egan reproduced in Hendricks, "The Kinetoscope," pp. 9–10.

Chapter Nine. Thomas A. Edison, Incorporated

1. Benjamin Hampton, *A History of the Movies* (New York: Arno, 1970), p. 18.

2. Balio, *American Film Industry*, p. 23. See also Charles Musser, *Before the Nickelodeon: Edwin S. Porter and the Edison Manufacturing Company* (Berkeley and Los Angeles: University of California Press, 1989).

3. Shop order #542 (1891) testing batteries for GE; #623 and 631 (1892) experimental battery for Edison; #646 (1893) experiments on recharging Lalande battery; #752 (Feb. 1894); #935 (June 1889) personal battery experiments TAE, ENHS.

4. *Edison Phonograph Monthly* (Aug. 1906).

5. Conot, *Streak of Luck*, p. 347; Edison claimed in an interview with *Harper's Weekly* that "there will not be a belt in the mill" (21 Dec. 1901), ENHS.

6. *Edison Phonograph Monthly* (Nov. 1906).

7. Vanderbilt calls the Edison battery "amazing" because of the complex machinery and processes that were developed for it. *Edison the Chemist*, p. 213; Silver Lake cleaning shop order #1140, 1124, ENHS.

8. Edison started by offering advice, encouragement, and financial assistance to Lansden; his motivation was described thus, "I am anxious that a practicable vehicle shall be put on the market." TAE annotation on J. M Lansden to TAE, 2 July 1903, ENHS. He then went on to encourage Weir, of the Adams Company, to build a copy of the Lansden wagon. TAE to Weir, Nov. 1904, ENHS.

9. In 1908 Edison bought out the Battery Supply Company of Newark, a competitor in the primary battery business. This union obtained the services of E. Hudson and F. J. Lepreau who managed TAE, Inc.'s primary battery business in the twentieth century. Frank J. Prial, *75 Years of Packaged Power* (Bloomfield, N.J.: McGraw Edison, 1964), p. 15, NJSL.

10. The merger movement in business reached a peak during the early years of the lab. Between 1888 and 1905, 156 trusts were formed. Stanley Lebergott, *The Americans: An Economic Record* (New York: W. W. Norton, 1984), p. 314.

11. "The Prophecy of the Famous Wizard," newspaper interviews, 16 July 1899, ENHS.; "Advantages of the Electric Car," promotional material in Storage Battery folder, ENHS.

12. "Edison's Most Important Discovery," *Harper's Weekly* (21 Dec. 1901).

13. Early years of the new technology and the Pope Company are described in James J. Flink, *America Adopts the Automobile* (Cambridge: MIT Press, 1970), pp. 238–40. Tricycle experiments in experimental list of 1 Aug. 1889, laboratory folder, ENHS.

14. Thomas P. Hughes, *Elmer Sperry: Inventor and Engineer* (Baltimore: Johns Hopkins Press, 1971), pp. 81–85.

15. James J. Flink, *The Car Culture* (Cambridge: MIT Press, 1975), pp. 15–16.

16. Edison car, TAE annotation on T. C. Weir to TAE, 11 April 1902, ENHS.

17. TAE to H. Byllesby, 3 June 1911, LB 110306, p.550, ENHS.

18. One headline in the 1899 news clippings in ENHS was, "It Will Run by Electricity, Will be Light, Simple Enough for a Child to Manage, and Will be Sold for No More than a Horse and Carriage." *New York Morning Journal*, 25 June 1899.

19. Robert Anderson, "The Motion Picture Patents Company," (Ph.D. diss., University of Wisconsin, 1983), p. 166.

20. Notebooks from 1887 detail experiments on rechargeable batteries (see N 870301, N 880610, N 921109). The first shop order for storage battery work is #28 (Jan. 1888). See W. B. Carlson, "Thomas Edison as a Manager of R&D: The Case of the Alkaline Storage Battery," *Technology and Society* 7 (Dec. 1988): 4–12.

21. The lab staff looked at patents and scientific literature from several European countries. Richard H. Schallenberg, *Bottled Energy: Electrical Engineering and the Evolution of Chemical Energy Storage* (Philadelphia: American Philosophical Society, 1982), pp. 353–54.

22. Meadowcroft claims that Edison said in 1882 that a better storage battery could be found. William Meadowcroft, *Edison and His Storage Battery* (West Orange, N.J.: TAE, Inc., 1935), p. 4, NJSL; Dyer and Martin, *Edison: His Life*, p. 554.

23. He admitted (in 1905) that he had not found the problem of the leaking cells and did not know when he would. Annotation on L. Weir to TAE, 20 Jan. 1905. The campaign to find the problem is recounted in Edison's letters to Bergmann: 9 Nov. 1904, 29 Nov. 1904, 9 Dec. 1904, ENHS.

24. *Motor Age* 12 (13 Feb. 1908): 12.

25. Trial balance, 30 April 1905, 1905 Storage Battery folder, ENHS.

26. By 1910 its indebtedness to him was $1.3 million, a sum that was more than the capital stock of the company. Financial memo, 9 May 1910, ENHS. A disgruntled Gilmore told his friends at Victor that Edison took $1.5 million in cash out of the Phonograph Company. A. Clark to ERJ, 31 Aug. 1908, box 4, AHC.

27. TAE to F. Comitot, 15 Nov. 1907, LB 070701, p. 178, ENHS.

28. TAE to F. T. Collver, 16 Oct. 1907, LB 070701, p. 114; Asst. Sec. to F. Scheffler, 5 Nov. 1908, LB 080919, p. 294, ENHS.

29. Gilmore memo, 1906 Phonograph folder, ENHS.

30. Roundtable letter, 1920 Eng. Dept. files, ENHS.
31. Shop orders #792 (Dec. 1894), #813 (April 1896), ENHS.
32. The first long-playing cylinders were introduced for the Twentieth Century Graphophone in 1905; they were followed by celluloid cylinders produced by the Indestructible Company and sold by Columbia. Welch and Read, *Tinfoil to Stereo*, p. 100. (The Indestructible Record Company was formed in 1906 around Thomas Lambert's celluloid patents.)
33. *Edison Phonograph Monthly* (Jan. 1910).
34. A. Westee to Dyer, 30 March 1910, ENHS.
35. Dyer to TAE, 30 March 1910, ENHS.
36. Dyer to TAE, 30 March 1910. By 1911 the Edison Storage Battery Co. was doing about $60,000 of business each month, but its start-up costs were so high (a result of years of R and D) that it had yet to record a profit. TAE to W. Sloane, 30 Jan. 1911, LB 101218, p.350, ENHS.
37. W. M. Lybrand to Dyer, 23 March 1910, ENHS. Lybrand was a partner in an accounting firm, who was sending information about company organization to Dyer.
38. Dyer to National Phonograph Co., 13 Sept. 1910, ENHS.
39. Press clippings in 1911 TAE, Inc. folder, ENHS.
40. The corporation was financially strong, with $306,069 cash in hand, $349,237 invested, and its accounts receivable exceeding accounts payable by over $750,000. Its surplus of assets over liabilities was $9,752,556. The corporation was to act as a financial umbrella, supporting the weak members with the earnings of the strong. In 1911 it was due $879,456 from its affiliated companies, an indication of the financial strain of supporting the weaklings. National Phonograph Co. to R. G. Dun Co., 31 July 1911, ENHS.
41. A. Clark to ERJ, 31 Aug. 1908, box 4, AHC. The talking machine business was a close fraternity in which managers moved from one company to another. Several managers left Edison for Victor and took valuable information and financial data with them.
42. As the *Edison Phonograph Monthly* (March 1908) admitted: "From an organizational standpoint the depression has been advantageous, forcing us to reorganize and reduce."
43. E. J. Berggren to Dyer, 11 May 1911, ENHS.
44. This involved taking away purchasing, advertising, accounting, and advertising functions from smaller companies and incorporating them into the departments of the main administration. It was estimated by the accountants that this would save approximately $50,000 each year. Several smaller organizations, such as the Bates Co., kept their corporate identities intact although TAE, Inc. took over their management. TAE, Inc. annexed certain functions, like advertising and purchasing, regardless of the legal independence of the various Edison companies. The reduced administration was seen as "the wisest and most generally accepted cure for decreased profits." Roundtable letter, 1920 Eng. Dept. files, ENHS.
45. Roundtable letter, 1920 Eng. Dept. files, ENHS.

46. Bylaw of 15 April 1911, ENHS.

47. Dyer to T. Graf, 20 March 1911, ENHS.

48. Batchelor diary, 1338, p. 51, quoting Aylsworth, 1 Sept. 1903; Columbia advertising in Primary Printed collection, ENHS.

49. "Try everything you can towards economy. No one is safe in this cold commercial world that can't produce as low as his greatest competitor." TAE to Insull, 10 May 1895, ENHS.

50. Read and Welch, *Tinfoil to Stereo,* p. 96.

51. J. Schermerhorn to Gilmore, 19 Feb. 1907; memo to Weber, 6 Feb. 1907, ENHS.

52. C. Wilson to Gilmore, 24 May 1907; Weber memo, 10 July 1907, ENHS.

53. Roundtable letter, 1920 Eng. Dept. files, ENHS.

54. Dyer memo, 25 March 1910, ENHS.

55. *Edison Phonogram* 2 (Dec. 1892): 263–74

56. Weber and his staff carried out improvements on machines, often at the request of dealer feedback. For example, they altered the horn of the phonograph to improve sound quality and give the machine a new look, redesigning the horn and its attachment brackets in 1907. N. Durand to Gilmore, 9 March 1907, ENHS.

57. Engineering Department History, by John Constable, found in 1919, TAE, Inc., undated but probably written about 1920 (henceforth cited as Constable History).

58. Dyer memo, 25 March 1910, ENHS. The seeds for this organization were sown in 1909. Peter Weber appointed an assistant (Charles Schiffl) to deal with the day-to-day engineering work of the phonograph product. Schiffl was transferred from the Works to the main laboratory building, and there he worked on the design of new phonographs. Schiffl's move to the lab was the first step in bringing many engineering functions under one centralized control. P. Weber to all foremen, 29 Jan. 1909, ENHS.

59. TAE to Dyer, 20 Jan. 1910, ENHS.

60. Dyer to T. Graaf, 21 March 1910, ENHS.

61. This powerful committee also approved expenditure on product development, supervised production engineering and the manufacture of tools, and decided on improvements in the manufacturing plant. Dyer memo, 3 June 1910, ENHS.

62. C. Wilson memo, 24 June 1910, ENHS.

63. Dyer to A. Westee, 26 Jan. 1910, ENHS.

64. Hounshell, *American System to Mass Production,* pp. 13, 237.

Chapter Ten. The Diamond Disc

1. Numbers from accounts sheet in 1912 Phonograph folder, ENHS; Victor numbers from B. L. Aldridge, "The Victor Talking Machine Company," reproduced in Ted Fagan and William R. Moran, *The Encyclopedic*

Discography of Victor Recordings (Westport, Conn.: Greenwood, 1983). Quote is in F. Dolbeer memo, 10 Jan. 1911, ENHS.

2. Bruce, *Alexander Graham Bell*, pp. 262, 350.

3. "An Outline of Facts Pertaining to the Talking Machine Industry," a typed document produced for advertising purposes by the Victor Company that got into Meadowcroft's hands. In 1913 Phonograph folder, ENHS.

4. Eldridge Johnson's personal recollections, as dictated to D. E. Wolff, 25 Oct. 1910, box 11, AHC. The first motor delivered to the company was too expensive to manufacture and it was followed by an altered model that went into production in August 1896. The reproducer was developed with Alfred Clark, and a joint patent was issued.

5. Edgar Hutto, "Emile Berliner, Eldridge Johnson, and the Victor Talking Machine Company," *Journal of the Audio Engineering Society* 25 (Oct. 1977): 666–73; "High Light History of Victor Phonographs," Victor file, Division of Mechanisms, Smithsonian Institution.

6. Johnson recalled that a melted down Edison cylinder was used to make up the disc in the experiments. E. R. Fenimore Johnson, *His Master's Voice Was Eldridge Reeves Johnson* (Milford, Del.: State Media, 1974), pp. 45–48. Johnson said that this process was the most important asset of the company. Recording remained virtually unchanged in principle until electrical recording in 1925. E. R. Johnson, "History of Victor"; "Sworn Statement by E. R. Johnson," June 1928, box 8, AHC.

7. Geoffrey Jones, "The Gramophone Company: An Anglo-American Multinational," *Business History Review* 59 (Spring 1985): 79–81.

8. Annual Reports of the Victor Co., box 32, AHC. Johnson began his construction program about 1905, a few years after Edison. He built a new office building, a new laboratory, and several multistory factory buildings.

9. The original patent was filed in 1903, number 143,060, and was supported by several others filed in 1903–6.

10. Johnson's patents on the Victrola covered both the mechanism and the design. The original patent, awarded in 1907 (and reissued later), claimed to improve the reproduction of sound and "the appearance of the machine as a whole, so as to provide an ornamental piece of furniture."

11. John Wesley Hyatt developed celluloid and gained the important patents. His brother, Isiah Smith Hyatt, coined the term and the brothers set up the Celluloid Manufacturing Co., which began operations in Newark in 1872. Robert Friedel, *Pioneer Plastic* (Madison, Wisc.: University of Wisconsin Press, 1983), pp. 12–17. Thomas Lambert gained a patent covering the use of the material as a recording medium. The Indestructible Record Co. was established around his patents in 1906. Read and Welch, *Tinfoil to Stereo*, p. 96.

12. Even with modern equipment and ears, one can hear a difference in the Blue Amberol recordings, which have a clarity and a sound of their own. One of the supporters of the Blue Amberols is George W. Childs, who at eighty-five still plays his Amberols on a converted Diamond Disc machine. He writes, "They were the finest acoustic records ever

made . . . playing them through the disc horn is wonderful!" Letter to the author, 3 Dec. 1987.

13. F. Dolbeer to Dyer, 23 Dec. 1910; J. Blackman to Dyer, 8 Dec. 1910, ENHS.

14. George L. Frow, *The Edison Disc Phonographs and the Diamond Discs* (Tunbridge Wells, U.K.: G. L. Frow, 1982), p. 16; ERJ to L. Douglas, 19 Sept. 1911, box 7, AHC.

15. There can be little doubt that the West Orange laboratory copied the Victrola; in 1911 Edison was designing cabinets for a new machine "like the Victrola, but far finer." TAE to Cunningham Piano Co., 19 Dec. 1911, LB 111204, p. 132, ENHS.

16. TAE to Legal Dept., 6 Dec. 1911, legal box 15, ENHS. After considering the patent position, the Legal Department anticipated litigation over the Diamond Disc violating Victor's patents but regarded the commercial advantages worth the risk. L. Hicks to Dyer, 12 April 1911, legal box 15, ENHS.

17. Vanderbilt, *Edison the Chemist*, pp. 131–35.

18. Frow and Selfi, *Edison Cylinder Phonographs*, pp. 138–47.

19. Read and Welch, *Tinfoil to Stereo*, pp. 192–93.

20. Edison's notes for Hutchison, 1912 Phonograph folder, ENHS.

21. L. McChesney to W. Miller, 24 Jan. 1915, ENHS.

22. Price lists and promotional material in Primary Printed collection, ENHS.

23. Maxwell memo, 21 July 1910, ENHS.

24. F. Babson to C. Wilson, 29 Dec. 1914, ENHS.

25. H. F. Miller to Goldsmith, LB 151216, p. 121, ENHS.

26. TAE to E. H. Johnson, 28 Oct. 1912, ENHS.

27. Frow, *Edison Disc*, p. 38; C. E. Fairbanks to Mambert, 31 May 1917, ENHS.

28. M. R. Hutchison To C. Edison, 15 April 1912, ENHS. Hutchison wrote, "Someone in the Works is trying to undermine me." Hutchison diary, 7 Jan. 1912.

29. TAE to H. Byllesby, 2 Dec. 1911, LB 110628, p. 695, ENHS.

30. Manufacturing Committee minutes, 26 Jan. 1911, ENHS.

31. TAE memo, 5 Nov. 1913, Eng. Dept. Files, ENHS.

32. Obituary, *New York Times*, 18 Feb. 1944; Hutchison was made chief engineer in August 1912. Diary, 17 July and 12 Aug. 1912.

33. P. Cromelin to Dyer, 17 Aug. 1911, ENHS.

34. "And have found, to my surprise, that there is very little originality in it . . . the fact is that musical composition is full of plagiarism." Interview in *Musical Trade Review* (20 Feb. 1911), ENHS.

35. TAE to T. Graf, 20 Nov. 1911, LB 110628, p. 622, ENHS.

36. TAE to Gallup and Alfred, 21 Feb. 1916, LB 160127, p. 387, ENHS.

37. Meadowcroft to C. Aston, 19 Feb. 1917, LB 170122, p. 182, ENHS.

38. Accounts records for kinetophone project, especially in shop orders, #2127, 1988, 2001 and 2153, billbooks 1899–1910, ENHS.

39. MPPC circular, 1909 Motion Picture folder, ENHS.

40. See Robert J. Anderson, "The Motion Picture Patents Company," (Ph.D. diss., University of Wisconsin, 1983).

41. Balio, *American Film Industry,* p. 114.

42. TAE to H. P. Weidy, 29 June 1899, LB 980131, p. 606, ENHS.

43. Newspaper clippings, 5 May 1910, Motion Picture folder, ENHS.

44. J. W. Farrell to Dyer, 15 June 1909, ENHS.

45. Meadowcroft to J. Eliot, 8 Oct. 1912, ENHS.

46. D. Bliss to W. Greene, 10 May 1912, ENHS.

47. Interview of 27 Aug. 1910; Hutchison to TAE, 19 Dec. 1912, ENHS.

48. Annotation on 1913 Hutchison memo, Motion Picture folder, ENHS.

49. H. F. Miller to W. French, 12 Sept. 1910, LB 100829, p. 55, ENHS.

50. Dyer to R. Cross, 22 April 1912, 8 July 1912, ENHS.

51. C. Wilson memo, 18 Dec. 1912, ENHS.

52. Hutchison diary, 28 April 1913, ENHS.

53. The story of this project is told by Art Shiffrin, "70th Anniversary of the Theatrical Release of Kinetophone," *Society of Motion Picture and Television Engineers Journal* 92 (July 1983): 739–51; and "Time Code and Mr Edison," *db* 16 (Dec. 1982): 30–37. See also Rosalind Roganoff, "Edison's Dream: A Brief History of the Kinetophone," *Cinema Journal* 15 (Spring 1976): 58–68.

54. Hutchison to TAE, 19 Jan. 1914, ENHS.

55. A. Shiffrin, "The Trouble with the Kinetophone," *American Cinematographer* (Sept. 1983): 50–54.

56. TAE to Hutchison, 14 Jan. 1914, ENHS.

57. Hutchison to TAE, 19 Jan. 1914, ENHS.

58. A. M. Kennedy to Hutchison, 20 May 1914, ENHS.

Chapter Eleven. The Rise of the Organization

1. Conot, *Streak of Luck,* p. 408.

2. Hutchison diary, 10 Dec. 1914, ENHS.

3. Dated 19 Dec. 1914, Phonograph folder, ENHS.

4. John D. Venable, *Out of the Shadow: The Story of Charles Edison* (East Orange, N.J.: C. Edison Fund, 1978), p. 76.

5. C. H. Wilson memo, 30 Dec. 1912, ENHS.

6. J. Constable, "Report on New Model," 9 Nov. 1914, ENHS.

7. R. H. Allen to IRS, 1 Oct. 1920, ENHS.

8. TAE to Bergmann, 24 Dec. 1910, LB 101208, ENHS.

9. Hutchison minutes of meetings of Engineering Department, 19 Nov. 1914, Eng. Dept. files, ENHS.

10. R. A. Bachman to Mambert, 31 Dec. 1914, ENHS.

11. Wear on Diamond Discs was immediately noticed by Victor's engineers when they tested them; E. R. Johnson to A. Clark, 3 Feb. 1913, Johnson papers, box 4, AHC.

12. Meadowcroft to Dickson, 30 Dec. 1915, ENHS.

13. Roundtable letter, 1920, Eng. Dept. files, ENHS.

14. A. Chandler, *Strategy and Structure: Chapters in the History of American Industrial Enterprise* (Cambridge: MIT Press, 1962).

15. Alfred P. Sloan, *My Years with General Motors* (New York: Doubleday, 1964), p. 42.

16. Dinwiddie memo, Cylinder Division Blue Book, 21 April 1917, ENHS.

17. Meadowcroft, box MB 68; Mambert memo, 22 Jan. 1916, ENHS.

18. Mambert to Musical Phonograph Division, 5 Feb. 1916, ENHS.

19. I. A. Ventres to C. Edison, 17 March 1914; Mambert to divisions, 26 April 1916, ENHS.

20. Roundtable letter, 1920 Eng. Dept. files, ENHS.

21. A memo stated that it was Edison's "personal desire . . . to effect greater economies and properly conserve our finances." Financial Executive memo #3602, 5 June 1916, ENHS.

22. Ibid.

23. John Venable, *Out of the Shadow,* ch. 4.

24. F. W. Taylor's book, *The Principles of Scientific Management* (New York: W. W. Norton, 1967), was first published in 1911. His calling card for his management consulting business announced: "Systematizing Shop and Manufacturing Costs a Specialty." Thomas Cochran, *American Business in the Twentieth Century* (Cambridge: Harvard University Press, 1972), p.75.

25. Taylor, *Scientific Management,* pp. 48–49.

26. Taylor spoke of enforcing standardization and adoption of his methods, Montgomery, *Fall of House of Labor,* p. 229.

27. Noble, *America by Design,* pp.263–68; Donald R. Stabile, "The Du Pont Experiments in Scientific Management: Efficiency and Safety, 1911–1919," *Business History Review* 61 (Autumn 1987): 366–69.

28. C. Wilson to all division heads, 14 Dec. 1914, ENHS.

29. Mambert to Edison Storage Battery Co., 22 Jan. 1916, ENHS.

30. Meadowcroft, "Edison Industries," box MB 78; C. Edison to managers, 25 June 1917, ENHS.

31. The years after 1900 saw an increase in the importance of professional accountants, especially in the wake of income and corporate taxation. See David F. Hawkes, "The Development of Modern Financial Reporting Practices among American Manufacturing Corporations," in Baughman, *American Management,* p. 116.

32. R. H. Allen to Mambert, 3 Years Report, 3 April 1919, ENHS.

33. Mambert memo, 11 Sept. 1916, ENHS.
34. Conot, *Streak of Luck,* p. 422.
35. Dyer to TAE, 21 Oct. 1912, ENHS.
36. Financial Executive memo #3497, 6 March 1916, ENHS.
37. Disc Record Division Blue Books (1916–19) contain the copious correspondence and calculations about disc record costs.
38. Mambert to C. Edison, 30 Jan. 1918, ENHS.
39. Mambert to W. Stevens, 26 Aug. 1918, ENHS.
40. Mambert to J. W. Robinson, 16 Nov. 1918, ENHS.
41. Constable History, 1919, ENHS.
42. His terms of employment were, "Devote your entire time and attention to perfecting a process of . . . reproducing . . . pictures in natural colors." J. White to A. Werner, 15 June 1900, ENHS.
43. N. Durand to C. Wilson, 16 Feb. 1910, ENHS.
44. Hutchison to C. Wilson, 8 Dec. 1914, ENHS.
45. Memo of 11 Sept. 1917, Laboratory Blue Book, ENHS.
46. Constable History, 1919, ENHS.
47. J. Constable to C. Edison, 2 April 1920, Eng. Dept. files, ENHS.
48. Hutchison to TAE, 5 Nov. 1913, ENHS.
49. Edison's lab to A. Fleming, 24 July 1916, LB 160630, p. 302, ENHS.
50. C. E. Mitchell, Jan. 1918, Employment ratings Blue Books, ENHS.
51. *Edison Phonograph Monthly,* Nov. 1906.
52. G. Clark to J. Constable, 16 July 1918, Disc Record Division files, ENHS.
53. Edison's exalted position had always justified his own draftsman, but the reorganization of drafting gave Edison's personal drafting to the main department. J. Constable to W. O. lab, 24 Oct. 1916, ENHS.
54. Lab floor plan, 25 Nov. 1918; J. Constable to C. Edison, 19 Sept. 1916, ENHS.
55. J. Constable to C. Hayes, 27 Nov. 1917, ENHS.
56. TAE to S.W. Cutting, 20 Oct. 1916, LB 160913, p. 516; Meadowcroft to Delaney, 8 Jan. 1916, LB 151216, p. 330, ENHS.
57. Meadowcroft to C. Goodwin, 20 Oct. 1916, LB 160913, p. 520; Lab floor plan, 25 Nov. 1918, ENHS.
58. Lab floor plan, 25 Nov. 1918, ENHS.
59. Meadowcroft to A. Fleming, 24 July 1916, LB 160630, p. 302, ENHS.

Chapter 12. Business and Technology: The Dictating Machine

1. TAE, "The Perfected Phonograph," *North American Review* 146 (June 1888): 641–50.
2. Durand in Ediphone minutes, 27 Dec. 1916. Charles Edison fund, East Orange, N.J. (henceforth cited as CEF).

3. Woodbridge, *Dictaphone,* pp. 12–13.
4. Edison, "The Perfected Phonograph."
5. Miller reminiscences, *New York Public Ledger,* 22 March 1925, ENHS.
6. Obituary, *New York Times,* 1 July 1949, p. 25.
7. Durand to Dyer, 17 March 1909, ENHS.
8. Durand in Ediphone minutes, 27 Dec. 1916, CEF.
9. Charles Hibbard came to the lab in 1906 to work on the business machine. He was placed under the supervision of Peter Weber. Gilmore to Durand, 26 March 1907, ENHS.
10. Dyer to F. Scribner, 18 April 1907, ENHS.
11. Durand to Dyer, 17 March 1909, ENHS.
12. N. Holland to Hutchison, 5 Dec. 1913, ENHS.
13. N. Holland to Hutchison, 31 Oct. 1912, ENHS.
14. Advertising circular, box D8, ENHS.
15. Ibid.; J. Constable to N. Durand, 1 April 1918, ENHS.
16. Woodbridge, *Dictaphone,* p. 14.
17. TAE to Gouraud, 30 May 1889, LB 890521, pp. 153–70; Comparison of Business Phonographs, 19 May 1905, Phonograph folder, ENHS.
18. Advertising material, 1907 Phonograph folder, ENHS.
19. Ediphone minutes, 27 Dec. 1916, CEF.
20. Edison business phonograph promotional material, 1911 Phonograph folder, ENHS.
21. Holland reports, 1918 Phonograph folder, ENHS.
22. *Scientific American* 111 (12 Sept. 1914): 216.
23. Advertising flyer of North American Phonograph Co., Primary Printed, ENHS.
24. Margery W. Davies, *A Woman's Place Is at the Typewriter: Office Work and Office Workers 1870–1930* (Philadelphia: Temple University Press, 1982), p. 31; Braverman, *Labor and Monopoly Capital,* pp. 294, 298.
25. Davies, *Woman's Place,* p. 79. Between 1870 and 1910 the number of women who worked for wages doubled. By 1930 women accounted for 25 percent of all wage earners and 96 percent of stenographers and typists. Green, *World of the Worker,* pp. 43, 105.
26. Advertising copy, 1909 Phonograph folder, ENHS.
27. Numbers for production are given in Durand's Tips pamphlets to dealers. Conrad Benesham has compiled lists of Ediphone serial numbers by year, which give a good guide to output. These sources are at ENHS.
28. Davies, *Woman's Place,* pp. 30–31.
29. Durand to Gilmore, 9 March 1907, ENHS.
30. Ediphone competition bulletin, #179, 6 April 1931, ENHS.
31. Ediphone Buyers Guide, Warshaw collection, Smithsonian Institution.
32. National Ediphone Sales Conference, 9 Jan. 1925, ENHS.
33. Charles Edison to division managers, 4 Sept. 1917, ENHS.

34. "A neatly dressed young business woman . . . but not one who thinks she is hired for the exhibition of her best clothes," Ediphone Tips #160, 1912, ENHS.

35. Memo to Constable, Development Work for Ediphone, 1 Nov. 1919, ENHS.

36. Durand to C. Edison, 31 Jan. 1918, ENHS.

37. "List of Suggestions . . . for Improvement of the Ediphone," 1918 Phonograph folder, ENHS.

38. Holland to C. Schiffl, 1910 Phonograph folder, ENHS.

39. Holland to M. R. Hutchison, 31 Oct. 1912, ENHS.

40. Meadowcroft to F. Kimball, 2 May 1911, LB 160314, p. 663, ENHS.

41. Ediphone Buyers Guide, Warshaw collection, Smithsonian Institution.

42. Durand to S. Langley, 2 April 1918, ENHS.

43. *Scientific American* 111 (12 Sept. 1914): 216.

44. Leffingwell's first of many books on this movement was entitled *Scientific Office Management* (1917). It was soon followed by Lee Galloway's *Office Management: Its Principles and Practice* (Chicago: A. W. Shaw, 1925). These books argued that the purpose of office management is control over the office and office work. Braverman, *Labor and Monopoly Capital*, pp. 304–7. Davies points out that "efficiency experts" in the office tried to eradicate traditional practices in the same way that industrial engineers undermined the "shop culture" on the factory floor. *Woman's Place*, p. 104.

45. Meadowcroft box MB 78; Ediphone business, 6 May 1918, ENHS.

46. Annual Ediphone Conference, 7 Jan. 1928, Charles Edison closing address, ENHS.

47. Green, *World of the Worker*, p. 105.

48. Ediphone competition bulletin #167, 14 Nov. 1929, TAE, Inc., Phonograph Division, ENHS.

49. Durand to J. Constable and C. Edison, 27 Aug. 1918, ENHS.

50. J. L. Meikle, *Twentieth Century Limited: Industrial Design in America 1925–1939* (Philadelphia: Temple University Press, 1979), p. 82.

Chapter Thirteen. The Impending Conflict

1. In a newspaper interview, 18 Nov. 1895, Edison described how high voltage alternating currents would make an excellent weapon. He thought that sprays of water would make the necessary contact. He also talked about torpedoes, dynamite guns, and bombs on balloons. 1895 Torpedo folder, ENHS.

2. Conot, *Streak of Luck*, p. 415.

3. R. Schallenberg, "The Anomalous Storage Battery: American Lag in Early Electrical Engineering," *Technology and Culture* 22 (Oct. 1981): 725.

4. Richard Compton-Hall, *Submarine Boats: The Beginnings of Underwater Warfare* (New York: Arco, 1984), p.76.

5. Ibid., pp. 96–97. Holland's story is told by Richard K. Morris, *John Philip Holland* (Annapolis, Md.: U.S. Naval Institute Press, 1966).

6. Hutchison to Nixon and Mannock, 25 March 1911, ENHS.

7. A. Preston, *Submarines* (London: Phoebus, 1975), p. 27.

8. Hutchison experiments, diary entries, 6 March 1911, 12 Aug. 1911, 26 Sept. 1911, 1 July 1914, and 15 July 1914, ENHS. Accidents of the early submarines were very common; between 1901 and 1914 there were about sixty-eight serious accidents worldwide. Compton-Hall, *Submarine Boats,* p. 162. Explosions due to battery gases exploding occurred in U.S. Navy submarines before and after the E2 disaster, including explosions on the *Octopus* and *Stingray* in 1907/08. Edwin P. Hoyt, *Submarines at War* (New York: Stein and Day, 1983), p. 24.

9. TAE to F. D. Roosevelt, 10 Sept. 1915, ENHS.

10. J. Daniels to TAE, 7 July 1915, Naval Consulting Board (NCB) records, box 1, ENHS.

11. In 1903 Admiral Melville pointed out the advances of Germany's fleet and the connection with industrial research. His arguments were summarized by Willis Whitney on 20 Dec. 1915, and a copy was given to Edison, NCB box 1, ENHS.

12. Newspaper interview, 18 Nov. 1895, ENHS.

13. TAE to *Chicago Examiner,* 6 Aug. 1914: "Civilization is under going a surgical operation. It is necessary to settle for all time that Dynastic Militarism shall disappear from the earth. That the people shall rule through constitutional governments . . . as they do now in the United States," ENHS.

14. TAE to Rollin Kirby, 10 May 1917, LB 170426, p. 90, ENHS.

15. J. Daniels to TAE, 7 July 1915, NCB records, box 1, ENHS.

16. Lloyd N. Scott, *Naval Consulting Board of the United States* (Washington, D.C.: GPO, 1920), p. 111.

17. TAE to Tillman, 1916, NCB box 4, ENHS.

18. TAE to J. Daniels, 4 March 1918, NCB box 22; TAE to C. Harwood, 6 April 1917, LB 170122, p. 571. Despite years of effort, Edison was never able to cut through the bureaucratic red tape of the navy to build a naval laboratory. As usual, the difficulty with establishing the lab was finance; Congress did not appropriate enough money to suit Edison's grand design—he estimated that the lab would cost $5 million fully equipped. He vigorously opposed the establishment of the lab near Washington, arguing that it could only be a success if it was far away from the capital. His own choice was Sandy Hook, New Jersey. NCB boxes 4 and 22 contain much of the correspondence. A laboratory was built for the navy in Annapolis, Md., in the 1920s.

19. Extract from Naval Appropriation Bill, H.R. 15947, NCB box 4, ENHS.

20. The *New York Sun*, 7 Feb. 1916, pointed out this contradiction in Edison's naval service and blamed him for installing a dangerous technology on the submarine. Hutchison diaries, 31 Dec. 1916, noted, "We were in no way to blame but the odium has gone all over and has hurt our business some."

21. Meadowcroft to N. Pratt, 11 June 1917, LB 170426, p. 226, ENHS.

22. Sir Arthur Hezlet, *Electronics and Sea Power* (New York: Stein and Day, 1975), pp. 147–50; "Submarines Betrayed by Sound Waves," *Scientific American* 113 (16 Oct. 1915): 333, 346.

23. List of experimental work, NCB box 21, ENHS.

24. TAE to B. F. Thompson, 22 Aug. 1916, NCB box 4, ENHS.

25. J. R. Hill, *Anti-Submarine Warfare* (Annapolis, Md.: Naval Institute Press, 1985), p. 42.

26. "Edison's work during the war," Meadowcroft manuscript, ENHS.

27. Fessenden's career and accomplishments are detailed in Hugh Aitken, *Continuous Wave,* Chs. 2,3,4, and 11.

28. Wise, *Willis Whitney,* pp. 187–91.

29. Hezlet, *Electronics and Sea Power,* pp. 151–53. Hezlet concludes that radio intelligence was the most important electronic factor in the defeat of the U-boats. Robert M. Grant (*U-Boat Destroyed* [London: Putnam, 1964], pp. 151–61, appendices) shows that the great majority of sinkings came from mines.

30. Edison went to Washington with three assistants to make the statistical survey, "Strategic Plans for Saving Cargo Boats from Submarines." "Edison's War Work," NCB box 21, ENHS.

31. Hutchison's diary notes Edison's growing concern with shortages of chemicals; H. T. Leeming to department heads, 18 Aug. 1918, tells of shortages causing layoffs, ENHS.

32. Meadowcroft to C. Aston, 19 Feb. 1917, LB 170122, p. 182, ENHS.; Vanderbilt, *Edison the Chemist,* pp. 248–51.

33. Meadowcroft, "Edison's work during the war," ENHS.

34. Newspaper cuttings, 1915 Benzol file; Meadowcroft to Emery, 31 Oct. 1916, ENHS.

35. TAE annotation on W. Maxwell to E. T. Gundlach, 27 Nov. 1915, ENHS.

36. Fenimore Johnson, *His Master's Voice,* p. 102.

37. R. Allen to Mambert, Three year report, 3 April 1919, ENHS.

38. As E. R. Johnson explained, "War excitement always creates a demand for music." Speech of 4 Oct. 1919, in collected work, box 11, AHC. Dane Yorke ("The Rise and Fall of the Phonograph," *American Mercury* 27 [Sept. 1932]: 1–12) notes a wartime demand for music, p. 8.

39. J. Constable to W. Maxwell, 27 Oct. 1919; J. Constable to TAE, Outline of lab's activities, 25 Nov. 1919, Eng. Dept. files; Mambert minutes, Executive Committee meeting, 16 May 1918. In addition to Edison's experiments the lab carried out production engineering for the manufacture

of government contracts. During the United States' participation in the war, total expenditures of $238,235 were made on experiments, the largest amount going to the submarine location project, H. H. Eckert to S. Mambert, 18 July 1919; cost sheets of WWI experimental work, NCB box 21, ENHS.

40. J. Constable to C. Luhr, 28 Dec. 1917, ENHS.

41. Allen, Three year report, ENHS.

42. Venable, *Out of the Shadow,* pp. 78–79.

43. *Edison Herald,* 25 Feb. 1919, ENHS.

44. Executive Committee minutes, 23 May 1918, ENHS.

45. Kellow memo, 1920, ENHS.

46. Unsigned draft of eight-page memo from Mambert's office to TAE, 29 April 1918, ENHS.

47. Undated copy in Eng. Dept. files, probably Dec. 1919, ENHS.

48. Executive Committee minutes, 23 May 1918, ENHS.

49. V. T. Stewart to J. Constable, 5 Feb. 1918, ENHS.

50. N. Durand to C. Edison, 4 Sept. 1918, ENHS.

51. Experimental costs on dictating machines rose from about $9,000 a year in 1913/14 to $15,000 a year in 1920. This sum represents the cost of one experimenter, half an assistant or Works engineer, and a skilled machinist. Ediphone minutes, CEF; experimental lists, accounts, ENHS.

52. Mambert memo, 1920, ENHS.

53. G. T. Owen to W. Maxwell, 3 July 1918, ENHS.

54. J. W. Robinson to Mambert, 2 Dec. 1918, ENHS.

55. J. Constable to C. Edison, 15 Aug. 1918; C. Luhr in Executive Committee minutes, 16 May 1918, ENHS. A check by the Park Service staff revealed that none of the machine tools in the lab is later than 1907.

56. TAE to J. Franklin, 12 Dec. 1916, LB 161107, p. 331, ENHS.

57. C. Huenlich to N. Durand, 3 May 1920, Eng. Dept. files, ENHS.

58. Lepreau to C. Wilson, Primary Battery Division Blue Book, memo of 28 June 1918, ENHS.

59. Comments quoted in J. Constable to C. Edison, 26 Nov. 1919, Eng. Dept. files, ENHS.

60. J. Constable memo, 31 July 1919, ENHS.

61. Meadowcroft to Applegate, 23 March 1917, LB 170122, p. 471, ENHS.

62. TAE to J. Franklin, 12 Dec. 1916, LB 161107, p. 331, ENHS.

63. G. M. Ryder to F. R. Shell, 2 Dec. 1920, ENHS.

64. W. Maxwell to J. Constable, 2 Feb. 1920, Eng. Dept. files, ENHS.

65. Kellow memo, 1920, ENHS.

66. J. Constable memo, 31 Oct. 1916, ENHS.

67. Executive Committee minutes, 16 May 1918, ENHS.

68. Meadowcroft to S. Stewart, 31 Dec 1915, LB 151216, p. 248, noting that Edison still worked eighteen to twenty hours a day. In response to a medical questionnaire in 1930, Edison said he maintained the same work day of eighteen to twenty hours, although he was sleeping more than

before (six hours per twenty-four compared with less than five before). Longevity Enquiry, 12 Oct. 1930, ENHS.

Chapter Fourteen. The End of an Era

1. "Business Activities of M. Jones," a personal history of TAE, Inc.'s personnel manager, Jones collection, box 13, ENHS.

2. Dyer to A. Jaynes, 16 March 1911, ENHS.

3. Anne Huber Tripp, *The IWW and the Paterson Silk Strike of 1913* (Urbana: University of Illinois Press, 1987), pp. 35–36.

4. Strike flyer, 1916, ENHS.

5. TAE to Mambert, 16 Dec. 1919, Mambert correspondence, vol 1; Edward Marshall, "Machine Made Freedom," *Forum*, Oct. 1926.

6. Anon. to TAE, Inc., 31 Dec. 1919, M. Jones collection, box 11; J. Constable to C. Edison, 2 April 1920, ENHS.

7. E. E. Hudson resigned from his position as division manager of primary batteries in 1916 and joined the Waterbury Battery Co. This company improved its product and dropped its prices. It soon took many customers from TAE, Inc. Hudson seems to have borne a grudge against his old employers and played a vital role in leading the Waterbury Co. The new division manager at West Orange, F. J. Lepreau, stated: "We are dealing with a competitor who knows all our purchasing, engineering, manufacture and sales" information. Lepreau to C. Wilson, 28 Jan. 1918, Primary Battery Blue Books, 1916–19, ENHS. F. K. Dolbeer was a general manager of the National Phonograph Co. and became sales manager of amusement phonographs in TAE, Inc. He contacted Victor in secret in 1912, shortly after his appointment as sales manager, because he was desperate to leave West Orange. Johnson was reluctant to employ him because he knew that Edison would retaliate by hiring away Victor's disc experts for his Diamond Disc campaign. ERJ to L. Douglas, 2 May 1912, box 7, AHC. Dolbeer joined Victor in 1914.

8. W. Maxwell to C. Edison, 13 Dec. 1918, ENHS.

9. F. J. Lepreau quote in minutes of a special meeting of board of TAE, Inc. with executives of Primary Battery Division, 9 April 1918, Primary Battery Division Minutes, vol. 1, CEF.

10. L. McChesney annotation on letter to TAE, 2 Feb. 1917, ENHS. In 1917, eighteen states were considering censorship and regulation of the industry.

11. This rough estimate of costs in TAE, Inc. from film lists of 1909 and 1912; accounts; H. Plimpton to Dyer, 24 April 1912; memo to H. Lanahan, Dyer correspondence, 2 Sept. 1916, ENHS.

12. C. Wilson to C. Edison, 23 and 25 March 1918, ENHS.

13. Mambert memo, 16 Feb. 1920, Mambert correspondence, ENHS.

14. Solomon Fabricant, *The Output of Manufacturing Industries, 1899–1937* (New York: National Bureau of Economic Research, 1940), p. 578.

15. Columbia went into receivership in 1921 and the different parts were later revived; the dictating machine business came under a new company, the Dictaphone Co., Woodbridge, *Dictaphone,* p. 14; Robert Lacey, *Ford* (Boston: Little Brown, 1986), p. 268; Chandler, *Strategy and Structure,* pp. 128–29.

16. The administrative staff was cut from 1,074 to 528. Anne Jardin, *The First Henry Ford: A Study in Personality and Business Leadership* (Cambridge: MIT Press, 1970), pp. 112–13; Carol Gelderman, *Henry Ford the Wayward Capitalist* (New York: Dial Press, 1981), p. 195. Six hundred telephones were sold and all pencil sharpeners were removed.

17. TAE to A. Williams, 5 Jan. 1921, ENHS.

18. Hutchison diary, 1 Jan. 1921. Hutchison also notes that Edison "clipped Charles' wings." When Edison returned to West Orange after the war he ordered cutbacks in the organization in anticipation of a deflationary policy. Minutes of Executive Committee, 1 May 1918, ENHS.

19. TAE to Mambert, 16 Dec. 1919, Mambert correspondence; business activities of M. Jones, MJ box 13, ENHS.

20. Executive Committee Minutes, 2 Feb. 1924, p. 8, RR 68. Edison later wrote a scathing criticism about Mambert only being good enough for a clerk. Annotation on letter for reference, 1924, ENHS.

21. C. Edison to TAE, 12 July 1926, ENHS.; John Venable, *Out of the Shadow,* p.80.

22. "Personal Audit," Jones collection, box 19, 1920; Maxwell resigned 8 March 1922, Edison Phonograph Works minutes, CEF.

23. "High Light History of Victor," Smithsonian Institution.

24. Mambert correspondence, vol. 1, 26 July 1920; W. Stevens to C. Edison, 23 Sept. 1918, Financial Cabinet Blue Books, ENHS.

25. Memos in 1925 Phonograph folders, ENHS, mention plans to get out of Amberola business because of "heavy losses."

26. Executive Committee Minutes, 21 Aug. 1921, RR, ENHS.

27. Wise, *Willis Whitney,* appendices; Reich, "Industrial Research," p. 182. The Research Branch became the Department of Physical Research Engineering in 1919. By 1912 Harold D. Arnold supervised its staff of 234 and those of the Chemical Research Engineering Department (48) and Transmission Engineering Department (131).

28. Accounts records and laboratory folders for 1920–22, ENHS.

29. Jenkins, *Images and Enterprise,* p.312; Rae, "Application of Science," in Oleson and Voss, *Organization of Research,* p.263.

30. David A. Hounshell and John K. Smith, *Science and Corporate Strategy: Du Pont R&D, 1902–1980* (Cambridge: Cambridge University Press, 1988), p. 100; Jeffrey L. Sturchio, "Chemistry and Corporate Strategy at Du Pont," *Research Management* 27 (Jan. 1984): 12.

31. Wise, *Willis Whitney,* appendices; Reich, "Industrial Research," pp.181–82; Hounshell and Smith, *DuPont R&D,* p. 120.

32. Ibid.

33. Minutes of Edison Phonograph Works directors, 9 March 1921, ENHS.

34. Edison claimed that there was no "so-called improvement" in talking machines from 1914–20 that was unfamiliar to the lab. While he might have anticipated phonograph developments, TAE, Inc. did not introduce any new ones. Message to the Edison phonograph dealers, 26 June 1919, Primary Printed, ENHS.

35. Yorke, "Rise and Fall of Phonograph," p. 10; Johnson thought the Aeolian Company was "a high class act." ERJ to Leon Douglas, 2 Sept. 1915, box 7, AHC.

36. Brunswick led the way in the new console cabinets which gave it a strong market position. Hutto, "Emile Berliner, Eldridge Johnson," p.672; TAE annotation on 17 March 1926 letter, ENHS.

37. Edison was involved in the development of an automatic changer in 1926. Annotation on letter of 17 March 1926, ENHS. Victor introduced theirs in 1927 with great effect while two models of changers sat in the West Orange laboratory. Frow, *Edison Disc,* pp. 73–74.

38. TAE to L. Follett, 4 Dec. 1916, LB 161107, stating that lab has been "working for a long time" on an automatic stop. An automatic stop was introduced and withdrawn in 1917; further trials were made in 1919 and the 1920s. Frow, *Edison Disc,* pp. 190–93.

39. A lateral cut reproducer was introduced and then withdrawn in 1914; finally, the Edisonic Line of the late 1920s had a universal reproducer capability. Frow, *Edison Disc,* p. 40.

40. W. Maxwell Report, Musical Phonograph Division, 1911, ENHS.

41. Annotated advertising copy in 1927 Phonograph folder, ENHS.

42. Mark Sullivan, *Our Times: Pre-War America* (New York: Scribners, 1930), pp. 369–93.

43. Questionnaire answered by TAE, 1926 Phonograph folder, ENHS.

44. W. Mallory to Meadowcroft, 7 Jan. 1922, ENHS.

45. Calvin Child was Victor's recording expert. He noted in 1917 that "this dance music changes from day to day and this may be something entirely new that we should get after at once." He asked Johnson to send over his son Fenimore, who knew something about "Jass" music and how it should sound. Child to ERJ, 13 June 1917, box 4, AHC.

46. Barnouw, *Tower in Babel,* p. 128.

47. Art Walsh to C. Edison, 25 April 1927, ENHS.

48. Comparison of talking machines and live recitals was not new in the industry. The first Edison tone tests were in 1915. They continued until the mid 1920s. Frow, *Edison Disc,* pp. 236–41.

49. List of suggestions for improving Ediphone, 1918 Phonograph folder, p. 5, ENHS.

50. "The general public look for a loud reproducer first and then quality," memo in 1911 lab folder, ENHS.

51. H. C. Brainerd to TAE, 16 Nov. 1926, ENHS.

52. Gradeon advertisement, *New York Times*, 2 July 1926, p. 9.

53. See Harold G. Bowen, *The Edison Effect* (West Orange: Thomas Alva Edison Foundation, 1951) for a history of Edison's role in wireless.

54. Aitken, *Continuous Wave*, p. 217.

55. J. Constable letter, 13 May 1919, ENHS.

56. TAE to L. DeForest, 17 June 1926, ENHS.

57. Reich, "Industrial Research," pp. 157–59.

58. Eng. Dept. files, Feb. 1917; J. Constable to E. Boykin, 25 Oct. 1918, ENHS.

59. Gleason L. Archer, *A History of Radio* (New York: American Historical Society, 1938), p. 112. Sarnoff thought that an ideal price would be $75, about the same price as the first popular phonographs (p. 113).

60. Meadowcroft annotation on "The Story of a Great Achievement" in Radio folder. Edison anticipated the day when wireless would be in general use. Edison annotation on 19 May 1898 letter, ENHS.

61. W. Miller to G. H. Mann, 21 May 1910, LB 100521, ENHS.

62. Fenimore Johnson, *His Master's Voice*, pp. 112–13.

63. TAE annotation on Sun Radio data sheets, "Speaker Distortion;" TAE to H. Phillips, 29 Dec. 1925, ENHS.

64. *New York Times*, 3 Oct. 1926.

65. Annotation on C. Edison to TAE, 21 June 1927, ENHS.

66. TAE to H. Phillips, 29 Dec. 1925, ENHS.

67. Archer, *History of Radio*, pp. 208–9; Aitken, *Continuous Wave*, p. 472.

68. C. J. Pusateri, *A History of American Business* (Arlington Heights, Ill.: Harlan Davidson, 1982), p. 243.

69. TAE to C. Edison in 1926 Radio folder, ENHS.

70. Venable, *Out of the Shadow*, p. 82.

71. Fenimore Johnson, *His Master's Voice*, p. 112. Estimates produced by the radio industry put sales in 1922 at $60 million. The figure for 1924 was $358 million. Barnouw, *Tower in Babel*, p. 125.

72. R. A. Allen to C. Edison, 22 April 1925; Art Walsh to Charles Edison, 25 April 1927, ENHS.

73. W. Miller to TAE, 25 Oct. 1925, ENHS.

74. Frow, *Edison Disc*, p. 72; Read and Welch, *Tinfoil to Stereo*, pp. 184–85.

75. Read and Welch, *Tinfoil to Stereo*, p. 256; Hutto, "Emile Berliner, Eldridge Johnson," p. 672. Johnson did not want customers putting things on the flat top of his machines because he thought that they would stop playing (and buying) records. He took all the blame for the decision to introduce "hump back" machines that the public rejected. ERJ to B. G. Royal, 26 Aug. 1932, box 8, AHC.

76. Fenimore Johnson, *His Master's Voice*, p. 114.

77. Kellow to TAE, 22 Sept. 1919; Kellow to G. Ryder, 12 Feb. 1920, ENHS.

78. Leon Douglas was experimenting with "telephone" recording as

early as 1916. He noted that a great deal of R and D was being carried out on wireless telegraphy at that time and advised Victor to keep a close watch on these developments in case the competition should gain an advantage. Douglas to B. G. Royal, 3 March 1916, box 7, AHC.

79. Maxwell's work and his theory of matched impedance is described in Read and Welch, *Tinfoil to Stereo,* pp. 240–54. The microphone diaphragm moved about one-tenth of the distance of the acoustic diaphragm to record the loudest sounds; its weight was one-twentieth of the Edison recorder. Edward C. Weinte, "General Principles of Sound Recording," *Bell Laboratories Record* 7 (Nov. 1928): 83.

80. Clark, "His Master's Voice," p. 223.

81. Hutto, "Emile Berliner, Eldridge Johnson," p. 673; "Excerpt from Mr. de la Chapelle's talk," script for tour of Victor Works, 15 Jan. 1927, ENHS. Annual profit for 1926 and 1927 was over $7 million. ERJ financial folder, box 11; annual reports, box 31, AHC.

82. Annotation on 1925 newspaper cutting in Radio folder, ENHS.

83. Phonograph advertising in Frow, *Edison Disc,* pp. 70–71; 1928 Edison Industries Products, promotional booklet, ENHS.

84. A. Walsh to C. Edison, 15 April 1927, ENHS.

85. Columbia made $2.5 million in 1927. Frow, *Edison Disc,* p. 74. Victor's profits were much higher. See note 81, this chapter.

86. Note from C. Edison to TAE, 1927 Phonograph folder.

Chapter Fifteen. The Last Years

1. Splitdorf Co. stockholder circular, 25 Jan. 1929, ENHS. At the time that RCA was formed there was an "implied promise" that Edison would be allowed to use the patents (Venable, *Out of the Shadow,* p. 82), but when TAE, Inc. tried to enter the pool, RCA refused, forcing TAE, Inc. to buy out a company that had the patent rights.

2. Venable, *Out of the Shadow,* p. 82.

3. Advertising notes to radio announcer, 10 June 1929; promotional material, Primary Printed, ENHS.

4. Splitdorf Co. stockholder circular, 25 Jan. 1919, ENHS.

5. Standard Daily Trade Services, 21 July 1925, ENHS.

6. Venable, *Out of the Shadow,* p. 83. The decision to leave the radio business was prompted by a deadline in 1931 to pay $100,000 for an RCA license. Conot, *Streak of Luck,* p. 447.

7. Minutes of TAE, Inc. board meeting, 8 Aug. 1928, RR 68, ENHS.

8. Stuart Leslie, *Boss Kettering: Wizard of General Motors* (New York: Columbia University Press, 1983), pp. 219–27.

9. Notes and sketches, box D26, ENHS.

10. Advertising material in Primary Printed, ENHS.

11. Press release, TAE, Inc., 22 Nov. 1928, ENHS.

12. Advertising material in Primary Printed, ENHS.

13. Studio Music Committee Minutes, 17 Dec. 1928, J. P. Buchanan folder, Phonograph Division Papers, ENHS.

14. RCA purchased the Victor Co. in 1929 after Johnson had sold the stock to a group of New York bankers. David Sarnoff of RCA had long coveted Victor's Camden plant, which had been built up into an impressive facility. Eugene Lyons, *David Sarnoff* (New York: Harper and Row, 1966), p. 145. Although all sources give the threat of radio as the reason for Johnson's sale of Victor, his version and those of his son and business associates point to his poor health as the main reason.

15. Advertising copy from proofs in 1907 advertising file, ENHS.

16. Yorke, "Rise and Fall of Phonograph," p. 7; Promotional material, 15 May 1912, ENHS.

17. P. Sutcliffe to H. M. Scott, 27 Nov. 1918. "A lot of people think that he (Edison) is still working at a small lab in Menlo Park." Several textbooks also keep Edison at Menlo Park his whole career. R. Current, T. H. Williams, F. Friedel, *The Essentials of American History* (New York: Alfred A. Knopf, 1972), p. 255. A high school text (Melvin Shwartz and John R. O'Connor, *Exploring American History* [New York: Globe, 1974], p. 351) places all his inventions at the invention factory in Menlo Park.

18. Obituary notice in the *New York Times*, cited by Josephson, *Edison: A Biography*, p. 481. See Wachhorst, *Thomas Alva Edison* for a complete analysis of the Edison legend. From the European point of view, Edison represented America and the Gilded Age. One historian sees him as the "most representative figure of the Age." Hugh Bryan, *The Pelican History of the USA* (London: Penguin, 1985), p. 460.

19. David L. Lewis, *The Public Image of Henry Ford* (Detroit, Mich.: Wayne State University Press, 1976), p. 223.

20. Josephson, *Edison: A Biography*, pp. 478–81. This account of the Jubilee is based on the memoirs of Ford's assistant, E. G. Leobold.

21. General Electric advertisement copy in *Scientific American* 113 (16 Oct. 1915): back cover.

22. Advertising circular in 1917 Motion Picture folder, ENHS.

23. Memo of 13 June 1917, ENHS.

24. 1911 *Annual Report of Victor Co.*, box 32, AHC.

25. Fenimore Johnson, *His Master's Voice*, p. 103.

26. Douglas was given an annual salary of $15,000 when he joined the company. Sworn statement of ERJ, 1928, box 8 experiments and history, AHC. Johnson considered that Douglas, Child, and their respective departments were main assets of the company.

27. ERJ, Sworn statement, 1928, box 8, AHC. Johnson had the same lack of faith in the patent system as Edison.

28. "The Victor selling organization is the most important and most expensive of the whole establishment." ERJ, "History of Victor."

29. Victor's advertising budget after the war exceeded $1 million each year; in 1924 the company spent $5 million, which made it one of the

largest advertisers of any American business. Yorke, "Rise and Fall of Phonograph," p. 11; Fenimore Johnson, *His Master's Voice*, p. 103.

30. "I have strong ideas about advertising," wrote Edison. Annotation on Amusement Phonograph Committee Minutes, 17 Jan. 1913; Meadowcroft to Mallory, 11 May 1922, on some new sales propositions Edison thought up.

31. The sales staff of TAE, Inc. were continually pleading for more advertising, which always lagged behind that of the competition. The Blue Books of the Ediphone Division, 1916–18, have several examples, including Durand's arguments to board of directors to increase advertising expenditure above 5 percent of sales (26 Dec. 1916, ENHS).

32. "There is a Future for the Talking Machine," ERJ speech, box 11, Historical file, AHC.

33. Advertising materials, 1926, box D8, ENHS.

34. History of Victor Co., box 15; ERJ to L. Douglas, 3 June 1913, box 7, AHC.

35. ERJ to L. Douglas, 1 June 1907, box 7, AHC.

36. ERJ to L. Douglas, 3 June 1913, box 7, AHC.

37. ERJ to A. Clark, 3 Feb. 1913, box 4, AHC.

38. "Edison opposition does not appear to amount to much." ERJ to L. Douglas, 5 Oct. 1914, box 7, AHC. "We are the only real manufacturer of talking machines" (1919). Fenimore Johnson, *His Master's Voice*, p. 102.

39. TAE, Inc. Minutes, 1 Feb. 1926, RR 68, ENHS.

40. The laboratory at Fort Myers is now a museum and open to the public. Edison maintained experiments at Fort Myers and remained in communication by mail with West Orange. H. Miller to H. Heifman, 24 Dec. 1909, LB 090927, p. 433, ENHS.

41. "Edison Hunting for Rubber in Weeds," *Science and Invention* 95 (26 Nov. 1927): 18–19; Edison Botanic Research Corp., Report to Shareholders, 15 Jan. 1932, ENHS.

42. N 880102, pp. 101–5, ENHS.

43. Vanderbilt, *Edison the Chemist*, pp. 281–82.

44. Robert C. Halgrim, *The Edison Record* (Fort Myers: Edison Museum, nd.), p. 12; Fort Myers press clippings, 2 Aug. 1927, ENHS.

45. *Brooklyn Times*, 1928 Rubber folder, ENHS.

46. Instructions for collecting and shipping latex bearing plants, 1 July 1927, ENHS.

47. Annotation on a letter of 2 Dec. 1929, ENHS.

48. In "Rubber from Weeds: My New Goal" (*Popular Science* 3 [Dec. 1927]: 9), Edison claims that this "is the most complicated problem I have ever tackled."

49. Vanderbilt, *Edison the Chemist*, p. 299.

50. Although Edison succeeded in this project, the strategic problem was solved when Ford acquired Brazilian rubber plantations. Lewis, *Ford*, p. 165.

51. TAE to Hood Wright, Nov. 1887, N 871115, ENHS.

52. TAE, Inc. had $5 million in war bonds. Conot, *Streak of Luck*, p. 448.

53. U.S. Congress, *Congressional Record*, vol. 71, no. 97, Resolution of 21 Oct. 1929.

54 This unlikely claim was made in an advertisement for Heinz barbecue sauce, *USA Today*, 24 May 1984.

55. Charles Edison to TAE, nd., D box, ENHS. In the first nine months of 1926, the combined loss of the phonograph business was $941,000. This was offset by a profit of $1,734,000 from batteries. The Dictating Machine Division made a small profit of $278,000. Edison accounts, ENHS.

56. RCA booklet #211 for dealers, box 15, AHC.

57. National Research Council, *Industrial Research Laboratories of the U.S.* (Washington, D.C.: National Research Council, 1931).

58. *Industrial Arts Index, 1926–1927* (New York: H. W. Wilson, 1928), pp. 1020–21; *Industrial Arts Index, 1931* (New York: H. W. Wilson, 1932), pp. 1395–96.

59. Testimony of John W. Lieb, 13 April 1925, ENHS. "The rule that has been adopted not to permit your name to be used in connection with any enterprise except those which you control and for which you feel a personal responsibility is a safe one to follow." Dyer to TAE, 11 July 1910, ENHS.

60. H. Miller to H. Tasker, 19 Aug. 1909, LB 090507, p. 521; promotional brochures in Primary Printed, ENHS.

Epilogue

1. TAE, Inc., *Annual Report*, 1952, CEF.

2. Venable, *Out of the Shadow*, pp. 237, 240–43.

Index

Books in the Series

The American Railroad Passenger Car
by John H. White, Jr.

Neptune's Gift: A History of Common Salt
by Robert P. Multhauf

Electricity before Nationalisation: A Study of the Development of the Electricity Supply Industry in Britain to 1948
by Leslie Hannah

Alexander Holley and the Makers of Steel
by Jeanne McHugh

The Origins of the Turbojet Revolution
by Edward W. Constant II (Dexter Prize, 1982)

Engineers, Managers, and Politicians: The First Fifteen Years of Nationalised Electricity Supply in Britain
by Leslie Hannah

Stronger Than a Hundred Men: A History of the Vertical Water Wheel
by Terry S. Reynolds

Authority, Liberty, and Automatic Machinery in Early Modern Europe
by Otto Mayr

Inventing American Broadcasting, 1899–1922
by Susan J. Douglas

Edison and the Business of Innovation
by Andre Millard

Designed by Chris L. Hotvedt

Composed by The Composing Room of Michigan, Inc., in Goudy Old Style with Lubalin Graph display

Printed by the Maple Press Company, Inc., on 50-lb. Glatfelter Offset A-50 Antique paper

The New Literary Criticism and the Hebrew Bible

edited by
J. Cheryl Exum
and David J.A. Clines

Journal for the Study of the Old Testament
Supplement Series 143

Published by JSOT Press
JSOT Press is an imprint of
Sheffield Academic Press Ltd
343 Fulwood Road
Sheffield S10 3BP
England

Typeset by Sheffield Academic Press
and
Printed on acid-free paper in Great Britain
by Biddles Ltd
Guildford

British Library Cataloguing in Publication Data

New Literary Criticism and the Hebrew Bible.—(JSOT Supplement Series, ISSN 0309-0787; No. 143)
I. Exum, J. Cheryl II. Clines, David J.A.
III. Series
220.4

ISBN 1-85075-424-1

Contents

ABBREVIATIONS

AB	Anchor Bible
ANVAO	Avhandlinger utgitt av det Norske Videnskaps-Akademi i Oslo
BEATAJ	Beiträge zur Erforschung des Alten Testaments und des antiken Judentums
BKAT	Biblischer Kommentar: Altes Testament
BR	*Biblical Research*
BZ	*Biblische Zeitschrift*
BZAW	Beihefte zur *ZAW*
CBQ	*Catholic Biblical Quarterly*
FOTL	The Forms of the Old Testament Literature
HAR	*Hebrew Annual Review*
HSM	Harvard Semitic Studies
HUCA	*Hebrew Union College Annual*
ICC	International Critical Commentary
IEJ	*Israel Exploration Journal*
IOSOT	International Organization for the Study of the Old Testament
JAAR	*Journal of the American Academy of Religion*
JBL	*Journal of Biblical Literature*
JSOT	*Journal for the Study of the Old Testament*
JSOTSup	*Journal for the Study of the Old Testament*, Supplement Series
JSS	*Journal of Semitic Studies*
KHAT	Kurzer Hand-Commentar zum Alten Testament
LD	Lectio divina
NCB	New Century Bible
NKZ	*Neue kirchliche Zeitschrift*
OTL	Old Testament Library
OTS	*Oudtestamentische Studiën*
PMLA	Proceedings of the Modern Language Association
RB	*Revue biblique*
RSR	*Recherches de science religieuse*
SJOT	*Scandinavian Journal of the Old Testament*
VT	*Vetus Testamentum*

VTSup	*Vetus Testamentum*, Supplements
WBC	Word Biblical Commentary
WMANT	Wissenschaftliche Monographien zum Alten und Neuen Testament
ZAW	*Zeitschrift für die alttestamentliche Wissenschaft*

CONTRIBUTORS TO THIS VOLUME

Alice Bach, Department of Religious Studies, Stanford University, California.
Robert P. Carroll, Department of Biblical Studies, University of Glasgow, Scotland.
David J.A. Clines, Department of Biblical Studies, University of Sheffield, England.
J. Cheryl Exum, Department of Biblical Studies, University of Sheffield, England.
Francisco García-Treto, Department of Religion, Trinity University, San Antonio, Texas.
David Jobling, St Andrew's College, Saskatechewan, Canada.
Francis Landy, Department of Religious Studies, the University of Alberta, Edmonton, Alberta.
Stuart Lasine, Department of Religion, Wichita State University, Wichita, Kansas.
Peter D. Miscall, St Thomas Theological Seminary, Denver, Colorado.
Robert Polzin, School of Comparative Literary Study, Carleton University, Ottowa, Ontario.
Hugh Pyper, Department of Theology and Religious Studies, the University of Leeds, England.
Ilona Rashkow, Deaprtment of Comparative Studies, State University of New York at Stony Brook, New York.

THE NEW LITERARY CRITICISM

David J.A. Clines and J. Cheryl Exum

Literary criticism, or what German biblical scholarship termed *Literarkritik,* has featured prominently in scholarship on the Hebrew Bible since the rise of the historical-critical method in the early nineteenth century. This literary criticism of the Bible had as its goal—since it was foundational for *historical*-critical study—the reconstruction of the *history* of the biblical literature. Its method was to analyse the stylistic and (to some extent) the ideological differences among the various writings of the Hebrew Bible—and especially *within* books like Genesis and Kings—in order to separate earlier from later, simpler from more elaborated, elements in the text. The magisterial four-document theory of the sources of the Pentateuch given classic formulation by Julius Wellhausen, and the still regnant hypothesis of a Deuteronomistic edition of the books of Joshua to 2 Kings proposed by Martin Noth, are showcases of the methods and results of traditional literary criticism.

The subject matter of this volume, the 'new' literary criticism of the Hebrew Bible, whatever form it takes, has almost nothing in common with that *Literarkritik.* It is not a historical discipline, but a strictly literary one, foregrounding the textuality of the biblical literature. Even when it occupies itself with historical dimensions of the texts—their origin or their reception—its primary concern is the text as an object, a product, not as a window upon historical actuality.

But exactly what is meant by the 'new' literary criticism? How new is 'new' depends upon how traditional a vantage point one takes up to begin with. To those still preoccupied with historical criticism (and they remain the majority in biblical studies), any focus on the text as a unitary object—any consideration, that is, of its style, its rhetoric or its structure—counts as a new tendency. But to those engaged in the newest of the 'new' literary criticisms—feminist, Marxist, reader-response, deconstructionist and the like—even stylistics, rhetorical criticism and structuralism and other formalist criticisms are no longer 'new'; they are, by some reckonings, already *passé*.

Thus, in conceiving this volume, we had to decide what would count as 'new' among literary criticisms. Others might define it differently, but, for us, 'the new literary criticism' signifies all the criticisms that are post-structuralist. That is what we think is 'new' in our discipline: the theoretical approaches that have come into the limelight in literary studies generally in the 70s and 80s, and that can be expected to influence the way we read the Hebrew Bible in the present decade. It is not surprising, nor even especially unfortunate, that Hebrew Bible studies should adopt the methods of general literary criticism only a decade or two after they are developed outside our own discipline. But what is certain is that by the end of this decade approaches that may now present some degree of novelty or even shock value to traditional biblical critics will be incorporated into the daily practice of mainstream scholars (just as the language and interests of the formalist rhetorical critics and narratologists of the 60s and 70s have become part of the common professional stock in trade).

The new literary methods have already started to make their mark in biblical studies, and the editors of this volume believe that the time is ripe to present to the discipline a sampler of the kind of work that they themselves take pleasure in and believe holds promise of an upsurge of intellectual creativity in the field.

What, then, are the characteristics of the new literary criticism as it has been applied to the Hebrew Bible? If the essays in this volume can be regarded as representative—and the editors certainly think they can—the first thing that strikes one is how eclectic the new literary criticism is. While some of the essays here can be characterized as one thing rather than another, 'feminist' perhaps, or 'psychoanalytic', most of them move freely from one critical approach to another, combining materialist with reader-response criticism, psychoanalytic with ideological criticism, and so on. In their diversity, these essays reflect the multidisciplinary nature of these new criticisms, their resistance to tidy classification (for example, in positing a woman reader, feminist criticism is also reader-response criticism, and in reading against the grain, it works like deconstruction). The experimental quality of these essays also suggests that what biblical study needs at this moment is not so much systematization as a spirit of exploration and methodological adventurousness, where every new way of looking at our familiar texts is to be eagerly seized upon and tested for all it is worth.

The second noticeable feature is that, in this interweaving of methods, there appears a spirit of goodwill, cooperation even. In general literary criticism, the opposite has frequently been the case, with a whole apparatus of gurus and disciples, feuds, misunderstandings, and political manoeuvrings. In Hebrew Bible 'new' literary studies, and certainly in this volume, on the other hand, there is no bad blood, no methodological purism, no 'school' mentality, no sneers at other approaches. Why this should be so is hard to guess, whether because there are not yet enough 'new' literary critics in Hebrew Bible studies to draw up battle lines, or because they all feel they are still making common cause against a common foe, an unselfconscious historical criticism. Whatever the reason, the message these essays convey, however subliminally, is that there are no holds barred, and no automatically inappropriate angles of vision upon our texts—and that even in centres of institutional power there are no longer any arbiters of what may and may not be legitimately and fruitfully said about our texts.

A third feature of these essays is their orientation to texts. The editors invited the contributors to offer an essay that represented their current work, and not one contributor wrote a truly theoretical piece. Perhaps this is not surprising, since most of the authors are professional biblical scholars and not literary theorists. But perhaps it is noteworthy all the same that the essayists never thought it necessary to set out their exact theoretical position, to distance themselves from similar-sounding approaches, or even to attempt to justify the theory they were exemplifying. It may not always be like that in the future, and perhaps a different group of authors even at this present juncture would do things differently; but the impression readers of this volume may quite properly gain is that the contributors decided there were so many interesting new things to say about texts from the perspectives of these new criticisms that they chose not to linger over the theoretical niceties.

A fourth characteristic of these essays is that they press beyond 'interpretation' to 'critique'. The traditional concern of biblical criticism has been understanding, interpretation, exegesis. In these essays, viewed collectively, on the other hand, there are distinct signs of the movement that biblical studies seems poised to make, that is, to evaluation of the biblical texts from standpoints outside the ideology of the texts themselves. Whether from a feminist standpoint, or some other intellectual or ethical position, biblical scholars are beginning to see the need to develop critiques of their material—in order to make their criticism truly critical. As long as we do not challenge the world views of our literature, as long as we limit our researches merely to questions of meaning and refuse to engage with questions of value, it will become increasingly hard for us to justify the place of biblical studies within the human sciences. Not all the essays here present such a critique explicitly; but it may be regarded as the tendency of the 'new literary criticism' in general, if only in its very plurality, to call into question the values embedded in the traditional scholarship. In biblical studies such values include an often unspoken privileging of the ideology set forth or assumed by the texts, which the new literary criticism will surely expose.

All the methodological diversity and eclecticism of these essays makes it impossible to label the contributions with one or another theoretical tag or even to arrange them in any logical order. What would a logical order be?, we asked ourselves. So the volume must stand as a witness to the plurality of criticisms in Hebrew Bible studies that seems now to be here to stay, a sign of the times, a marker of directions in which study of the Hebrew Bible is likely to develop in the present decade. It may, of course, for some readers signal the further fragmentation of the field that makes it impossible to 'keep up' with what is being written about these texts that we have in common. But perhaps that once laudable desire to keep up, to 'master' and 'control' the field, is ripe for suppression, or at least sublimation, now that it can be recognized for what it is, as yet another manifestation of the academic will to power (or, alternatively, of scholarly insecurity). In a post-modern world, there is no centre, no standing-place that has rights to domination, no authority that can manage or control what is to count as scholarship. No one, therefore, need lose any sleep over the existence of methodologies that are unfamiliar, uncongenial or questionable, nor about their incapacity to participate in every new approach. Our hope rather is that some of the trends represented in these essays will prove fascinating, stimulating, intriguing, or even enraging, to readers—who will share our sense of excitement about the intellectual challenges our texts present us with at this moment in history.

Finally, since the literary criticism presented in this volume is by its own profession 'new', and the contributors would prefer to be addressing not just one another but also a scholarly audience in general that may as yet be somewhat unfamiliar with the shape of the new literary criticisms, it may be helpful to sketch here the main outlines of the theoretical positions and approaches that may be referred to as the 'new literary criticism'. But first, a word about literary approaches that we would not now classify as 'new'.

Literary Criticisms No Longer 'New'

New Criticism

New criticism stands for an attitude to texts that sees them as works of art in their own right, rather than as representations of the sensibilities of their authors. Against the romantic view of texts as giving immediate access to the ideas and feelings of great minds, the new criticism regards texts as coherent intelligible wholes more or less independent of their authors, creating meaning through the integration of their elements. And against a more positivistic scholarship of the historical-critical kind, new criticism emphasizes the literariness of literary texts and tries to identify the characteristics of literary writing.

In biblical studies the term 'new criticism' has been rarely used, but most work that is known as 'literary'—whether it studies structure, themes, character, and the like, or whether it approaches the texts as unified wholes rather than the amalgam of sources, or whether it describes itself as 'synchronic' rather than 'diachronic', dealing with the text as it stands rather than with its prehistory—can properly be regarded as participating in this approach.

Rhetorical Criticism

Rhetorical criticism, sharing the outlook of new criticism about the primacy of the text in itself, and often operating under the banner of 'the final form of the text', concerns itself with the way the language of texts is deployed to convey meaning. Its interests are in the devices of writing, in metaphor and parallelism, in narrative and poetic structures, in stylistic figures. In principle, but not often in practice in Hebrew Bible studies, it has regard to the rhetorical situation of the composition and promulgation of ancient texts and to their intended effect upon their audience. But, like new criticism, its primary focus is upon the texts and their own internal articulation rather then upon their historical setting.

Structuralism

Structuralist theory concerns itself with patterns of human organization and thought. In the social sciences, structuralism analyses the structures that underlie social and cultural phenomena, identifying basic mental patterns, especially the tendency to construct the world in terms of binary oppositions, as forming models for social behaviour. In literary criticism likewise, structuralism looks beneath the phenomena, in this case the texts, for the underlying patterns of thought that come to expression in them. Structuralism proper shades off on one side into semiotics and the structural relations of signs, and on the other into narratology and the systems of construction that underlie both traditional and literary narratives.

The New Literary Criticisms

The literary criticisms that have been sketched above, and have been typified as no longer 'new', have by no means outlived their usefulness, nor have they been invalidated by the appearance of the criticisms yet to be discussed. The essays in this volume are proof enough that contemporary biblical critics are delighted to have in their repertory a vast array of methods; their skill is often to know which criticisms best to deploy with a given text. For every one scholar who rigorously explores the ramifications of a single method there seem to be ten who owe no methodological allegiances. Even the historical-critical method, the precursor of them all, can have its place in the most 'advanced' work, as these essays witness. Here, however, we consider the newer literary criticisms that have been the occasion for this volume.

Feminist Criticism

Feminist criticism can be seen as a paradigm for the new literary criticisms. For its focus is not upon texts in themselves but upon texts in relation to another intellectual or political issue; and that could be said to be true of all the literary criticisms represented in this volume. The starting point of feminist criticism is of course not the given texts but the issues and concerns of feminism as a world view and as a political enterprise. If we may characterize feminism in general as recognizing that in the history of civilization women have been marginalized by men and have been denied access both to social positions of authority and influence and to symbolic production (the creation of symbol systems, such as the making of texts), then a feminist literary criticism will be concerned with exposing strategies by which women's subordination is inscribed in and justified by texts. Feminist criticism uses a variety of approaches and encourages multiple readings, rejecting the notion that there is a 'proper way' to read a text as but another expression of male control of texts and male control of reading. It may concentrate on analysing the evidence contained in literary texts, and showing in detail the ways in which women's lives and voices have in fact been suppressed by texts. Or it may ask how, if at all, a woman's voice can be discovered in, or read into, an androcentric text. Or it may deploy those texts, with their evidence of the marginalization of women, in the service of a feminist agenda, with the hope that the exposing of male control of literature will in itself subvert the hierarchy that has dominated not only readers but also culture itself.

Materialist or Political Criticism

In a materialist criticism, texts are viewed principally as productions, as objects created, like other physical products, at a certain historical juncture within a social and economic matrix and as objects that exist still within definite ambits constituted by the politics and the economics of book production and of readerships. More narrowly, materialist criticism analyses texts in terms of their representation of power, especially as they represent, allude to or repress the conflicts of different social classes that stand behind their composition and reception.

Psychoanalytic Criticism

A psychoanalytic criticism can take as its focus the authors of texts, the texts themselves, or the readers of the texts. Since authors serve their own psychological needs and drives in writing texts, their own psyches are legitimate subjects of study. It is not often we have access to the psyche of a dead author, but even if little can be said about the interior life of real authors, there is plenty to be inferred about the psyches of the authors implied by the texts. Just as psychoanalytic theory has shown the power of the unconscious in human beings, so literary critics search for the unconscious drives embedded within texts. We can view texts as symptoms of narrative neuroses, treat them as overdetermined, and speak of their repressions, displacements, conflicts and desires. Alternatively, we can uncover the psychology of characters and their relationships within the texts, and ask what it is about the human condition in general that these texts reflect, psychologically speaking. Or we can turn our focus upon empirical readers, and examine the non-cognitive effects that reading our texts have upon them, and construct theoretical models of the nature of the reading process.

Reader-Response Criticism

The critical strategies that may be grouped under the heading of reader-response criticism share a common focus on the reader as the creator of, or at the very least, an important contributor to, the meaning of texts. Rather than seeing 'meaning' as a property inherent in texts, whether put there by an author (as in traditional historical criticism) or somehow existing intrinsically in the shape, structure and wording of the texts (as in new criticism and rhetorical criticism), reader-response criticism regards meaning as coming into being at the meeting point of text and reader—or, in a more extreme form, as being created by readers in the act of reading.

An obvious implicate of a reader-response position is that any quest for determinate meanings is invalidated; the idea of 'the' meaning of a text disappears and meaning becomes defined relative to the various readers who develop their own meanings. A text means whatever it means to its readers, no matter how strange or unacceptable some meanings may seem to other readers.

Reader-response criticism further raises the question of validity in interpretation. If there are no determinate meanings, no intrinsically right or wrong interpretations, if the author or the text cannot give validation to meanings, the only source for validity in interpretation has to lie in 'interpretative communities'—groups that authorize certain meanings and disallow others. Validity in interpretation is then recognized as relative to the group that authorizes it.

Deconstruction

Deconstruction of a text signifies the identifying of the Achilles heel of texts, of their weak point that lets them down. As against the 'common sense' assumption that texts have more or less clear meanings and manage more or less successfully to convey those meanings to readers, deconstruction is an enterprise that exposes the inadequacies of texts, and shows how inexorably they undermine themselves. A text typically has a thesis to defend or a point of view to espouse; but inevitably texts falter and let slip evidence against their own cause. A text typically sets forth or takes for granted some set of oppositions, one term being privileged over its partner; but in so doing it cannot help allowing glimpses of the impossibility of sustaining those oppositions. In deconstruction it is not a matter of reversing the oppositions, of privileging the unprivileged and vice versa, but of rewriting, reinscribing, the structures that have previously been constructed. The deconstruction of texts relativizes the authority attributed to them, and makes it evident that much of the power that is felt to lie in texts is really the power of their sanctioning community.

SELECT BIBLIOGRAPHY

Literary Theory

Catherine Belsey, *Critical Practice* (London: Methuen, 1980).

Art Berman, *From the New Criticism to Deconstruction: The Reception of Structuralism and Post-Structuralism* (Urbana, IL: University of Illinois Press, 1988).

Ralph Cohen (ed.), *The Future of Literary Theory* (New York: Routledge, 1988).

Terry Eagleton, *Literary Theory: An Introduction* (Oxford: Basil Blackwell, 1983).

Ann Jefferson and David Robey (eds.), *Modern Literary Theory: A Comparative Introduction* (London: Batsford, 1982).

Frank Lentricchia, *After the New Criticism* (London: Athlone Press, 1980).

David Lodge (ed.), *Modern Criticism and Theory: A Reader* (London: Longman, 1988).

—(ed.) *20th Century Literary Criticism: A Reader* (London: Longman, 1972).

Literary Theory and the Bible

Mieke Bal, *Death and Dissymmetry: The Politics of Coherence in the Book of Judges* (Chicago: University of Chicago Press, 1988).

David Jobling and Stephen D. Moore (eds.), *Poststructuralism as Exegesis, Semeia* 54 (1991).

Terence J. Keegan, *Interpreting the Bible: A Popular Introduction to Biblical Hermeneutics* (New York: Paulist Press, 1985).

Edgar V. McKnight, *The Bible and the Reader: An Introduction to Literary Criticism* (Philadelphia: Fortress Press, 1985).

Mark Minor (ed.), *Literary-Critical Approaches to the Bible: An Annotated Bibliography* (West Cornwall, CT: Locust Hill Press, 1992).

Stephen D. Moore, *Literary Criticism and the Gospels: The Theoretical Challenge* (New Haven: Yale University Press, 1989).

Mark Allan Powell, *The Bible and Modern Literary Criticism: A Critical Assessment and Annotated Bibliography* (New York: Greenwood Press, 1992).

Regina Schwartz (ed.), *The Book and the Text: The Bible and Literary Theory* (Cambridge, MA: Blackwell, 1990).

Anthony C. Thiselton, *New Horizons in Hermeneutics* (London: Marshall Pickering, 1992).

Charles E. Winquist (ed.), *Text and Textuality, Semeia* 40 (1987).

Deconstruction

Harold Bloom *et al.*, *Deconstruction and Criticism* (London: Routledge & Kegan Paul, 1979).

Christopher Butler, *Interpretation, Deconstruction, and Ideology: An Introduction to Some Current Issues in Literary Theory* (Oxford: Clarendon Press, 1984).

Jonathan Culler, *On Deconstruction* (Ithaca, NY: Cornell University Press, 1982).

Howard Felperin, *Beyond Deconstruction: The Uses and Abuses of Literary Theory* (Oxford: Clarendon Press, 1985).

Christopher Norris, *Deconstruction: Theory and Practice* (London: Methuen, 1982).

Deconstruction and the Bible

David J.A. Clines, 'Deconstructing the Book of Job', in *The Bible as Rhetoric: Studies in Biblical Persuasion and Credibility* (ed. Martin Warner; Warwick Studies in Philosophy and Literature; London: Routledge, 1990), pp. 65-80, and in *What Does Eve Do to Help? And Other Readerly Questions to the Old Testament* (JSOTSup, 94; Sheffield: JSOT Press, 1990), pp. 106-23.

—'Haggai's Temple, Constructed, Deconstructed and Reconstructed', *SJOT* 7 (1993), pp. 51-77, and in *Second Temple Studies* (ed. Tamara C. Eskenazi and Kent H. Richards; JSOTSup; Sheffield: JSOT Press, forthcoming).

Edward L. Greenstein, 'Deconstruction and Biblical Narrative', *Prooftexts* 9 (1989), pp. 43-71.

David Jobling, 'Writing the Wrongs of the World: The Deconstruction of the Biblical Text in the Context of Liberation Theologies', *Semeia* 51 (1990), pp. 81-118.

Peter Miscall, *The Workings of Old Testament Narrative* (Semeia Studies; Philadelphia: Fortress Press; Chico, CA: Scholars Press, 1983).

Hugh C. White, 'The Joseph Story: A Narrative which "Consumes" its Content', *Semeia* 31 (1985), pp. 49-69.

Feminist Criticism

Hélène Cixous and Catherine Clément, *The Newly Born Woman* (Minneapolis, MN: University of Minnesota Press, 1986).

Mary Eagleton (ed.), *Feminist Literary Criticism* (London: Longman, 1981).

—(ed.) *Feminist Literary Theory: A Reader* (Oxford: Basil Blackwell, 1986).

Elizabeth A. Flynn and Patrocinio P. Schweickart (eds.), *Gender and Reading: Essays on Readers, Texts, and Contexts* (Baltimore: Johns Hopkins University Press, 1986).

Luce Irigaray, *This Sex Which is Not One* (Ithaca, NY: Cornell University Press).

Mary Jacobus (ed.), *Women Writing and Writing about Women* (London: Croom Helm, 1979).

Nancy Miller (ed.), *The Poetics of Gender* (New York: Columbia University Press, 1986).

Elaine Showalter (ed.), *The New Feminist Criticism* (New York: Pantheon Books, 1985).

Susan Rabin Suleiman (ed.), *The Female Body in Western Culture: Contemporary Perspectives* (Cambridge, MA: Harvard University Press, 1986).

Toril Moi, *Sexual/Textual Politics: Feminist Literary Theory* (London: Routledge & Kegan Paul, 1988).

Robyn R. Warhol and Diane Price Herndl (eds.), *Feminisms: An Anthology of Literary Theory and Criticism* (New Brunswick, NJ: Rutgers University Press, 1991).

Feminist Criticism and the Bible

Alice Bach, 'The Pleasure of Her Text', in *The Pleasure of Her Text: Feminist Readings of Biblical and Historical Texts* (ed. Alice Bach; Philadelphia: Trinity Press International, 1990), pp. 25-44.

—'Reading Allowed: Feminist Biblical Criticism at the Millennium', *Currents in Research: Biblical Studies* 1 (1993).

Mieke Bal, *Lethal Love: Feminist Literary Readings of Biblical Love Stories* (Bloomington, IN: Indiana University Press, 1987).

—*Murder and Difference: Gender, Genre, and Scholarship on Sisera's Death* (trans. M. Gempert; Bloomington: University of Indiana Press, 1988).

— (ed.), *Anti-Covenant: Counter-Reading Women's Lives in the Hebrew Bible* (Bible and Literature Series, 22; JSOTSup, 81; Sheffield: Almond Press, 1989).

Phyllis Bird, 'The Harlot as Heroine: Narrative Art and Social Presupposition in Three Old Testament Texts', *Semeia* 46 (1989), pp. 119-39.

—'Israelite Religion and the Faith of Israel's Daughters: Reflections on Gender and Religious Definition', in *The Bible and the Politics of Exegesis: Essays in Honor of Norman K. Gottwald on his Sixty-Fifth Birthday* (ed. David Jobling, Peggy Day and Gerald T. Sheppard; Cleveland, OH: Pilgrim Press, 1991).

Athalya Brenner, *The Israelite Woman: Social Role and Literary Type in Biblical Narrative* (The Biblical Seminar, 2; Sheffield: JSOT Press, 1985).

— (ed.), *The Feminist Companion to the Bible* (Sheffield: Sheffield Academic Press, 1992–), 10 vols. (when complete).

Athalya Brenner and Fokkelien van Dijk-Hemmes, *On Gendering Texts: Female and Male Voices in the Hebrew Bible* (Leiden: Brill, 1993).

Claudia V. Camp, 'What's So Strange about the Strange Woman?', in *The Bible and the Politics of Exegesis: Essays in Honor of Norman K. Gottwald on his Sixty-Fifth Birthday* (ed. David Jobling, Peggy Day and Gerald T. Sheppard; Cleveland, OH: Pilgrim Press, 1991).

Claudia V. Camp and Carole R. Fontaine (eds.), *Women, War, and Metaphor: Language and Society in the Study of the Hebrew Bible, Semeia* 61 (1993).

Peggy L. Day (ed.), *Gender and Difference in Ancient Israel* (Minneapolis: Fortress Press, 1989).

J. Cheryl Exum, *Fragmented Women: Feminist (Sub)versions of Biblical Narratives* (JSOTSup, 163; Sheffield, JSOT Press; Valley Forge, PA: Trinity Press International, 1993).

J. Cheryl Exum and Joanna W.H. Bos (eds.), *Reasoning with the Foxes: Female Wit in a World of Male Power, Semeia* 42 (1988).

Esther Fuchs, 'The Literary Characterization of Mothers and Sexual Politics in the Hebrew Bible', *Semeia* 46 (1989), pp. 151-166 (also in *Feminist Perspectives on Biblical Scholarship* [ed. Adela Yarbro Collins; Chico, CA: Scholars Press, 1984], pp. 117-36).

—'Marginalization, Ambiguity, Silencing: The Story of Jephthah's Daughter', *Journal of Feminist Studies in Religion* 5 (1989), pp. 35-45.

—'Contemporary Biblical Literary Criticism: The Objective Phallacy', in *Mappings of the Biblical Terrain: The Bible as Text* (ed. Vincent L. Tollers and John R. Maier; Lewisburg: Bucknell University Press, 1990), pp. 134-42.

Danna Nolan Fewell and David M. Gunn, 'Controlling Perspectives: Women, Men, and the Authority of Violence in Judges 4 and 5', *JAAR* 58 (1990), pp. 101-23.

Pamela J. Milne, 'The Patriarchal Stamp of Scripture: The Implications of Structuralist Analyses for Feminist Hermeneutics,' *Journal of Feminist Studies in Religion* 5 (1989), pp. 17-34.

Ilana Pardes, *Countertraditions in the Bible: A Feminist Approach* (Cambridge, MA: Harvard University Press, 1992).

Political Criticism

Fredric Jameson, *The Political Unconscious: Narrative as a Socially Symbolic Act* (Ithaca, NY: Cornell University Press, 1981).

Pierre Macherey, *A Theory of Literary Production* (London: Routledge & Kegan Paul, 1978).

Political Criticism and the Bible

K. Fussel, 'Materialist Readings of the Bible: Report on an Alternative Approach to Biblical Texts', in *God of the Lowly: Socio-Historical Interpretations of the Bible* (ed. Willy Schottroff and Wolfgang Stegemann; trans. Matthew J. O'Connell; Maryknoll, NY: Orbis Books, 1984), pp. 13-25.

Norman K. Gottwald, 'The Theological Task after the Tribes of Yahweh', in *The Bible and Liberation: Political and Social Hermeneutics* (Maryknoll, NY: Orbis Books, 1983), pp. 190-200.

—'Literary Criticism of the Hebrew Bible: Retrospect and Prospect', in *Mappings of the Biblical Terrain: The Bible as Text* (Bucknell Review; ed. Vincent L. Tollers and John R. Maier; Lewisburg: Bucknell University Press, 1990), pp. 27-44.

— 'Social Class as an Analytic and Hermeneutical Category in Biblical Studies', *JBL* 112 (1993), pp. 3-22.

David Jobling, 'Feminism and "Mode of Production" in Ancient Israel: Search for a Method', in *The Bible and the Politics of Exegesis: Essays in Honor of Norman K. Gottwald on his Sixty-Fifth Birthday* (ed. David Jobling, Peggy Day and Gerald T. Sheppard; Cleveland, OH: Pilgrim Press, 1991), pp. 239-51.

—'"Forced Labor": Solomon's Golden Age and the Question of Literary Representation', *Semeia* 54 (1992), pp. 57-76.

Jorge Pixley, *Biblical Israel: A People's History* (Minneapolis: Fortress Press, 1992).

Psychoanalytic Criticism

Shoshana Felman (ed.), *Literature and Psychoanalysis. The Question of Reading: Otherwise* (Baltimore: Johns Hopkins University Press, 1982).

Jane Gallop, *The Daughter's Seduction: Feminism and Psychoanalysis* (Ithaca, NY: Cornell University Press, 1982).

Daniel Gunn, *Psychoanalysis and Fiction: An Exploration of Literary and Psychoanalytic Borders* (Cambridge: Cambridge University Press, 1988).

Edith Kurzweil and William Phillips, *Literature and Psychoanalysis* (New York: Columbia University Press, 1983).

Simon O. Lesser, *Fiction and the Unconscious* (Chicago: University of Chicago Press, 1957).

Shlomith Rimmon-Kenan, *Discourse in Psychoanalysis and Literature* (London: Methuen, 1987).
J.P. Muller and W.J. Richardson (eds.), *The Purloined Poe: Lacan, Derrida and Psychoanalytic Reading* (Baltimore: Johns Hopkins University Press, 1988).
Meredith Anne Skura, *The Literary Use of the Psychoanalytic Process* (New Haven: Yale University Press, 1981).

Psychoanalytic Criticism and the Bible

Adrian Cunningham, 'Psychoanalytical Approaches to Biblical Narrative (Genesis 1–4)', in *A Traditional Quest: Essays in Honour of Louis Jacobs* (ed. Daniel Cohn-Sherbok; JSOTSup, 114; Sheffield: JSOT Press, 1991), pp. 115-36.
—'Type and Archetype in the Eden Story', in *A Walk in the Garden* (ed. Peter Morris and Deborah Sawyer; JSOTSup, 136; Sheffield: JSOT Press, 1992), pp. 290-309.
Francis Landy, *Paradoxes of Paradise: Identity and Difference in the Song of Songs* (Bible and Literature, 7; Sheffield: Almond Press, 1983).
Anna Piskorowski, 'In Search of her Father: A Lacanian Approach to Genesis 2–3', in *A Walk in the Garden* (ed. Peter Morris and Deborah Sawyer; JSOTSup, 136; Sheffield: JSOT Press, 1992), pp. 310-18.

Reader-Response Criticism

David Bleich, *Subjective Criticism* (Baltimore: Johns Hopkins University Press, 1980).
—*The Double Perspective: Language, Literacy, and Social Relations* (New York: Oxford University Press, 1988).
Umberto Eco, *The Role of the Reader: Explorations in the Semiotics of Texts* (Bloomington, IN: Indiana University Press, 1979).
Judith Fetterley, *The Resisting Reader: A Feminist Approach to American Fiction* (Bloomington, IN: Indiana University Press, 1978).
Stanley E. Fish, *Is There a Text in This Class? The Authority of Interpretive Communities* (Cambridge, MA: Harvard University Press, 1980).
—*Doing What Comes Naturally: Change, Rhetoric, and the Practice of Theory in Literary and Legal Studies* (Durham, NC: Duke University Press, 1989).
Elizabeth Freund, *The Return of the Reader: Reader-Response Criticism* (New York: Methuen, 1987).
Norman Holland, *The Critical I* (New York: Columbia University Press, 1992).
Wolfgang Iser, *The Act of Reading: A Theory of Aesthetic Response* (London: Routledge & Kegan Paul, 1978).
Susan R. Suleiman and Inge Crosman (eds.), *The Reader in the Text: Essays on Audience and Interpretation* (Princeton: Princeton University Press, 1980).
Jane P. Tompkins (ed.), *Reader-Response Criticism: From Formalism to Post-Structuralism* (Baltimore: Johns Hopkins University Press, 1980).

Reader-Response Criticism and the Bible

David J.A. Clines, *What Does Eve Do to Help? and Other Readerly Questions to the Old Testament* (JSOTSup, 94; Sheffield: JSOT Press, 1990).
Robert Detweiler (ed.), *Reader Response Approaches to Biblical and Secular Texts*, *Semeia* 31 (1985).

Robert M. Fowler, 'Who is "The Reader" in Reader-Response Criticism?', *Semeia* 31 (1985), pp. 5-23.
—'Mapping the Varieties of Reader-Response Critical Theory', *Biblical Interpretation* (forthcoming, 1994).
Edgar V. McKnight, *Post-Modern Use of the Bible: The Emergence of Reader-Oriented Criticism* (Nashville: Abingdon Press, 1988).
J. Severino Croatto, *Biblical Hermeneutics: Toward a Theory of Reading as the Production of Meaning* (Maryknoll, NY: Orbis Books, 1987).

GOOD TO THE LAST DROP: VIEWING THE SOTAH (NUMBERS 5.11-31) AS THE GLASS HALF EMPTY AND WONDERING HOW TO VIEW IT HALF FULL*

Alice Bach

> R. Joshua b. Karhah said: Only two entered the bed and seven left it. Cain and his twin sister, Abel and his twin-sisters. 'And she [Eve] said: I have gotten a man...' R. Isaac said: When a woman sees that she has a child she exclaims, 'Behold, my husband is now in my possession'.
>
> *Bereshit Rabbah* 22.2

> It is important for us to guard and keep our bodies and at the same time make them emerge from silence and subjugation. Historically, we are the guardians of the flesh; we do not have to abandon that guardianship, but to identify it as ours by inviting men not to make us 'their bodies', guarantors of their bodies. Their libido often needs some wife-mother to look after their bodies. It is in that sense that they need a woman-wife [*femme*] at home, even if they do have mistresses elsewhere.
>
> Luce Irigaray, 'The Bodily Encounter with the Mother'

The most tempting aspect of producing feminist readings of biblical texts is to implicate readers in the act of resisting a stable set of attitudes about male representations of women as constructed in the literature of ancient Israel. The strategy I am adopting comes from one of the generals of the French Resistance: Jacques Derrida. His model of reading is undergirded by a desire to resist two complementary beliefs about

* I am grateful to my colleagues, Arnold Eisen and Howard Eilberg-Schwartz in the Department of Religious Studies at Stanford, for reading successive drafts and encouraging me toward writing the article I wanted to write. A.J. Levine of Swarthmore College offered much food for thought and diet soda.

texts: (1) a text has identifiable borders or limits; (2) a text exists within a stable system of reference to other texts of 'information' (its context) that can be represented, for example, by appending scholarly notes.[1] Much biblical critical theory has remained tightly locked within the borders of texts that are determined by constructions of provenance, dating and canon, and any other limits that scholars have assigned to the work.[2] Limits are considered here as 'everything that was to be set up in opposition to writing (speech, life, the world, the real, history, and what not, every field of reference—to body, mind, conscious or unconscious, politics, economics, and so forth)' (Derrida 1979: 257). By selecting a pre-text[3] or source whose central concern is the control of women's sexuality, Num. 5.11-31, the ritual of the Sotah, I intend to focus the reader's attention upon traditional readings that have preserved patriarchal values while containing woman as the object of male anxieties. The Sotah narrative invites a departure from traditional interpretations, which figure the woman as social and material reproducer of children.

The challenge here is to stir up a new brew, where men's attempts to control women's bodies are reread as male vulnerability—the fear of woman engorging male power through her enveloping sexuality. The mysterious water that the woman is forced to drink is contained within a vessel handed to her by the priest. This ritual vessel is metonymic for the womb containing semen, for a sexually pure wife guarantees her husband

1. Kamuf's introduction (1991: 255) to Derrida's article, 'Living On: Border Lines', which I have quoted here, serves as a description of biblical institutional resistance to deconstructive thinking. Derrida's article invites readings that overflow the possibilities of borders and of complete reference. Thus, I follow the leader in resisting biblical critics' concern with staying within con/textual limits, canonical, linguistic, temporal.

2. For an article that suggests one way in which indeterminate readings of the deconstructive kind may be applied to biblical texts, see Greenstein 1989.

3. I adopt this term from Bal, who suggests that a text's 'double meaning keeps reminding us of the active work on preceding texts, rather than the obedient repetition of them' (1991: 430). This article owes much to the genie-like character that drives the work of Bal, who refuses to be content inside the container of biblical literary convention.

a womb vessel filled solely with his seed. This ritual is necessary because the purity of the womb vessel is in doubt. Similarly, on a literary level, readings have been contained within institutional 'vessels', or canons, immobilizing feminist readers as surely as the liquid in the Sotah vessel maintains the wife under the husband's control.

Encamped outside traditional textual borderlines, I have escaped the boundaries of Num. 5.11-31 and its mishnaic expansion, Tractate Sotah, into modern commentaries, creating a narrative of Sotah, a text that permits a husband to accuse his wife of adultery, without having the two witnesses traditional in Israelite law in cases of capital crimes. My text presents additional characters: ancient sages, who made no pretense of covering up their desire to inflict pain upon errant women, assuming the guilt of the woman brought before the priest, and recent interpreters of the ritual, who share an agenda of normalizing the text. My Sotah text reflects a Derridean concern with the relation between texts once their borders have been blurred. A borderline perspective allows the reader to pose questions that historical investigations have not asked about the impact of these texts upon a woman reader.[4]

To further blur borders, to demonstrate how one part of a text may be relevant to others, I have added con-texts of biblical sexual politics: Genesis 39, a narrative in which a woman attempts to initiate sexual activity with the male hero; Proverbs 5 and 7, texts of warning in which an *'ishshah zarah* stands

4. Perhaps it would be helpful to remind the reader that my reading is not intended to replace or dominate earlier interpretations. Rather I pose different questions. The case is analogous to a gendered reading of *Cinderella*. In analyzing the folktale, tradition critics would focus upon the ritual of the prince placing the glass slipper upon the woman's foot (is the slipper always glass? is the incantation formulaic?); archaeologists might provide the shape of the slipper and suggest it wasn't glass but linen; philologists will attempt to provide a link between the ancient word *xxx*, 'pumpkin' and its etymological cognate, *yyy*, resulting in the modern word *coach*. These scholarly investigations have no impact upon the forceful moral codes that keep women waiting for the prince. *Cinderella* has been all too clear to women. In fact, the straight line from women's obedience to the salvific arrival of the prince was not broken until questions were raised about the ideological biases of the storytellers and their interpreters.

poised to seduce the male reader; and Deut. 22.13-30, in which laws governing sexual activity reflect male attempts to control female sexuality. To these con-texts of women's improper behavior, the Sotah ritual stands as an antidote.

Feminist biblical scholars in the past decade have used literary strategies of reading to point out that women are defined in relation to their family roles: they are daughters, wives, and mothers.[5] Although literary feminists studying the ancient world have struggled with the difficulties of reading male-authored texts, which do not provide access to women's inner thoughts, or tell us much about their daily lives, most of the readings have not attempted to break out of the institutional containers, which view biblical texts as discrete measured works without a context.[6] By stressing the borderless nature of texts, I hope to dissolve barriers that have prevented readings that extend beyond the verses of a biblical passage. In spite of my desire to dissolve borders, I do not claim to recover women's lived reality in my reading. While the account of the Sotah may not add material details about women's lives in ancient Israel, I think it reveals a lot about what women had to put up with. Political theorist Susan Okin provides a set of questions that are tied to such a concern. Her sharp distinction in looking at the ways men and women have habitually been defined by social and political philosophers will be useful in my analysis of the Sotah narrative. She writes: 'Philosophers who, in laying the foundation for their political theories, have asked, "*What are men like?* What is man's potential?" have frequently in turning to the female sex, asked "*What are women for?*"' (Okin 1979: 10, italics mine).

5. Some of the most subtle and helpful of these works have been produced by Bird, Exum, Fuchs, Meyers.

6. I suggest to the reader two notable exceptions to encased readings. Bal's biblical studies (1985, 1986, 1988) illustrate the benefits of trans-disciplinary readings of codes in order to break disciplinary borders. *Reading Rembrandt* (Bal 1992) is a startling performance in which a literary critic reads visual works and shows how they both fill the gaps of the literary texts they augment and produce further questions about those texts. Exum (1992) examines the traditional views of Greek tragedy and suggests untraditional ways in which biblical texts can be read as tragic without being dependent upon the classical model.

The problem of how a reader survives the ritual of Sotah, when she cannot swallow the implicit threats to women stirred up in the ritual, can be solved by posing Okin's political question, *What are men like?*, as a psychoanalytic one. This strategy is particularly well suited to biblical texts, since they are male-authored, and thus contain the symptomatic utterances of the author/narrator. From a psychoanalytic perspective, the text operates as a pair of male doubles: the male narrator and the ideal reader, who is also male. Thus, while the woman is the object of the text of Numbers 5, she is excluded from the male dialogue. The reader is allied with the male author/narrator by ignoring the sexual desire of the female character and by joining the author in a critique of her sexual behavior. As I found in my reading of modern interpreters, the gender of the reader does not determine the reading. Frymer-Kensky, for example, like her male counterparts, is not concerned with forming an alliance with the female character in the text. Trying to imagine a female reader contending with male subjectivity, a critic reading with a feminist-psychoanalytic strategy can subvert a text's desire so as to hear what it does not wish to say. 'Son and father agree to write the mother out of the text, for to desire her is not to have the phallus. They conspire both to rid the text of her and to entrap her in it; she is immured' (Segal: 169). Reading the text against its demand, however, *reading as a woman*, allows the subversion of male doubling and allows the figured woman in the text to communicate with a feminist reader. Reading through the sexual codes, a feminist reader charts the literary coercion traditional institutions have used to define the female as other.

Sotah with a Twist

The Sotah is unique in biblical law: it is the only trial by ordeal; it is the only occasion on which a person can be accused of a capital crime without two witnesses. The half-disrobed woman with dishevelled hair, appearing as though she has been caught in an intimate act, is not even permitted to utter the self-incriminatory oath: the ritual oath is put in the mouth of the priest. Only the potion is put in the mouth of the woman. She is forced to swallow what she knows. In my view, then, the Sotah is a unique vehicle for envisioning what is denied, repressed, and silenced in ancient Israelite culture.

The Sotah is not unique in making a woman's fate be determined by men. It is not unique among biblical descriptions of ritual in its textual ambiguities that make it difficult to determine what actually occurred. Interpretations of the Sotah are not unique in having focused upon elements of historicity in the text: is the ritual indeed a trial by ordeal? what are the components of the drink? what occurs physiologically to the woman after swallowing the brew? does she die? does she become sterile? is the Sotah a divine forerunner of RU 486, a chemically induced abortion? Num. 5.11-31 is not the only text that interpreters have failed to read as a political text expressing the fears of its male authors toward woman and their colonization of the female body. Symbolically the woman becomes the currency of the exchange between males. This transaction bears similarities to the Deuteronomic laws concerning undesirable sexual acts. Under the dictates of a phallic economy the father or husband can demand reparation for the damaging of the woman's body. Female sexuality uncontained deflates the phallic economy in which all gains accrue to the master (Benstock 1991: 95).

In the ancient Near East, whose cultures demonstrated a flourishing phallic economy, a man could buy his way out of an adulterous situation by compensating the husband and accepting a discounted wife. Unlike its neighbors, Israel had no provision for a husband to mitigate the death penalty for a wife and her partner convicted of adultery.[7] The fact that the crime of adultery was incorporated into the Sinaitic covenant guaranteed its fateful consequences. 'Unless it [adultery] was punished with death,' Milgrom argues, 'God would destroy the malefactors and indeed the entire community that had allowed it to go unpunished'(1990: 349).

But what if the duplicitous wife is not caught?

Because the concern with ensuring paternity was so strong in Israel, a ritual was devised to further protect the husband from the possibility of a 'wandering wife', and its attendant loss of

7. Hammurabi §129; Middle Assyrian Laws §§14-16; Hittite Laws §§192-93; *ANET* 171, 181, 196.

prestige. The Sotah ritual described in Num. 5.11-31 is constructed around suspicion of adultery, rather than proof of the crime in which two witnesses were required in order to pass sentence of death. The horror of trial by ordeal applied to the woman accused indicates the social view of adultery. Further, it reflects the patriarchal attempt to assure a husband that his honor could be restored if he had so much as a suspicion that his wife had been fooling around. Female erotic desire, then, was understood as erratic, a threat to the social order. By drowning such desire, the traditional order was assured of continuing dominance over women's bodies.

Fateful In/fidelities

As I have stated, the ordeal of Sotah described in Num. 5.11-31 is unique in the Bible, but there are other biblical texts that reflect male anxiety about female sexuality. Genesis 39 evinces the dangers of rampant female sexuality, and, in Proverbs 5 and 7, the *'ishshah zarah* is the paradigm for the woman who uses her sexuality to ensnare men. I shall use these two texts as examples of male-authored warnings about women's sexuality, warnings that are textual defenses of the trial by ordeal. Both the Egyptian woman and the *'ishshah zarah* are examples of woman's sexuality out of control. The Sotah stands as an antidote.

Numbers 5.3 permits a suspicious husband to accuse his wife of adultery without fear of punishment. The figure of the lascivious wife in Genesis 39 supports a husband's suspicion of adultery. Potiphar is told by his wife that his servant has attempted to rape her. A measure of uncertainty salted with suspicion must exist in Potiphar's mind, since he throws Joseph into prison rather than ordering him killed. Both the Sotah and Genesis 39 indicate the presence of a smoking gun, but no body. In each case suspicion of women's sexual impurity results in loss of honor for the husband. Since the sexual activity described by the wife of Potiphar has not been witnessed by any other character—in other words, she has not been caught *in flagrante delicto*—her situation bears certain similarities to that of the woman accused in Numbers 5. Genesis 39, however, presents witnesses who have heard about the sexual invitation and its rejection. Joseph knows a story different from the one told by his mistress; he remains silent. The narrator and reader know that no sexual crime has been committed. Thus they convict the woman for letting her sexual desire flame out of control. If, however, the reader chooses to place the woman in the subject position, and to question the anxiety of the male narrator and of the character Joseph, she can produce a reading that transforms the female character from the mute figure silenced under the terms of phallocentric discourse. This act of reading is what French feminists have termed 'producing an alternative female imaginary' (Irigaray: 197). As I have argued elsewhere, reading to recover the suppressed story of the wife of Potiphar can result in a story of fatal attraction, female obsession with the male love object (1993).

In Numbers 5, like Genesis 39, no crime at all need be committed. The vivid images in the husband's imagination are all that is necessary to bring his wife to the tabernacle to drink the bitter water. On the basis of suspicion of her activities, the Israelite husband could bring his wife before the priest, who would administer the *me hammarim*, 'bitter water',[8] to determine her guilt or innocence. As in rituals of this sort, the punishment was incorporated within the act (or ordeal) itself. An innocent woman survived drinking the potion; a guilty one suffered some sort of punishment related to her sexuality. Interpretations of exactly what the woman's punishment was have varied widely from the time of the Tannaim to the present day.

In Genesis 39 the wife's sexual fantasy condemns her to narrative humiliation; in Numbers 5 the husband's sexual fantasy

8. Scholars have debated the meaning of this difficult term. Sasson has suggested that *mrr* is connected to the Ugaritic root, 'to bless', with a resulting merismus, 'waters that bless and waters that curse'. Also imbuing the term with powers of judgment, Brichto argues that one cannot derive *marim* from the verb *mrr*, to be bitter. He supports his suspicion with the contention that neither the dirt from the tabernacle floor nor a few drops of ink could account for bitterness. He has provided an intriguing suggestion that one read the term *marim* (as derived from the verb *yrh* 'to teach') *mei hammarrim* as a construct with a hiphil plural of abstraction, understood as 'spell-inducing water' (59). In addition, a reading that understands *me hammarim* as spell-inducing waters that would 'teach' the guilt or innocence of the woman is compelling.

condemns his wife to drink the bitter water and be publicly humiliated. Even if the woman is found to be innocent and survives the ordeal (both the biblical text and its expansions emphasize the possibility of the woman's guilt, as I argue below), she has been shamed in front of the community. The priest, the male mediator figure representing her husband's rights, unbinds her hair, an act that evokes a picture of female sexuality unbound: the loosed hair of the loose woman. Holding her husband's jealousy offering in her hands, the woman stands submissively while the priest acts as her mouthpiece, reciting her self-condemning oath. Having been revealed in the presence of the community, even an innocent wife will have difficulty regaining status and respect, since the husband's suspicion has been transmitted to the community. Verse 31 attempts to stabilize the husband's position. He is exonerated regardless of her guilt or innocence. The very ambiguity of the wife's position—did she or didn't she?—separates her from the usual position of the wife-woman, who receives her social identity from her husband.

The husband free from blame differs from the case of the newly married man who falsely accuses his bride of sexual impurity.[9] Deuteronomic law states that if the father can present proof of his daughter's virginity, the husband receives a dual punishment: he must pay reparations to her father (100 shekels of silver) and he may not divorce his wife, who has been slandered/degraded (Deut. 22.19). If, however, the father cannot produce the evidence to clear his daughter, she is assumed to be impure and is stoned to death in front of her father's door. The execution carried out at the father's door provides a vital clue to the integral connection in both cases between the father and the husband: if the girl is guilty, the father has either knowingly or not offered for sale to the husband damaged goods. In the case of the girl's innocence, the husband must pay damages to the father, whose good name has been damaged. Thus the law reflects the men as subjects of the concern and the woman as the object of male ownership.

9. Phillips argues, I think convincingly, that the question in this case is not paternity so much as the husband's eagerness to recover the *mohar*, 'bride price' (1981: 13).

Only another man can verify the husband 's accusation. In the Sotah ritual, the priest functions as intermediary, acting on the husband's suspicion. According to traditional interpretations, the father God enters the bitter waters to determine the woman's guilt or innocence. The physical evidence confirming her innocence is a clean functional womb revealed to the community of witnesses after the deity's 'inspection'. In the Deuteronomic law, the father produces the bloody evidence of his daughter's sexual purity, again assuring a clean functional womb. As in the case of the Sotah, the woman's version of her own story is not considered. Thus, even when a husband can be punished for falsely accusing his wife of sexual impurity (Deut. 22.19), she is not to be believed. The law reflects the concerns of the phallic economy: it protects her father if she is innocent; it protects her husband if she is guilty.

The textual emphasis on the woman's secrecy in Num. 5.13 undergirds the author's concern with the difficulty of discerning female sexual purity. Four times within the indictment, the 'fact' of the woman's secrecy is repeated: 'without the knowledge of her husband', 'she keeps secret', 'without being forced', 'and there was no witness against her' (v. 13). Like a too-rapid heartbeat, the repetition is a telling clue about the power of male fears and fantasies about women's secrets. It is not surprising that the patriarchal society has fashioned a law that protects men's suspicion of women and their dark secrets.

Tractate *Sotah* in the Mishnah elaborates the biblical case law in Num. 5.11-31. The sages describe even more pain and suffering in store for the bad wife. It is worth looking in some detail both at the passage in Numbers and at the Mishnah's interpretation of it in order to understand the fierce reaction in biblical as well as postbiblical Israel to the act of adultery as a crime both against the husband and against the larger community. According to Jewish law, a wife faced a punishment of death if she willingly had sexual relations (*wayyitten 'ish bak 'et-shekobto*, Num. 5.20) with a man other than her husband (*shakab 'otak*, Num. 5.19; *shikbah 'immi* is the invitation of the wife of Potiphar to Joseph in Gen. 39.7). If there is no witness to the act ('none of the men of the house was in the house', Gen. 39.11), it is assumed the woman was not taken by force, but was a willing participant. Deut. 22.23 states that a woman who is taken in the city and does not cry out for help is equally guilty with the man who lay with her (*shakab 'immah*). Hittite law even more sharply defines the woman's culpability: '[I]f [a man] seizes her in (her) house, it is the woman's crime and the woman shall be killed' (§197; *ANET*, 196). Clearly there was an ancient connection between the territory of the woman (inside her house or the city) and her ability to control any situation occurring within 'her' borders. If, however, the act occurs in the countryside, the Deuteronomic lawgivers understood the crime differently. Away from the structured life of the town, the woman's (assumed) screams would not have been heard. She is exonerated, but the man is put to death, his crime equated with the act of someone who attacks and murders his neighbor (Deut. 22.27). One can infer from this crime against a woman's sexual purity, a capital crime, as is murder, that a sexually ravaged woman had no more future than a dead woman. Thus, a woman who participated voluntarily in her own defilement (allowing another man access to her husband's private place) would invoke the same death penalty: the swallowing of the bitter Sotah.

After the husband has accused his wife of adultery, he is enjoined to bring her to the priest for the trial by ordeal, after bringing to the tabernacle a cereal offering for his own 'jealousy'. Later sages indicate that the *torat sotah* is in effect even if the lover or the husband is a castrate. Thus, even if her unfaithfulness could not have resulted in progeny and even if the husband could not have been concerned about the paternity of his subsequent children, she would still be required to drink.[10] For some of the rabbis, then, the protection of paternity becomes secondary to the protection of male honor and integrity of the household. This reading of *torat sotah* would indicate the male desire to compensate a castrated husband by assuring him the same rights in respect to his wife as a potent man. Its inclusion in

10. *Bemidbar Rabbah* 9.17. *Sidrah Naso*, where this interpretation appears, is thought by some scholars to be based upon the ancient Tanhuma, which frequently preserves original readings not found in Buber's edition. See the introduction to *Bemidbar Rabbah* (trans. J.J. Slotki; London: Soncino Press, 3rd edn, 1983).

Sidrah Naso implies that no man may threaten the position of the husband, even a man without procreative organs. What a vivid illustration of the woman as mode of exchange in the phallic economy!

The root *sth* is used in Numbers 5 to describe the activity of the adulterous woman as a 'turning aside', from the marriage path. In three of the four examples (vv. 19, 20, 29) the verb is used in connection with the wife 'turning aside from under her husband's control'; extramarital sexual relations for a wife are understood as her breaking out of her proper place (*tahat 'ishek* 'underneath your husband' (*'ishah*, 'her husband', in v. 29). Some standard translations do not acknowledge the vivid verbal portrait of sexual activity that is suggested by the Hebrew: RSV reads 'under your husband's authority'; JPS reads 'while married to your husband'. Neither allows for a possible sexual allusion.

In addition to its use in Numbers 5, the root *sth* is found in only one other biblical book, Proverbs, where the young man is warned to 'turn aside' from the way of evil men (4.15) and 'not to allow his heart 'to turn aside to the path' of the seductive woman (7.25) since her house is the way to Sheol, going down to the house of death (7.27).[11] Indeed, the victim is compared with an ox headed for the slaughterhouse, a deer bounding toward a noose, a bird winging into a snare (7.22-23). Extending the text's animal metaphors produces a reading in which the young man with the eager innocence of an animal rushes exuberantly toward a predetermined death. Female sexuality is a trap baited by a predatory female hunter. If the youthful male reader of Proverbs stumbles on the paths of wicked men (*resha'im*), there is no indication that he will end up in the dire shape predicted if he sets his foot on the path of the *'ishshah zarah*. The connection of turning aside (*sth*) for sexual purposes

11. Fishbane (1974: 44) reads the connection between the motif of female seduction and Prov. 6.20-35 as an inner-biblical midrash on the Decalogue. According to Fishbane, 'what makes this case significant is that the various prohibitions are presented in the light of a general warning against adultery—or, more specifically, in the light of the seduction of false wisdom in contrast with divine wisdom, Prov. 8–9'. Thus, the tension is between the adulterous woman and the good woman.

with punishment occurs only in the instance of misreading the scented trap of the *'ishshah zarah.*

In spite of the conventional interpretation in English of *'ishshah zarah* as a foreign woman, the Hebrew word *zarah,* meaning 'strange', does not necessarily equate such 'foreignness' with ethnicity. The word can imply otherness, as reflected in the woman who is depicted by the RSV as a 'loose woman' or 'adventuress'. Her otherness is understood in contradistinction to the good woman, *'eshet hayil,* who is not described in terms of her sexuality. Thus, *'ishshah zarah* is foreign to goodness, to wisdom. Scholars continue to dispute whether she is actually a foreigner or, as I suspect, a woman whose explicit sexuality made her a social outcast and therefore an outsider.[12] In each case, nevertheless, the turning aside is clearly on to the path of illicit sexual relations since the verbal root is *sth* rather than *swr.* The far more common root meaning 'turn aside', *swr,* is not always understood in an explicitly sexual manner even when used in connection with women, e.g., by the author of Proverbs of the beautiful woman 'turning aside (*sarat*) from the paths of discretion' (11.22). One assumes that these are improper paths, but the text does not indicate that they are necessarily sexual ones.

In the Proverbs account, the *'ishshah zarah* is firmly rooted on the evil path, indeed her house leads to Sheol. There is no warning for a female reader not to stray into these paths; rather the warning is presented to her potential male victim. It is the vulnerable young male, a nameless parallel to the chaste hero Joseph, who must be warned against 'turning aside,' or turning *toward* the *'ishshah zarah,* a parallel to the character of the Egyptian wife. The roles of the two women in Proverbs, the *'eshet hayil* and the *'ishshah zarah,* are fixed; the author of Proverbs expects no textual engagement between a wife-woman and her sexual twin. He is not concerned with exploring possible shadings in either woman's character. Nor is he

12. Bird (1974: 87 n. 44) designates *'ishshah zarah* as the 'other' woman, contrasting her with the wife. De Vaux (1965: 36) considers the term to contain nothing more loaded than 'the wife of another man'. For further discussion, see Camp 1985; Humbert 1937; McKane 1970: 285, 287; Snijders 1954: 103-104.

worried about stones in the paths of young women, causing them to stumble. It is male readers (the sons) to whom the author (the father) addresses his collection of maxims and warnings. Once again the woman is the object of male anxiety: subduing her sexuality is the key to his safety.

In the prohibition in Numbers 5, God instructs Moses (v. 11) to present the case of a woman who is suspected of wandering (*tisteh*, v. 12) from the authority of her husband. Thus the crime and its ritual punishment are seen to be devised by the deity, not by the community. If there is a warning to the woman, it is in the description of her startling punishment: 'when the LORD makes your thigh fall away and your body swell' (RSV). The Hebrew text is more vivid than the English translation: literally, 'when God causes your thigh (*yarek*) to droop and your womb (*bitnek*) to swell' (v. 21). These terms are suggestive of the sexual act. The word *yarek* is a commonly understood euphemism for sexual organs (e.g. Gen. 24.2, 9; 46.26; 47.29; Exod. 1.5; Judg. 8.30). In these other biblical usages, the word refers to the male 'seat of procreative power', according to BDB, although in Num. 5.21, 22, 27 BDB considers *yrk* as parallel to *btn*. The word *beten* is often understood to refer to the womb (e.g. Gen. 25.23, 24; 38.27; Hos. 12.4; Job 10.19, 31.15; Qoh. 11.5; Ps. 139.13). The parallelism (*yarek // beten*) in vv. 21, 22, 27 suggests strongly that *yrk* does not mean 'thigh' but 'reproductive organs' (as against BDB) and thus emphasizes the wife's role as bearer of the husband's legitimate heirs. It is her place of procreation (*yarek // beten*) that has been violated, and thus will be deformed or destroyed by the priestly potion, a magical brew of holy water and the dust from the floor of the tabernacle (v. 17). If these terms tell us what women are for, they also make it clear what women are not for. The male fantasy imagines the woman as possessing the *yarek*, the seat of procreative power, and thus threatening to 'reverse the body symbolism on which the father's authority is established' (Newsom: 153). A similar version of this pervasive fantasy occurs in the Proverbs description of the *'ishshah zarah* as 'sharp as a two-edged sword' (5.4).

Unbinding the woman's hair, and placing the husband's

jealousy offering (*minhat qena'ot*, v. 15)[13] into her hands, the priest functions as proxy of the offended male, the husband, and of the deity whom the woman's sexuality has taunted. Yet, as the male unbinding another man's woman, he is also the mirror of the lover, touching the forbidden woman. The *minhat qena'ot* held in the wife's hands symbolizes her potential danger, as the one holding and possibly controlling his sexuality. It can also echo the secret lover, whom she held instead of her husband, the sex that resulted in jealousy. Then the priest pronounces the terms of the trial by ordeal, the no-win situation for the woman. In my opinion a strong subtexual suggestion of sexual language exists in the Hebrew text. I provide below an interpretation that intentionally teases out these nuances.

> If no man has *profaned your body,*
> if you have not *turned aside to uncleanness*
> while *you should have remained underneath your husband,*
> be free from this *bitter water that brings forth the agony* (v. 19).
>
> But if *you have turned toward your lover,*
> though under your husband's power,
> if *some man other than your husband*
> *has placed his seed inside your house,*
> *then let the water that brings this curse*
> *pass into your bowels*
> *and make your womb swell*
> *and your thigh fall open* (vv. 20, 22).

The most remarkable aspect of the priest's speech, as I have interpreted it, is the extent of 'guilty' language, shown here in italics. The emphasis is placed on the woman's sexual acts and the agony that results from her turning aside from her husband. If she is innocent, none of these wrenching pains will occur. But as they are all detailed, her possible purity is drowned, or at least diluted, by the volume of curse that issues from the priest's mouth. Thus, the stream of language acts to accuse and punish as much as the priestly potion streaming into the woman. If the woman is innocent, the water will pass through her, and she will

13. The connection between the offering of jealousy (*qn'*), which becomes the *torat haqena'ot* in Num. 5.29, and the husband's sole and complete rights to his wife, is emphasized through a linguistic play. The husband has a legal right to protect jealously (*qn'*) his acquired property (*qny*).

continue to function as a wife, to produce her husband's children (v. 28).[14] A sympathetic interpretation states that if the husband is innocent, the wife will be tested by the bitter water; if the husband has accused his wife wrongly, the wife will not be harmed (Phillips 1981). By picturing so vividly the woman having had sex with a man other than her husband, the text makes it difficult to remember that she might be innocent and not have to undergo the punishment that is described in such detail. By having these words pronounced about her, the woman is verbally punished even if the bitter water does not punish her physically. There is no incantation mentioned that will give equal time to her innocence. A reading that assigns guilt to the husband's accusation would switch the focus of the text from his fears, simultaneously switching power to the wife. The dominance of the husband over his wife is reflected, then, in the text's emphasis on her guilt. The husband's honor is further restored by his dominance over the shadow of the unknown lover, whose intimacy with the wife has been both recalled and repudiated by the priest. If the woman has committed the acts of which she has been accused, then YHWH's judgment shall transform the water:

> the water that causes the agony shall stream into [enter] her
> and shall cause her bitter pain
> and her belly/womb shall swell
> and her thigh/womb shall sag (v. 27).

Instead of her lover's semen entering her, it is the water of judgment that streams into the woman. The poison will cause her belly/womb to swell with pain and in torment her sexual organs will collapse.[15] A most arresting allusion to the sexual act

14. *Bemidbar Rabbah* assures the woman that if she is pure, the water will not affect her, 'for this water is only like dry poison placed upon healthy flesh and cannot hurt it' (9.33). Characterizing the water as poison certainly makes clear its deleterious effect upon the one who must swallow it.

15. Frymer-Kensky supplies a medical explanation for the result of the flooding of the woman's sexual organs. According to her interpretation, the woman suffers the 'collapse of the sexual organs known as a prolapsed uterus...Conception becomes impossible, and the woman's procreative life has effectively ended (unless, in our own time, she has corrective surgery)' (1984: 20-21). What I find most interesting about this description is its

gone wrong. The string of verbs in vv. 20-21—*'et-yerekek nophelet ve'et-bitnek tsavah*, 'your thigh/sexual organs shall *sag/fall away* and your womb *swell/distend'*—echoed in v. 27, focuses attention upon the sexual act. Thus, a connection is made between the husband's loss of prestige through his wife's adulterous act and the loss of erection. The wife's sagging (and therefore empty) womb becomes a symbol of measure for measure punishment meted out for the husband's loss of prestige. Through the punishment that drains her of sexuality and power, he regains his authority.

What Are Men Like?

Many rabbinic sages assume that the result of the woman's drinking the bitter water is death. *Bemidbar Rabbah* records an *aggadah* (Naso 9) that illustrates the magical or divine nature of the bitter water, which can discern the difference between a good woman and an evil one. Two married sisters look very much alike but live in different towns. The one who lives in Jerusalem is 'clean'. The other is 'defiled', and goes to her good sister and pleads with her to take her place in the ritual of the bitter water. The good sister agrees, drinks the water, and is unharmed. Returning home, her sister, who has played the harlot, comes out to embrace her. As they kiss, 'the harlot smelled the bitter water and instantly died'. While the story supports the view that a clean woman will be untouched by the water, as she has been untouched by a man other than her husband, it also makes clear that death caused by the bitter water is the just punishment for an adulterous woman.

There are no recorded cases of the administration of the *torat sotah*, although the *aggadah* quoted above gives the rabbinic view of unavoidable death to the guilty woman—even if she has not actually swallowed the potion. Proximity to the judgmental drink is sufficient to cause punishment. There are no *aggadoth* that record a happy ending for the innocent woman. Jewish tradition maintains that Rabbi Johanan ben Zakkai, shortly after

offering of analgesia for what the text describes as 'bitter pain'. Put another way, the interpreter seems intent on slowing the pulse of the passage, rendering it safe for a modern reader.

the destruction of the Second Temple, abolished the ordeal, because he felt that divorce was sufficient to separate the husband from his possibly adulterous wife (*m. Soṭah* 9.9). The Tosefta (*Soṭah* 14.2) offers a less romantic explanation. 'The ritual of bitter waters is performed only in cases of suspected [unprovable, without witnesses] adultery, but now there are many who fornicate in public (with witnesses)'. In any case, the fact that the sages may have rejected the ordeal *in principle*, Romney-Wegner observes, does not allow us to assume that 'it constitutes a rejection of the double standard that assigned women far less sexual freedom than men' (Romney-Wegner 1988: 54).

In the Mishnah's elaboration of the law of the *sotah*, the sages separate the wife's sexuality into two parts. 'By paying bride-price the husband acquired both the sole right to intercourse with her and (still more important to the sages) the sole right to utilize her reproductive function' (Romney-Wegner 1988: 52-53). This second aspect of the husband's property rights is emphasized in the Mishnah tractate *Soṭah*, chs. 1–6: A wife who is sterile, past menopause, or for any other reason unable to bear children, does not have to drink the priestly potion. But if the husband divorces such a wife on his suspicion of her sexual impurity, Rabbi Meir says she does not receive her *kethubah*, 'marriage settlement'. Rabbi Eliezer, who clearly knows what women are for, adds that the husband is justified in marrying another woman and having children with her (*Soṭah* 4.3). Thus, even being suspected of sexual impropriety has its price.

If the wife is unable to bear children, the threat to the husband is more symbolic than real. A Talmudic passage links the Sotah with the instance of Maacah, the mother of Asa the king, who is punished with the loss of her status as queen because she has made an 'abominable image'. Rabbi Judah defines *miphlezeth*, the 'abominable image', as an object which 'intensifies licentiousness (*maphli lezanutha*), as R. Joseph taught: It was a kind of phallus with which she had daily contact' (*Avodah Zara* 44a). Touching the phallus, like touching the golden calf, puts the woman in contact with the locus of male power. While Queen Maacah's crime involved holding a symbolic phallus, anxiety was also raised at the possibility of the woman touching a penis.

Deuteronomy presents the case of a woman whose husband is wrangling with an opponent in the marketplace; the wife goes to his aid. In trying to defend her husband, she touches the crotch of the other man, an offense that shames her husband, but, more alarming, brings her into contact with the male organ. A woman seizing a man's genitals will have her hand cut off (Deut. 25.12).

The sages linger over the image of the adulterous woman, the woman who has enticed the wrong man:

> If she were clothed in white garments, he [the priest] covered her in black ones. If she had upon her ornaments of gold, necklaces, earrings, and rings on her fingers, they take them from her in order to disgrace her; and after that he brings an Egyptian rope[16] and ties it above her breasts. Everyone who wants to behold her comes to gaze at her (*m. Soṭah* 1.6).

After undressing her in the text, the ancient rabbis embellish the violent destruction of the woman alluded to in the biblical account:

> Hardly has she finished drinking before her face turns yellow and her eyes bulge and her veins swell, and they say, 'Take her away! take her away! that the Temple Court be not made unclean' (*m. Soṭah* 1.7).

A later rabbinic description is even more graphic:

> She painted her eyes for his sake, and so her eyes bulge. She braided her hair for his sake, and so the priest dishevels her hair. She beckoned to him with her fingers and so her fingernails fall off. She put on a fine girdle for his sake, and so the priest brings a common rope and ties it above her breasts. She extended her thigh to him and therefore her thigh falls away. She received him upon her womb, and therefore her belly swells. She fed him with the finest dainties; her offering is therefore the food of cattle. She gave him to drink choice wine in elegant flagons, therefore the priest gives her to drink the water of bitterness in a piece of earthenware (*Bemidbar Rabbah* 9.24).

16. The text reads *hevel mitsra'*, literally 'rope made from rushes', which was a contemptuous name for a slave (who had presumably been forced to make the rope in Egypt) and was considered a badge of shame. One is tempted to connect this mention of shameful rope with the Egyptian courtier's wife who shamed her husband with her adulterous longings. A rabbinic play on *hbl* is tempting since the noun also means 'birth pangs'.

In the rabbinic view, the punishment is not extraordinary. The principle of measure for measure that opens *Soṭah* 7, *bammidah sh'adam 'oded bah moddin,* states that the punishment fits the crime: since she 'adorned herself for transgression', God undressed her:

> with her thigh (*bayarek*) did she first transgress,
> and then with the belly (*beten*),
> therefore shall the thigh be stricken first
> and then the belly
> and the rest of the body shall not escape (*m. Soṭah* 1.7).

The sages then provide other examples of measure for measure: Samson, who looked at women with lust in his eyes, has his eyes gouged out by Philistines; Absalom was vain about his hair so he was suspended by his glorious hair. Absalom (again) had copulated with the ten secondary wives (*pilagshim*) of his father; thus, ten javelins are thrust into him. Clearly there is intentional sexual imagery in these biblical examples cited as parallels to the ritual of the *Sotah*. For the ancient rabbis, measure for measure acted as a control against sexual transgressions, by men or women. An adulterous person lost whatever merit she or he may have achieved throughout the rest of their life. Rabbi Judah haNasi ruled that merit held in suspense the immediate effects of the bitter water, but the 'woman would not bear children or continue in comeliness, but *she will waste away by degrees and in the end will die the self-same death*' (*Soṭah* 3.5). Thus, ancient readers probably believed that if a woman drank the ritual water she would not survive.

The Glass Half Empty

Modern interpreters seem intent on mopping up the bitter waters and downplaying their deleterious effects on the suspected wife. In doing so, however, they do not recognize their own interest in normalizing the Sotah as a Jewish ritual. Their cool medical explanations of a prolapsed uterus or false pregnancy stand in stark contrast to the hot fantasies of the ancients. One reading suggests that among the horrible physical effects that take place upon drinking the bitter water for the adulterous wife who has conceived through that union, the fetus will be aborted. If, on the other hand, the woman is innocent and has conceived with her husband, she will 'retain the seed' (v. 28) and bear her husband's child (Romney-Wegner 1988: 52).

After referring to Numbers 5 as 'a harrowing ordeal' (55), Brichto argues that the dangers in trial by ordeal are physical, and 'the danger in the potion is hypothetical—and at that, explicitly nonexistent if the woman is innocent' (56). Frymer-Kensky rejects the category of trial by ordeal as 'unwarranted and misleading'. She prefers to consider the Sotah as an example of the classic purgatory oath. Milgrom wavers, claiming that the genius of the Sotah ritual is that it removes the ability to punish from human hands and gives it into the divine realm, which would indicate trial by ordeal. In his Numbers commentary, he refers to the ritual as 'the ordeal'. Nonetheless, because Milgrom does not understand the resulting punishment of the woman as death, but merely sterility, he views the ritual of Sotah as lacking the critical element of a classic trial by ordeal: death of the guilty person. In contrast, Fishbane assumes its status as trial by ordeal. He offers a form-critical analysis of similarities between the 'draught-ordeal' ritual described in Numbers 5 and the Babylonian parallel of a case of suspected adultery in *Code of Hammurabi* 131-32. What all these analyses overlook is that the ritual of Sotah is initiated by the husband's suspicion of his wife's adulterous activity. The biblical text echoes the fear of female secrecy four times in one verse: 'it is hidden from the eyes of her husband, she is undetected, since she was not caught, and there is no witness against her' (v. 13). This fear of what another man might be doing inside his wife's house (or body) results in a protection of that house by the husband through the ritual of Sotah.

These historical critics are concerned with the extent of the woman's physical punishment—is it miscarriage, sterility, or death?—and whether the trial was actually carried out. What I find of central interest in Numbers 5 is not its degree of historicity, but rather what its existence tells us about men's fear of women's sexuality. As I have shown, concern with sexual politics allows the reader to see what is at stake in patriarchal guarding and regarding the female body. The existence of the

Sotah within the biblical corpus functions as a means of social control over wives who might ignite their husband's anger. The ritual shames her, even if she is found innocent. Accepting the text's construction of a situation in which the husband and community must be able to determine a woman's sexual purity, the contemporary scholars under review here have produced a unified picture of woman as threat to her husband's status. Remaining within the framework of belief that accepts suspicions about women's sexual lives, their techniques do not disrupt the fixed binary oppositions that categorize sexuality and gender.

Milgrom argues that the trial actually protects the woman from the 'lynch-mob mentality' of the angered community (1990: 348-50). Because the ritual has been assigned to the priest, and thus, the opportunity of dealing with the errant wife has been removed from the hysterical mob, an innocent woman would be protected from the wrath of her accusers. Brichto goes even further in transmuting the ordeal into a balm by asserting that the Sotah protects 'the woman as wife in the disadvantaged position determined for her by the mores of ancient Israel's society'. While the argument may seem attractive to those concerned with preserving the woman's life, it does not seem to have been one professed by ancient interpreters. Each time they refer to the woman put to the ordeal of bitter waters, they describe calamitous physical results. While they linger textually over the destruction of the guilty woman's body, they create no such parallel about the preservation of the innocent woman's body. While the interpreter as observer can gaze at a guilty woman's body, it would be a crime against the husband to gaze at an innocent wife's body. Describing her physically would be equivalent to undressing her. The guilty woman has already been observed in her nakedness, her husband already shamed.

That one could in theory assume an ordeal that at the least causes sterility as a way of protecting the woman is difficult to support. A sterile woman in a culture in which women function as child-bearers does not have a salutary future. Protection seems to be constructed for the husband, who, even upon his narrowest suspicion of his wife's infidelity, can force her to submit to this ordeal. The text even provides for the safety of the suspicious husband in the event that the wife is proved innocent. He is completely exonerated even if his suspicions are proved false (v. 31).

Fishbane has delineated a complex and elegant inner-biblical exegesis that connects the prophetic use of the unfaithful wife motif (as illustrated by Hos. 1–2; Isa. 50.1; 51.17-23; 57.3-14; Ezek. 16; 23) as a metaphor for Israel's infidelity to YHWH with the divine judgment exercised in the ordeal in Numbers 5 (Fishbane 1974: 40-45). Certainly connecting the motif of an adulterous woman betraying her husband with the people Israel's betrayal of YHWH is a striking example of the vitality of midrashic technique at work. What is troublesome is that Fishbane, like the early midrashists, assumes the guilt of the woman in Numbers 5. Reading with the ideology of the text, he raises no suspicions about the possible motives of the accuser.

In overlooking the indeterminate nature of the crime as presented in the Sotah, Fishbane produces a new text, one in which the woman is known to be guilty of the crime. In the prophetic view, Israel's harlotry is overt; the people have been plainly worshiping other gods. It would seem that the legal texts upon which the prophets are playing would be those that refer to such a *witnessed* offense of adultery (Exod. 20.14; Lev. 20.10; Deut. 22.22), not to the *suspected* adultery in Numbers 5.

A narrative text that contains motifs of both the trial by ordeal ritual and of Israel as harlot is Exod. 32.19-20, where Moses (acting as priest) makes a potion from the golden calf that the people of Israel are required to drink. While this text has been connected since Talmudic times with the Sotah ritual (*Avodah Zarah* 44a), there are two noteworthy differences. In the Exodus text, after drinking, all the guilty people of Israel are struck down by a divine plague. Their communal act is met with communal punishment, whereas the Sotah sets the isolated woman apart from the community. Most important, Moses has seen their act of infidelity. 'When he approached the camp and saw the calf and the dancing, Moses' anger burned hot...' (v. 19). The priest in the Numbers ritual, on the other hand, has no proof of the woman's guilt at the time he makes her swallow the drink. Moses plays the role of the priest mixing up the deadly brew, although his actions at the outskirts of the camp appear to be impulsive, stemming from his fury at the people's disloyalty to YHWH. He is not carrying out part of a formal ritual. In the Exodus text YHWH claims the dual role of shamed husband and divine vindicator. Both biblical recipes contain divine ingredients. Moses' brew contains the powdered remains of other gods, and the Sotah is made viable by the presence of YHWH.

The 'when' and 'how long' of the ritual and its resultant punishment concerns Frymer-Kensky. She envisions the woman going home 'to await the results at some future time' (1984: 22), yet the text does not imply any passage of time from the point of swallowing the potion to its devastating effect. While Frymer-Kensky's reading provides the woman an element of privacy—the woman would suffer her punishment at home—such a thoughtful emphasis on privacy directly counters the publicness of the ritual undergone upon the altar of the tabernacle. As I understand the measure-for-measure principle upon which this text is based, at best the woman would be rendered sexually dysfunctional for her sexual philandering. A parallel element requires a public punishment for a public sexual display. Since the wife did not remain at home alone, she will be publicly punished. Since the wrong man saw her body, everyone will see her sexually humiliated.

Brichto and Milgrom consider the effect of the bitter water upon the guilty woman to be more sinister, and more permanent, than abortion: sterilization. While I am more persuaded by an interpretation that embraces long-term effects than by one that supposes spontaneous abortion, I am skeptical that a husband would deprive himself of a fecund wife on the basis of his suspicions. In its favor, a barren or menopausal wife was exempt from the ordeal—which leads one to conclude that the potion had to affect the woman's ability to bear children. In this light Milgrom's 'cleansing of the womb' theory, which would return to the husband a wife able to conceive, has merit. The ordeal shares with many ancient laws involving women the overriding concern with protecting paternity, assuring the husband that any child born to his wife is his.

Another element of the readings of these scholars (Fishbane, Brichto, Frymer-Kensky, Milgrom) that I find curious is that none of them is struck by the fact that the woman is condemned to undergo the ordeal on the basis of her husband's suspicion, not on proof. And that if the woman was proved guilty, as reflected by some womb-shaking punishment, they do not imagine that death would be the result. Cases of 'proved' adultery were treated as capital crimes; it would seem that the swollen womb and sagging thigh would be all the evidence one would need that the woman was guilty. Indeed the proof would be of divine origin, since the magical (or divine) nature of the potion is that once inside the woman it discerns the purity or defilement of her body.

Milgrom argues that God has taken the punishment out of the hands of the outraged community, and thus the husband and community are forbidden to cause the death of the woman. YHWH has punished her through sterilization (1990: 349). If one accepts Milgrom's softer interpretation, the punishment remains in patriarchal hands: the deity avenges the crime against the husband. The woman is deprived of speech and action. *What is the woman for* in Milgrom's interpretation? She is the vessel through which the male-concocted brew flows. If she is a proper vessel, the liquid fills her like semen. She will bear the children of her husband. If she is an improper vessel, the liquid redefines her. She will no longer bear children.

Each of these modern interpretations remains within the borders of the biblical ritual. While the analyses describe or reflect the husband's existing suspicion of his wife, they all serve to augment the sense of suspicion about women that is produced by the biblical text. The concentration on suspicion of the woman also leaves unexamined the biblical constraint upon a wife not to behave in a suspicious manner, not to arouse her husband's anxieties. What I find missing in the recent analyses is any acknowledgment of the consequence to the woman of shaking up the sexual/gender system. At the same time there is no attempt to challenge, or even comment upon, the institutional structure of patriarchy that used the ritual of Sotah to put a woman in her place.

The Glass Half Full

From the sampling of midrashic texts imagining the fate of the wandering wife, it seems clear that the ancient rabbis were not embarrassed about creating violent sexual images of punishment for the wife who might have double-crossed her husband. In their attempt to limit the negative impact of the Sotah text, modern interpreters generally ignore the violent language in these texts. The readings of Milgrom and Brichto defend the practice of the ritual of Sotah as a strong means of protection of the woman against her irate husband. But the possibility of vengeance as the husband's motive is ignored, as are the images of shrinking genitalia and distended womb. Such considerations would provide the reader with an alternative possibility to the woman's guilt.

As more scholars apply feminist theories to the Sotah, the presumption that the male point of view is universal or normative will be dissolved. The scant or perfunctory examination of women's responses to this corrosive text is an example of how much work needs to be done in the area of feminist analysis of biblical texts. Romney-Wegner has recently added to the literature of the Sotah by producing such a critical analysis of the text as legal document. She notes two prongs of discrimination reflected by the ritual of the Sotah: (1) there is no corresponding ritual for an errant husband, since adultery is defined as a crime committed by a wife against her husband, not a husband against his own wife. (2) The Sotah is the only case in either the Bible or the Mishnah that circumvents the normal rules of evidence, in which two witnesses are necessary in capital cases; the result is a double standard of due process. The wife's personal rights are diluted by the husband's property rights.

Romney-Wegner's insights are important to a feminist analysis of ancient legal texts that kept women contained. Since her interests are legal and not literary, however, the powerful language of the text, and the rabbinic fantasies that expand upon it, are out of her purview. She does not wonder about the effect upon the image of women when a society creates its only trial by ordeal in order to punish their improper sexual behavior. Clearly the integration of research on women from many different fields is needed to circumvent the borders of our particular disciplines. Each analysis of the Sotah will present a partial picture until we use an interdisciplinary analysis that encompasses many partial views.

The crucial element of the Sotah text, regardless of whether one wishes the accused woman to suffer a horrible death or merely to sip a noxious cocktail, is that it reflects the potency of male imaginings. As surely as the innocent bird eagerly wings toward the tempting snare, the husband imagines his wife as luscious Eve, the source of trouble and the root of desire. A tamed Eve pleases men, a wild one frightens them, but in neither aspect does she serve the needs of women. The Sotah both reflects and supports the patriarchal social system that cannot accept the woman without seeking to offset the threat that she represents, a threat of dissolution, anarchy and antisocial disorder.

BIBLIOGRAPHY

Bach, Alice

1993 'Breaking Free of the Biblical Frame-Up: Uncovering the Woman in Genesis 39', in *A Feminist Companion to Genesis* (ed. Athalya Brenner; The Feminist Companion to the Bible, 2; Sheffield: Sheffield Academic Press): 318-42.

Bal, Mieke

1985 *Narratology: Introduction to the Theory of Narrative* (Toronto: University of Toronto Press).

1987 *Lethal Love: Feminist Literary Readings of Biblical Love Stories* (Bloomington: Indiana University Press).

1988 *Murder and Difference: Gender, Genre, and Scholarship on Sisera's Death* (trans. Matthew Gumpert; Bloomington: Indiana University Press).

1988 *Death and Dissymmetry: The Politics of Coherence in the Book of Judges* (Chicago: University of Chicago Press).

1992 *Reading Rembrandt: Beyond the Word–Image Opposition* (Cambridge: Cambridge University Press).

Bird, Phyllis

1974 'Images of Women in the Old Testament', in *Religion and Sexism* (ed. R.R. Ruether; New York: Simon & Schuster): 41-88.

1989a 'The Harlot as Heroine: Narrative Art and Social Presupposition in Three Old Testament Texts', *Semeia* 46: 119-39.

1989b 'Women's Religion in Ancient Israel', in *Women's Earliest Records: From Ancient Egypt and Western Asia* (ed. Barbara S. Lesko; BJS; Atlanta: Scholars Press): 283-98.

Benstock, Shari
1991 *Textualizing the Feminine: On the Limits of Genre* (Norman: University of Oklahoma Press).

Brichto, H.C.
1975 'The Case of the *Sota* and a Reconsideration of Biblical Law', *HUCA* 46: 55-70.

Camp, Claudia
1985 *Wisdom and the Feminine in the Book of Proverbs* (Sheffield: Almond Press).

Derrida, Jacques
1979 'Living On: Borderlines' (trans. James Hulbert), in *Deconstruction and Criticism* (ed. Harold Bloom *et al.*; New York: Seabury Press).

Exum, J. Cheryl
1985 '"Mother in Israel": A Familiar Figure Reconsidered', in *Femininist Interpretation of the Bible* (ed. Letty M. Russell; Philadelphia: Westminster Press): 73-85.
1992 *Tragedy and Biblical Narrative: Arrows of the Almighty* (Cambridge: Cambridge University Press).

Fishbane, Michael
1974 'Accusations of Adultery: A Study of Law and Scribal Practice in Numbers 5.11-31', *HUCA* 45: 25-45.

Frymer-Kensky, Tikva
1984 'The Strange Case of the Suspected Sotah (Numbers v 11-31)', *VT* 34: 11-26.

Fuchs, Esther
1987 'Structure and Patriarchal Functions in the Biblical Betrothal Type-Scene: Some Preliminary Notes', *Journal of Feminist Studies in Religion* 3: 7-13.
1988 'For I Have the Way of Women: Deception, Gender and Ideology in Biblical Narrative', *Semeia* 42: 68-83.

Gallop, Jane
1990 'Why Does Freud Giggle When the Women Leave the Room?', in *Psychoanalysis and...* (ed. R. Feldstein and H. Sussman; New York: Routledge): 49-54.

Greenstein, Edward L.
1989 'Deconstruction and Biblical Narrative', *Prooftexts* 9: 43-71.

Humbert, Paul
1937 'La femme étrangère du livre des Proverbes', *Revue des études sémitiques* 6: 40-64.

Irigaray, Luce
1985 *This Sex Which Is Not One* (trans. Catherine Porter; Ithaca: Cornell University Press).
1991 'The Bodily Encounter with the Mother', in *The Irigaray Reader* (ed. Margaret Whitford; Oxford: Basil Blackwell): 34-46.

Jackson, B.S.
1975 'Reflections on Biblical Criminal Law', in his *Essays in Jewish and Comparative Legal History* (Studies in Judaism in Late Antiquity, 10; Leiden: Brill): 25-63.

Jacobus, Mary

1982 'Is There a Woman in This Text?', *New Literary History* 14: 117-41.

Kamuf, Peggy (ed.)

1991 *A Derrida Reader: Between the Blinds* (New York: Columbia University Press).

McKane, William

1970 *Proverbs* (OTL; Philadelphia: Westminster Press).

McKeating, Henry

1979 'Sanctions against Adultery in Ancient Israelite Society, with Some Reflections on Methodology in the Study of Old Testament Ethics', *JSOT* 11: 57-72.

Milgrom, Jacob

1981 'The Case of the Suspected Adulteress, Numbers 5.22-31: Redaction and Meaning', in *The Creation of Sacred Literature* (ed. Richard F. Friedman; Berkeley: University of California Press): 69-75.

1990 *Numbers* (The JPS Torah Commentary; Philadelphia: Jewish Publication Society of America).

Neufeld, E.

1944 *Ancient Hebrew Marriage Laws* (London).

Newsom, Carol A.

1989 'Woman and the Discourse of Patriarchal Wisdom: A Study of Proverbs 1–9', in *Gender and Difference in Ancient Israel* (ed. Peggy L. Day; Minneapolis: Fortress Press): 142-60.

Okin, Susan Moller

1979 *Women in Western Political Thought* (Princeton: Princeton University Press).

Phillips, Antony

1973 'Some Aspects of Family Law in Pre-Exilic Israel', *VT* 23: 349-61.

1981 'Another Look at Adultery', *JSOT* 20: 3-25.

Romney-Wegner, Judith

1988 *Chattel or Person? The Status of Women in the Mishnah* (New York: Oxford University Press).

Sasson, Jack

1972 'Numbers 5 and the Waters of Judgment', *BZ* 16: 249-51.

Slotki, Judah J. (trans.)

1983 *Bemidbar Rabbah* (London: Soncino Press, 3rd edn).

Snijders, L.A.

1954 'The Meaning of *zar* in the Old Testament,' *OTS* 10: 97-105.

Vaux, Roland de

1965 *Ancient Israel: Its Life and Institutians* (New York: McGraw–Hill).

INTERTEXTUALITY AND THE BOOK OF JEREMIAH: ANIMADVERSIONS ON TEXT AND THEORY

Robert P. Carroll

> Yet, what appears as a lack of rigour is in fact an insight first introduced into literary theory by Bakhtin: any text is constructed as a mosaic of quotations; any text is the absorption and transformation of another. The notion of *intertextuality* replaces that of intersubjectivity, and poetic language is read as at least double.
>
> Julia Kristeva[1]

> Every text, being itself the intertext of another text, belongs to the intertextual, which must not be confused with a text's origins: to search for the 'sources of' and 'influence upon' a work is to satisfy the myth of filiation. The quotations from which a text is constructed are anonymous, irrecoverable, and yet *already read*: they are quotations without quotation marks. The work does not upset monistic philosophies, for which plurality is evil. Thus, when it is compared with the work, the text might well take as its motto the words of the man possessed by devils: 'My name is legion, for we are many' (Mark 5.9).
>
> Roland Barthes[2]

> Literature is not exhaustible, for the sufficient and simple reason that no single book is. A book is not an isolated being: it is a relationship, an axis of innumerable relationships.
>
> Jorge Luis Borges[3]

1. J. Kristeva, 'Word, Dialogue and Novel', in *The Kristeva Reader* (ed. T. Moi; Oxford: Basil Blackwell, 1986), p. 37 (ET by A. Jardine, T. Gora, and L.S. Roudiez of 'Let mot, le dialogue et le roman', in *Semiotiké* [Paris; Editions du Seuil, 1969], pp. 143-73 [146]).

2. R. Barthes, 'From Work to Text', in *Textual Strategies: Perspectives in Post-Structuralist Criticism* (ed. J.V. Harari; Ithaca, NY: Cornell University Press, 1979), p. 77 (ET by J.V. Harari of 'De l'oeuvre au texte', *Revue d' Esthétique* 3 [1971]).

3. J.L. Borges, 'A Note on (towards) Bernard Shaw', in his *Labyrinths:*

I

Having written copiously on the book of Jeremiah, I still find that that book eludes my reading of it. It resists all my reading strategies and other scholars' reading strategies do not persuade me at all that they have got the measure of Jeremiah. I can cope with being mystified and even defeated in my reading of the Bible because its alienness is both self-evident and generally acknowledged by contemporary writers on the Bible. I am unable to accept many of my colleagues' readings of Jeremiah because they tend to demystify it to the point of domestication. Its alienness—of time, place, culture, ideology, etc.—is charmed by such reading strategies and often becomes the underwriting of their own programmes. Such domestications I wish to eschew, while leaving myself open to the criticism that I have failed to milk the text of its obvious theological wealth.[4] That failure is not all that great because in my opinion a success here would be a betrayal of the text. In this paper I shall attempt to put forward a rather different approach to the understanding of the book of Jeremiah, using insights from modern literary theory known as 'intertextuality'. In discussing the book of Jeremiah in terms of my own approaches to it in conjunction with those of other commentators on the book I have already begun to *practise* an intertextual approach to Jeremiah.

Selected Stories and Other Writings (London: Penguin Books, 1970), pp. 248-49.

4. This I take to be the main point of W. Brueggemann's criticism of my work on Jeremiah in his review of recent commentaries on Jeremiah: 'Jeremiah: Intense Criticism/Thin Interpretation', *Int* 42 (1988), pp. 268-80. Brueggemann has now contributed two fine volumes of commentary on Jeremiah himself, so his own theological reading of the text can be scrutinized: *To Pluck Up, To Tear Down: A Commentary on the Book of Jeremiah 1–25; To Build, To Plant: A Commentary on Jeremiah 26–52* (International Theological Commentary; Grand Rapids: Eerdmans; Edinburgh: The Handsel Press, 1988, 1991). I have offered some reflections on the Jeremiah commentaries of Carroll, Holladay and McKane in my articles, 'Radical Clashes of Will and Style: Recent Commentary Writing on the Book of Jeremiah', *JSOT* 45 (1989), pp. 99-114; and 'Arguing about Jeremiah: Recent Studies and the Nature of a Prophetic Book', in *Congress Volume, Leuven 1989* (ed. J.A. Emerton; VTSup, 43; Leiden: Brill, 1991), pp. 222-35.

The term 'intertextuality' (or *intertextualité*) is taken from the seminal 1966 essay by Julia Kristeva. This essay is a presentation and development of the central ideas of the Russian literary critic Mikhail Bakhtin.[5] In her discussion of Bakhtin she writes:

> Writer as well as 'scholar', Bakhtin was one of the first to replace the static hewing out of texts with a model where literary structure does not simply exist but is generated in relation to *another* structure. What allows a dynamic dimension to structuralism is his conception of the 'literary word' as an *intersection of textual surfaces* rather than a point (a fixed meaning), as a dialogue among several writings: that of the writer, the addressee (or the character) and the contemporary or earlier cultural context.[6]

She develops her explication of Bakhtin's work further:

> The addressee, however, is included within a book's discursive universe only as discourse itself. He thus fuses with this other discourse, this other book, in relation to which the writer has written his own text. Hence horizontal axis (subject–addressee) and vertical axis (text–context) coincide, bringing to light an important fact: each word (text) is an intersection of word (texts) where at least one other word (text) can be read. In Bakhtin's work, these two axes, which he calls *dialogue* and *ambivalence*, are not clearly distinguished. Yet, what appears as a lack of rigour is in fact an insight first introduced into literary theory by Bakhtin: any text is constructed as a mosaic of quotations; any text is the absorption and transformation of another. The notion of *intertextuality* replaces that of intersubjectivity, and poetic language is read as at least *double*.[7]

These two quotations from a complex essay will have to serve as points of definition. The term 'intertextuality' defines the literary object/event/word as an 'intersection of textual surfaces' and as 'a mosaic of quotations'. In other words, a text is always both pretextual and contextual, as well as being textual. It is not simply generated by a writer, but is a complex production

5. She focuses mainly on Bakhtin's *Rabelais and his World* (ET 1965) and his *Problems of Dostoevsky's Poetics* (ET 1973); but the reader interested in his dialogism theory might be better recommended to read M. Bakhtin, in M. Holquist (ed.), *The Dialogic Imagination: Four Essays* (trans. C. Emerson and M. Holquist; Austin, TX: University of Texas Press, 1981).

6. Kristeva, 'Word, Dialogue and Novel', p. 37 (emphases original).

7. Kristeva, 'Word, Dialogue and Novel', p. 37 (emphases original).

formed by prior textual events and the interaction of writers/ redactors/readers with such a contexting textuality.

Further intertextual definitions of intertextuality may be added to Kristeva's initial analysis. Among many definitional statements, the following by John Frow are given here to aid understanding of the nature of the intertextual:

> Texts are therefore not structures of presence but traces and tracings of otherness. They are shaped by the repetition and the transformation of other textual structures.
>
> These absent textual structures at once constrain the text and are represented by and within it; they are at once preconditions and moments of the text.
>
> Texts are made out of cultural and ideological norms; out of the conventions of genre; out of styles and idioms embedded in the language; out of connotations and collective sets; out of clichés, formulae, or proverbs; and out of other texts.[8]

The notions of otherness and repetition are fundamentally important in defining the nature of intertextuality. They point to the codedness of textuality and emphasize the fact that a text reflects a system (or code) of other textual factors (or structures). Every text makes its readers aware of other texts. It insists on an intertextual reading.

Turning from the definitional to the biblical text, it is possible to demonstrate the self-evident nature of the intertextuality of the Bible by referring to what is already known about that collection of many books that we call 'the Bible'. This in turn will allow me to introduce an intertextual reading of Jeremiah as a natural follow-on from the collection of books in which the book of Jeremiah now has its place. For once it may be worthwhile pointing out the obvious so as to remind readers of the Bible of what they may be forgetting when reading that book.

Whatever our ideological holdings and however we may favour reading the Bible, the presentation of the books constituted by the different canons of the Bible (Hebrew, Greek, Christian, etc.) represents certain narratological arrangements

8. J. Frow, 'Intertextuality and Ontology', in *Intertextuality: Theories and Practices* (ed. M. Worton and J. Still; Manchester: Manchester University Press, 1990), pp. 45-55 (45).

that tell a story (e.g. from Genesis to 2 Kings or from Genesis to 2 Chronicles/Ezra–Nehemiah). This story (or these stories) depend very much on all the books being arranged in sequence and therefore the story that emerges from the Bible is fundamentally intertextual. Each textual unit (or book) within the larger collection of books depends on the books preceding and succeeding it for its place in and contribution to the story. Some canon-conscious communities may prioritize certain elements of the narrative and thereby make all the other books in the collection dependent on or reflective of the prioritized books—as happens in Orthodox Judaism, where the Torah of Moses is the main focus of revelation and everything else in the canon is (mere) commentary on Torah. Intertextuality becomes metacommentary.[9]Jameson The relationship then between Torah and the rest of the books ('the law and the prophets') is an intertextual one. In various Christian communities (whether Catholic, Orthodox or Protestant) such intertextualities are taken up into a greater intertextuality created by the addition of the (Greek) New Testament to the (Greek version of the) Hebrew Bible. The storyline of the New Testament depends very much on the (meta)narrative of the Hebrew Bible and would be meaningless without it. Much of what is in the New Testament is generated by an intertextual dialogue with the older collection of writings (the letter to the Hebrews and the Apocalypse make the intertextual point without remainder!). Within the New Testament itself there is a strong intertextual factor in the production of the Synoptic Gospels, where the Gospels feed on and off each other. The Gospel of Mark signals its intertextual nature immediately by identifying the 'beginning of the gospel of Jesus Christ, *as it is written* in Isaiah the prophet...' (1.1-2a) in intertextual terms. Matthew's genealogy of Jesus reflects the Gospel as book (1.1, 'the book of the genealogy of Jesus Christ') and is modelled on the *toledoth* literature of the Hebrew Bible. Luke's Gospel equally signals its

9. On metacommentary see F. Jameson, 'Metacommentary' in his *The Ideologies of Theory: Essays 1971–1986*. I. *Situations of Theory* (London: Routledge, 1988), pp. 3-16 (article originally published in *PMLA* 86.1 [1971], pp. 9-18). David Clines and Robert Carroll are currently editing a volume on the Hebrew Bible under the general title of *Metacommentary*.

intertextual construction by acknowledging the many who had 'undertaken to compile a narrative' before he set out 'to write an orderly account' of the matter (1.1-4).

The intertextual nature of the Bible cannot be gainsaid. Whether we read the Hebrew Bible, the literature from Qumran, the New Testament or the Bibles of the various Christian churches, we are always reading texts that have been generated intertextually. I state the obvious in order to put it behind us. The whole Bible (whichever one is used) is a mosaic of mosaics (a mosaic of Mosaics, also). In classical critical theory about the Bible, the Documentary Hypothesis about the generation of the Pentateuch from four documents (JEDP) is a primitive form of an intertextual account of the writing of the Bible. However much that theory (in whatever version) may be under review in current biblical scholarship, and whatever theory of biblical composition may (or may not) replace it eventually in the guild of biblical studies, any account of how the primary narratives of the Bible came to be written will have to have an intertextual basis.[10]

It would take a chapter in itself to outline the dominant intertextual relations between the books in the Hebrew Bible. The so-called Deuteronomistic History of Joshua–Judges–Samuel–Kings (with the book of Deuteronomy as its prologue) is edited so as to be quite dependent on the Pentateuch, and it regularly cites it. Throughout that History (and also in Chronicles) there are many references to books that form intertextual connections with what is in the biblical text. Any examination of the books making up the Prophetic Collection (Isaiah–Jeremiah–Ezekiel–the Twelve) will discover a whole intertextual world where each individual book will be found to contain a considerable amount of material common to other books (the most obvious example may be Isa. 2.2-4 = Mic. 4.1-3, but further similar examples could be multiplied a hundredfold). To read and understand Isaiah 40–55 it is necessary to know the book of Psalms; to read the book of Jeremiah requires a deep knowledge of the

10. My own view of the composition of the Pentateuch, which tends to follow Rendtorff's notion of *Bearbeitungen*, is merely academic here, and the intertextuality of the Pentateuch is a perception about the five books independent of any particular theory of composition (in or out of vogue).

Deuteronomistic History. The examples could be multiplied, but the argument would not be made any firmer by statistical information. The books of the Bible are interwoven by and from each other and no account of their composition that avoids addressing their intertextual nature can be an adequate account of anything in the Hebrew Bible.

II

To turn to the book of Jeremiah after this most general of introductions is to see just how intertextual the biblical books really are. The major, and in my opinion now classic, commentary on Jeremiah by Bernhard Duhm (1901), with its explanation of the composition of the book of Jeremiah in terms of the poems of Jeremiah, the book of Baruch and supplementation of these two documents by later writers, indicates the fundamentally intertextual nature of Jeremiah without using such terminology.[11] The notion of supplementers working on prior documents and producing the book as we know it already contains in it the basic idea of intertextuality. Texts are generated by prior texts. So the material for an intertextual account of the book of Jeremiah is there in the work of the commentators of this century (Duhm's work was mostly done in the last century, but the publication of his commentary in the first year of this century makes his work a twentieth-century book). Whether we develop Duhm's work or enhance it by modification or expansion using the subsequent work of Sigmund Mowinckel and William McKane does not materially affect Duhm's fundamentally important contribution to the modern understanding of the book of Jeremiah.[12] Mowinckel may favour the 'streams

11. B. Duhm, *Das Buch Jeremia* (KHAT, 11; Tübingen: Mohr [Paul Siebeck], 1901), pp. xi-xx.

12. S. Mowinckel, *Zur Komposition des Buches Jeremia* (Videnskapsselskapets Skrifter, 4; Hist.-Filos. Klasse, 1913, 5; Oslo: Jacob Dybwad, 1914), and *Prophecy and Tradition: The Prophetic Books in the Light of the Study of the Growth and History of the Tradition* (ANVAO, 2; Hist.-Filos. Klasse, 1946, 3; Oslo: Jacob Dybwad, 1946); W. McKane, *A Critical and Exegetical Commentary on Jeremiah*. I. *Introduction and Commentary on Jeremiah I–XXV* (ICC; Edinburgh: T. & T. Clark, 1986). For a brief survey of the work of Duhm, Mowinckel and others on the composition of Jeremiah, see R.P. Carroll,

of tradition' approach rather than Duhm's 'supplements', and McKane may advocate a 'rolling corpus' notion of the book's generation, but all these commentators see Jeremiah as, in very important senses, being the product of development and supplementation (that is, as an *Ergänzungstext*). And the fundamental feature of any *Ergänzungstext* is its intertextual construction.[13]

Our knowledge of the processes that gave rise to the book of Jeremiah in the first place is absolutely nil. Everything we know (or imagine we know) is based on a highly interpretative account of what we may imagine is 'information'. We posit certain texts and narratives in Jeremiah as being prima facie an account of how the book was written (e.g. Jer. 36). This judgment is always open to question because it is analogous to lifting ourselves up with our own boot-straps—it is a boot-strapping operation because we extrapolate from information contained within a book when we have yet to demonstrate that such information can be used reliably for the purposes we have in mind. Commentators who rely on their own reading of Jeremiah 36 as *the* account of how *the book* of Jeremiah was written, so that 'the book is largely the work of the scribe Baruch', read more into the text than can be warranted by any prior argument.[14] On the other hand, treating the text as an *Ergänzungstext* allows us to recognize the obvious and then permits us to produce an intertextual account of the book's production. Jeremiah 36 allows a glimpse of this possibility because it presents an account of the transformation of Jeremiah's spoken oracles into a written document. When the king has the scroll of Jeremiah's words burned, Baruch the scribe *re*writes the scroll and 'many similar words were added' to the words of Jeremiah (36.32).[15]

Jeremiah: A Commentary (OTL; London: SCM Press, 1986), pp. 38-50. My articles referred to above in n. 4 offer further comment on the composition of Jeremiah, especially in relation to the work of Holladay and McKane.

13. On the connections between Jeremiah and 2 Kings, see C.C. Torrey, 'The Background of Jeremiah 1–10', *JBL* 56 (1937), pp. 193-216; on Jeremiah as an *Ergänzungstext* see my 'Arguing about Jeremiah', cited in note 4 above (pp. 229-31).

14. The brief quotation is from W.L. Holladay, *Jeremiah. 2. A Commentary on the Book of the Prophet Jeremiah, Chapters 26–52* (Hermeneia; Philadelphia: Fortress Press, 1989), p. 24.

15. I have written at greater length on the narrative of ch. 36 and also on

The intertextuality of that story must be obvious, even to incompetent readers. Words are transformed into writing and the first written version is replaced with a rewritten and longer second version. Incorporated into the second scroll is the first scroll, but it is now supplemented or even transformed—we do not know what the relations between the two may have been, other than whatever is conveyed by the term 'added'—and in that change you have a testimony to the scroll's intertextuality. The intertextual is the pretextual further inscribed. Baruch's second scroll has as its pretext the first scroll (its pretext is the words of Jeremiah), and its incorporation into the second scroll demonstrates the intertextual nature of that second scroll.

The fundamental intertextuality of the book of Jeremiah may be demonstrated using an approach different from the *Ergänzungstext* thesis. Perhaps the first and most obvious thing to notice about the book of Jeremiah is the fact (and fact it is!) that its final chapter (52) is also the final chapter of 2 Kgs (25). How intertextual can you get? Whatever the reasons for incorporating 2 Kings 25 into the book of Jeremiah and thereby forming an inclusio or closure between the Deuteronomistic History and Jeremiah, the intertextual nature of Jeremiah is strongly indicated. The shared chapter is declarative of the strong relationship between Jeremiah and the Deuteronomistic History. A general reading of Jeremiah will show that it is a book in dialogue with the History and also dependent on it. Modern scholarship on Jeremiah often talks about a Deuteronomistic *edition* of Jeremiah.[16] Whatever may be indicated by this point of view, and whatever justification there may be for it, it does point to an important aspect of the book of Jeremiah. The language, discourse analysis, topoi and other concerns of the

the role of writing in the book of Jeremiah: see my IOSOT paper 'Manuscripts Don't Burn—Inscribing the Prophetic Tradition: Reflections on Jeremiah 36' (Paris 1992; to be published in the BEATAJ volume of papers given at that Congress), and my G.W. Anderson Festschrift contribution, 'Inscribing the Covenant: Writing and the Written in Jeremiah' (to be published by JSOT Press in the *JSOT* Supplement Series).

16. The fullest account of the matter is undoubtedly W. Thiel, *Die deuteronomistische Redaktion von Jeremia 1–25* and *Die deuteronomistische Redaktion von Jeremia 26–45* (WMANT, 41, 52; Neukirchen–Vluyn: Neukirchener Verlag, 1973, 1981).

Deuteronomistic writers are also to be found in Jeremiah. Even allowing for a general theory of a Deuteronomistic edition of the Prophetic Collection, there is a stronger element of Deuteronomistic writing in Jeremiah than in any other prophetic book in the Hebrew Bible.[17] The relationship between Jeremiah and Deuteronomism (whatever is built into that catch-all term) is a highly intertextual one. The colophon of Jer. 1.1-3 apart (most of the colophons to the prophetic books reflect deuteronomistic influence), a number of narratives show Deuteronomistic traces throughout (e.g. 7.1–8.3; 11.1-13; 25.1-14; 26; 44), and the cycle of material on the 'house of the king of Judah' (21.11–23.6) in particular. In 22.10–23.6 the collection of poems is given a series of prose commentaries that owe much to the Deuteronomistic History. Without the linking commentary the poems could not be understood in the ways suggested by the prose explanations. The intertextuality of the cycle is glaringly obvious. The poems with their commentary are a prime example of the intertextuality of Jeremiah as 'an intersection of textual surfaces' where poems about anonymous persons (apart from 22.28-30) intersect with the History's list of the last kings of Judah to form an intertextual account of their fates. Jeremiah's poems become commentary on the kings and thereby supplement the History's account of them. Such intertextual supplementations help to incorporate the book of Jeremiah into the Deuteronomistic literature (hence the closure of Jeremiah with 2 Kgs 25).

In much more general ways the intertextuality of Jeremiah can be demonstrated to the reader of the book. There are so many intertextual elements *within the book itself* that it is difficult to know where to start in the argument. The inclusios, the chiastic structures, the repeats of pieces of text within the book, the editorial rearrangements of such repeats—all testify to the intertextual nature of the book. Much of the material contained in Jeremiah has been used in so many different ways to create the book that it is itself already intertextually generated without our having to go outside of the book to demonstrate its intertextual

17. On general features of a Deuteronomistic editing process in the prophets, see for example W.H. Schmidt, 'Die deuteronomistische Redaktion des Amosbuches. Zu den theologischen Unterschieden zwischen dem Prophetenwort und seinem Sammler', *ZAW* 77 (1965), pp. 168-93.

nature from other pretextual sources. The differences, especially of arrangement and placement, between the Hebrew and Greek editions of Jeremiah may force any reader (or commentators) into some intertextual account of the book. If the Greek versions of Jeremiah are scrutinized, say, in the Göttingen Septuagint edition, where each and every page is constituted by a quarter or less of text and three-quarters or more of alternative readings, then the intertextuality of Greek Jeremiah is simply beyond dispute.[18] Joseph Ziegler has provided us with the means of producing a first-class intertextual account of just one reception history of Jeremiah, but a comprehensive commentary on the Greek texts of Jeremiah remains to be written. Perhaps with the increased interest in newer approaches to the Bible—especially in terms of literary, intertextual and *Rezeptionsgeschichte* approaches to it—study of the Hebrew Bible will come to include a serious treatment of the LXX.[19]

Much of the book of Jeremiah is pieced together by the manipulation of fragments and snatches of text, so that the construction and production of the book must be regarded as having followed various intertextual routes. Pieces of text are brought together as topoi to form collections of related material: for example, the cycle of material on the cult (7.1–8.3); the drought cycle (14.1–15.4); material on the royal house (21.11–23.6); the cycle on the prophets (23.9-40) and a further collection of material held together by the topos of prophets (chs. 27–29). The construction of such cycles indicates an intertextual focus whereby editors, redactors, writers—I think these are overlapping rather than synonymous terms in current biblical scholarship, though I would not like to have to take the witness stand and swear on oath as to what the differences were between them—brought together bits of texts that they regarded as

18. J. Ziegler (ed.), *Jeremias, Baruch, Threni, Epistula Jeremiae* (Septuaginta. Vetus Testamentum Graecum, auctoritate academiae scientiarum Gottingensis editum, 15; Göttingen: Vandenhoeck & Ruprecht, 2nd edn, 1976).

19. Apart from various monographs on the LXX of Jeremiah, McKane's ICC volume should be recognized for the importance of its treatment of the LXX in relation to the Hebrew text of Jeremiah. A full-scale commentary on the actual Greek texts of Jeremiah is still a desideratum of Jeremiah studies.

belonging together in order to make them form a unitary statement about certain matters. Juxtaposed, these originally separate elements are now intertextually constituted, and their meanings are shaped and reshaped by their present conjunctions in the book. Intertextuality creates meaning by these writerly means. Repeated uses of the same pieces of text (e.g. 6.13-15 = 8.10-12; 11.20 = 20.12; 16.14-15 = 23.7-8; 23.19-20 = 30.23-24) suggest construction processes in the book that indicate a highly intertextual reflectivity going on in the creation of the book of Jeremiah. Such intertextuality inevitably moves us away from original authorship—whatever that may mean in terms of the sources behind the Bible—to contemplate the production processes whereby *the text took this shape* (i.e. the form we have it in). Whatever account we may wish to offer for the production of the book of Jeremiah, it will have to take its intertextual nature into consideration. But then, as Roland Barthes states, 'The theory of the Text can coincide only with the activity of writing'.[20] We need some account of how the book came to written form and any such account is necessarily focused on techniques of ancient writing in the Near East.

Many other intertextual features of the book may be noted. The writer of the prologue in ch. 1 has gathered together a set of figures or *Leitmotiven* (*Leitwörter*) in 1.10 that reflect the uses of these words throughout the book itself. This extrapolation from the text reveals the intertextual engagement involved in the book's production. The fractured chiasmus of 1.10, with its six terms 'pluck up', 'break down', 'destroy', 'overthrow', 'build', 'plant', directs the reader how to read the book. The terms themselves appear in various combinations in 12.14-17; 18.7, 9; 24.6; 31.28, 38, 40; 42.10; 45.4 (with 31.28 using all six terms). A more complex pattern of usage could be suggested if every occurrence of one or more of these terms in the book were scrutinized. Readers who wish to follow this intertextual *Holzweg* may pursue their own reading of Jeremiah.

A much more interesting and complex intertextual feature of the book of Jeremiah may be seen in some of the narratives. A reading of these narratives will demonstrate how intertextuality functions to overflow the text's boundaries and to force the

20. Barthes, 'From Work to Text', p. 81.

reader into encountering other texts. This combination of intertextuality and narrativity in Jeremiah has the added bonus of moving the discussion away from more traditional textualist issues to more modern narratological matters in relation to the biblical text. In 7.1-15 there is what may now be called the famous 'temple sermon' (famous because so many preachers when stuck for a text always retreat to it for inspiration). The text is complex and has a double representation of proclamation in the temple precincts. It is introduced by one of the many standard 'reception of the divine word' formulas so dominant in the book of Jeremiah.[21] Thereafter the section is all sermon. No further contextualization information is given. Its present context in the cycle of material directed against cultic practices gives it a certain troping and suggests an outcome consonant with 8.1-3. In ch. 26, however, the sermon—or part of it—reappears, with much greater contextualization, and is presented as part of a narrative mostly taken up with the reception of the sermon. The whole chapter serves to introduce the second half of the book of Jeremiah—a book that has come to an end in 25.30-38, as it were. Thus ch. 26 is both significant as the *restart* of the tradition and pregnant with possibilities. As the introduction to the block of material in chs. 26–36, it tropes the earlier version of the sermon in various ways. I use the word 'earlier' here to mean 'earlier in the book' rather than in historical terms. The reader (or hearer) of the book will have read (or heard) 7.1-15 before hearing (or reading) ch. 26. This way of talking (writing) about the narrative already acknowledges the highly intertextual nature of the book of Jeremiah, but I know no other way of taking the text of Jeremiah seriously than to recognize its intertextual nature.

Structural elements in the narrative of ch. 26 link it with ch. 36, which closes the section of chs. 26–36. So ch. 26 has to be read in conjunction with ch. 36 in order for the circle of signification in the narratives to be closed. Chapter 26 sets up ch. 36 and ch. 36 concludes a matter left open-ended in ch. 26. The two narratives are intertextually bound together. In both narratives the words

21. On this feature of Jeremiah, see P.K.D. Neumann, 'Das Wort, das geschehen ist... Zum Problem der Wortempfangsterminologie in Jer. i–xxv', *VT* 23 (1973), pp. 171-217.

of the prophet, whether spoken or written, are responded to by various social strata of Judaean society. Problems of and questions about representation in the Bible cannot be dealt with here or this essay will run on too long, but readers should be reminded of Jacques Derrida's point about representation in order not to be led astray by the narratives of chs. 26 and 36 in relation to Judaean history. Derrida writes:[22]

> the authority of representation constrains us, imposing itself on our thought through a whole dense, enigmatic, and heavily stratified history. It programs us and precedes us...

So I shall not deal with what these narratives represent in terms of social interaction between prophet and people or even with the representation of Jeremiah in them. Only their intertextual features interest me here.

In ch. 26 various groups react to what Jeremiah says, both for and against his point of view. In a highly stratified narrative, different social strata hear him out, defend or attack him and, eventually, he escapes by the agency of a member of an important family (26.24). In the course of various debates about the legitimacy of what Jeremiah has to say, rural elders cite Mic. 3.12 in his defence (26.17-19). This incorporated citation in a narrative is very rare in the prophetic texts—quotations are all too common, but 26.18 is set into an ongoing narrative—yet it underlines the highly intertextual nature of the story. It is not only intertextual in relation to the book of Jeremiah (the argument of these paragraphs), it is intertextual in relation to Micah. The text of ch. 26 turns paradigmatic at vv. 17-19 and 20-23 before indicating the outcome of the debate in v. 24. In ch. 36, set a few years later than ch. 26, the prophet again attempts to influence the community (and the whole state: 36.9; cf. 26.2) in a particular direction. This time he does not preach to the worshippers, but the scroll of his words written by Baruch is read by Baruch on various occasions to different groups of people. The fate of the scroll indicates the final rejection of the prophet's words, and the story left open-ended in ch. 26 is closed in ch. 36. However, once ch. 36 is read into the story of ch. 26, the

22. J. Derrida, 'Sending: On Representation', *Social Research* 49.2 (1982), pp. 294-326 (304) (ET by P. and M.A. Caws).

intertextuality of the matter becomes more complex because 2 Kings 22 is implicated in ch. 36. Jeremiah 36 and 2 Kings 22 are bound up together intertextually. They reflect one another and together constitute a paradigm of how to hear prophecy read and how not to hear it read. These narratives mirror each other. Jer. 7.1-15 may be only the content of the prophetic word, but chs. 26 and 36 and 2 Kings 22 are all about the reception of that word. Their intertextuality proclaims itself in every line. It is a moot point whether Jeremiah 36 or 2 Kings 22 is the primary narrative on which the other is based.[23] As the book of Jeremiah is fundamentally dependent on the Deuteronomistic History, it may be concluded that Jeremiah 36 reflects 2 Kings 22 rather than the other way around, but from an intertextual point of view it hardly matters which came first. Both belong together and have to be read together intertextually.

The overstrong, intertextual relationship between Jeremiah and the Deuteronomistic History has already been stressed throughout this essay. It may be further spelled out to underline the point. For Deuteronomy—Deuteronomism includes the History and Deuteronomy—the figure of Moses, especially *Moses as prophet*, is absolutely fundamental. The most important texts for this viewpoint are Deut. 18.15-22 and 34.10-12—but the whole story of Moses, especially in Exodus 2–7, is told in terms of the formal aspects of the commissioning narratives of prophets. In Jeremiah 1 there are various elements that reflect connections between Jeremiah and Moses. Whether influence is from Moses to Jeremiah or vice versa is again a moot point.[24]

23. The vexed question of whether 2 Kings or Jeremiah came first cannot be dealt with here. Someday soon biblical scholarship will have to rethink all these matters and develop much better theoretical bases for reading the Hebrew Bible. On the relative order of Jer. 36 and 2 Kgs 22, see C. Minette de Tillesse, 'A reforma de Josias', *Revista Biblica Brasileira* 6 (1989), pp. 41-61, and his contribution to IOSOT 1992, 'Josias et Joiaqim: 2 R 22/Jer 36' in the BEATAJ volume referred to in n. 15 above.

24. On Jeremiah and Moses from the conventional point of view, see the many works of W.L. Holladay, esp. his 'The Background of Jeremiah's Self-Understanding: Moses, Samuel, and Psalm 22', *JBL* 83 (1964), pp. 153-64; 'Jeremiah and Moses: Further Observations', *JBL* 85 (1966), pp. 17-27; *Jeremiah*. I. *A Commentary on the Book of the Prophet Jeremiah, Chapters 1–25* (Hermeneia; Philadelphia: Fortress Press, 1986), pp. 26-31; *Jeremiah*, II,

Moses is directly referred to in 15.1, and in 11.1-13 Jeremiah is represented as being a preacher of the covenant. They are intertextual reflections of each other and the wise reader of the text today will avoid the folly of trying to relate either to history or to each other, except in textualist terms.

There is a vast number of intertextual elements in the book of Jeremiah, many of which are beyond the length of this study to encompass. A few further points may be brought to the reader's attention before concluding remarks are in order. A close reading of the cycle in 3.1–4.4 will convince even a sceptic of the book's dependence on other texts (hence its intertextual nature). Jer. 3.1 begins the cycle with a somewhat tendentious citing of Deut. 24.1, 4, so immediately the interpretation becomes an intertextual reading of Jeremiah and Deuteronomy. The reference to King Josiah in 3.6-8 picks up a minor element in the book of Jeremiah (cf. 1.3; 22.11; 25.1, 3; 36.2) and necessarily makes the reader refer to the Deuteronomistic History, where the story of Josiah has its proper place. All that the book of Jeremiah does with Josiah is midrashic in nature, and biblical midrash is inevitably intertextual. In 4.3 there is an echo of Hos. 10.12c (cf. the echo of Hos. 7.4 in Jer. 23.10 and the citation of Hos. 3.5a in Jer. 30.9). Intertextual relations between Hosea and Jeremiah are well known in the standard works on Jeremiah, though not everybody would want to offer an intertextual account of the matter. There are also strong intertextual relations between Jeremiah and Ezekiel which certainly warrant us taking a very serious intertextual approach to understanding how these texts (Hosea, Jeremiah, Ezekiel) came to be written.[25]

pp. 38-39. See also L. Alonso Schökel, 'Jeremías como anti-Moisés', in *De la torah au messie: Etudes d'exégèse et d'herméneutique bibliques offertes à Henri Cazelles pour ses 25 années d'enseignement à l'Institut Catholique de Paris (Octobre 1979)* (ed. M. Carrez, J. Doré and P. Grelot; Paris: Desclée, 1981), pp. 245-54.

25. Some of the materials for this approach can be found in C. Hardmeier, *Prophetie im Streit vor dem Untergang Judas: Erzählkommunikative Studien zur Entstehungssituation der Jesaja- und Jeremiaerzählungen in II Reg 18–20 und Jer 37–40* (BZAW, 187; Berlin: de Gruyter, 1989); T.M. Raitt, *A Theology of Exile: Judgment/ Deliverance in Jeremiah and Ezekiel* (Philadelphia: Fortress Press, 1977); C.R. Seitz, *Theology in Conflict: Reactions to the Exile in the Book of Jeremiah* (BZAW, 176; Berlin: de Gruyter, 1989). On

It is of the essence of intertextuality that it consists of 'an intersection of textual surfaces' and 'as a dialogue among several writings'—to echo Kristeva again—and therefore an intertextual approach to the understanding of the prophetic books is necessary in order to account for the ways in which the different books seem to echo each other all the time.

Jer. 25.30 reflects Amos 1.2 and Joel 3.16. That is, all three references are essentially the same text, with local variations in each book. That example can be multiplied throughout the book of Jeremiah, especially in the material generally known as 'the oracles against the nations' (chs. 46–51). Jer. 51.58b is identical to Hab. 2.13b. The material on Moab in ch. 48 has many similarities to the material on Moab in Isaiah 15–16 (e.g. 48.34-36 as a conglomerate formed from Isa. 15.2-6; 16.11-12). Jer. 49.12-22 is expanded by material from Obad. 1-4, among other places (see the standard commentaries). In the second volume of his major commentary on Jeremiah William Holladay offers a great deal of data on this common material in Jeremiah and other biblical books.[26] He uses an extremely old-fashioned approach which is rather theoretically uninformed and writes about 'Jeremiah's dependence on...' without drawing any conclusions about the intertextuality of all this 'dependence'. He does, however, provide a great amount of data from which any careful reader may be able to deduce a theory of intertextuality for the composition of the book of Jeremiah. I, for my part, cannot see how the inspired prophet of Holladay's account is going to go around quoting everybody else's words in order to make up his own words. If it is Baruch who is translating Jeremiah's words into the words of all the other prophets, then a different theory of prophecy is required from Holladay. I would prefer to see developed an intertextual account of the matter that would take full cognizance of the fact that so much of the book of Jeremiah is made up of intertextual elements—and that therefore any theory of the book's composition would

the Hosea–Jeremiah connection see K. Gross, 'Hoseas Einfluss auf Jeremias Anschauung', *NKZ* 42 (1931), pp. 241-56, 327-43. I would prefer an intertextual account of the matter rather than an 'influence' (shades of Harold Bloom!) approach to these biblical books.

26. Holladay, *Jeremiah*, II, pp. 44-53.

have to allow for some distance between the 'historical Jeremiah' and the written words we now read. That distance would have to be considerably further than appears to be allowed for in Holladay's voluminous writings on Jeremiah.

One final set of examples may conclude the textualist part of my essay: 6.22-24 = 50.41-43 (with minor variations). All these double texts in Jeremiah force the modern reader to think about redactional compositions of the work and make us abandon particularity of reference in understanding what a prophet may have said. I have singled out this example of a double text because it demonstrates the strong intertextual relationship between the cycle of oracles in 4.5–6.26 and the cycle of poems in chs. 50–51 in the cycle of 'the oracles against the nations' (chs. 46–51). Many of the same terms, figures, images, metaphors and metonyms occur in both sets of texts. Ostensibly chs. 4–6 represent Jeremiah's preaching against Jerusalem-Judah and chs. 50–51 represent his preaching against Babylon. The same dominant motif of 'the foe from the north' figures in both cycles. But if in chs. 4–6 the foe from the north is Babylon, who then is that foe from the north in chs. 50–51 when Babylon itself is the target? How can the same material serve such different purposes? The question may be more acutely addressed to the Greek edition, which has the material against the nations placed in the middle of the book, whereas in the Hebrew edition the oracles of chs. 46–51 are at the end of the book, thus creating a fine symmetry between chs. 4–6 and 50–51. The infinite adaptability of the text points in an interesting direction: any statement in the prophetic texts may be made up of fairly conventional, clichéd material and therefore can have as its referent (if referent there be!) a wide range of possible meanings. Stock phrases can be loaded with precise reference by variation: e.g. 'against you, O daughter Zion' in 6.23 easily becomes 'against you, O daughter Babylon' in 50.42. An examination of the book of Jeremiah will reveal it to be made up of multitudinous clichés, conventionalized speech, commonplaces, quotations, proverbs, and all the other forms of expression so characteristic of intertextual productions. Whether we need to revise our image of the prophet to one who went around adapting old sayings and updating well-tried routines of oracular expression or should pursue our intertextual readings in a rather different direction in their bearing on the theory of the text's composition is a matter for much debate.

III

The basic data that I have chosen for consideration here are only part of the book of Jeremiah. More and different examples could have been given, but I do not think that they would have made much difference to my argument about the intertextuality of Jeremiah. The production of the prophetic books in the Hebrew Bible is very much an intertextual matter. There may have been original speakers behind the traditions represented by the books, but I do not think that in the books we are dealing directly with them. Their original work—this assumption may be challenged in many ways—has been troped by writers with other agendas and turned into the texts we know. What a theory of intertextuality does for the prophets is quite a complicated matter. It does not necessarily rule out authorship, for, as Barthes notes about the notion of the intertextual, it 'must not be confused with a text's origins' (see epigraphic introduction), but it probably means that we cannot work with old-fashioned notions of speaker–writer–text simplicities. The force of the theory is that it makes us recognize the intertextuality of all language. We already knew that! But what modern theory does is both to remind us of what we thought we knew (but have often forgotten) and to force us to acknowledge it in praxis. The intertextual points to the pretextual textuality that governs texts and that sets up webs or networks in which texts are coded and by which the author of any text is turned into a 'guest' in that text.[27] The long occupation with the author of texts in biblical studies, the almost obsessional concern to identify who wrote what and to attribute every fragment of a text to a specific author or to assign each layer of a text to the genuine, the secondary or the gloss—these are concerns that have been in the process of being abandoned in recent decades. An intertextual approach to biblical texts should assist that

27. Barthes, 'From Work to Text', p. 78. The word Barthes uses is *réseau*, which is better translated as 'network' than 'web' (the translation preferred by the translator of his article).

abandonment further. The author must be abandoned simply because the biblical texts are hardly 'authored' in the modern sense of an author as the actual writer of a text. Intertextuality goes much further than that, but even that would be a start in some circles. Jeremiah studies would certainly benefit greatly from the abandonment of the search for either 'the historical Jeremiah' or 'the author of the book of Jeremiah'. I believe both quests to be doomed to utter failure and also to be a waste of time and energy. An intertextual approach to the book of Jeremiah which sees it as a network of pretextual, contextual textualities would focus our attention on the text rather than on data to which we have no access.

Intertextuality, among so many other things, means that no text can ever be seen as existing as a closed system or as a hermetic or self-sufficient text. It always exists in terms of and over against other texts. Other texts helped to create it. Its writers are always readers of other texts. So it always exists in reference to other texts. A different intertextual approach to Jeremiah could have been taken by looking at Jeremiah in relation to the book of Psalms (e.g. 17.5-8 and Ps. 1), but that would not have established anything other than what I have written.[28] What I think is now required in Jeremiah studies is an intertextual approach which will not only investigate the whole range of intertextual elements in the construction of the book of Jeremiah, but will also turn its attentions to imagining the conditions of the production of the book. If this is how the book looks, how did it come to be this way? Its intertextuality raises many questions which may be dealt with in different ways. The historical–critical approach to the Bible always likes to work with the 'historical' because it believes specific historical occasions generate texts. That may well be the case, though I do not envy anybody who imagines that they have access—or can gain access—to the occasions that produced the biblical texts. If, in

28. Again the question of which way the influence should be understood vis-à-vis the Psalms and Jeremiah is a matter of debate; see P.E. Bonnard, *Le psautier selon Jérémie: Influence littéraire et spirituelle de Jérémie sur trente-trois Psaumes* (LD, 26; Paris: Cerf, 1960); cf. Holladay, *Jeremiah*, II, pp. 64-70. The matter of the lament poems in Jer. 11–20, commonly called 'the confessions of Jeremiah', is far too complex to discuss here.

Holladay's approach to Jeremiah, the prophet is shown to be dependent on all those other books, then we must ask about the nature of such a prophet and also ask whether all those writings were in existence *and* authoritative to the point that Jeremiah would use them to construct his own work. In a book that contains such a scathing dismissal of the written (Jer. 8.8), it may not be so obvious that the speaker would depend so much on other scribal contributions.[29] On the other hand, perhaps he did and therefore we, the readers, need to distance ourselves from this Jeremiah character because he behaves like somebody who lacks self-awareness.

There are many aspects of the theory of intertextuality that I have not taken into account in this brief survey of its application to the book of Jeremiah. That is always one of the dominant problems of using modern theory to explicate ancient texts. The metaphysics that shores up modern theories is not always appropriate for ancient writings or, for that matter, itself secure from serious criticism (this seems to me to be very much the case with any Marxist or marxisant theory used in reading the Bible). In offering here an account of the intertextuality of Jeremiah I recognize that I am using a fairly simple model of the theory in order to offer some illustrations of how such a theory can assist in reading the Bible. I am no stranger myself to theory or its use, so I do know what its shortcomings can be.[30] Whatever lack of theoretical sophistication may be charged against the user of theory, there is also always the charge that the theory involves a long, complicated way of getting to where a simpler, more traditional approach has already taken its anti-theory devotees. Such animadversions against theory are inevitable. Yet they should not hide from us the fact that most of us tend

29. My articles referred to in n. 15 above offer some observations on Jer. 8.8.

30. In my book *When Prophecy Failed: Reactions and Responses to Failure in the Old Testament Prophetic Traditions* (London: SCM Press, 1979), I used the theory of cognitive dissonance, developed from social psychology, as a reading strategy for analysing the prophets. Critical reaction to that book was generally hostile to the theoretical sections and more approving of the less theoretical material. Getting the blend of text and theory right in current biblical studies is a very difficult task. This essay illustrates some of the problems.

to use some theory of composition or writing, whether or not we are as conscious of it as we might be. New theories come along in order to displace older theories because the yield from the older approaches has become steadily less and less. It is the failure of old theories *and* the death of old theoreticians that give new theories their opportunities to perform.

Intertextuality describes a number of phenomena that are very old, very common and remarkably well known. More traditionalist approaches might talk about 'echo', 'influence', 'borrowing', 'quotation', etc., though, to be fair to the concept of intertextuality, it goes much further than these terms do and covers a much wider metaphysical range too. In biblical studies much work has already been done on intertextual matters, though without calling it by such a name.[31] So any intertextual approach to the Bible can forge links with work in progress and with so much solid research already completed. McKane's 'rolling corpus' approach to Jeremiah still seems to me to be a very promising excavation of the compositional modes of Jeremiah 1–25. It is intertextuality *within* the book of Jeremiah. Other approaches will take up the intertextual as the relations between texts from different books.[32] That approach will generate a very large body of work and reflects what is beginning to appear in biblical scholarship. The more old-fashioned terminology of 'borrowing' or whatever should not conceal from us the similarity of the enterprise. We all know how musicians 'borrow' or 'steal' from each other—can Johann Sebastian Bach or Mozart be understood except in relation to the works from which they have taken so much and transformed it? In twentieth-century music the practice of deliberately taking from others or using their styles of music to construct new music is too well known to require much comment. The music of Gustav Mahler is perhaps most noted for this blending of styles taken

31. I have in mind here the very solid work of M. Fishbane, *Biblical Interpretation in Ancient Israel* (Oxford: Clarendon Press, 1985). Others, too many to name, also belong here.

32. E.g. D. Boyarin, *Intertextuality and the Reading of Midrash* (Indiana Studies in Biblical Literature; Bloomington: Indiana University Press, 1990); R.B. Hays, *Echoes of Scripture in the Letters of Paul* (New Haven: Yale University Press, 1989).

from other composers, but from Monteverdi to Mahler the intertextuality of music has been part of the given in European music. The modern composer Alfred Schnittke (of German-Russian Jewish extraction) describes his own music in terms of its 'polystylism' (an aspect of stylistic pluralism). Such 'polystylism' is simply a form of stylistic manipulation whereby a composer takes a style from another composer (in Schnittke's case the main influences include Mahler, Schönberg and Berg) and reworks it for purposes other than those used by the original musician.[33] What polystylism describes in music, intertextuality describes in writing.

An intertextual approach to Jeremiah brings to the fore the vexed old question of the original author, which seems to plague biblical studies. While there is nothing in the theory that would rule out some carefully specified notion of 'author' (original writer, editor, redactor or final writer, editor, redactor or what?), it does in general diminish the role of the author as unique intersubjective personality. Since so much of the theological reading of the Bible wants to remain within striking distance of 'scripture as sacred writing with authoritative implicatures', an intertextual account of Jeremiah can raise serious problems of authenticity, legitimacy and authority. In these senses, much of current debate about 'canonical criticism' is a contested site of struggle about the authority of the Bible, and 'intertextuality' can be seen as an enemy of the canonic. There are certainly family resemblances between an intertextual reading of the Bible and the approach of 'canonical criticism'. Both approaches wish to read the text in its totality of textual and intertextual networks, codes and systems. But the canonic approach to the Bible wishes to rush to closure and to control the ownership of the Bible in ways unimaginable to the intertextualists.[34] Old and new theoretical positions do battle here. An intertextual account

33. For those who do not know the music of Alfred Schnittke, see the work of John Webb on Schnittke, esp. his 'Schnittke in Context', *Tempo* 182 (September 1992), pp. 19-22; and the work of the music critic Ronald Weitzman, who is an ardent follower of Schnittke's music.

34. This is my perception of the function of the work of the Yale theologians, especially that of Brevard Childs and also the Lindbeck–Thiemann school (including the work of Hays referred to above in n. 32).

of any biblical text (or of the Bible itself) will not necessarily serve any particular ideological position, whereas the approach of 'canonical criticism' inevitably tropes the discussion in ecclesiastical directions. I would have to say then that intertextuality holds better promise for biblical scholarship as a whole, as a discipline involving the 'community of scholars'. The canonic approaches will then favour the 'community of believers' better than they will the larger world of scholarship. That need not mean a rift between intertextual approaches and canonic approaches. An intertextual study of the Bible (or parts of it) need not confine itself to questions about canonic intentions or may not even address them at all. Different interests determine these matters. Yet both approaches recognize certain fundamental points: the texts are intertextual because texts are always in dialogue with other texts and also in dialogue with other readers, which is another way of saying the same thing.[35]

35. A shorter and earlier version of this paper was read to the Postgraduate Seminar of the Glasgow University Faculty of Divinity's Department of Biblical Studies. Something of the long, engaged discussion that followed the paper has been taken into account in the rewriting of this chapter. I am grateful for all the responses, though I will not swear to having amended my life or views adequately yet.

A World Established on Water (Psalm 24): Reader-Response, Deconstruction and Bespoke Interpretation

David J.A. Clines

Let's talk of readers' response. Or, since I am doing the talking, let me talk of *this* reader's response.

There are things about this fine and famous psalm a reader like me cannot swallow. There is, for instance, the idea of the world being founded upon seas and rivers. The poet, for his part, actually believes (does he not?) that underneath the rocks and dirt of the earth's surface there is an underworld sea, fed by rivers, upon which the world floats. And I do not believe that. Or rather, to put it more strongly but more exactly, I know that that view is wrong.

But this is not the only point on which I cannot buy the ideology of the psalm. For me, this cosmological misapprehension is only the outcropping of a larger seismic fault that runs hidden beneath the whole surface of the psalm.

I will be arguing that the psalm is riddled with religious ideas as unacceptable as its cosmology, and further, that it is not even internally coherent. At the end I will suggest an answer to the question of what is to be done with a piece of sacred literature that is so ideologically and religiously alien today, even to a person of goodwill toward it (like myself), and that speaks with so uncertain a voice. I will, in other words, deploy three strategies: an ideologically slanted reader-response criticism, a deconstructionist critique, and a new proposal for a goal-oriented hermeneutic, which I call 'bespoke' or 'customized' interpretation.

1. *A Reader-Response Criticism*

Let me first speak of the reader that I am. Toward the poem as a whole I find myself ambivalent. All my life I have found the poem powerful and uplifting. This is partly due to the background music I inevitably hear when I read the poem, the singing of it by the Scottish Male Voice Choir, all the vogue in my religious neck of the woods in the fifties. But there is also something grand and elevated about its tone that attracts me—at least, that attracts a romantic and soulful part of me.

I also recognize and accept that the poem has been, for two and a half millennia or more, a vehicle for worship in Jewish and Christian communities; and however unlovely those communities may have been, I have no urge to sniff at their religious experience. In short, I want to be able to say something positive about this poem.

The other side of it is that the poem is built upon two ideologies that I deplore: the first a notion that 'holiness' attaches to places, the second an idea of victory in war as glorious.

a. *Holiness*

According to the poem, only those who live blameless lives are entitled to enter the temple of the Lord—it is those who have clean hands and a pure heart who 'shall', or 'should', ascend the hill of the Lord and stand in his holy place (vv. 3-4).[1] No doubt there is a sense of fit here, an idea that pure people and things belong in holy places, and that outside the temple, *pro fano*, is the place for the profane. But there is equally plainly a sense that the holiness that exists in the 'holy place' is in need of protection from the impure, that it is open to contamination by unholiness.[2]

In such an account, holiness is being understood both in a religious-cultic and in an ethical sense: holy places clearly cannot

1. Is this a prediction of who in fact shall enter the holy place, or who it is who is entitled to enter it?

2. Holiness is 'defined on the one hand as that which is consistent with God and his character, and on the other as that which is threatened with impurity' (D.P. Wright, 'Holiness (OT)', in *The Anchor Bible Dictionary* [ed. D.N. Freedman; New York: Doubleday, 1992], III, pp. 237-49 [237]).

be holy in an ethical sense, but are holy only because they have been marked out as such by a divine signal.[3] Humans, on the other hand, need to match the holiness of the holy place by the kind of holiness that they can acquire, which is ethical purity (and not, of course, religious-cultic designation, unless they happen to be priests). In the language of the poem, the place is 'holy' and the entrants to it are 'clean'. Ethical 'uncleanness' is unsuitable for a 'holy' place.

My question to myself, as a reader checking all the time on my responses to texts, is: Can I tolerate a notion of holiness that sees it as contaminatable? If the world contains relatively small pockets of holiness, like a hill of the Lord or a temple, surrounded by vast areas of unholiness, like (presumably) everywhere else, and if the unholy has the power to contaminate the holy but the holy does not have the power to infect the unholy, what future, I ask myself, is there for the holy? The holy is rather under threat, is it not, if it has to be protected from the unholy by the exclusion of unrighteous people from visiting the sanctuary. For if impure people are supposed to be kept out of the shrine, or keep themselves out, in order to protect its holiness, what happens if impure people are inadvertently allowed in? Does the holy thereby become unholy?

In a word, Is the holy to be at the mercy of doorkeepers? Would it not be better, I say to myself, to think of holiness, as a symbol of the divine, as incapable of being damaged by humans? If it is worth the name of holy, must it not in any case be more powerful than its opposite, whatever *its* name? Why not think of the divine presence as a powerful purifying influence that can quite easily cope with sinners and can in some way annihilate their impurity? A temple, then, if it is to be conceived

3. 'We cannot make shrines and cannot select their "positions", but can never do more than merely find them' (G. van der Leeuw, *Religion in Essence and Manifestation* [trans. J.E. Turner; New York: Harper & Row, 1963], p. 398). Typically the holy place in Israelite religion is 'the place which Yahweh your God shall choose to put his name there' (Deut. 12.5), that is, the place of theophany. And Israel is to 'take care' that it does not choose its own holy places (Deut. 12.13). See also my paper, 'Sacred Space, Holy Places and Suchlike', *Trinity Occasional Papers* (Festschrift for Han Spykeboer) 12/2 (1993) (forthcoming).

of as a dwelling of the divine presence, would be a place where the unrighteous were confronted by the contrast between their badness and divine goodness, and thus it would function as a locus of ethical transformation. Holiness would be viewed, not defensively as it is here, as a substance in need of protection, but as a force for positive change in the community.

But if I 'buy' the psalm, I 'buy' its ideology of holiness, and I had better be aware of what I am doing.

b. *War*

The second ideology sustaining this poem that I find myself unable to accept is of the glory of war, or rather, of victory in war. It is not that the humans are warlike, but that the deity himself is. This only makes it worse, from my ethical perspective at least.

It comes, in fact, as something of a shock to the first-time reader of the poem (or, shall we say, to the curious and close reader) that it moves in that direction. For in its first strophe the poem has breathed a pacific air of stability and constructiveness. At its beginning, there is a creative act of 'founding' and 'establishing' that has overridden any cosmic tendency to instability, and there is not a hint of conflict in the world order that results. And in the second strophe, there are no real villains or any sign of organized opposition to the forces of good that needs to be put down by force. It is in this context of world stability and personal goodness that we encounter what is the principal truth, for this poem, about the God who dwells on the holy hill and whose face the generation of the righteous is seeking. This God is celebrated, not for his creative powers (strophe 1) nor as the fount of human goodness (strophe 2), but because he is 'mighty in battle' (v. 8) and 'Yahweh of armies' (v. 10). What makes him 'glorious' is that he is 'strong and mighty' enough to achieve military victories. There is no glory, in this poem, in creating the world, there is no glory in being the object of worship by clean-living toilers up the steep ascent of Zion. The glory that gains him the right of access through the ancient gates is his glory gained on the field of battle.

Now, as we all know, glory and honour in war is nothing other than victory. The victors always retire in honour, the defeated in disgrace. But what makes victory, and what makes defeat? Not the rightness of the cause, not the gallantry of the combatants, not the prayers of the faithful. Victories are won by superior numbers, by alliances, by tactics, and by chance. And a victor deserves praise for nothing other than winning. This is not my idea of glory, and the fact that someone says military prowess is what makes God glorious does not impress me.

We had better know what we are doing. In subscribing to Psalm 24, we are writing a blank cheque for war, for the validity of war imagery to describe the deity's activities, and for the unexamined assumption that war solves problems. If I 'buy' the psalm, I 'buy' its ideology of war.

A reader-response approach to this psalm, then, highlights elements in it, quite fundamental elements, that raise uncertainties, if not hostilities, in the mind of the modern reader, this one at least. These have proved to be uncertainties about whether *we* can affirm what it is the psalm seems to be affirming.

2. *A Deconstructive Critique*

The problems with this psalm are greater than those, however. We next must consider, not whether we can affirm the psalm, but, whether the *psalm itself* affirms what it affirms. Are there aspects in which it is at odds with itself, perhaps even to the extent of undermining what it is professing? Does it deconstruct itself at all?

Yes, in these four respects.

1. *Although the whole world belongs to the Lord (v. 1), it is not all 'holy'.*

Now according to the cultural conventions in which our text participates, the 'holy' is defined as what belongs to the deity. A temple, heaven, priests are 'holy' because of their attachment to the deity. It follows that if the whole earth is 'the Lord's', the whole earth is 'holy'.

This view affirmed by the poem in its opening lines is subsequently undermined by the reference to the 'holy place' belonging to the Lord, presumably upon the 'hill of the Lord' (v. 3). If all the world belongs to the Lord, in what sense can *one hill* 'belong' to the Lord? And if all the world is holy by virtue of his possession of it, in what sense can *one* place be 'holy'?

I conclude that while the poem wants to maintain that the world as a whole is undifferentiatedly the Lord's possession, it cannot sustain this view, but allows v. 3 to deconstruct v. 1.

2. *Although all those who live on the earth 'belong' to the Lord (v. 1), some of them must be his enemies.*
Again, the two affirmations undermine one another. For in what sense could it be said that the deity 'owns' his enemies? If he finds it necessary to engage them in battle, and if battle against them is so difficult that any victory over them is 'glorious', how could they already be said to be 'his'?

So the reference to warfare deconstructs the assertion of the Lord's ownership of and lordship over all the earth's inhabitants—and vice versa.

3. *Although ascending the hill of the Lord proves one's innocence, those who who ascend are in need of 'vindication' from God.*
Those who ascend the hill of the Lord are promised 'vindication' from God. The implication is that at present they lack such vindication and stand in need of it.

In the eyes of whom do they stand in need of vindication? Presumably both God and themselves are well aware of their moral virtue, so it must be in the eyes of others that they need to be vindicated. But where are the people who are refusing them recognition, and before whom their virtue must be demonstrated? There is nothing in this poem about any assaults on the integrity of the righteous by the wicked, nor any complaint that these people of clean hands and pure hearts are being persecuted or otherwise maltreated by those less upright than themselves.

So the poem craves vindication for the innocent worshippers, but, deconstructively, cannot find any respect in which they might need it.

Furthermore, since it is only those of clean hands that are permitted to ascend the hill of the Lord, the very act of participation in worship is sufficient testimony of their uprightness.

They already have their vindication, and so the promise of a future vindication becomes nugatory.

4. *Those who worship on the hill are expected to have clean hands and not to have lifted up their soul to vanity. But the deity is not.*
A double standard in ethics is in operation here.

The worshippers must have clean hands or they will contaminate the holiness of the hill. But the deity ascends it straight from the battlefield, his hands dripping with blood. Does 'lift up your heads, O gates' then mean 'Look the other way'?

The worshippers must not have lifted up their souls to vanity, but the deity has been soldiering away, seeking the bubble reputation even in the cannon's mouth. 'Reputation' is nothing but Shakespearian for 'glory', and the quest for glory in war is surely a quintessential lifting up of the soul to vanity.

In short, the qualities demanded of the worshippers are deconstructed by the qualities praised in the deity they worship. And vice versa.

These are not the only places in which this poem deconstructs itself, but they are pretty central. The question arises: What is to be done with such a text?

3. *Bespoke Interpretation*

In the rest of this essay I want to offer a framework for dealing with such a question. I call it a goal-oriented hermeneutic, an end-user theory of interpretation, a market philosophy of interpretation, a discipline of 'comparative interpretation'. This framework has two axes.

First, there is the indeterminacy of meaning. Second, there is is the authority of the interpretative community.

First, then, comes the recognition that texts do not have determinate meanings. Whatever a text may mean in one context, it is almost bound to mean something different in a different context. 'Bus stop' will mean one thing when attached to a pole at the side of the road, another thing when shouted by an anxious parent to a child about to dash into that road. 'Jesus saves' will have one meaning when it stands by itself, but another meaning when it is followed by 'Moses invests'.

We may go further. Nowadays we are recognizing that texts not only do not have determinate meanings, they do not 'have' meanings at all. More and more, we are coming to appreciate the role of the reader, or the hearer, in the making of meaning, and recognizing that, without a reader or a hearer, there is not a lot of 'meaning' to any text. Psalm 24 means whatever it means to its various readers, and if their contexts are different, it is likely that it will mean different things to different readers. There is no one authentic meaning that we must all try to discover, no matter who we are or where we happen to be standing.

The second axis for my framework is provided by the idea of interpretative communities. If we ask who it is that authorizes or legitimates an interpretation, who it is that says something may count as an interpretation and not be ruled out of court, the answer can only be: some group, some community. Solipsistic interpretations may be fun for their inventors—you meet a better class of reader that way—but if there is no group who will accept them, they don't survive. Some interpretations are authorized by the SBL, some by the ecclesiastical community, but most by little sub-groups within these communities—the Intertextuality in Christian Apocrypha Seminar and the like. The market for interpretations is getting to be very fragmented these days, and I sometimes count myself lucky if I can sell an interpretation to six people.

What we call legitimacy in interpretation is really a matter of whether an interpretation can win approval by some community or other. There is no objective standard by which we can know whether one view or other is right; we can only tell whether it has been accepted. What the academic community today decides counts as a reasonable interpretation of Psalm 24 *is* a reasonable interpretation, and until my community decides that my interpretation is acceptable, it *isn't* acceptable.

Of course, what one community finds acceptable, another will find fanciful or impossible. The local Faculty of Divinity will not approve of the interpretations of our psalm made by St Augustine and his community, neither would St Augustine think much of the interpretations of the Faculty of Divinity. There are no determinate meanings and there are no universally agreed upon legitimate interpretations.

What are we exegetes then to be doing with ourselves? To whom shall we appeal for our authorization, from where shall we gain approval for our activities, and above all, who will pay us?

The simplest answer for academics has long been that we will seek the approval of no one other than our fellow academics. If our papers get accepted by *Vetus Testamentum* and *New Testament Studies* they are valid, and if they don't they're not.

This safe answer has started to fall apart, though. We are beginning to realize that what counts as a valid interpretation in Cambridge does not necessarily do so in Guatemala City or Jakarta or Seoul—and certainly not vice versa. The homogeneity of the 'scholarly world' is proving fissiparous, and many smaller interest groups are taking the place of a totalitarian *Bibelwissenschaft*. More and more scholars are seeking their legitimation from communities that are not purely academic.

Where does that leave us?

If there are no 'right' interpretations, and no validity in interpretation beyond the assent of various interest groups, biblical interpreters have to give up the goal of determinate and universally acceptable interpretations, and devote themselves to producing interpretations they can sell—in whatever mode is called for by the communities they choose to serve.

This is what I call 'customized' interpretation. Like the bespoke tailor, who fashions from the roll of cloth a suit to the measurements and the pocket of the customer, a suit individually ordered or bespoken, the bespoke interpreter has a professional skill in tailoring interpretations to the needs of the various communities who are in the market for interpretations. There are some views of Psalm 24 that the church will 'buy' and 'wear', and others that only paid up deconstructionists, footloose academics and other deviants will even try on for size.

There is nothing unethical in cutting your garment not only according to your cloth but also according to your customer's shape. Even in a market economy, no one will compel you to violate your conscience, though it may cost you to stick to your principles. As a bespoke interpreter responding to the needs of the market, I will be interested, not in 'the truth', not in universally acceptable meanings, but in eradicating shoddy interpretations that are badly stitched together and have no durability, and I will be giving my energies to producing attractive interpretations that represent good value for money.

In such a task interpreters of today do not have to start from scratch. For this programme has a green angle too. It is ecologically sound, because it envisages the recycling of old waste interpretations that have been discarded because they have been thought to have been superseded. In this task of tailoring to the needs of the various interpretative communities, interpreters can be aided by the array of interpretations that have already been offered in the course of the history of the interpretation of the Bible. In fact, what has usually been called the 'history of interpretation' is ripe for being reconceived as a discipline of 'comparative interpretation', providing raw materials, methods, critiques and samples for the work of designing intelligible and creative interpretations for end-users. For too long the interpretations of the past have been lumped together under the heading of the 'history' of interpretation, with the unspoken assumption that what is old in interpretation is out of date and probably rotten and the hidden implication that what is new is best.

Recycling Christian interpretations is a good way to start the programme of comparative interpretation. For the first thing we notice is that among the readings of the patristic period no one is striving for a *correct* interpretation. Here the only fixed point is that the king of glory is Christ, and the exegesis is driven by the question, When then did Jesus Christ enter these gates? Any moment in the history of Jesus Christ to which these words can attach themselves will yield an acceptable interpretation.

For example, in the fourth-century *Gospel of Nicodemus* the gates are the gates of hell, which Christ breaks, freeing its inhabitants—the harrowing of hell. In Augustine, the king of glory is ascending after the resurrection, and the scene is one of welcome into the heavenly courts. For Gregory of Nyssa, on the other hand, the scene is the descent of Christ to earth in the incarnation.

The poem has, in Christian interpretation, then, transcended its original significances in the history of ancient Israel, whatever they were, and has become multivalent.

And every new interpretation creates an access of meaning for the poem. Here is a brand-new interpretation, fresh from your friendly corner bespoke interpreter. Come and buy. It is a non-religious interpretation that attends to the connotations rather than the denotations of the language, and it doesn't require you to give up any other favourite interpretation you may already have.

Let's say Psalm 24 is about world-building and world-orienting, about locating oneself at the centre (the Lord's hill), up it (ascending) and in it (entering the gates). And let's say the world that is being built is the world of meaning, and the poem concerns making a world of meanings, meanings secure enough to be going on with.

In Psalm 24, then, we are celebrating a world that is founded, established—a world where we can find the direction to the Lord's hill, for example, a world where Wittgenstein could say, Now I can go on. It has orientation and it has elevation: it is three-dimensional space—which is to say, a world for living in.

Now in the world of meaning there is undifferentiated space—the earth at large—and there is a particularity of space—a specific hill, the hill we seekers for meaning are interested in ascending. And in order to ascend the mountain of particular meaning—that is, to establish *the* meaning of the text—we need a pure heart, of course, because purity of heart is to will one thing—and no swearing deceitfully by the false gods of theory. Now each of us sets out on the quest for meaning alone: 'Who (singular) shall ascend the hill of the Lord?...The one that has clean hands...' We ascend the mountain in our singularity; but when we attain the blessing, which is the vindication of our quest, we find ourselves in the company of a whole generation of seekers for meaning, a veritable Fishian interpretative community: 'This is the generation of those who seek your face...'

The one who ascends the hill is, himself or herself, personally a king of glory. There is nothing glorious in itself; glory signifies the esteem of others. Glory is the recognition by a public who acclaim success in the quest for meaning. Yes, it *is* a struggle, though a demilitarized one, against the intractability of experience and the bewildering array of interpretations already in the field.

Centring ourselves, knowing which way to turn, is a construction of a reality, a world-ordering enterprise. But if we even ask for a moment how firm a foundation we saints of the Lord have laid for ourselves in this world-ordering enterprise, we recognize that the world we have established is founded not upon pillars but upon seas and rivers. We float on a raft of signifiers under which signifieds slide playfully like porpoises; but we have to live *as if* the foundations were solid all the way down to bedrock. We cannot peer too long into the deconstructive underworld waters.

I have often wondered what one should do after deconstructing a text. A true deconstructionist would say, Start deconstructing the deconstruction. But there is another answer, which is truer, I think, to the experience of readers who have performed, or witnessed, a deconstruction. It is very difficult to forget a deconstruction; it is hard to get it out of your head. But the mind demands more order than deconstruction will leave us with, and will go on wilfully constructing, inventing new connotations, new contexts, new interrelationships that will shore up the text, even if only temporarily.

That is what I feel the course of this essay has done. I wanted to expose the fragility, the volatility of the text, its weakness and its incoherence. It was not in order to recommend its abandonment or replacement by some other stronger and less questionable text, but to point up the fragility of texts in general, the inconclusiveness of interpretations, and the impulse nevertheless to stitch them together again no matter how. Weaving and interweaving of interpretations that mean something to someone, that meet with a cry of recognition or at least a grudging assent from some interpretative community—that resolidifies texts. It is the best we can hope to do. It is something like building a universe, intelligently knit together but resting ultimately on unpredictable and ever shifting underground waters. Which was itself an interpretation of Psalm 24.

WHO'S AFRAID OF 'THE ENDANGERED ANCESTRESS'?

J. Cheryl Exum

> Who's afraid of the big bad wolf, the big bad wolf, the big bad wolf?
>
> The three little pigs
>
> Let's take a look: we shall find illumination in what at first seems to obscure matters...
>
> Jacques Lacan

A Thrice-Told Tale

Three times in Genesis the patriarch, the eponymous ancestor of Israel, travels to a foreign country, where he passes his beautiful wife off as his sister because he fears the locals will kill him on her account if they know he is her husband. Abraham and Sarah are the ancestral couple in the primal scene (Gen. 12, where their names are Abram and Sarai) and in the first repetition (Gen. 20, by which time their names have been changed to Abraham and Sarah). Sarah is taken to be the wife of the foreign ruler (the pharaoh of Egypt in Gen. 12, and Abimelech of Gerar in Gen. 20) and then returned to Abraham when the ruler learns of the ruse. The third version (Gen. 26) concerns Isaac and Rebekah; the foreign ruler is again Abimelech of Gerar; and the matriarch is *not* taken. In all three cases, the patriarch prospers, the foreign ruler is (understandably) upset, and the matriarch has no voice in the affair.

It is generally agreed that the tales are variants on the same theme. The characters change and details vary, but the fabula remains the same. Within biblical scholarship, this thrice-told tale is often referred to as 'the Endangered Ancestress' or 'the

Ancestress of Israel in Danger'.[1] The widespread use of this label raises the question, What kind of danger do scholars think the matriarch is in? If, as is generally accepted, these stories represent in some way a threat to the threefold promise to Abraham of land, descendants, and blessing, then the threat is to the promise, and it follows that the patriarch, not the matriarch, is in danger. The promise, after all, was made to him—not to her or to the two of them (see Gen. 12.1-3)—and without his wife how can he have descendants?

Or is the danger faced by the matriarch the loss of honor? This could be said to be an issue in Genesis 20, where the narrative is at pains to assure us that nothing of a sexual nature took place between Abimelech and Sarah. Here the omniscient narrator tells the audience:

> Now Abimelech had not approached her (Gen. 20.4).

He then gives the statement divine authority by placing it in the mouth of God, who speaks to Abimelech in a dream:

> Therefore I did not let you touch her (Gen. 20.6).

Finally, by having Abimelech publicly justify Sarah's reputation, he ensures that all the characters in the story share in this knowledge.

> To Sarah he said, 'Look, I have given a thousand pieces of silver to your brother; it is your vindication in the eyes of all who are with you; and before everyone you are righted'[2] (Gen. 20.16).

It is not so clear that nothing of a sexual nature happened in the primal scene, Genesis 12, where we hear that 'the woman was taken into the pharaoh's house' (v. 15) and the pharaoh says, 'I took her for my wife' (v. 19). Interestingly, what did or did not happen to Sarah in the royal harem receives more attention from scholars than it does from Abraham. Bernhard Anderson, in his annotations to the Revised Standard Version, would

1. E.g. Keller 1954; von Rad 1961: 162-65, 221-25, 266; Koch 1969: 111-32; Polzin 1975; Westermann 1985: 159; Coats 1983: 109, 149, 188; Biddle 1990.

2. Following the RSV. The translation of the obscure Hebrew is problematic, but this seems to be the sense; see Westermann 1985: 328; von Rad 1961: 224; Skinner 1910: 319.

apparently have us believe that the story is less explicit and shocking than it actually is, for he explains that Sarah 'was *almost* taken into Pharaoh's harem' (italics mine). (Does this mean she got only to the door?) Koch, Polzin, Miscall, and Coats, in contrast, assume that Sarah did have sexual relations with the pharaoh.[3] Koch's judgment, incidentally, is as ethnocentric as it is androcentric: 'There is one feature of the story missing which would be natural to us: there is no reluctance to surrender the woman's honour'. To support his conclusion that the earliest form of the story did have Sarah committing adultery, Koch appeals to what he believes other women would do: '[I]t seems obvious that the Bedouin women are so devoted to their menfolk that to protect a husband's life they would willingly lose their honour'.[4]

What is this honor anyway but a male construct based on the double standard, with its insistence on the exclusive sexual rights to the woman by one man? The scene in Genesis 16, where the situation is reversed, is comparable and illuminating. Genesis 12 and Genesis 16 raise the issue of the matriarch or the patriarch having sexual relations with someone else. In Genesis 12, Abraham tells Sarah to let herself be taken by another man 'in order that it will go well with me because of you and I may live on your account' (v. 13). In Gen. 16.2, Sarah tells Abraham to have sexual intercourse with Hagar ('go in to my maid') so that she may obtain a child through Hagar. Neither Abraham nor Sarah is concerned with what this intimate encounter might mean for the other parties involved, but only with what he or she stands to gain. In Genesis 16, we are told specifically that Abraham had sexual intercourse with Hagar ('he went in to Hagar and she conceived', v. 4), but such specific detail is omitted from Genesis 12 (we shall return to this point below). Significantly, no one speaks of Abraham's loss of honor in Genesis 16, nor is there much concern for Hagar's honor—a fact that indicates 'honor' is not only a male construct but also a class construct. Abraham, who as a man is not required to be

3. Koch 1969: 125; Polzin 1975: 83; Miscall 1983: 35; Coats 1983: 111.

4. Koch 1969: 127; cf. Abou-Zeid 1966: 253-54, 256-57. For discussion of honor and its relationship to the politics of sex, see Pitt-Rivers 1977: esp. 113-70.

monogamous, cannot be dishonored by having sex with Hagar at Sarah's urging. Neither can Hagar be dishonored, since a slave has no honor to lose.

It is not the woman's honor so much as the husband's property rights that are at stake. Still, we might expect the patriarch to show some concern for his wife's well-being. It is thus curious that in all three cases the patriarch does not consider that the matriarch might be in danger. On the contrary, he thinks *he* is in danger:[5]

> I know that you are a beautiful woman. When the Egyptians see you, they will say, 'This is his wife'; and they will kill me and let you live (Gen. 12.11-12).
>
> It was because I thought, There is surely no fear of God in this place, and they will kill me because of my wife (Gen. 20.11).
>
> When the men of the place asked about his wife, he said, 'She is my sister', for he feared to say 'my wife', thinking, 'lest the men of the place kill me because of Rebekah, for she is beautiful' (Gen. 26.7).

Whether or not the patriarch's fear is justified—whether or not he really is in danger or whether his fear is simply displaced—is a question we shall explore. If the patriarch does not suppose that the matriarch is in danger, neither is there any evidence that the *matriarch* thinks she is in danger. In fact, we do not know what she thinks about *anything*, which is a very good indication that the story is not really about the matriarch at all. She neither acts nor speaks in any of the versions, though in the second version speech is indirectly attributed to her: Abimelech tells God that Sarah told him that Abraham was her brother (Gen. 20.5). If her only speech is one reported by another character in the narrative, the matriarch can hardly be said to become a narrative presence in any real sense. She is merely the object in a story about male relations (and we shall inquire below how the two men respond in relation to the object). What, then, is the danger, and to whom? More important, why do we hear about it three times?

Most studies of Genesis 12, 20, and 26 are concerned with the relationship between the three stories: how are they alike and different, and how are the differences to be accounted for

5. Clines 1990: 67-68.

(which often means, how can the repetition be explained away)? Now what happens in Genesis 12, 20, and 26 is very disturbing. A man practically throws his wife into another man's harem in order to save his skin. Yet the questions one most often encounters about this text are generally along the lines of: What is the oldest form of this story?[6] Or, Are the three accounts oral or written variants?[7] Are Genesis 20 and 26 more ethical than Genesis 12?[8] The disturbing issues raised by the story are sometimes deplored[9] but then set aside in favor of disengaged discussion of the growth of the tradition, the relative dates of the versions, and such historical questions as whether or not the stories reflect customs of 2000 to 1500 BCE (the so-called patriarchal period), or whether a man could or should marry his half-sister (the controversial evidence of Nuzi).

A few scholars have inquired into the role of these stories in the context of the larger narrative.[10] A sustained contextual reading of the three stories is offered by David Clines, who concludes that the patriarch is more of a danger to foreigners than they are to him.[11] But reading the three tales in their context also exposes problems. For example, in Genesis 20 Sarah would be over ninety years old, and we might wonder why Abraham thinks other men would take such an interest in her. Moreover, Abraham has now been told by God that Sarah will be the mother of his heir, which makes it even harder to understand why he would let another man take her (it may even be

6. See Van Seters 1975: 167-91; Koch 1969: 111-32; Noth 1972: 102-109; Westermann 1985: 161-62.

7. On the issue of literary dependency, see Van Seters 1975: 167-91; Westermann 1985: 161-62; cf. Alexander 1992. For an argument that the pentateuchal sources use the same (wife-sister) motif to develop different themes, see Petersen 1973. For discussions of the stories as oral variants, see Culley 1976: 33-41; and the more recent folkloristic approach of Niditch 1987: 23-66.

8. Most commentators agree with Koch (1969: 126), who thinks that 'moral sensitivity becomes gradually stronger'; Polzin (1975: 84) argues that Gen. 12 is as sensitive to ethical issues as are chs. 20 and 26.

9. Von Rad (1961: 162) calls Gen. 12 'offensive', and speaks of the 'betrayed matriarch' (p. 164); see also Vawter 1977: 181.

10. Clines 1990: 67-84; Fox 1989; Rosenberg 1986: 70-98; Steinberg 1984; to a lesser degree, Polzin 1975; Miscall 1983: 11-46.

11. Clines 1990: 67-84.

the case that Sarah is already pregnant with Isaac).[12] In Genesis 25, Esau and Jacob are born to Isaac and Rebekah, and by the end of the chapter they are already hunting and stealing birthrights respectively. Thus in Genesis 26, when Isaac says of Rebekah, 'She is my sister', we might wonder, what has become of the twins? These are only some of the difficulties a contextual reading must engage. I mention them not because I intend to offer a contextual reading here, but rather to underscore how puzzling and uncanny the tale is both in context and in isolation. We encounter one set of problems when the three versions are read in their larger context and other problems when they are considered in their own right. In fact, one might say that this tale in its three forms calls attention to itself by virtue of the surplus of problems it poses to interpretation. I propose that a different kind of approach to the repeated tale in Genesis 12, 20, and 26 could provide new insights into some recurrent difficulties. Specifically, I want to offer a psychoanalytical alternative to previous, largely form and tradition-historical, approaches.

By proposing a psychoanalytic-literary reading as an alternative, I am not claiming that this approach will 'solve' the problems posed by these chapters whereas other approaches do not. On the contrary, I maintain that posing questions and opening up new dimensions of a text are as fruitful an enterprise as the traditional critical approach of seeking answers as if answers were objectively verifiable. Like psychoanalysis, psychoanalytical criticism is neither externally verifiable nor falsifiable. We can only follow it, as Freud says about analysis, to see where it will lead,[13] and, in the process, hope to illuminate a hitherto uncharted textual level, the narrative unconscious. My approach appeals to the multiple levels on which stories function; like dreams, they are overdetermined. As Freud points out in comparing texts to dreams, which, he argues, require over-interpretation in order to be fully understood, 'All genuinely creative writings are the product of more than a single motive and more than a single impulse in the poet's mind, and are open to more than a single interpretation'.[14]

12. So Vawter 1977: 245; Miscall 1983: 32; Clines 1990: 75-76.
13. Freud 1961: 4.
14. Freud 1965: 299. I see little difference in my suggesting below that

To anticipate my argument: a psychoanalytic-literary approach takes as its point of departure the assumption that the story in Genesis 12, 20, and 26 encodes unthinkable and unacknowledged sexual fantasies. Because there is something fearful and attractive to the (male) narrator about the idea of the wife being taken by another man, a situation that invites the woman's seizure is repeated three times. The tale would thus appear to illustrate Freud's *Wiederholungszwang*, the repetition compulsion—the impulse to work over an experience in the mind until one becomes the master of it—whose locus, according to Freud, is the unconscious repressed.[15] The text is a symptom of the narrator's intra-psychic conflict. But whereas the repetition compulsion is neurotic and an obstacle to awareness, telling the story of the patriarch's repetitive behavior offers the occasion for a 'working out' of the neurosis.

> Repetition is both an obstacle to analysis—since the analysand must eventually be led to renunciation of the attempt to reproduce the past—and the principal dynamic of the cure, since only by way of its symbolic enactment in the present can the history of past desire, its objects and scenarios of fulfillment, be made known, become manifest in the present discourse.[16]

Repeating the story, working over the conflict until it is resolved, provides a semiotic cure for the neurosis. By the charmed third time the cure is effected; that is to say, it is believed.

In approaching the text from a psychoanalytic-literary perspective, I am not proposing to psychoanalyze the characters. Rather than treat characters in a story as if they were real people with real neuroses, I want to examine the world view these literary creations represent. Taking a cue from psychoanalytical theory and building upon the similarities between interpreting dreams and interpreting texts, I shall consider all the characters in the text as split-off parts of the narrator. When a dream is analyzed in psychoanalysis, the analysand is brought

Abraham behaves as he does because of fear and desire that his wife gain sexual knowledge of another man and, say, Westermann's contention (1985: 164) that Abraham behaves this way because of insufficient trust in the divine promises. For insightful remarks about the way traditional scholarship disguises its subjectivity, see Miscall 1983: 40-42.

15. Freud 1961: 16-25 *passim*.

16. Brooks 1987: 10.

to recognize aspects of herself or himself in the various characters of the dream. In our thrice-told tale we will consider the characters in the story as aspects of the narrative consciousness. Thus not just the female characters but the male characters also are expressions of male fantasies, anxieties, etc. When I say, 'Abraham fears for his life', I refer to Abraham not as if he were a real human being but rather as a vehicle for the androcentric values and the androcentric world view of the biblical narrative. It bears pointing out that I am not proposing to psychoanalyze the author either, in the sense that the author, any more than Abraham, is a real person. I assume, with most biblical scholars, that these ancient texts are a communal product, and, further, I assume they received their final redaction at the hands of men. The narrative thus does not reflect an individual's unconscious fantasies, but rather, we might say, it owes its creation to a kind of collective androcentric unconscious, whose spokesperson I shall call simply 'the narrator'.

Features Obscure and Obscuring

In a recent study of the Abraham traditions, Joel Rosenberg remarks that 'the "wife-sister" motif, considered as an item of history and tradition, is an obscure and suggestive theme whose full meaning will probably continue to elude us'.[17] As my epigraph from Lacan indicates, I want to look for illumination in what at first glance seems to obscure matters.[18] The tales exhibit many puzzling features. Why, for example, does the patriarch fear that he will be killed for his wife? Why doesn't he consider the possibility that she might simply be taken from him? He could be overpowered and robbed of his wife, or sent away without her, or an attempt could be made to buy him off. He assumes, however, a moral code according to which the foreign men in question will *not* commit adultery but they *will* commit murder. And when he says, in Gen. 12.13 and 20.11, '*They* will kill me', does he imagine that they would all attack him at once (and if so, who would get the woman)? Or, by assuming many men will want his wife, is he simply accepting in advance that

17. Rosenberg 1986: 77.
18. Lacan 1988: 41.

there is nothing he can do to save both his wife and his life? He is not concerned about what might happen to his wife in another man's harem, and clearly not interested in protecting her. In fact, by claiming that the beautiful woman is his (unmarried) sister, the patriarch guarantees that his wife *will* be taken.

Having taken the woman (in Gen. 12 and 20), the foreign ruler, upon learning that she is Abraham's wife, gives her back to her husband. He does not kill Abraham, as Abraham had feared, even though now he has good reason, since Abraham's lie about Sarah's status has both placed him in an unacceptable position and brought trouble upon his land (plagues in Gen. 12 and barrenness in Gen. 20). In Genesis 26, Abimelech is incensed at what *might have happened* and takes measures to ensure that it will not happen in the future. What the patriarch seems to fear, and says explicitly that he fears in Gen. 20.11—lack of morality ('there is surely no fear of God in this place')—is proved by events to be not the case. Moreover, he already attributes a certain morality to the foreign men when he assumes they will kill him rather than commit adultery with a married woman.

The crucial question is, Why does the patriarch—twice in the person of Abraham and once Isaac—repeat his mistakes? Why does he need to set things up so that another man will seize his wife not once, but three times? To answer that the threefold repetition is the result of three different pentateuchal sources or of three variants in the oral tradition behind the text is to beg the question.[19] As recent literary criticism of the Bible recognizes, the final form of the text is not a haphazard product but rather the result of complex and meaningful redactional patterning. If the androcentric tradition keeps repeating this story, we can assume that the story fills some need.

The Repetition Compulsion

We begin with what is apparent. The story is about fear and desire: desire of the beautiful woman and fear of death because

19. Indeed, one of the early arguments of source criticism for multiple authorship of the Pentateuch was the fact that the patriarch, and his son after him, would hardly have been so foolish as to repeat the ruse three times.

of her. In all three versions the patriarch considers his wife desirable to other men, and in the first two, he is right: the woman is desired, as is witnessed by the fact that she is taken as a wife by another man. In all three instances, the matriarch's desirability makes the patriarch afraid for his life, though his fear turns out to be unjustified. In assessing the patriarch's behavior in response to the perceived threat, Clines remarks that 'the danger is all in the patriarch's mind to begin with'.[20] This being the case, a psychoanalytical approach should prove especially useful. But it is not just what might or might not be going on in the patriarch's mind that will concern us. As I have indicated, all the characters in this repeated story are vehicles for the narrative neurosis.

Each of the stories, the primal scene and its repetitions, is preoccupied with the *same unconscious fantasy*: that the wife have sex with another man. Psychoanalysis tells us that this must be the unconscious desire because this is precisely what the patriarch sets up to happen. It is important to keep in mind that the desire is unconscious; what Freud says about Oedipus's desire is applicable here: in reality it would likely cause him to recoil in horror.[21] What is unconsciously desired is also unconsciously feared; as I hope to show, the story is repeated in an effort to envision and simultaneously to deny the possibility of such a sexual encounter taking place between the wife and another man. Psychoanalysis draws attention to the close relationship between desires and fears. Am I afraid of heights because unconsciously I desire to jump? Is homophobia in reality a fear of one's own repressed sexual urges? Fear in Genesis 12, 20, and 26 is conscious but displaced. The patriarch fears for his life, the assumption being that the foreign man will want the woman all to himself. Abraham is willing to let the other man have her, since the woman must belong to one man or the other but cannot be shared; she cannot belong to both. This is the familiar double standard, according to which men may have sexual relations with more than one woman, but a woman cannot have sexual knowledge of a man other than her husband. The

20. Clines 1990: 68.

21. Freud, Letter to Wilhelm Fliess of Oct. 15, 1897, cited by Felman 1983: 1022.

remarkable thing about the patriarch's ruse is that it ensures that his wife *will* gain sexual knowledge of another man. Certainty is better, more controllable, than doubt.

Since we are dealing with a text, and not with an analysand who can contribute actively to the psychoanalytical process, we can only speculate about what lies behind the fear and desire. It could be the need to have the woman's erotic value confirmed by other men, what René Girard describes as the mechanism of triangular desire.[22] Having chosen a particular woman as the object of his desire, the man needs other men's desire to validate his choice, and even to increase his desire. Or, losing the woman to another man is desirable because he will be free of the woman and the responsibility she entails. This is the male fantasy of sex without commitment; he will be free to have other women, unhampered by the domesticity that the wife represents. There may be deeper, more distressing, desires as well. The same object (originally, according to much psychoanalytical theory, the mother's body) evokes both reverence and hostility. Thus the fascination with the notion of the woman being taken by another man may mask a fear and hatred of woman that desires her humiliation (there is no question that the story objectifies the woman). Other explanations might be sought in what Freud calls 'the mysterious masochistic trends of the ego'.[23] Losing the woman to another man is also threatening, because sexual knowledge of another man would provide the woman with experience for comparison. Other men might be 'better', or know some things about sex he does not know, and perhaps she will enjoy with them what she does not experience with him. This takes us back to the patriarch's displaced fear. His fear for his life at the hands of other men disguises the fact that it is really the woman's sexual knowledge that is life-threatening for him. It is 'safer' for him to fear other men than to acknowledge his fear of the woman's sexuality.

22. See Girard 1965: esp. 1-52.

23. Freud 1961: 12. We might also keep in mind that the repetition complex is related to the desire for death and the delaying of it, which is reflected in the patriarch's fear of death because of the woman.

Patriarchy's Talking Cure

The fabula in which the wife is, in effect, offered to the other man is repeated until the conflict revolving around the woman's feared and desired sexual knowledge has been resolved. By managing fear and desire within an ordered discourse, the narrative functions as a textual working-out of unconscious fantasies, a semiotic cure for the neurosis.

Let us consider first the fundamental similarities between the three tales. All three raise the possibility that the matriarch have sex with a man other than her husband. The patriarch is not only willing for his wife to commit adultery; he invites it. The foreign ruler, on the other hand, will not willingly commit adultery. The patriarch might thus be viewed as a cipher for the unconscious desire, the foreign ruler as the embodiment of fear, and the story as the locus of the tension. The *difference* in the three tales is significant for resolving the conflict. In the first, Sarah is taken into the royal harem, and restored when the pharaoh learns that she is already another man's wife. But did she have sexual relations with the pharaoh? We cannot be sure, for this version of the story does not satisfactorily resolve the issue. It must, therefore, be repeated. The second time around, matters are different. In Genesis 20, Sarah is again taken, but Abimelech does not lay a hand on her. It is no doubt reassuring that what is unconsciously desired and feared does not take place, but the situation remains potentially threatening as long as the woman is allowed to enter another man's household. In the third version, Genesis 26, the possibility of what is both desired and feared taking place is ruled out from the start: Rebekah is not even taken into Abimelech's house.

In the working out of the neurosis, the realization of the fantasy is precluded. To describe this process as it is actualized in the narrative, I shall borrow some terms from Freud, without applying them in a strictly Freudian sense.[24] Instead I shall use a

24. I am offering neither a Freudian reading nor suggesting the superiority, or even validity, of Freudian analysis (in recent years there have been numerous important feminist critiques of Freudian theory). For basic distinctions between the ego, the id, and the super-ego, see Freud 1960; Freud used these terms differently and sometimes indiscriminately, and he changed his usage over time.

fundamental Freudian concept as a metaphor in order to clarify the contradictory impulses in the text. The foreign ruler, who expresses moral outrage at the deception Abraham has perpetrated, is a kind of super-ego, an enforcing, prohibiting agency, to Abraham's id, unconscious desire ready to give over the woman. In other words, the positions occupied in Freudian theory by the super-ego and the id, i.e. the self-observing, self-critical agency in the ego and the libidinous unconscious desire, are fantasized as characters in the story. The text is metaphorically in the position of the ego, where these contradictory impulses are finally resolved.

In the first version, the pharaoh is upset, but his response does not crystallize the moral issue; the super-ego is not yet highly developed.

> What is this you have done to me? Why did you not tell me that she was your wife? Why did you say, 'She is my sister', so that I took her for my wife? (Gen. 12.18-19).

In the second version, in contrast, we find a virtual obsession with issues of sin and guilt, all signs of a highly active conscience. The pharaoh's 'What is this you have done to me?' becomes Abimelech's

> What have you done to us? How have I *sinned* against you that you have brought on me and my kingdom a great *sin*? *Deeds that are not done* you have done to me (Gen. 20.9).

This super-ego, however, needs external moral support, and thus the narrative begins with a lengthy dialogue between Abimelech and God in a dream.[25] God, as symbol and overseer of the moral order, passes judgment: 'You are a dead man because of the woman you have taken; she is another man's wife' (v. 3). With continued emphasis on the issue of innocence

25. On the legal character of the dialogue, see Westermann 1985: 322-23. Interestingly, the locus for dealing with the conflict here is a dream. Freud saw dreams as fulfillments of unconscious wishes. Even anxiety dreams and punishment dreams, such as this one, perform this function, 'for they merely replace the forbidden wish-fulfillment by the appropriate punishment for it; that is to say, they fulfill the wish of the sense of guilt which is the reaction to the repudiated impulse' (Freud 1961: 37).

versus guilt, Abimelech protests his innocence before the law, appealing to his ignorance of Sarah's status:

> Lord, would you slay a *righteous* people? Did he himself not say to me, 'She is my sister'? And she herself said, 'He is my brother'. In the *integrity* of my intentions and the *innocence* of my hands I have done this.

Abimelech is 'innocent' because God, the moral law, prevented him from 'sinning': 'It was I who kept you from sinning against me; therefore I did not let you touch her' (v. 6). Fear of punishment provides powerful motivation for adherence to the law: 'If you do not return her, know that you shall surely die, you and all that is yours' (v. 7).

This ethical rationalization is carried through on every level of the narrative in Genesis 20. Just as Abimelech (in the position of super-ego) justifies himself to God (external moral law), so also Abraham (in the position of the id, the unconscious desire) justifies his deceit to Abimelech (super-ego):

> It was because I thought, There is surely no fear of God in this place, and they will kill me because of my wife. Besides she is indeed my sister, the daughter of my father but not the daughter of my mother; so she could be my wife.

Subtly he tries to shift the blame by implicating God:

> When God caused me to wander from my father's house, I said to her, 'This is the kindness you must do me: at every place to which we come, say of me, "He is my brother"'.

Abraham's protestations of innocence are like psychoanalytical negations: if he were innocent he would not need to protest so much. He undermines his defense—that he feared the lack of morality 'in this place'—by adding that he told Sarah to claim he was her brother 'at every place to which we come', indicating compulsive behavior and not a single aberration. This 'Freudian slip' is a sign of a guilty conscience, the need to be caught in the lie—and commentators have caught him.[26] The libido still feels the need to be held in check against its own powerful impulses.

By the third time (Gen. 26), the super-ego functions independently of external restraints; it rejects the very notion of the

26. E.g. Miscall 1983: 15; Westermann 1985: 326; Coats 1983: 150.

woman having sex with another man. The moral issue is generalized. 'One of the people', not the Self who no longer feels threatened, 'might have lain with your wife'—but nothing happens. We are informed in v. 7 that the men of Gerar asked Isaac about Rebekah, so we know they have noticed her. We are also told (v. 8) that Isaac and Rebekah were in Gerar for a long period of time, so we also know they are not interested. The fascination with the fantasy has been abandoned. As on the previous occasions, the id is held accountable to the super-ego, but it is no longer viewed as threatening: 'You'—the fascination with the woman's desired and feared sexual knowledge—'would have brought guilt upon us', Abimelech tells Isaac (v. 10), but (so the implication) I—the admonitory, judgmental agency in the ego—prevented it. In this version, the super-ego does not need God, the external source of morality, to tell it what to do. It makes its own law: 'Whoever touches[27] this man or his wife shall be put to death' (*mot yumat*). In the Bible, this kind of apodictic formulation appears in the legal material. In psychoanalysis, the ability to internalize moral standards is a sign of maturity.

It can hardly be fortuitous that once the story ceases to entertain the fantasy of another man having the woman, the patriarch is pictured enjoying the woman sexually, and the other man witnesses it. Abimelech looks out his window and sees Isaac 'fondling' (NRSV) or 'caressing' (Westermann) Rebekah. Whatever the precise meaning of the verb *metsaheq*, a pun on Isaac's name, it has to refer to some form of sexual intimacy, since, on the basis of this activity, Abimelech recognizes that Isaac and Rebekah must be man and wife. In this final version of the tale, the fantasy of the woman's having sex with another man is rejected in favor of the (also fantasized) assurance that her sexuality belongs exclusively to the patriarch.

And what of the other man's watching? According to Girard's theory of triangular desire, the relation between the

27. The verb *ng'* was used of approaching the woman sexually in 20.6. Here it has a double meaning, since it is also applied to the man in its more general sense of harming. The inclusion of 'this man' in the edict may be taken as a sign of acceptance of the dangerous impulses as no longer capable of jeopardizing the Self.

rivals in an erotic triangle is as important as their relationship to the object of desire.[28] Using the Girardian triangle as a model, I suggested above that the desiring subject (the position occupied in our narratives by the patriarch) needs the desire of other men to confirm the excellence of his sexual choice. The patriarch sees the matriarch as an object of beauty, and thus an object of desire ('I know that you are a beautiful woman', 12.11; cf. 26.7), but he needs to know that other men desire her too; so he sets up a situation that will elicit their desire: he presents her as an available woman.[29] The prestige of his rival only serves to affirm that the woman he has selected is worthy of desire.[30] The rival who takes the matriarch has the ultimate social prestige—he is the pharaoh or the king—and he has sexual prestige because he has a harem; he can have any woman he likes, and one assumes he chooses only the best. He is also willing to pay a high price for the woman, either to possess her (12.16) or as restitution (20.14, 16)—further testimony to her value. Girard examines stories, like ours, where the hero appears to offer the beloved wife to the rival, and concludes, 'He pushes the loved woman into the mediator's arms in order to arouse his desire and then triumph over the rival desire'.[31] Having Abimelech, the rival, witness his sexual activity with the matriarch is the patriarch's ultimate turn-on, his incontestable victory over rival desire. In

28. Girard (1965) proposes that our desire for something does not really come from ourselves, nor does it lie in some kind of intrinsic worth in the object of our desire; rather it is based on looking at what other people find desirable. Other people become our models, 'mediators of desire' in his theory, whose desire we copy. The positions in Girard's metaphorical triangle are: the desiring subject; the mediator of desire, who defines the subject's desire for him or her; and the object of the desire.

29. White (1991: 180-83) makes a similar point about the beautiful woman as an object of desire in Gen. 12, but he evaluates Abraham's desire differently, as different from and superior to that of his rivals.

30. Girard 1965: 50.

31. Girard 1965: 50. Girard also argues that 'the impulse toward the object is ultimately an impulse toward the mediator' (p. 10) and that the desiring subject wants to become his mediator/rival (p. 54). The patriarch becomes like his wealthy, powerful rival when he becomes wealthy at the foreign ruler's expense (12.16, 20; 20.14, 16; cf. 26.12-14, where the envy theme is continued), and when the ruler recognizes him as more powerful—for example, as a prophet who can pray for him, or simply as 'much mightier than we are' (26.16).

this version of the fantasy, the roles are here reversed. The patriarch is no longer in the position of the fearing/desiring subject; the other man is. Fear of the woman's knowledge of other men is transformed into other men's envy of him.

Not a Woman's Story

I have argued that Genesis 12, 20, and 26 deal with an unacknowledged and unthinkable male fantasy. In the patriarch–matriarch–foreign ruler triangle, the matriarch never becomes a narrative presence. Though addressed by men—Abraham says, 'Say you are my sister' (12.13); Abimelech says, 'Look, I have given your brother a thousand pieces of silver; it is your vindication...' (20.16)—the matriarch never speaks and only once is she reported to have spoken (20.5). The woman has no voice in determining her sexual status and no control over how her sexuality is perceived or used. Susan Niditch calls Sarah in Genesis 12 a 'tacit accomplice'.[32] Sharon Pace Jeansonne considers her less an accomplice than a silent object.[33] In my reading, she is both accomplice and object because she, like the other characters, is a creation of the narrative unconscious. The male fantasy that created her character is not interested in the woman's point of view—her reaction to Abraham's suggestion, her willingness to be exchanged for her husband's well-being, or her experience in the harem of a strange man. The question of force versus consent, crucial for constructing the woman's perspective, is not raised.[34]

The woman is only an object in a story about male fears and desires. The possibility of the wife having sex with another man is taken out of the control of the woman and made solely an

32. Niditch 1987: 59.

33. Jeansonne 1990: 17. Jeansonne maintains that Sarah's silence is not evidence of complicity but rather a sign of her powerlessness; similarly Rashkow 1992. This is quite literally an argument from silence, and it too easily leads us into a victim–victimizer dichotomy that ignores women's complicity in patriarchy. On this point I agree with Niditch (1987: 59), but for a different reason: Sarah is an accomplice because her character is the creation of an androcentric narrator. Sarah is not, as White (1991: 185) would have it, an 'innocent victim', because she is complicit.

34. This is also the case with Hagar in Gen. 16; see above.

affair between men. This is the only way androcentric ideology can conceive of it, unless, as in the case of Potiphar's wife, the woman is a 'bad woman',[35] which, of course, the matriarch cannot be or else she would not qualify to be the matriarch. As it is posed in Genesis 12, 20, and 26, the question is not, Will the woman commit adultery, but, Will the other man commit adultery? The patriarch thinks not: he thinks the other man will kill him rather than commit adultery with a married woman. The foreign ruler also rejects the thought of adultery. The result is a kind of gentlemen's agreement about the other man's property, which reflects the biblical understanding of adultery as less a matter of sex than a violation of another man's property rights.[36] Legislating the husband's exclusive sexual rights to his wife is an effective way of controlling women's desired and feared sexuality. That the patriarch, the foreign ruler, and God all recognize the seriousness of adultery with a married woman is crucial to the ideology of all three versions (what the woman thinks is irrelevant).

'She Is Indeed My Sister'

Scholars generally deal with Abraham's claim that Sarah really is his half-sister in Gen. 20.12 by asking whether or not it is a lie. Clines and Miscall think Abraham is lying;[37] Westermann, von Rad, Speiser, and Skinner think he is telling the truth.[38] Some apologists call Abraham's claim that Sarah is his sister a 'white lie'.[39] Regardless of whether or not Sarah and Abraham are sister and brother, we know it is not true of Isaac and Rebekah. From a psychoanalytic-literary perspective, the important issue is not the veracity of Abraham's claim but the fact that in all

35. See Bach 1993.

36. See Westbrook 1990. For an interpretation of Gen. 12 that sees the taboo against sex with a married woman exploited by Abraham to set up the pharaoh, see White 1991: 174-86. For an anthropological perspective, see Pitt-Rivers 1977: 159, who suggests the stories are about 'sexual hospitality', where women are used to establish relations among groups of men; see pp. 113-70.

37. Clines 1990: 76; Miscall 1983: 14-15.

38. Westermann 1985: 326; von Rad 1961: 222; Speiser 1964: 92; Skinner 1910: 318.

39. Anderson, annotations to the RSV; Fox 1989: 32.

three versions the brother–sister relationship is imagined. All three accounts raise the issue of consanguinity simply by having the patriarch tell the foreigners that the matriarch is his sister. Might we not see in this latent incest fantasy a desire to achieve unity with the other? In the Song of Songs, for example, the man uses the epithet 'sister, bride' to refer to the woman as sign of intimacy. Clearly the matriarch's kinship ties to the patriarch are important to these stories in Genesis 12–36; she must come from his own people, his own kind.[40] As a sibling, the matriarch is more 'self' than 'other'—more like the patriarch than different. Fantasizing her as his sister may represent a narcissistic striving toward completeness or wholeness, whose realization can only be imagined in his mirror-image from the opposite sex (she is what he would be if he were a woman). Oedipal desire, of which, according to Eve Kosofsky Sedgwick, the Girardian triangle is a schematization,[41] may be at work here as well. As his close female relative, the sister is a stand-in for the mother as object of desire (and Sarah is the arch mother). In this case, Abraham will have married a girl as much like the girl who married dear ol' dad as possible. Fear of the father's wrath may explain his willingness to give her back, symbolically, to the father—the subject position held in our tale by the powerful, foreign ruler–authority figure. In the end, his relationship to his mother-substitute is legitimized by the father. This is the significance of the fact that Abimelech *sees* Isaac and Rebekah engaged in sexual play: it represents the father's acknowledgment that this woman rightfully belongs to the 'son' and the father's permission for him to have sex with her.

40. For anthropological readings of the three accounts as representing a movement from incest to the preferred form of marriage, see Pitt-Rivers 1977: 154-55; Donaldson 1981. Pitt-Rivers offers a suggestive reading of these accounts in relation to the story of the rape of Dinah, Gen. 34; see pp. 151-71. On the matriarchs' role in Gen. 12–36, see also Exum 1993: 94-147.

41. Sedgwick 1985: 22. See her discussion (pp. 21-27), which, in contrast to Girard, takes gender into account as a constituent factor. Interestingly, Freud saw the repetition complex as going back to some period of infantile sexual life, to the Oedipus complex; see Freud 1961: 19.

Who's Afraid of 'The Endangered Ancestress'?

We have looked at the thrice-told tale in Genesis 12, 20, and 26 as a symptom of the narrative's intra-psychic conflict, a conflict between the unconscious desire that the wife gain sexual knowledge of another man and the fear that this could happen. The conflict appears in disguised and distorted form: the patriarch fears for his life because of his beautiful wife, and passes her off as his sister, thereby allowing another man to take her into his harem. In reality, the fear is of the woman's sexuality, which is both desired and feared. There is a compulsive need to repeat the story until the conflict is resolved. In Genesis 12, the super-ego (the pharaoh) is subject to the id (Abraham); he takes the woman. In Genesis 20, the super-ego (Abimelech) has external moral support (God). He is subject to the id (Abraham) in that he takes the woman, but subject to external law (God) in that he does not touch her. But morality based on external authority is not the best solution for the patriarchal neurosis. In the third version (Gen. 26), the moral code is internalized; the fascination with the woman's desired and feared sexuality no longer poses a threat; the neurosis is cured; the cure is believed.[42]

In the children's refrain, 'Who's afraid of the big bad wolf, the big bad wolf, the big bad wolf?', we find a denial of fear that, as such, is also a recognition of fear. The thrice-told tale in Genesis 12, 20, and 26 functions similarly. It says, in effect, 'Who's afraid of the woman's sexual knowledge?' And it answers by reassuring the patriarch that there is no need to fear. But it betrays itself, for, like the ditty about the big bad wolf, it acknowledges that there is something to be feared. If the danger in these three stories is woman's sexuality and woman's sexual knowledge, who or what is in danger? To the

42. Later retellings of these stories continue the process of filling gaps, thereby resolving some of the anxiety-provoking ambiguities (for example, Did Abraham lie about Sarah's being his sister?; What happened to Sarah in the harem?; Did Abraham know what happened in the harem?) and some give Sarah a greater role (for example, Sarah prays for protection, and the ruler is afflicted 'because of the word of Sarai' [*'al debar sarai*, Gen. 12.17]). On later versions of the tale in Jewish and Islamic sources, see Firestone 1991.

question, 'Who or what is afraid of the woman's sexual knowledge?', the answer is, 'Patriarchy'.

BIBLIOGRAPHY

Abou-Zeid, Ahmed
1966 'Honor and Shame among the Bedouins of Egypt', in *Honour and Shame: The Values of Mediterranean Society* (ed. J.G. Peristany; Chicago: University of Chicago Press): 245-57.

Alexander, T.D.
1992 'Are the Wife/Sister Incidents of Genesis Literary Compositional Variants?', *VT* 42: 145-53.

Bach, Alice
1993 'Breaking Free of the Biblical Frame-up: Uncovering the Woman in Genesis 39', in *A Feminist Companion to Genesis* (ed. A. Brenner; The Feminist Companion to the Bible, 2; Sheffield: Sheffield Academic Press): 318-42.

Biddle, Mark E.
1990 'The "Endangered Ancestress" and Blessing for the Nations', *JBL* 109: 599-611.

Brooks, Peter
1987 'The Idea of a Psychoanalytic Literary Criticism', in *Discourse in Psychoanalysis and Literature* (ed. S. Rimmon-Kenan; New York: Methuen): 1-18.

Clines, David J.A.
1990 *What Does Eve Do to Help? And Other Readerly Questions to the Old Testament*. JSOTSup, 94; Sheffield: JSOT Press).

Coats, George W.
1983 *Genesis, with an Introduction to Narrative Literature* (FOTL, 1; Grand Rapids: Eerdmans).

Culley, Robert C.
1976 *Studies in the Structure of Hebrew Narrative* (Semeia Supplements; Philadelphia: Fortress Press; Missoula, MT: Scholars Press).

Donaldson, Mara E.
1981 'Kinship Theory in the Patriarchal Narratives: The Case of the Barren Wife', *JAAR* 49: 77-87.

Exum, J. Cheryl
1993 *Fragmented Women: Feminist (Sub)versions of Biblical Narratives* (JSOTSup, 163; Sheffield: JSOT Press; Valley Forge, PA: Trinity Press International).

Felman, Shoshana
1983 'Beyond Oedipus: The Specimen Story of Psychoanalysis', in *Lacan and Narration: The Psychoanalytic Difference in Narrative Theory* (ed. R. Con Davis; Baltimore: Johns Hopkins University Press): 1021-53.

Firestone, Reuven
1991 'Difficulties in Keeping a Beautiful Wife: The Legend of Abraham and Sarah in Jewish and Islamic Tradition', *JJS* 42: 196-214.

Fox, Everett
1989 'Can Genesis be Read as a Book?', *Semeia* 46 (*Narrative Research on the Hebrew Bible*, ed. M. Amihai, G.W. Coats and A.M. Solomon): 31-40.
Freud, Sigmund
1960 *The Ego and the Id* (trans. Joan Riviere; rev. and ed. James Strachey; New York: W.W. Norton).
1961 *Beyond the Pleasure Principle* (trans. and ed. James Strachey; New York: W.W. Norton).
1965 *The Interpretation of Dreams* (trans. and ed. James Strachey; New York: Avon Books).
Girard, René
1965 *Deceit, Desire, and the Novel: Self and Other in Literary Structure* (trans. Y. Freccero; Baltimore: Johns Hopkins University Press).
Jeansonne, Sharon Pace
1990 *The Women of Genesis: From Sarah to Potiphar's Wife* (Minneapolis: Fortress Press).
Keller, Carl A.
1954 '"Die Gefährdung der Ahnfrau". Ein Beitrag zur gattungs- und motivgeschichtlichen Erforschung alttestamentlicher Erzählungen', *ZAW* 66: 181-91.
Koch, Klaus
1969 *The Growth of the Biblical Tradition: The Form-Critical Method* (trans. S.M. Cupitt; New York: Charles Scribner's Sons).
Lacan, Jacques
1988 'Seminar on "The Purloined Letter"' (trans. J. Mehlman), in *The Purloined Poe: Lacan, Derrida and Psychoanalytic Reading* (ed. J.P. Muller and W.J. Richardson; Baltimore: Johns Hopkins University Press): 28-54.
Miscall, Peter D.
1983 *The Workings of Old Testament Narrative* (Semeia Studies; Philadelphia: Fortress Press; Chico, CA: Scholars Press).
Niditch, Susan
1987 *Underdogs and Tricksters: A Prelude to Biblical Folklore* (San Francisco: Harper & Row).
Noth, Martin
1972 *A History of Pentateuchal Traditions* (trans. Bernhard W. Anderson; Englewood Cliffs, NJ: Prentice–Hall).
Petersen, David L.
1973 'A Thrice-Told Tale: Genre, Theme, and Motif', *BR* 18: 30-43.
Pitt-Rivers, Julian
1977 *The Fate of Shechem: Or, the Politics of Sex* (Cambridge: Cambridge University Press).
Polzin, Robert
1975 '"The Ancestress of Israel in Danger" in Danger', *Semeia* 3 (*Classical Hebrew Narrative*, ed. R.C. Culley): 81-98.
Rad, Gerhard von
1961 *Genesis* (trans. J.H. Marks; Philadelphia: Westminster Press).

Rashkow, Ilona N.
1992 'Intertextuality, Transference, and the Reader in/of Genesis 12 and 20', in *Reading between Texts: Intertextuality and the Hebrew Bible* (ed. D.N. Fewell; Louisville, KY: Westminster/John Knox): 57-73.

Rosenberg, Joel
1986 *King and Kin: Political Allegory in the Hebrew Bible* (Bloomington: Indiana University Press).

Sedgwick, Eve Kosofsky
1985 *Between Men: English Literature and Male Homosocial Desire* (New York: Columbia University Press).

Skinner, John
1910 *A Critical and Exegetical Commentary on Genesis* (Edinburgh: T. & T. Clark).

Speiser, E.A.
1964 *Genesis* (AB, 1; Garden City, NY: Doubleday).

Steinberg, Naomi
1984 'Gender Roles in the Rebekah Cycle', *Union Seminary Quarterly Review* 39: 175-88.

Van Seters, John
1975 *Abraham in History and Tradition* (New Haven: Yale University Press).

Vawter, Bruce
1977 *On Genesis: A New Reading* (Garden City, NY: Doubleday).

Westbrook, Raymond
1990 'Adultery in Ancient Near Eastern Law', *RB* 97: 542-80.

Westermann, Claus
1980 *The Promises to the Fathers: Studies on the Patriarchal Narratives* (trans. David E. Green; Philadelphia: Fortress Press).
1985 *Genesis 12–36* (trans. J.J. Scullion; Minneapolis: Augsburg Press).

White, Hugh C.
1991 *Narration and Discourse in the Book of Genesis* (Cambridge: Cambridge University Press).

A Reader-Response Approach to Prophetic Conflict: The Case of Amos 7.10-17

Francisco O. García-Treto

The new prominence given to the reader in literary interpretation of the Bible—the stimulus for the first form of this essay[1]—has led me to explore and to attempt to demonstrate what a 'reader-oriented' approach can produce, using the report of Amaziah's confrontation with Amos[2] as an example. The approach and the method spring from the work of contemporary 'reader-oriented' critics[3] who have emphasized the central importance of the reader in the production of the meaning of texts. Among these, I find most useful the well-known work of Stanley Fish, and particularly his notion of 'interpretive communities',[4] as a general frame of reference. Such communities, Fish says,

> are made up of those who share interpretive strategies not for reading (in the conventional sense) but for writing texts, for constituting their properties and assigning their intentions. In other

1. A paper presented to the Prophetic Literature Section of the Society of Biblical Literature at its 1991 Annual Meeting in Kansas City, Missouri.
2. Amos 7.10-17.
3. Among the principal theorists of what is broadly known as 'reader-response' criticism are Wolfgang Iser, Umberto Eco, Norman Holland, Stanley Fish and David Bleich. For an excellent recent discussion of the area, see Michael Steig, *Stories of Reading: Subjectivity and Literary Understanding* (Baltimore: Johns Hopkins University Press, 1989), ch. 1, 'Theories of Reading: An End to Interpretation?', pp. 3-16.
4. Stanley Fish, *Is There a Text in This Class? The Authority of Interpretive Communities* (Cambridge, MA: Harvard University Press, 1980). See especially pp. 167-73.

> words, these strategies exist prior to the act of reading and therefore determine the shape of what is read rather than, as is usually assumed, the other way around.[5]

Fish postulates that there is no such thing as 'simply reading', a hypothetical activity that would imply 'the possibility of pure (that is, disinterested) perception'.[6] Rather, a reader—or a reading—always proceeds from the basis of certain 'interpretive decisions' which in turn lead to the adoption of the 'interpretive strategies' that produce or determine the reading. Interpretive strategies, in fact, 'are the shape of reading, and because they are the shape of reading, they give texts their shape, making them rather than, as it is usually assumed, arising from them'.[7] This concept provides a helpful resolution to the fear of 'interpretive anarchy', which even for readers who have given up the 'impossible ideal' of 'perfect agreement' on texts with a 'status independent of interpretation' remains an obstacle to reader-oriented approaches. That fear, says Fish,

> would only be realized if interpretation (text making) were completely random. It is the fragile but real consolidation of interpretive communities that allows us to talk to one another, but with no hope or fear of ever being able to stop.[8]

Within a given interpretive community, Fish concludes, 'the only "proof" of membership is fellowship, the nod of recognition from someone in the same community, someone who says to you what neither of us could ever prove to a third party: "we know"'.[9] Post-modern literary critics, or feminist interpreters, or 'the guild' of Society of Biblical Literature members, or Brazilian Ecclesial Base Communities,[10] or fundamentalist protestants,[11]

5. Fish, *Is There a Text?*, p. 171.
6. Fish, *Is There a Text?*, p. 168.
7. Fish, *Is There a Text?*, p. 168.
8. Fish, *Is There a Text?*, p. 172.
9. Fish, *Is There a Text?*, p. 173.
10. See the work of Carlos Mesters, in particular *Defenseless Flower: A New Reading of the Bible* (trans. from the Portuguese by Francis McDonough; Maryknoll, NY: Orbis Books, 1989).
11. For a helpful analysis, see Kathleen C. Boone, *The Bible Tells Them So: The Discourse of Protestant Fundamentalism* (Albany, NY: State University of New York Press, 1989), especially ch. 5, 'For Correction: The Interpretive

can all be recognized—or, more precisely, recognize themselves—as more-or-less strictly defined interpretive communities. These communities are engaged in reading the Bible, if not 'in different epochs', most certainly 'with different world-views'. I end this prologue by quoting Michael Steig's third premise for a model of reading and understanding texts, which adds an important corollary:

> Understanding, in the act of reading literature, is a temporary condition of satisfaction arrived at subjectively and, in the dialectical sense of the term, intersubjectively; it is not directly related to 'meaning' in its narrowest sense—the signification of small language units, such as words and sentences, or the propositional 'message' of a text.[12]

In order to begin to sketch a reading of Amos 7.10-17, I would like to specify more precisely the interpretive community of this particular reader, a community largely congruent with, but partly different from, that of the general class of academic biblical scholars active at this time in North America. Because of my ethnic background, cultural allegiance, personal history and experience and numerous other factors, I see myself—and others see me—as a Hispanic-American. I find myself in sympathy with Justo González's recent plea for a conscious effort on the part of Hispanics to read the Bible 'in Spanish', that is, a reading of people—whether biblical scholars or unsophisticated folk—who read the Bible 'as exiles, as members of a powerless group, as those who are excluded from the "innocent" history of the dominant group'.[13] That reading will approach the Bible as a political book, asking first

> not the 'spiritual' questions or the 'doctrinal' questions—the Bible is not primarily a book about 'spiritual' reality, except in its own

Community', pp. 61-75. Boone also uses Fish's concept as a theoretical base.

12. Steig, *Stories of Reading*, p. xiv.

13. Justo L. González, *Mañana: Christian Theology from a Hispanic Perspective* (Nashville: Abingdon Press, 1990), p. 85. By 'innocent history', González means 'a selective forgetfulness, used precisely to avoid the consequences of a more realistic memory', and he identifies it as characteristic of the dominant group's world view.

> sense, nor is it a book about doctrines—but the political questions: Who in this text is in power? Who is powerless? What is the nature of their relationship? Whose side does God take?[14]

The first predisposition that will inform my reading, then, will be to read for the signs of power in the discourse of the characters, in their actions, in their relationship—in short, to read politically.

There is a second predisposition in my reading, and that is to pay particular attention to the performative, as over against the purely logical, aspects of the discourse. That is to say, that as a product of a Caribbean culture at whose very heart is the tension between—rather than the synthesis of—Africa and Europe,[15] this reader is conditioned to expect and to appreciate improvisational performance, particularly in situations of controversy. By improvisational performance I mean a strategy of discourse that resolves tension and 'displaces centers without displacing them' by turning to the other in an attempt to overwhelm, to seduce, or to decenter, rather than to convince with logical argument. Arguing that Caribbean culture does not stop at the shores of the Gulf of Mexico, but that its characteristics are present as well in North American black expression, Antonio Benítez-Rojo recognizes in Martin Luther King one of its exemplars:

> His African ancestry, the texture of his humanism, the ancient wisdom in his words, his improvisatory nature, his cordially high tone, his ability to seduce and be seduced, and above all, his vehement status as a 'dreamer' (*I have a dream*...) and performer,

14. González, *Mañana*, p. 85.

15. The 'mestizaje' model, which Virgilio Elizondo has proposed as a hermeneutic key for Mexican-American culture—see his *Galilean Journey: The Mexican-American Promise* (Maryknoll, NY: Orbis Books, 1983)—must be modulated for the Caribbean by voices such as that of Antonio Benítez-Rojo: 'In fact, this *mestizaje* is a concentration of conflicts, an exacerbation brought about by the closeness and density of the Caribbean situation. Then, at a given moment, the binary syncretism Europe–Africa explodes and scatters its entrails all around: here is Caribbean literature. This literature should not be seen as anything but a system of texts in intense conflict with themselves.' See A. Benítez-Rojo, 'The Repeating Island', in *Do the Americas Have a Common Literature?* (ed. Gustavo Pérez-Firmat; Durham, NC: Duke University Press, 1990), pp. 85-106.

> all make up the Caribbean element of a man who is unquestionably idiosyncratic in North America. Martin Luther King occupies and fills the space in which Caribbean thought (L'Ouverture, Bolívar, Martí, Garvey) meets North American black discourse; that space can also be filled by the blues.[16]

If 'the Caribbean' in this sense extends into North America, then the analysis of the performative forms of African-American discourse that Henry Louis Gates, Jr presents in his recent work,[17] particularly the concept of 'Signifyin(g)',[18] can be as relevant to my reading of Amos as those forms are to my world view and to my culture. By 'Signifyin(g)', Gates means a form of intertextuality, traditional in African-American discourse, in which 'repetition with revision, or repetition that signals difference',[19] as one scholar has recently characterized it, is valued in performative discourse. As Drewal observes, 'Signifyin(g) can include any number of modes of rhetorical play. "To signify" is to revise that which is received, altering the way the past is read, thereby redefining one's relation to it.'

In what follows, I want to read Amos 7.10-17, giving primacy to my two 'predispositions', that is, to use Fish's terminology, I have made the 'interpretive decision' of reading from my 'difference', which leads me to an 'interpretive strategy' that, for the purposes of this essay, I will simplify to two elements. These are, first, reading for expressions of power relations and characterizations of power, and second, reading for the sort of

16. Benítez-Rojo, 'The Repeating Island', p. 103.

17. Henry Louis Gates, Jr, *The Signifying Monkey: A Theory of African-American Literary Criticism* (New York: Oxford University Press, 1988). See also his *Figures in Black: Words, Signs, and the 'Racial' Self* (New York: Oxford University Press, 1987).

18. See especially ch. 2 ('The Signifying Monkey and the Language of Signifyin(g): Rhetorical Difference and the Orders of Meaning', pp. 44-88) of *The Signifying Monkey*. When referring to Gates's term in this essay, I will, as he does, capitalize the term and enclose the final *g* in parentheses.

19. Margaret Thompson Drewal, *Yoruba Ritual: Performers, Play, Agency* (Bloomington, IN: Indiana University Press, 1992), p. 4. The West-African Yoruba are the major ancestral group of Caribbean and North American blacks.

performative intertextuality which, borrowing Gates's term, can be called 'Signifyin(g)'.[20]

Power Games

Amaziah comes on the scene accompanied by narratorial fanfare. In the first half of v. 10 full titles are in order, and the narratorial voice utters them in a jingling rhyme: *wayyišlaḥ 'ᵃmaṣyâ kōhēn bêt-'ēl 'el-yārob'ām melek-yisrā'ēl*, as if to emphasize the inflated self-importance of the character. Amaziah in fact has a way with a title himself, as his direct discourse demonstrates in v. 13: Bethel is 'the king's sanctuary' and 'a temple of the kingdom'. I cannot help seeing here a parodic portrayal of someone for whom title and political authority are everything, to the extent that Beth-*el*, the 'house of *God*', becomes indeed '*the king's* sanctuary' and 'a temple of the kingdom'. Highly ironic also is the lack of any reference to God in Amaziah's discourse, whether he is addressing his superior Jeroboam, or his troublesome trespasser Amos. The priest of Bethel does not invoke divine authority, but rather he speaks as a functionary of the state. In the light of the preceding, Amaziah's much-commented use of the term *ḥōzeh* in addressing Amos can be put into context. The term, as David Petersen and others have amply demonstrated, is indeed a relatively common designation of southern prophets,[21] equivalent to *nābî'* in meaning and in function, the main difference being precisely that it is a *Judean* title. Title-conscious Amaziah may therefore have even sounded exaggeratedly polite when he addressed Amos as *ḥōzeh*, a 'southern' term, but what results is an emphasis on his certainty that Amos does not belong in the North. He uses *ḥōzeh* as an ironic trope on *nābî'*, a trope that epitomizes the point of his attack on Amos: Amos does not 'belong' in Bethel, Amos has no

20. A related concept in Cuban culture is 'choteo', but for the purposes of this essay it does not appear necessary to engage in detailed cultural analysis. See Gustavo Pérez-Firmat, *Literature and Liminality: Festive Readings in the Hispanic Tradition* (Durham, NC: Duke University Press, 1986), for an introduction to 'choteo'.

21. David L. Petersen, *The Roles of Israel's Prophets* (Sheffield: JSOT Press, 1981), especially pp. 51-69.

authority, as a southerner, with which to challenge Amaziah and his northern state cult. Amos may be a prophet, but he is no prophet *here*! Many Hispanics can speak from experience for a similar use of the Spanish title 'Señor' by some Southwestern Anglos—the word in itself may be polite, but the intent is to underline the otherness, the foreignness, and therefore the unequal status of the addressee. As a priest, Amaziah is a custodian of a system, which Mary Douglas so well described, that equated purity with order, and uncleanliness with 'matter out of place'.[22] By underlining, no matter how politely in external formality, the otherness of Amos, Amaziah is quite literally 'treating him like dirt', as a preface to 'putting him in his place'.

When, besides considering Amaziah's use of titles, we move to consider his actions, what we find first is a strategy of maximizing Amos as a political threat toward the king by presenting Amos's words as extremely hostile and dangerous to Jeroboam and to the nation—

> Amos has conspired against you in the very center of the house of Israel; the land is not able to bear all his words—[23]

and certainly of misrepresenting Amos's words about the king:

> For thus Amos has said,
> 'Jeroboam shall die by the sword, and Israel must go into exile away from his land'.[24]

The intent is clearly to make Amos seem a worse threat than he actually is, to elicit precisely the swift and drastic action from the political authority about which his words implicitly warn Amos. To put it bluntly, there is a double-dealing duplicity in Amaziah that is clear to the reader in the two messages, the first to the king and the second to the prophet. To the latter, addressed as we have seen by the patronizingly polite *ḥōzeh*, Amaziah directs his famous rebuke, in which he orders him to leave immediately—whether *lēk bᵉraḥ-lᵉkā 'el-'ereṣ yᵉhûdâ* is a command to 'flee' or to 'make haste' is not really the central issue—and then casts the insulting insinuation that Amos's first interest in prophesying

22. Mary Douglas, *Purity and Danger: An Analysis of the Concepts of Pollution and Taboo* (London: Routledge & Kegan Paul, 1966).

23. Amos 7.10 (NRSV).

24. Amos 7.11 (NRSV).

is making a living: 'earn your bread there, and prophesy there'.[25] Finally, as has been pointed out, Amaziah utters those ironic words in which the priest of the 'House of God'—who obviously knows what he is saying—calls it 'the king's sanctuary' and 'a temple of the kingdom', giving that as sufficient reason for telling the prophet to leave.

Signifyin(g)

I propose that we read Amos's reply to Amaziah as an example of something akin to what Gates calls 'motivated Signifyin(g)', a 'rhetorical transfer...which serves to redress an imbalance of power, to clear a space, rhetorically'.[26] Amos needs to revise Amaziah's exclusionary text in order precisely to remain in the power game, to 'clear a space' for himself, to 'have the last word'. Gates uses Mikhail Bakhtin's typology of narrative discourse, in particular what Bakhtin calls 'double-voiced discourse', and, within that, the two subcategories of parody and of hidden polemic as a basic construct for elaborating his theory of 'Signifyin(g)'.[27] It is the last of these subcategories, that of 'hidden polemic', that I think is particularly illuminating of Amos's reply. Bakhtin, who characterized hidden polemic as 'barbed words...words used as brickbats', defines this kind of speech as discourse that, besides its orientation toward a referential object, brings to bear a polemical attack 'against another speech act, another assertion, on the same topic'.[28] If,

25. Obviously, since Amaziah's own living was derived from his religious function, there is the possibility that the comment is to be seen as an attempt to let Amos know that, in some way, they share a common interest, and that Amaziah understands Amos's interest in prophesying at a major sanctuary, even if he does not approve and cannot permit Amos's activity at Bethel. Even if this is his intention, Amos clearly chooses to interpret the comment as a patronizing insult and to react accordingly.

26. Gates, *The Signifying Monkey*, p. 124.

27. Gates, *The Signifying Monkey*, pp. 110-11.

28. Mixail Baxtin (Mikhail Bakhtin), 'Discourse Typology in Prose', trans. Richard Balthazar and L.R. Titunik, in *Readings in Russian Poetics: Formalist and Structuralist Views* (ed. Ladislav Matejka and Krystyna Pomorska; Ann Arbor: Michigan Slavic Publications, 1978), pp. 176-96 (187-88) ('Tipy prozaiceskogo slova', in *Problemy tvorcestva Dostoevskogo* [Leningrad, 1929],

as Gates says, borrowing a term from the language of jazz to express the method of Signifyin(g), 'revision proceeds by riffing upon tropes',[29] I would read Amos's reply to Amaziah first as a series of 'riffs' on the latter's use of *ḥōzeh* as what I have called above 'an ironic trope on *nābî'*'. Amos pointedly ignores the term *ḥōzeh* but launches into a 'riff'—or a variation, if you will—in which he denies his '*nabî*hood'—'I am not a *nābî'*, nor the son of a *nābî'*', apparently going Amaziah one better in denying his own status, at the same time that he turns Amaziah's irony inside out by embracing the lack of status and title which the priest's ironic *ḥōzeh* had implied. He doesn't stop 'riffing', however, but launches next into a series of self-identifications as a rustic which can be read as yet another quick-witted elaboration on Amaziah's attack. Amos goes on apparently agreeing with Amaziah's implication that he does not belong where he is—'I am a herdsman, and a dresser of sycamore trees', also claiming that the Lord found him 'following the flock'. This language is a well-known headache for interpreters who try to take it literally and determine from it what was Amos's 'real' occupation. Bakhtin, indeed, warns readers that there are 'double-voiced' texts, in which 'discourse maintains a double focus', that is, texts that aim both at the overt referential object of speech and 'simultaneously at a second context of discourse, a second speech act by another addresser'. To ignore this and thus to treat the speech act as if it were ordinary, single-referent discourse, is to fail to 'get it', or, as Bakhtin puts it, in that case 'we shall take stylization for straight style and read parody as poor writing'.[30] Amos seems in effect to be telling Amaziah, 'I am *anything but* a prophet! I am nothing but a cowboy, a clodhopper, a sheepkicker!' The rhetorical, performative strategy of seeming to agree with the one who 'puts you down' in order to 'put him on' is what appears to be at work here, rather than an otherwise awkward attempt on Amos's part to present some

pp. 105-35). For a different translation of the same material, see Mikhail Bakhtin, *Problems of Dostoevsky's Poetics* (trans. Caryl Emerson; Minneapolis: University of Minnesota Press, 1984), pp. 185-203.

29. Gates, *The Signifying Monkey*, pp. 110-11.

30. Cited by Gates, *The Signifying Monkey*, p. 176.

sort of *curriculum vitae* as a 'self-justification speech'.[31] The real impact of this parodic self-deprecation appears immediately, however, when Amos reveals to Amaziah that the latter's attack on him in fact has gone right through Amos to impinge on the One behind him: '*Yahweh* took me from following the flock, and *Yahweh* said to me...' The repetition of Yahweh's name, which appears here for the first time in the exchange between priest and prophet, is the turning point in the power game in which the two have been engaged.[32] Amaziah's peremptory 'go, flee away to the land of Judah' is now 'trumped' by the report of Yahweh's 'go, prophesy to my people Israel', and thus negated. Negated, also, by that 'my people', is the claim of royal and national supremacy with which Amaziah had sought to expel Amos from Bethel. Amaziah is stripped of his assumed power, left naked as it were, to face the 'word of Yahweh'. His indictment is clearly put: 'You say, "Do not prophesy against Israel, don't drivel against the House of Isaac"', and his punishment is rudely and savagely stated—like the preceding, in Yahweh's name. These terms show just how thorough is to be Amaziah's loss of status—his wife will be defiled, his children slaughtered,[33] his inheritance lost, and he himself will be cast out of the land from which he had sought to expel Amos, to die in an unclean land. Finally, what Amaziah had reported to Jeroboam as *Amos's* 'unbearable' words

31. The term 'self-justification speech' comes from Shalom M. Paul, *Amos: A Commentary on the Book of Amos* (Minneapolis: Fortress Press, 1991), p. 249.

32. The 'signifying monkey' (a trickster figure in African-American folklore) typically taunts or tricks the lion—the loud and oppressive self-proclaimed 'king'—into attacking the powerful elephant, and therefore being the instrument of his own destruction. See Gates, *The Signifying Monkey*, ch. 2, 'The Signifying Monkey and the Language of Signifyin(g): Rhetorical Difference and the Orders of Meaning' (pp. 44-88).

33. The priest's actual wife and children, as I read this scene, are not represented literally in Amos's discourse. They are brought into play as abstract extensions of Amaziah, as tokens that Amos uses to extend and compromise his adversary's vulnerability. There is a striking parallel in this to the use of 'your mama' in the variety of Signifyin(g) called 'playing the Dozens' (for which see Gates, *The Signifying Monkey*, pp. 72-73 and *passim*) and in similar varieties of Caribbean discourse.

concerning Israel is returned to the priest as Yahweh's inexorable decree: Israel *will* go into foreign exile.

Conclusion

In a recent book, Walter Brueggemann calls Amos 7.10-17

> a clear moment when the monopoly of throne and temple is threatened, and then maintained. The priest Amaziah banishes Amos, the voice of an alternative imagination, with the ideological judgment: 'But never again prophesy at Bethel, for it is the king's sanctuary, and it is a temple of the kingdom'.[34]

Reading Amos 7.10-17 with power relations in mind—the first part of my strategy for reading—yields results that agree with Brueggemann's conclusion. Amaziah is a representative of institutional power, who acts to maintain the monopoly of power by seeking to exclude 'Amos, the voice of an alternative imagination' from Bethel. I am not so certain, however, that the text is unambiguously 'a clear moment when the monopoly of throne and temple is threatened, and then maintained', as Brueggemann says. Amos does not slink away defeated. The 'alternative imagination' he represents in fact wins the day, in a performance that decenters and overwhelms Amaziah, more importantly, a performance that persuades the reader that ultimate power, far from being 'a monopoly of throne and temple', remains with Amos's God.

34. Walter Brueggemann, *Interpretation and Obedience: From Faithful Reading to Faithful Living* (Minneapolis: Fortress Press, 1991), p. 188.

RUTH FINDS A HOME: CANON, POLITICS, METHOD

David Jobling

This essay has emerged as an unanticipated but necessary extension of my work on the process of canonization that created the literary entity '1 Samuel'—defined, that is, within a much larger narrative, a 'book' with precisely this beginning and end.[1] Such a book is defined, in fact, only in one of the two received canonical traditions, namely the LXX tradition, continued in Christian Bibles; the alternative Masoretic canon defines a single book of Samuel.

My initial purpose was to focus attention on the beginning of 1 Samuel, and to suggest that our reading of it *as a new beginning* has a profound effect on how we read the larger narrative. At first sight, the difference between the two canonical traditions is of limited relevance to this purpose, since the beginning of 1 Samuel is in both the beginning of a new 'book'. Doubtless because of my historical-critical conditioning, it took me a long time to realize that another difference between these canons is of direct relevance to my purpose, namely the inclusion in the LXX-Christian canon, but not in the Masoretic, of the book of Ruth between Judges and 1 Samuel. The canonical tradition which defines 1 Samuel as a separate book places immediately before it not Judges but Ruth. Thus the immediate context of the beginning of 1 Samuel in the Christian Bibles is in this respect different from the immediate context of the beginning of Samuel in the Jewish Bible.

I shall first discuss briefly, without regard to Ruth, some

[1] 1. For an account of this work, which is preliminary to a book on 1 Samuel, see my 'What, if Anything, Is 1 Samuel?', *SJOT* 7 (1993), pp. 17-31.

aspects of the beginning of 1 Samuel.[2] I shall then turn to the book of Ruth, suggesting significant ways in which the narrative that includes it differs from the narrative that excludes it; one of the effects of its inclusion, I shall argue, is to make us read the beginning of 1 Samuel as even more definitely a new beginning. I shall conclude with methodological reflection on my version of 'the new literary criticism'.

Reading the Beginning of 1 Samuel

There is no *canon* in which the story of Hannah is not the beginning of a new book. For contrast, one could simply propose a new division of the narrative, on some basis or other. But I prefer to try out an existing proposal, made by Martin Noth and developed by Dennis McCarthy. Working within the hypothesis of the 'Deuteronomic History', they of course take no account of Ruth. McCarthy's sections (which might perfectly well be called 'books') include ones which correspond to our Judg. 2.11–1 Samuel 12 (I shall call this 'the extended book of Judges') and 1 Samuel 13–2 Samuel 7.[3] I shall, in fact, keep in play *three* alternative divisions—Masoretic canon, LXX-Christian canon, and Noth–McCarthy.

Judgeship and Kingship

The major issue with which this part of the biblical narrative is dealing, I believe, is the assessment of the relative merits of judgeship and kingship as forms of government for Israel,[4] and I shall begin with the slant that the Noth–McCarthy scheme gives to this issue. Each of its 'books' concludes with a covenant-like passage: 1 Samuel 12, 2 Samuel 7 (1 Kgs 8, etc.). 1 Samuel 12, on the face of it, ratifies Saul's kingship and incorporates it into Israel's system. But how real is this kingship? Its reality is

2. This section is a summary of parts of 'What, if Anything, Is 1 Samuel?'

3. Martin Noth, *The Deuteronomistic History* (Sheffield: JSOT Press, 1981 [1957]), pp. 4-11 and *passim*; D.J. McCarthy, 'II Samuel 7 and the Structure of the Deuteronomistic History', *JBL* 84 (1965), pp. 131-38.

4. David Jobling, *The Sense of Biblical Narrative: Structural Analyses in the Hebrew Bible II* (JSOTSup, 39; Sheffield: JSOT Press, 1986), pp. 44-87.

subject to doubts of various kinds, of which I mention two. First, the literary form of 1 Samuel 12 is that of a *valedictory*, the last words of a great figure, which appropriately mark the moment of transition to something new—just like the last words of Joshua or of Moses. But it proves to be a *false* valedictory; Samuel remains alive for many chapters. So has 'something new' really emerged? The earlier statement that 'Samuel judged Israel all the days of his life' (1 Sam. 7.15) implies that, despite the *supposed* monarchy, judgeship is still in place.

Second, a problem that came more and more to dominate the presentation of the time of the judges was that of unworthy leaders who, far from restoring Israel to the faithfulness to Yahweh on which its very existence depended, themselves threatened that existence by their own unfaithfulness.[5] 1 Samuel 12 does nothing to alter this threat, since it brings kingship under the same conditional covenant that ruled the time of the judges (see vv. 14-15). But the ending of the *next* Noth–McCarthy 'book', 2 Samuel 7, *does* solve the problem, by excluding the kings from the conditionality of the covenant. We finally have here a kingship that has separated itself *theologically* from judgeship. Is a kingship that has not so separated itself a kingship at all?

The Noth–McCarthy division, through the endings of its 'books', foregrounds this problematic of a kingship that isn't one. The Masoretic canon's division between Judges and Samuel, by contrast, puts the issue of the systemic nature of kingship into the background. What it foregrounds is the present ending of Judges, which asserts, without consideration of the nature of kingship, that the lack of a king has negative consequences. So we begin to read the book of Samuel not only expecting a monarchy, but expecting monarchy to improve things. Samuel loses his rooting among the judges, and becomes a 'John the Baptist' of monarchy.

The LXX-Christian canon's creation of '1 Samuel' seems to carry further this tendency of the Masoretic canon. 1 Samuel is best described as 'The Book of Samuel and Saul'. It exactly covers the lifetime of both characters, beginning with the birth

5. Jobling, *Sense of Biblical Narrative II*, pp. 55-56.

of the older and ending with the death of the younger. It has many dramatic scenes of interaction between them, including the first appearance of Saul and the last, posthumous appearance of Samuel. It even presents their names as somehow related, and confusable. The literary effect of the creation of the book is to exploit the intertwining of these two lives in order to foreground their relation *to each other*, and to *background* not only Samuel's identification with the judges but also Saul's identification with the kings.[6]

Hannah and the Presentation of Women

The overwhelming tendency in scholarly literature is to read Hannah's song, in 1 Sam. 2.1-10, in relation to what follows rather than what precedes it. I believe this is the consequence of our instinctive reading of her story as a beginning, rather than of any real indicators in the song. The notion that v. 10 anticipates monarchy seems to me dubious, since the wording suggests celebration of an *existing* monarchy. This celebration in any case constitutes a major problem, since it stands in stark contrast to most of the rest of the song, which rejoices in Yahweh's liberation of the oppressed, including women, in terms compatible with a theology of revolution.

Whether one reads the song forwards or backwards makes an enormous difference to one's sense precisely of this issue. Polzin reads it as one of the songs of the Masoretic Samuel (cf. 2 Sam. 1.19-27; 22.2-51), and in particular he alleges extensive parallels between it and 2 Samuel 22.[7] Reading Hannah's song along with these songs of David leads Polzin, not surprisingly, to affirm strongly the monarchical aspect of Hannah's song. But

6. I have devoted a major essay to each of these issues: for the first, see n. 4; for the second, *The Sense of Biblical Narrative: Structural Analyses in the Hebrew Bible I* (JSOTSup, 7; Sheffield: JSOT Press, 2nd edn, 1986), pp. 12-30. Robert Polzin (*Samuel and the Deuteronomist: A Literary Study of the Deuteronomistic History. Part Two: 1 Samuel* [San Francisco: Harper & Row, 1989], pp. 26-30) seems to identify essentially these issues—whether there is adequate justification for kingship, and how Yahweh could abandon *Saul's* kingship—as the most critical in 1 Samuel.

7. *Samuel and the Deuteronomist*, pp. 31-35. I have critiqued these parallels in 'What, if Anything, Is 1 Samuel?'

to read it as one of the songs in the Noth–McCarthy 'extended book of Judges' results in a diametrically opposite tendency. In Judges 5, the song of another woman, Deborah, celebrates the sweeping away of the vaunted power of kings by the waters of Kishon. And in Judges 9, Jotham's anti-monarchical fable, though a man's song, finds its fulfilment in a woman's assassination of a king! The focus of our reading of Hannah's song, in this context, will be on the social revolution that *gets rid of* kings, and particularly on the revolutionary role of women.

Besides her song, another aspect of Hannah's story requires our attention. She seems at first sight very much defined by her family situation as the favoured but barren wife; not only her trouble, but also her eventual reward, are within this framework—we last hear of her as the mother of six (2.21). But *within* the family she acts with a striking independence, above all in her assumption of complete control over her firstborn son. At no point is any question raised as to her right to make her own vow to Yahweh regarding this child, and to carry it out.

These two issues—family and monarchy—are taken up in a good deal of recent feminist work bearing upon the role of women in the transition from judgeship to monarchy. Some of the work sees this transition primarily in literary terms—more or less, the transition from Judges to Samuel—while some of it suggests the historical framework of an actual transition to monarchy in Israel.[8] For present purposes, I shall confine myself to the primarily literary approaches.

Mieke Bal's thesis, in *Death and Dissymmetry*, is that the book of Judges is the literary product of a struggle over the transition from one pattern of kinship/marriage to another—from 'patrilocal' (the husband moves to the wife's father's house) to 'virilocal' (the wife moves to the husband's house; Bal adjusts the customary anthropological terms 'matrilocal' and 'patrilocal' to ones based on the wife's perspective).[9] She suggests that

8. For a review of this work, see my 'Feminism and "Mode of Production" in Ancient Israel: Search for a Method', in *The Bible and the Politics of Exegesis: Essays in Honor of Norman K. Gottwald on his Sixty-Fifth Birthday* (ed. D. Jobling, P. Day and G.T. Sheppard; Cleveland, OH: Pilgrim Press, 1991), pp. 239-51.

9. *Death and Dissymmetry: The Politics of Coherence in the Book of Judges*

patrilocal marriage provides women with *relatively* wider options than virilocal—not that it is beneficial for women, but that, among the various forms of 'patriarchy', there are relative differences which it is meaningful and necessary to analyse.

Regina Schwartz has recently discussed the role of stories featuring women and sexuality in the accounts of the incipient monarchy, arguing that power over women expresses political power.[10] She concentrates on three stories of David's women (Abigail, Michal, Bathsheba), noting in each case how it is a man (Nabal, Paltiel, Uriah) who appears as the 'victim', the woman being only a pawn in the male power-struggle.

It is fascinating to read Hannah in relation to the alternatives offered by Bal and Schwartz. Hannah seems fully integrated into the *virilocal* household, her life rotating about her husband, so that one might see her as the forerunner of Schwartz's women, mere pawns in patriarchal/monarchical games. Yet her assumption of control over her son—much more compatible with patri- than with virilocality—invites us to read her story as another episode in the struggle Bal perceives in Judges. In this connection, of course, we will not miss the link between Hannah's story and that of 'the Levite's concubine', so central to Bal's case. Both stories—a mere three chapters apart in the absence of Ruth—begin with 'a certain man of the hill country of Ephraim' (Judg. 19.1; 1 Sam. 1.1).

Ruth in the LXX*-Christian Canon*

The inclusion of the book of Ruth seems to me to extend the tendency I have discerned in the separation of Judges and Samuel, and in the creation of 01 Samuel: namely, to legitimize kingship, and specifically David's kingship.

Judgeship and Kingship

Both the opening and the closing words of Ruth confirm this tendency. The opening, 'In the days when the judges ruled', makes that time seem remote, a quite separate era from that of

(Chicago: University of Chicago Press, 1988), esp. pp. 85-86.

10. 'Adultery in the House of David: The Metanarrative of Biblical Scholarship and the Narratives of the Bible', *Semeia* 54 (1991), pp. 35-55.

the books to follow. The closing reference to David stands in sharp contrast to the closing of Judges, with its 'no king in Israel'; in fact the inclusion of Ruth and the creation of 1 Samuel together result in a series of books that *end* with (a) the urgent need for monarchy (Judges), (b) the announcement of the coming of David, founder of the 'true' monarchy (Ruth), and (c) the resolution (through Saul's death) of the complication of an *alternative* monarchy to the true one (1 Samuel).

But Ruth seems to me to have a profounder impact on the canon than this. For this book achieves *exactly the same journey* that 1 Samuel does—from 'the days when the judges ruled' to 'David'. Before we ever hear of Samuel or Saul, we know that the ground covered by their joint story can be covered without these two figures, much more briefly and also much more pleasantly; instead of the dark theological intricacies of a conditional covenant, there can be a sweet pastoral story that passes from famine to plenty, from death to birth.

This idea of a canonical alternative, or short-cut, is one I wish to pursue further. Ruth alludes (in 4.12) to the story in Genesis 38. The point of the allusion is not just the general theme of fertility, but the contribution *both* stories make to the genealogy that proceeds from Judah to David and his house. When one thinks of these two stories together, one thinks readily of 'intrusion'. Just as Genesis 38 famously intrudes into the Jacob–Joseph story, so is Ruth intruded, in one canonical tradition, into the story of Israel. In the Genesis story, ch. 38 comes immediately after the taking of Joseph to Egypt (37.36), which is the first intimation of the main theme of chs. 39–50, the descent of Jacob and his family into Egypt. Narratologically, there is no way that ch. 38 can be made to fit into this larger story,[11] and thematically it has to do with settling in Canaan. I believe that it represents a canonical *alternative* to 'going down to Egypt'; and I further suggest that Ruth confirms this alternative, belonging to a view of the past that does not include having been 'brought up from Egypt'. These 'intrusive' elements *subvert* the main *Heilsgeschichte* story, hinting that we could do without exodus,

11. It shows Judah as the independent head of a household, and easily old enough to be a grandfather, whereas in the Egypt story he is still part of his father's household.

conquest, Moses, and all the elaborated (conditional) theology associated with this story, and make do perfectly well with a canon consisting of Genesis up to ch. 38, and then jumping to David via Ruth. They seem to me to betray the sense of an 'alternative' past, consisting of the first ancestors, the dimly remembered judges, and the genealogy of David's house.[12]

The Presentation of Women

It is of particular interest for feminist reading that what separates Judges from 1 Samuel, and Bal's women from Schwartz's, is a *woman's* story, and even a woman's *book*. But feminist exegetes are divided about Ruth. Some are enthusiastic about the book, seeing Ruth as a strong, independent character, and her relationship to Naomi as an example of voluntary female bonding unique in the Bible; others see Ruth and her book as subserving male agenda.[13] I am more convinced by the latter. I have suggested that the pastoral cast of the story diverts attention from the issues being dealt with, and I also suspect that some features attractive to a certain feminist reading are a skilful

12. It is, of course, widely held that the Jerusalem theology, centered in the David–Zion complex of traditions, paid little attention to the exodus tradition; also that the Davidic tradition has an affinity to Genesis, at least the parts ascribed to J. My interest here lies neither in such historical hypotheses, nor in the question of literary sources that may have fed into the development of the canon, but rather in the literary effects whereby the canon seems both to uphold and to undermine the exodus theology.

Perhaps other similar elements are to be found. Judg. 1.1-21, likewise concerned with the Judahite genealogy, is also strikingly intrusive, belonging as it does to material set off by resumptive repetition—the events of Josh. 24.28-31, including the death of Joshua, are repeated in Judg. 2.6-9.

Is not some such sense of the past implied in the early chapters of Chronicles? 1 Chronicles begins its history, as opposed to genealogy, precisely at the point when 'Yahweh...turned the kingdom over to David' (10.14)—the account of Saul's death serves merely to close the book on any alternative history. The preceding genealogies give first place to that of Judah (chs. 2–4), within which, despite the parsimony of the narrative allusions in these chapters, the Gen. 38 story rates two full verses (2.4-5)!

13. 'It is the reader's task to determine whether this book affirms Ruth or ultimately erases her' (Amy-Jill Levine, 'Ruth', in *The Women's Bible Commentary* [ed. C.A. Newsom and S.H. Ringe; Louisville, KY: Westminster/John Knox, 1992], p. 79). Cf. Levine's whole introduction.

sugaring of the pill; the whole shape of the book allows women to take the initiative and make the plans—but only until the time comes for the real decision-making, when men take over (contrast ch. 4 with chs. 1–3).

In Bal's terms, the book reads like an apology for virilocal marriage. The choices of the two daughters-in-law in ch. 1, when placed directly after Judges, function to mark the dividing of the ways between patrilocal and virilocal marriage. Orpah chooses to stay in her own place, rather than be associated with her late husband's family—indeed, it is to her *mother's* house (1.8)[14] that she returns—and disappears from the canonical story. Ruth associates herself with her husband's family, and the *result* of her successful insertion into the virilocal system is that she participates in establishing the monarchy.[15] The ownership of women, as a legal issue, is central to the book, not only in ch. 4, where it is incidental upon the ownership of land,[16] but also in 2.5, where the question of ownership is the first that needs to be asked about any woman. In canonical terms, this stress on the legalities stands in contrast to the lawless (though sanctioned) acquisition of women in Judges 21.

Ruth's bond to Naomi, which forms the backbone of the book, and which is presented as voluntary in the book's most memorable words (1.16-17), in fact valorizes a relationship—mother-in-law to daughter-in-law—on the success of which the peace of the virilocal household depends (so that it is perhaps a just historical irony that Ruth's words have traditionally been used to express the bond of virilocal marriage). In this connection, note the charming micro-dialogue that the LXX-Christian canon sets up between Ruth 4.15 and 1 Sam. 1.8 (only a dozen verses apart). Ruth is more to Naomi than seven sons,

14. Cf. Carol Meyers, '"To her Mother's House": Considering a Counterpart of the Israelite *bêt 'āb*', in *The Bible and the Politics of Exegesis* (ed. Jobling, Day and Shephard), pp. 39-51.

15. Cf. Naomi Steinberg, 'The Deuteronomic Law and the Politics of State Centralization', in *The Bible and the Politics of Exegesis* (ed. Jobling, Day and Shephard), pp. 161-70.

16. Boaz's mentioning of the land inheritance to the next-of-kin before he mentions Ruth (vv. 3-5) may be perceived as a rhetorical trick. But note that he puts the land first also in his formal declaration in vv. 9-10.

while Elkanah claims to be more than ten sons to Hannah. Taken together, these two verses convey a message that neither alone adequately conveys: that the virilocal *system*, summarized in the triangle of head of household, his wife, and his surviving mother, is more important than mere fertility, which any system achieves.

On such a reading, the book of Ruth strikingly enacts the transition from Balian to Schwartzian woman. The relatively fluid situation in which Bal finds the women of Judges to live their lives is continued in the fluid situation in which Naomi and Ruth find themselves, and the range of options that seems open to them. But, through the very decisions that these women make, the book strives towards the relatively more fixed position of women under virilocality and monarchy. The decision of 'the Levite's concubine' to leave her husband's house was made when 'there was no king in Israel; all the people did what was right in their own eyes' (Judg. 17.6, 21.25; I use, not without irony, the non-sexist translation of NRSV). Orpah's decision was an analogous one, but it took her out of Israel and out of the story. Ruth's decision to 'find security...in the house of [her] husband' (1.9) made her the ancestor of Davidic kings. The decision of canonizers to find a home for her book between Judges and Samuel has helped make *Hannah* into merely the herald and facilitator of Davidic kings.

Political Reading: A Postscript on Method

This essay assumes a literary criticism much under the impact of feminism, Marxism and psychoanalysis. To be adequately political, a reading needs to reflect politically at every level of interpretation, and above all to be self-reflective, reading *itself* politically. So I shall here reflect on a series of issues that have arisen in the preparation of this essay, and on the way I have tried to deal with them.

What draws me to Judges and Samuel is their overt raising of political issues, and there has recently been, in the context of liberation and feminist interpretation and theology, a considerable quantity and variety of political readings of these books.[17] For me, such readings form the essential context for any work I do with these texts; yet most literary interpreters, even when they attend to the political themes *in* the text, are largely oblivious to political readings *of* the text.[18]

A great deal has been written, particulary within the Marxism *versus* deconstruction debates, on the politics of the *methods* one adopts within the general framework of 'literary' approaches.[19] In the great wealth of recent literary work on 1 Samuel,[20] there are two main trends; but both are responses to the same perception, that this text is a particularly complex and even self-contradictory one. One response is to try to exert control, to solve the text's problems by showing that the diversity expresses a single, though complex, ideological perspective (often equated with the narrator's point of view). The other is to accept the problems, even rejoice in them as creating the interest or fun of the text—to delineate, but not to 'solve', them.[21] This debate

17. For representative readings, in addition to the feminist ones discussed in this essay, cf. George V. Pixley, *God's Kingdom: A Guide for Biblical Study* (Maryknoll, NY: Orbis Books, 1981), pp. 20-24; Bruce C. Birch, *Let Justice Roll Down: The Old Testament, Ethics, and Christian Life* (Louisville, KY: Westminster/John Knox, 1991), pp. 204-12; Alice L. Laffey, *An Introduction to the Old Testament: A Feminist Perspective* (Philadelphia: Fortress Press, 1988), pp. 93-96, 105-107.

18. Polzin, for example, sees a battle going on in 1 Samuel between the ideologies of judgeship and kingship, and throws himself fully into the problems of the ideological commitments of the characters, the narrator, and sometimes the reader; but this 'reader', like Polzin's own authorial voice, has no particular location, and feminist or liberation criticism goes unheard (cf. e.g. *Samuel and the Deuteronomist*, p. 96; but such free-floating 'ideological' discussion occurs throughout the book).

19. On these debates, see David Jobling, 'Writing the Wrongs of the World: The Deconstruction of the Biblical Text in the Context of Liberation Theologies', *Semeia* 51 (1990), pp. 81-118.

20. For a review, see Robert Polzin, '1 Samuel: Biblical Studies and the Humanities', *RSR* 15 (1989), pp. 297-306.

21. To the first line belong Polzin, *Samuel and the Deuteronomist*, and Lyle M. Eslinger, *Kingship of God in Crisis: A Close Reading of 1 Samuel 1–12* (Bible and Literature, 10; Sheffield: Almond Press, 1985); to the second, Peter Miscall, *1 Samuel: A Literary Reaing* (Bloomington: Indiana University Press, 1986), and, to a lesser extent, James S. Ackerman, 'Who Can Stand

calls to mind the final verse of Judges, which contrasts having a king with 'everyone doing what is right in their own eyes'; but this contrast is expressed from a monarchical perspective, so that 'everyone doing what is right in their own eyes' will probably not be a fair description of the alternative to monarchy. Likewise, there are those who tendentiously read the current plurality of biblical methods as a situation of 'anything goes', and who yearn for some controls, for a 'king in Israel'. One's choice of method, it seems, may imply a stance towards the political matter of the text.

I do not in this essay employ specific techniques of structuralism (or of deconstruction, which I regard as a radicalization of structuralism); but I assume and work with its basic tenet, namely that a text has meaning only in its *difference* from other texts. First of all, I am reading the difference between two existing literary works, the two canons. But in setting up a system of differences, structuralism does not confine itself to *existing* literary works; it also posits possible but non-existing ones. If—and the proposition scarcely needs arguing—the given *divisions* of a literary work are a significant part of the work, then the biblical narrative *not* divided into books is a non-existing literary work, as is the Noth–McCarthy 'Deuteronomic History'. I posit these works as a way of saying things about the existing ones. This method, which I find perpetually fruitful, is related to Marxist approaches which interrogate the text for what it *fails* to say.[22]

The most troublesome methodological problems continue to lie in the relationship of literary reading of the Bible to historical hypotheses about it. This essay makes what must seem, to historical-critical sensibility, an extreme claim on behalf of the literary autonomy of the text; I refer to my basing lines of argument on detailed literary effects of the placement of Ruth in the LXX-Christian canon. For from a historical perspective there is a perfectly adequate explanation of this placement: the desire of the canonizers to put as many books as possible (Genesis to

before YHWH, This Holy God? A Reading of 1 Samuel 1–15', *Prooftexts* 11 (1991), pp. 1-24.

22. The classic statement is Pierre Macherey, *A Theory of Literary Production* (London: Routledge & Kegan Paul, 1978 [1966]).

Esther) in chronological order at the beginning of the Bible. So there was no specific *intention*, for example, to put Ruth 4.15 and 1 Sam. 1.8 near each other. A literary approach must insist that this collocation in the existing literary work has a status equal to that of any other literary datum, and that textual history and intention have, at this level, nothing to do with it.

On the other hand, my work relates itself at various points to historical hypotheses, and it is important to bring these relationships to consciousness, and to try to give an account of what is going on. I mention two issues. First, when dealing with the shift in the presentation of women between Judges and Samuel, suggested by Bal and Schwartz, I rather skirt the issue of the relationship between this *literary* observation and the *historical* hypothesis of scholars like Carol Meyers and Naomi Steinberg (based on models from the social sciences) that a shift in Israel from a less to a more statist form of government tended to restrict the options of women.[23] Second, I speak freely of a 'process' of canonical development, and of a 'tendency' in this process; this *historical* thesis, in fact, is what I am mainly arguing in this essay, though my literary observations could stand on their own in some other framework.

In the first case, I betray my anxiety over the possibility that the literature might be considered to 'reflect' the history. This anxiety is justified to the extent that notions of literature as a reflection of reality still reign, with various degrees of sophistication, in biblical studies; on the other hand, there must be something better to do with the literary and historical observations than just juxtapose them. In the second case, I fail to conceal my *desire* for a certain myth of the canonization process; the myth, namely, of an 'original' narrative very dubious towards monarchy, but whose true character the process obscured. In introducing the Noth–McCarthy scheme, I go beyond its value for purely synchronic comparisons, to the possibility that it represents diachronically an earlier way of dividing the narrative, in relation to which the canonical developments constitute a

23. Carol M. Meyers, *Discovering Eve: Ancient Israelite Women in Context* (New York: Oxford University Press, 1988), esp. pp. 189-96; Steinberg, 'The Deuteronomic Law Code'.

'tendency'. I desire this myth for its potential political impact on biblical studies.

I have no general answer to the problem of literature and history. Certain Marxist critics suggest models that I find usable.[24] More importantly, certain feminist critics empower me to develop a style that resists taking the problem *too* seriously, since it belongs to a male model of specialization; Bal and Schwartz, for example, seem not to share my anxiety, and work out ways of studying the biblical literature in significant relation to historical hypotheses about it. What changes the shape of the problem for politically engaged readers is their insistence that we look at everything from the perspective of *our own* historicity, including the historicity of what we do with the Bible, and why. The Bible is a political reality in the present, and nothing we do with it is separate from present reality. The Noth–McCarthy division, for example, which we are conditioned to assess by its ability to account for things we think we know about the *past*, is first of all a *present* thing; my use of it to establish a monarchical tendency in the development of the canon can be stood on its head by asking whether the forming of such a hypothesis does not indicate an anti-monarchical tendency in modern biblical scholarship! It is via a thorough immersion in the problems of the relationship between *our* history and *our* 'biblical text' (the text constituted by the variegated presence of the Bible in our culture) that we need to approach the problem of the relation between *past* history and the *past* existence of the biblical text. This is not a formula to solve everything; the problems of historical analogy between present and past remain immense. But the present essay assumes that this is the right approach in principle.

My very decision to take up the issue of canon has political aspects. Any literary study of the Bible must deal in some way with the issue of the whole and the parts; my inclination, here and elsewhere, to give priority to 'a sense of the whole', rather than to close reading, goes against the main trend. But is it merely a matter of taste? In the case of the Bible, 'the whole' is in some sense the canon, and it is surely as *canon* that the Bible

24. Especially Fredric Jameson, *The Political Unconscious: Narrative as a Socially Symbolic Act* (Ithaca, NY: Cornell University Press, 1981).

has its unique cultural power. Yet there has been very little serious examination of canons as literary works.[25]

The part played by canonical *sequence* in this exercise of cultural power needs further study, but it is hard to believe that it is not a basic aspect of a canon as it enters consciousness and habit. It is probably of greater significance for the Christian canons, which (including the New Testament) purport to tell one story from beginning to end, than for the Jewish canon.[26] So my assumption in this essay, that the placement of Ruth in the Christian canon has an impact on the reading of it and the books around it, seems justified.[27]

Finally, both my decision to pursue the canonical issues and my findings are probably related to the fact that canonization *as such* is a negative category in feminist discourse. In general literary-critical discourse, feminists have pointed out the political dimensions of the creation of a 'canon of great books', which becomes, for example, the authorized scope of university literature curricula.[28] Feminist scholars of the New Testament suggest that its canonization tended to exclude literature by, for, and about women.[29] Establishing a canon of scripture is a major exercise of power, and power characteristically works to further entrench itself.

25. The work of Northrop Frye is the most obvious exception. Cf. also Gabriel Josipovici, *The Book of God: A Response to the Bible* (New Haven: Yale University Press, 1988).

26. I shall take up this issue in an essay on 'The Canon of the Jewish Bible as a Literary Work' in a forthcoming collection (*No King in Israel: Post-Structural Essays on the Jewish Bible* [Sheffield: JSOT Press]).

27. Before we can even consider literary effects, we should not underestimate, given Josipovici's reminder of the importance of the sheer physicality of the Bible (*The Book of God*, pp. 29-36), the effect of Ruth simply as a physical barrier of some pages between Judges and 1 Samuel.

28. E.g. Sydney Janet Kaplan, 'Varieties of Feminist Criticism', in *Making a Difference: Feminist Literary Criticism* (ed. G. Greene and C. Kahn; New York: Methuen, 1985), pp. 37-58.

29. D.J. Good, 'Early Extracanonical Writings', in *The Women's Bible Commentary* (ed. Newsom and Ringe), pp. 383-89.

TRACING THE VOICE OF THE OTHER: ISAIAH 28 AND THE COVENANT WITH DEATH

Francis Landy

'We have made a covenant with death...we have concealed ourselves in illusion' (Isa. 28.15).

1. Woe, O crown of pride of the drunkards of Ephraim
and the fading flower [m.] of the beauty of his splendour,
which is at the head of the valley of fat things,
those hammered with wine.
2. Behold one strong and mighty to my Lord,
like a flood of hail, a storm of destruction,
like a flood of waters, powerful, overflowing,
he has cast down to earth by hand.
3. With feet they tread down/are trodden down,[1]
crown of pride of the drunkards of Ephraim.
4. And the fading flower [f.] of the beauty of its splendour,
which is at the head of the valley of fat things,
shall be like a first fig when it is not quite summer
which, as soon as the one who sees it sees it,
no sooner is it in his hand than he swallows it up.

5. In that day shall YHWH of Hosts be as a crown of beauty and a
    diadem of splendour for the remnant of his people.
6. And as a spirit of justice for the one who sits on justice,
and as power (for) those who turn back war at the gate.
7. And also these have raved with wine,
have tottered with drink,
priest and prophet have raved with drink,
are swallowed up by wine;
they have tottered from drink,

1. I read the grammatically anomalous *b*^{e}*raglayim tērāmasnâ* as a conflation of 'feet shall tread down' and 'with feet it (the crown, etc.) is trodden down', thereby achieving an ellipsis, a collision of active and passive experiences. For discussion of the phrase and emendations thereof, see Wildberger 1982: 1043 and Watts 1985: 360.

they have raved in vision,
they have uprooted judgment.
8. For all tables are full of vomit, shit,
without cease.

The Covenant with Death

A covenant with death is the ultimate absurdity, since death alone brooks no compromise; yet every post-edenic human endeavour is an attempt to make a deal with death, to postpone it, to render it malleable, to humanize it. The motif of the game with death, from Gilgamesh to Ingmar Bergman's *Seventh Seal*, is both a symbol for all human transactions with death, and, as play, a displacement, into wish-fulfilment.[2] We are drawn into the game not just because Death might be defeated, but because the game itself offers a space for fascination, for the suspension of closure. Into the mutual pleasure of the game is invested, not only the hope of immortality, but an invitation, that death lose its otherness. Thereby the relationship with death enters human reflection.

The covenant with death is antithetical to the covenant with YHWH, inscribed in the flesh but also in the text of the Torah and in a traumatic history. Much biblical polemic is dedicated to sustaining this opposition. It is not so much my intention to subvert it, to show that YHWH is a God of death as well as of life, as to investigate the effect of the opposition and convergence of the two covenants on writing, and with it prophetic writing.

The covenant with death is paradoxical as well as absurd, in that it is a bond with death that frees one from death. It entails, so the parallel passage in 8.19 suggests, a turning to one's ancestors, to the past, against a terrifying future. The past is the realm of memory, of the textual subconscious, whose revanchism is expressed in whispers and sighs (*hamᵉṣapṣᵉpîm wᵉhammahgîm*), in half-erased traces of language. The dead both

2. Symptomatic of scholarly discomfort with the metaphor in Isa. 28.16 is the attempt to find a concrete reference for it, either in terms of an actual cult of death or as an allegorical designation for an alliance with Egypt. For a general discussion, see Clements 1980: 229 and Wildberger 1982: 1073-75. See most recently van der Toorn 1988.

refuse to die, haunting our dreams and our imagination, and they tell us that we will die. We all have a compact, or at least a date, with death.

Poetry

Poetry plays with alternative worlds, with the infinite combinations of sounds and images, with the transition between narcissistic omnipotence and the terror of finitude. It is a game with language and the world that constitutes preeminently a 'transitional object', transitional between mother and child but also between union and separation (Winnicott 1972). The spoken or unspoken other player in this game is death, not only in that poetry tries to make sense of the world despite death, nor in that it seeks immortality for our voices and our lived experience, but in that it passes between being and non-being, what can and cannot be said, the thought of being and the unthought.[3]

Poetry, as player, is the antagonist of death. Perhaps it alone makes no covenant, refuses to compromise, with death. Writing otherwise is technology, *techne*, and thus, according to Derrida, an instrument of totalitarian control and impersonality.[4] 'We are all in peril of becoming thing' (Owen 1989: 150). Poetry fills or at least marks the gap between human being and thing with its possibilities of metamorphosis. For Derrida, a poet is a

3. For Heidegger, poetry—which for him encompasses every work of art and language—opens the cleavage, the difference, in the thought of Being to what cannot be thought therein, to the Unsayable. This is the site of the Holy, beyond the Givenness, or 'There isness' (*Es Gibt*) of Being. But if poetry marks a trace of the holy, it also sounds the knell of the philosophical subject (Taylor 1987: 37-58 [58]).

4. For Derrida, writing as *techne* is a *relation* between life and death (1978: 227). But the machine, by which Derrida means the representation of the psychical apparatus, is dead. As representation, writing is death. Derrida also reverses this: 'Death is representation'. For Derrida, writing consists of traces, each one of which is the site of the disappearance of the self. The erasure likewise is death (1978: 230). For a reflection on the possibility of non-totalitarian language, see his essay on Levinas, 'Violence and Metaphysics' (1978: 79-153 [148]).

metaphorical Jew,[5] who crosses the Jordan, from death to life. The river-crossing is the date of composition, of circumcision, the wound in the flesh that enables one to join the community of poets and Jews, whom Derrida terms 'autochthons' of language (1978: 66). For Celan, the primary event, that date that recurs always in his poems, is the trace of that which is now nothing, no one, 'No one's rose', ash; its God (*du*) likewise can only be experienced as smoke, as an intimate disappearance.[6]

The covenant with God is also distance from God, a distance the poet fills with ambiguous language. The enigma of poetry conceals the mystery of God, and everything else. The dissimulation of God's face allows us to speak (Derrida 1978: 67); the pleroma is disrupted to open up a space for self-questioning, for thought.[7] The fragmentation of the parousia that, according to Merleau-Ponty (1968: 152), results in the folding over and invagination of being, permits a dialectic of death and life, absence and presence, as the movement or trace of poetry.[8] According to Derrida (1978: 68), poetry originates in the breaking of the tablets, in a primal catastrophe, which is also a catastrophe within God. Between the shards of the tablets, the

5. Derrida quotes Marina Tsvetayevna: 'All the poets are Jews' (1986: 338). Similarly, in his essay on Jabès, Derrida writes that for Jabès 'the situation of the Jew becomes exemplary of the situation of the poet' (1978: 65).

6. References are to Celan's 'Psalm' to God as *Niemandsrose*, in the collection also called *Die Niemandsrose*, and to his poem *Am weissen Gebetriemen*, in *Atemwende* (Celan 1980: 142, 196). See the discussion in Derrida 1986: 333-34.

7. Derrida suggests, in his essay on Levinas, that within philosophical thought God is named within difference, and as difference. God, in this discourse, is both Life and Death, All and Nothing (1978: 115-16). To the objection that Levinas opposes philosophical discourse, Derrida responds that nonetheless he engages in it, to go beyond it, to achieve 'a certain silent horizon of speech' (1978: 117). One is reminded of Barthes' contrast between the ceaseless polemic engaged in by 'texts of pleasure' and the peace afforded by texts of *jouissance*. It might be noted that in Kabbalah—one of Derrida's many occult resources (1978: 74)—*'elohim* is the self-questioning, differentiating *sefirah*, Binah.

8. For Merleau-Ponty and Blanchot, death is absent presence or present absence (Taylor 1987: 96). For Blanchot (1982), all writing is *écriture du désastre*. For the trace as the movement between absence and presence that constitutes the world of sense, see Taylor 1987: 88.

possibilities of interpretation and recombination ramify. In those spaces, poetry becomes polysemous, inexhaustible and discontinuous. With the breaking of the tablets, according to the Midrash, death re-entered the world;[9] poetry is not only a resistance to that death, but always limited, fissured and impressed with it. Its ambiguity, as a sign of the covenant and of its breach, is also that of its success or failure. It marks the traces of that which has already vanished, but can do so only through the displacements and opacity of language. Its subject, as Derrida (1986: 332) says, is the unreadable: 'The unreadable is readable as unreadable, this is the madness or fire that consumes a date from within...' The indecipherability of the poem—its function as caesura—the ash, for example, at the centre of Celan's poems, threatens its words with illusoriness, with alienation. In Celan's poem, *Fadensonnen*, songs are to be found paradoxically only on the other side of that divide, leaving the words of the poem itself voiceless.[10]

The relation with death is perhaps more intimate. Poets are attracted to death: 'Now more than ever seems it rich to die'. Eros and Thanatos, according to Freud, are inextricable (Taylor 1992: 15); the erotic desire to unite with the other concludes in the non-differentiation of death. The poetry that articulates the loveliness and order of the world, and the pathos of its disintegration, verges, beyond Freud's pleasure principle, with the poetry that seeks regression, into the song of the nightingale, and relief from the pressure to make sense and the narcissistic play of mirrors between self and other.[11] The ambiguity of the

9. Cf. the opinion of R. Jose in *Mekilta de R. Ishmael*, Bahodesh IX (II, 276), and Nachmanides on Exod. 32.6. The loss of immortality as a result of the sin of the Golden Calf is a recurrent motif in the Zohar.

10. '[E]s sind/noch Lieder zu singen jenseits/der Menschen' ('there are/still songs to be sung on the other side/of mankind' (trans. Hamburger).

11. For Freud, that which lay beyond the pleasure principle was in fact its logical extension, since the pleasure principle consists in the resolution of tension, culminating in the homeostasis of death (Taylor 1992: 13-14). Lovers endlessly see themselves reflected in each other (Owen 1989: 126-27); the interplay of projection and introjection is not only a Freudian cliché, but the dynamic of Hegel's speculative philosophy, from which Bataille sought to escape through his cultivation of radical heterogeneity (see Taylor's essay on Bataille [1987: 115-48] and his discussion of Bataille's

poet corresponds to Lacan's split in the subject, between the conscious and the unconscious, the particularity of the individual, attached erotically to the world, and the universality of the matrix, neither being nor non-being, that is both its irrecoverable past and its inevitable future (Taylor 1987: 89-96; cf. Kristeva 1982).

The community needs poets to tell it the truth and to sustain its illusions; poets are accused of inventing fabulae, and of a radical critique that threatens social foundations. The other side of this double bind is that poets can tell the truth only through illusions, through the intricate art of replacement and opening gaps in the texture of language.[12] In poetry, truth and illusion, the real and the fantasmic, are interdependent, in the playspace composed of transitional objects.

Isaiah, Poetry and Death

Isaiah's poetry preeminently concerns death, the defences against death, and some opening of the horizons beyond death. This death, as for Rilke[13] and Celan, is universal; the turning to the ancestors is a reflex of the fear that there will be no descendants. From the vanishing future one buries oneself in the past.[14] The imagination of collective death is also a collapse of the poet's world. The problem in Isaiah is to find a language for the failure of the symbolic order, which is also an inversion of the covenant, and for the new voices that he hears, a poetry that will transmit in our language the 'other language' of God (28.11). The difficulty is compounded by the contradiction between the desire to communicate and the prohibition of comprehensibility, established paradigmatically in the call vision

celebration of the death-instinct [1992: 25-29]).

12. Owen's study *Mi-lou* is a sustained modern attempt at a Defence of Poetry against Plato's charge of immorality; I borrow greatly from it.

13. For the phenomenon of mass death, in the First World War, as an unbearable problematic for Rilke, and through Rilke, for Heidegger, see Wyschogrod 1985 (esp. ch. 1).

14. According to Lacan, in the repetition complex the past is projected into the future, whose ultimate horizon is death (1966: 318). The complementary movement is from the future to the past. See also Taylor 1987: 96.

(6.10). Isaiah is a poet who must not succeed, whose success is failure. Hence the alternation of a poetic idiom that is traditional, sophisticated and compressed, with one that is strange, naive and diffuse. Isaiah combines poetry of extraordinary density and polysemy with exorbitant repetition and syntactic fragmentation. This results not only in extreme difficulties of interpretation but in a dialectic of structure and anti-structure.[15]

The dialectic corresponds to that between texts of pleasure and texts of jouissance, to employ Roland Barthes' terminology (1973; cf. Landy 1991).[16] My interest in this essay is in the points of transition between jouissance and pleasure; where the poetic excitation is engendered; the resistance to significance; whether jouissance and pleasure are congruent or antithetic. Texts of pleasure reinscribe a culture; texts of jouissance are anti-cultural, and only irrupt in the interstices, in the gaps of texts of pleasure. An ideal analysis would describe the pleasure given by the text, its order, its accumulation of sensory and hermeneutic touches, and demarcate the excitement, the discharge of tension, of jouissance. This is what is mystical in poetry, the fusion with the voice beyond any particular significance or subjectivity, in which intoxication is also peace, the cessation of polemic.[17] This point of fusion is mysterious, the residue that remains when the text has been interpreted.

15. For a good discussion of the complexity and instability of Isaiah's similes, see Exum 1981.

16. Barthes' opposition between pleasure and jouissance conforms to Nietzsche's distinction between Apollonian and Dionysian discourses. Dionysus, the god of ecstasy, represents the shattering of individuation and the union of Eros and Thanatos. For a discussion of the Apollonian–Dionysian polarity in relation to the Song of Songs, see Landy 1983. Taylor (1992: 18-33) notes the parallel between Nietzsche's interrelated opposites and those of Freudian psychology (ego/id, conscious/unconscious, eros/thanatos), and the influence of Nietzsche on Bataille and Heidegger, and thus the whole modernist movement.

17. Barthes 1973: 15, 49 and *passim*. The search for a discourse that is not violent, that does not seek to impose itself on the other, characterizes Levinas's work and Derrida's essay on him ('Violence and Metaphysics', 1978: 79-153). For Heidegger, the 'most venturesome' poets take us to the realm of the holy, that is 'nothing human', but this is also extreme passivity, openness, to the traces of stillness beyond sound (1975: 141, 206-207).

Jouissance and pleasure, vocalic play and meaning, constitute the split voice of the poet, who speaks for the society he condemns, who speaks in his own voice, as part of the human community, as well as that of God.[18] But the latter voice is also ambiguous; it condemns him and his world to death, and is the voice of life. The prophet's response to the words he speaks may combine horror, and thus align him with the community's rejection of reality, with the desire for knowledge, no matter how terrible. This may be exemplified in v. 22, where the destruction (*kālâ wᵉneḥerāṣâ*) is counterbalanced by the privilege of hearing, containing the reverberations of catastrophe, among the deaf. Knowledge of God is both an ultimate horizon (as in Isa. 11.9) and transgressive. In our chapter, knowledge is always in question: 'To whom will he teach knowledge?' (v. 9), and what will he teach?

There are two especially difficult problems. The first is the relation of the power of the text to its clumsiness. Alongside poetry of very great sophistication we find ponderous vacuity, as in v. 21: 'to work his work, strange is his work; to perform his labour, peculiar is his labour'. Such 'bad' poetry seems integral to the poetics of Isaiah, to the breakdown of symbolic order. The crudity marks the encounter with the Real, beyond aesthetic construction.[19] The issue then is of the relationship of beauty and ugliness in establishing the tone of Isaiah,[20] duplicated, at the symbolic level, in the trajectory between beauty

18. The ambiguity is amplified in Celan's lecture, *Meridian*, cited in Derrida's essay, 'Shibboleth': '[The poem] speaks always in its own, inmost, concern...But I think...that it has always belonged to the hopes of the poem...to speak in the concern of an Other—who knows, perhaps in the concern of a *wholly* Other' (1986: 311-12).

19. See Lacan's lecture, 'Tuché and Automaton' (1977: 53-64). Lacan insists that psychoanalysis is essentially concerned with the encounter with the Real that is behind all signs, all psychic repetition, and with the real as it is experienced as encounter, or trauma. For Freud, the quest for the Real was also for death, in the form of the chthonic Diana, or the mother goddess (Lacan 1966: 412). Lacan contrasts the quest for the Real in psychoanalysis with the apprehension that psychoanalysis is a form of idealism. See also Taylor 1987: 83-90.

20. See the discussion of the relationship between tone and tension in Derrida 1982: 68-69.

and excrement, sense and non-sense, that we find in our chapter.

The second problem is more elusive. We are used to the notion of holistic reading, yet it is impossible to read Isaiah except in fragments.[21] The familiar accommodation, that we are dealing with the final form of the text, with Isaiah as a retrospective composition, preserves academic peace at the price of both coherence and fragmentation as inherent processes. The reductionism that assigns everything discordant to a different redactional level is troubling, because it results in such inferior poems. Little snippets, struggling under the weight of accretions and annotations, clash briefly with other snippets in a textual mêlée, congealing under the wintry gaze of a final editor. If the metaphorical power of poetry results from the interplay and juxtaposition of different linguistic levels and experiences, including different genres, then the power of poetry, its jouissance, is a priori excluded.

There is a further consequence. We have become accustomed to regarding the composition of the poetic corpus as a collective endeavour. Indeed, it is impossible not to posit communities of reception, supplementation, deletion. The scribal community, especially if linked to sacred authority, is an instrument of critical conservatism. The redaction-critical model proposes tidy poems produced by tame poets, each contributing to the canonical nest. What interests me is the voice of radical alterity,[22] which cannot be reduced to a tradition or political conformity, the powerful and utterly distinctive voice that I hear when I read the text. This is a fact of the reading experience that must be accounted for, especially if, as I suspect, I am not alone in my experience. The individual voice is responsible for Isaiah's status as one of the world's great poems/poets.[23] It is a voice that

21. This, however, is true of any text, as Roland Barthes points out throughout his oeuvre, most notably in *S/Z* and *Le plaisir du texte*.

22. The term *alterity* is germinal to Levinas's challenge to western totalitarian thought (cf. Taylor 1987: 194, and *passim*).

23. It could be objected that anthologies, such as Psalms, the Manyoshu, or the Greek Anthology, also become literary classics. Even in anthologies, however, there are individual voices; a collection, such as the Manyoshu, indeed often seeks out the best and most strikingly individual poems of an age. Even in a corpus as conventional as Psalms, there are self-questioning,

surfaces explicitly from time to time in the text, (e.g. in v. 22), foregrounding the poet's experience as one focus of attention.[24] Only thus can one account for the strangeness of the poetry. Conformist poetry, produced by pressure groups, would not be incomprehensible. Unless one were to suppose a surrealist or dadaist collective.

Illustration: Verses 1-8

My point about the reductive nature of redaction criticism may be illustrated in exemplary fashion by vv. 1-8.[25] Critics universally separate vv. 5-6 from vv. 1-4, and regard them as a very late insertion. Even Exum, who shows how closely integrated they are together, concedes this position; this is because it does not really concern her. She is interested in the final form of the text, and is prepared to be agnostic about its development (1982: 109, 116-17). In my view, however, if one eliminated

critical, and personal voices. I discussed some examples in an unpublished paper delivered at the International SBL in Vienna, 1990, entitled 'Deconstruction in Psalms', and in Landy 1991: 57-58.

24. L. Alonso Schökel (1987: 150) describes Isaiah as a classic writer in that, in contrast to Jeremiah, he does not insert himself into his poems. It seems to me that this judgment must be qualified.

25. The choice of vv. 1-8 might need some defence, since most commentators group vv. 7-8 with the next section. This is probably the consequence of another piece of received wisdom, namely that vv. 5-6 are a later insertion. The only grounds for a division at the end of v. 6 is that vv. 7-8 seem to refer to Jerusalemites, while the subject of vv. 1-4 is Ephraim (in vv. 5-6, 'the remnant of his people' presumably also signifies Judahites, but that's a different story). However, there are no verbal or thematic links between vv. 7-8 and 9-13, and there is a clear syntactic break. Indeed, it requires the invention of a completely fanciful story to connect vv. 7-13. My reasons for reading vv. 1-8 as a poetic unit are: (a) that they form a syntactic unit; (b) that they share the motif of drunkenness, which disappears for the rest of the chapter; (c) there is at least one metaphorical link, the verb *bl'*, 'swallow'. As Exum (1982: 109-10) points out, Jerusalemites do not become the explicit addressees until v. 14; vv. 7-13 (or 5-13) thus have a transitional function. It is with some discomfort that I write this footnote, since I am a believer in the Barthesian principle of the reader's responsibility for dividing the text into manageable bites; however, it seems to me that the breaks in our text are unusually clearly marked.

vv. 5-6 one simply would not have a poem. One moves from the false crown of vv. 1-4 to the true crown of vv. 5-6, and thence to the dissipation of sacred authority in vv. 7-8. Verses 1-4 and 7-8 match each other; at the centre of present disintegration is a glimpse of a different order, a different reality. The elimination of the centre, moreover, creates a different image of the poet(s)/ prophet(s) responsible for the text. One whose catastrophic vision is transposed into its opposite is clearly more interesting, complex and exciting than two poets who are monochromatically positive or negative.[26] A construct of the poet as interesting will invite more engaged readings than a construct of the poet as boring or uniform.[27] If the construct is of a liminal personality, such as a prophet,[28] it may lead to an experience of jouissance.[29]

The primary symbol in the passage is drunkenness. Drunkenness in Isaiah is a paradigmatically inane defence against death, as the *carpe diem* motif, 'Eat and drink, for tomorrow we die' (Isa. 22.13), suggests. Drink fends off but also anticipates death, anaesthetizing fear and rendering the subject unconscious. In Isa. 5.14, the company of drunkards dances into death. Drink is a symbol, however, for symbolic reversal: through alcohol, the symbolic order is breached; linguistic and social regression becomes the condition of bliss. Individual boundaries blur, as do those of class and value, depression and mania. Dionysus is

26. That the division is motivated by a disbelief that ancient people could be complex is indicated by the paucity of arguments adduced for making it. Petersen (1979: 107), for example, holds that the key argument for regarding vv. 5-6 as secondary is that they disrupt the continuity of discourse between vv. 1-4 and 7-8. The most detailed discussion is that of Vermeylen (1977: 388), for whom the use of the same vocabulary for negative and positive visions suggests different origins. The circularity of both these arguments is evident.

27. It should not be necessary to argue that every reading of a text is a construction—the semiotic process is always circular (see generally, Eco 1979). The imputation of a different author for every point of view implies, however, a second degree of construction: not only do we construct the author(s), but also the text(s).

28. For a classic description of liminal personalities, see Turner 1977, esp. the essay 'Liminality and Communitas' (94-130).

29. Barthes (1973: 67), however, proposed that absolute boredom may be conducive to jouissance.

cultivated at the centre of the society he threatens to destroy.

The passage is about beauty on the edge of destruction. The beauty of the splendour of Ephraim in the first verse (*ṣᵉbî tipa'rtô*) is transformed into the excrement and vomit of the last. The feast has become faeces and regurgitation; the ironic recycling of food combines with the retching of the stomach to suggest not only circularity but a turning of the inside out. Kristeva argues that beauty and disgust are the lining of the narcissistic space in which the baby separates itself from its mother; disgust heralds the approach of the *abject*, the object, which cannot yet be conceived as such, which is cast out (ab-jetted) of the self so that the self can be autonomous. Corpses, faeces, vomit are all symbols of the abject. The ultimate source of abjection, according to Kristeva, is the mother, whose power is also a capacity to destroy. Total dependence on the mother is infinitely threatening; in rejecting the mother, the infant rejects also the past. The other side of abjection, then, is desire.[30] The abject is constituted by repression, breached by jouissance (1982: 9-14).[31] In the centre of itself, an intoxicated society—intoxicated presumably metaphorically as well as literally—discovers abjection. This is especially fraught in the case of a sacred people, whose code of purity and impurity repeats the drama of abjection, casting out the defiling other, associated with death and the fertilizable feminine body, in order to establish its boundaries.[32] Coprophilia foreshadows the overture to death in v. 15; the orgy erases the differences between life and death, food and waste, conspicuous consumption and destruction. If the inside is turned out, the inner lining of the 'glorious beauty' of v. 1 is the archaic mother we thought we had excluded.[33] But this is also a

30. Kristeva (1982: 9-10) stresses the ambiguity of the *abject*: 'a composite of judgment and affect, of condemnation and yearning, of signs and drives'. The other side of the abject, according to Kristeva, is the sublime, achieved through sublimation: 'The abject is edged with the sublime. It is not the same moment on the journey, but the same subject and speech bring them into being' (1982: 11).

31. The repression is in fact 'primal repression', the principle of repression itself.

32. Kristeva devotes a chapter of *Powers of Horror* to 'The Semiotics of Biblical Abomination' (1982: 90-112). Cf. esp. p. 100.

33. The symptom of abjection is that one becomes abject oneself

reversion from sense to non-sense: from the symbolic order, which assigns things their place, making unmentionables unmentionable, to the anarchic play of the liberated body.

Where is the poet's jouissance? Condemnation might cloak complicity, a prurient indulgence at one remove. Or it might be the jouissance of righteous anger, that destroys the deceptive beauty of a brilliant but perverse culture for the sake of true aesthetics-ethics. That would be the familiar prophetic and divine self-justification. But this might be rationalization, for delight in violence for its own sake. On the other hand, the poet may be allied with the world he condemns. The relationship between alcoholic and poetic intoxication is long and terrible. It might be mimetic, as in v. 7, in which poetic rhythm ludicrously replicates the staggering hierophants. They may converge, as when drunken babble and hallucination stimulate verbal delirium. Poetry may be ascetic, its discipline requiring an attentiveness exclusive of any competition. Isaiah might represent this extreme in his call vision, when he is granted a pure word, in contrast to the impurity of language in which he is embedded (6.5-7).

Displacement prevents the resolution of these contraries. The ideal dominion is displaced into the future, subsequent to the expenditure of violence. The temporal disjunction permits both satisfactions. The beauty of Ephraim is celebrated and ironized, but only through metaphor, synecdoche and repetition. Stylistically, the first four verses are a set of sidetracks, deferments, and syntactic dislocations. The poetic pleasure that plays with images and sounds, imitating the hedonistic insouciance of the world it imagines, is hedged, in the intervals between its tableaux, by the anguish of disaster, and by the flight of the signifiers from the reality they portend. As we will see when we discuss v. 1 in detail, successive phrases, such as 'crown of pride of the drunkards of Ephraim', enable us to envisage the doomed world; each snapshot, each 'fading flower of the beauty of its splendour', is also a sign of closure, complete in its perfection and disintegration. The description distracts us from annihilation, and indirectly alludes to it. In it both moralistic anger, for

(Kristeva 1982: 5, 11). Socially, abjection confronts us with our animality; in our personal archaeology, with our earliest attempts to free ourselves 'of maternal entity' (1982: 12-13).

example at drunkenness, and sensuous delight are diffused inextricably in the intricacies of verbal texture.

Only close reading, paying attention to the patterning of sound as well as meaning, will reveal the interplay of jouissance and pleasure, fracture and articulation, in this passage. It begins with poetic art at its most perfect, with a description of the beauty of Ephraim that is exhaustive, polysemic and self-negating. The long list of epithets is both celebratory, like a throne name, and subversive: 'Crown of pride of the drunkards of Ephraim and fading flower of the beauty of its splendour, which is at the head of the valley of fat things, those hammered with wine'. In this sequence, 'the beauty of its splendour' (*ṣᵉbî tipa'rtô*) alliteratively matches and contrasts with 'the fading flower' (*ṣîṣ nōbēl*); the 'pride' of the 'crown' (*'ᵃṭeret gē'ût*) is implicitly undone by the drunkards over whom it reigns.[34] Drunkenness ill fits a crown, as we know from the words to Lemuel (Prov. 31.4-5); such a king is liable to be a lord of Misrule, and to exemplify carnivalesque inversion.[35] Paronomastically, the pride (*gē'ût*) of Ephraim is neutralized by the 'valley' (*gê'*) of 'fat things' (Exum 1982: 115),[36] while the violence of 'hammered with wine' (*hᵃlûmê yāyin*) induces stillness. Meanings proliferate: 'crown', for example, may be a metonym both of king and personified pride; it may represent the arrogant euphoria of the drunkards or the pretensions of Ephraim.

34. A number of critics hold that the 'crown' is a wreath worn by the drunkards (Wildberger 1982: 1047; Clements 1980: 225) as well as Samaria, surrounded by its fortifications. It is not apparent to me why these possible connotations should exclude reference to a king, except for distrust of polysemy, and a reductive desire to find a particular and concrete meaning for each image. If it speaks of 'the crown of the drunkards', it must, it seems, refer to an actual crown worn by them. Even if it does signify a crown worn by drunkards, or an an attitude of mind of theirs, it would still retain its emotive and symbolic aura. In Prov. 4.9, to which Wildberger turns for evidence, the 'crown' is clearly a metaphor for Wisdom's sovereignty.

35. For a carnivalesque reading of a prophetic narrative, see Garcia-Treto 1990.

36. Various critics eliminate the word play by reading *gê'* as *ge'ê*, following 1QIsa; cf. the discussion in Wildberger 1982: 1042. This seems motivated by little more than a dislike for complexity.

Ṣîṣ may mean 'diadem' as well as 'flower', while *nōbēl*, alongside its primary meaning of *fading*, connotes *folly* (*n*[e]*bālâ*), *drunkenness* (*nēbel*, 'bottle'), and *music* (*nēbel*, 'harp, lute'), all presumably associated with a feast.

Meanings also interfuse, for example into the undifferentiated impressiveness of *ṣ*[e]*bî tipa'rtô*, 'the beauty of its splendour,' which, as Exum (1982: 115) points out, combines the consonants of *'*[a]*ṭeret*, 'crown', and 'Ephraim' (*'eprayim*). The burst of beauty, at the centre of the line, summarizes the total aesthetic/sensual experience of Ephraim; it is expressed poetically in verbal excess and through its lack of an objective referent. It represents a moment of the sublime, which for Kristeva (1982: 11-12) is the other side of the abject. Here, where the lines of metaphor and metonymy meet, structure becomes unstructured. Its counterpart, however, is prophetic anger, in such loaded phrases as 'drunkards of Ephraim' and 'the valley of fat things/ones'; conspicuous consumption is implicated in a greedy and oppressive social system, and non-productive complacence (e.g. 'those hammered with wine'). The transition between *gē'ût*, 'pride', and *šikkōrê*, 'drunkards', produces a momentary shock, that the grandeur is in fact inebriation, allied with phonetic contrast; the harsh texture of the fricative (*š*) and plosive (*k*) in *šikkōrê*, 'drunkards,' is imbued with the intensity and structurelessness of prophetic rage. A similar effect is produced by the contrast between the soft liquids and labials of *nōbēl*, 'fading', and the initial affricate (*ṣ*) of *ṣ*[e]*bî*, 'beauty'. My point is not simply that metaphorical onomatopoeia traverses the boundaries of the semiotic and the symbolic, but that the text is a composite of structures and fissures, that the intricate verbal artistry, and the civilization it indicates, is threatened with *bouleversement*.

A perfect world, carefree and inviolate, is evoked and exposed as a nexus of emotional and symbolic tensions, poised between pathos, condemnation, and prospective nostalgia. It is shadowed by the introductory *hôy*, 'Woe', an inarticulate word on the threshold between the divine wish to express itself and its manifestation in human speech. *Hôy* introduces the genre of lament for the fragility of culture. Across the threshold is another world that mirrors our own,[37] whose desirability, e.g. in the image of the early fig of v. 4, arises both from the imagination of bliss, and the utter impurity, the freedom from repression, with which it is invested.

Verse 2 accomplishes the shattering of the world of v. 1, an explosion of violence demarcated by the deictic *hinnēh*, 'Lo'. *Hinnēh* matches *hôy*, 'Woe', as a formulaic anticipation of doom, and as a transcription of a paralinguistic gesture that breaks the continuity of the text. It is a sign of revelation, of a divine emissary ('One strong and mighty to my Lord') who is both contiguous with the prophet and is his *alter ego*, the 'other' whose advent he announces. What is curious, though, is that this emissary never appears.[38] No sooner do our eyes open, to truth, than they are distracted, by similes and other rhetorical sidetracks. The formidable attributes of the adversary, *ḥāzāq wᵉ'ammiṣ*, 'strong and steadfast', herald identification and action; instead, an eight-word double simile ('like a flood of hail, a storm of destruction, like a flood of waters, powerful, overflowing') intervenes between subject and predicate.[39] The simile is cleverly interwoven with the metaphor of v. 1, restoring its fictional/allegorical landscape; its power comes not only from the intensity of the storm, with its concatenation of heavenly and earthly disasters—hail, wind and flood—but from its irruption into the text. Simile provides homologies, but also opportunities for infinite regress; one goes through the looking glass of likeness into a different world—in other words, into fantasy.

The accumulated power of the epithets *ḥāzāq wᵉ'ammiṣ*, 'strong and steadfast', debouches in the simile, in the surge of

37. Ephraim is a symbolic as well as political entity in this context. The position of a prophecy against Ephraim at the beginning of a cluster of texts about Judah has been subject to some discussion. My view is that Ephraim functions as a 'transitional object', mediating between Judah and other nations, an other who is yet the same.

38. There is, of course, no shortage of attempts to compensate for this aporia. Irwin (1977: 8), for example, suggests that the prepositional *lamed* of *la'dōnāi*, 'to my Lord', is emphatic, and that the real subject is YHWH.

39. In fact, it is not clear where the simile ends. *Hinnîaḥ lā'āreṣ bᵉyād*, 'he has cast down to earth by hand', could either be the predicate of the main clause, as I assume, or the continuation of the simile, in which case the sentence remains incomplete.

water (*mayim kabbîrîm šōṭᵉpîm*, 'waters, powerful, overflowing'). The fantasy is of imperial phallic desire and its jouissance, in the service, however, of death and not of life. The object of desire is devastated; the subject is missing.[40] Non-relation substitutes for relation. The expenditure of violence in fantasy, in the encapsulated space of the simile, is a displacement of horror, that derives its energy from the real death it cannot say;[41] it is a refuge in play, in the possibility of reconstructing the world, making death tractable; but also it reflects the need, greed and fantasy of the other, the conqueror. What is his desire? According to Lacan (1977: 29 and *passim*), desire stems from a *manque à être*,[42] from narcissistic emptiness. All desire is for the Real, for Being, which is constituted in the archaic mother. The fantasy of the flood is a metaphorical transcription of unstructured drive-energy, submergence in the Real, that sweeps self and other away in pure kinesis. This is the basis of jouissance in the passage, overladen, however, with elements of anal fantasy, in which the accumulation of possessions, of being, is a defence, a screen, against the desire to be spendthrift, for loss of being.[43]

The rest of the description of the doom of Ephraim consists of disintegrating attempts at reparation. The mode is ironic/ pathetic, but it also transforms the fantasy. As if attempting to restore the past, virtually the whole of v. 1 is repeated in vv. 3-4. Exorbitant repetition is as characteristic of Isaiah as polysemic compression. The effect is mantric; one cannot let go of those lovely phrases, the perfect world. They are counterpointed, however, by the rhythm of the trampling feet of v. 3, and the single transformative word, *wᵉhāyᵉtâ*, 'And it shall be', at the beginning of v. 4. Like a broken record, the duplication of

40. This is compounded by the syntactic indeterminacy of the verb *hinnîaḥ*, 'caused to rest', whose subject could either be the Lord, or the one strong and steadfast, and whose object is equally uncertain (cf., for example, Petersen 1979: 105).

41. Owen (1989: 150-53) illustrates this in a brilliant analysis of Sylvia Plath's poem 'Cut', where 'the wit of substitution becomes the violent defense of words against physical violence to the self, against being transformed into thing'.

42. See also the discussion in Taylor 1992: 100-101.

43. A basic statement of his thesis that the anal object is a gift to the mother is to be found in Freud's Rat Man case history (1979: 93ff.).

language freezes time at the moment of dissolution. Instead of moving to the other side of that moment, we escape into yet another simile, that of the first fruit.

Coupling a simile to a metaphor results in a second-order figure of speech; the strangeness is compounded by the inappropriateness of likening a fading flower to an early fig.[44] If a simile takes us to a different domain, we find ourselves in familiar and somewhat clichéd surroundings.[45] Two elements save the simile from banality. The first is that of gender. The flower, on its reappearance in v. 4, is feminized; *ṣîṣâ*, 'flower' (f.), replaces *ṣîṣ*, 'flower' (m.). Likewise *bikkûrāh*, 'first fruit', is feminine. Feminizing its victims is a frequent prophetic tactic for exacting sympathy.[46] The desire of v. 2, which presumably is the ultimate referent, has been romanticized, has acquired a legitimate erotic façade. Moreover, the subject of desire has switched, from the unseen other to ourselves, the male Israelite reader, in our vernal perambulations.[47] The pleasant bucolic scene replaces the scene of disaster; the joy of eating, with its sexual suggestiveness,[48] is a fantasy

44. Exum (1981: 333-36) discusses an example of a comparison within a comparison that intensifies the poetic effect. Here the problem is that of the dissonance between the original metaphor and the simile. Exum notes the difficulty of the mixed metaphors (1982: 113). The Masoretic insertion of a *mappîq* in *bikkûrāh*, 'its first fruit,' subordinates the simile even more explicitly to the metaphor, while rendering it even more obscure.

45. Amos 8.1-3 uses the vision of summer fruit, and the pun *qēṣ/qāyiṣ*, 'summer/end', similarly to fuse metaphorically alimentary satisfaction and death. A similar metaphor, with God as subject, is to be found in Hos. 9.10: Israel is God's *bikkûrâ*, 'first fig', discovered by surprise in the wilderness.

46. Habitually, for instance, cities and peoples are figured as daughters. The daughter image may be lined with sadism, the sentimentality stirred in order to be shattered, as in the accounts of the daughter of Babylon in Isa. 47 and Ps. 137. Nevertheless, the cloying gesture is part of the emotive repertoire.

47. The reader/observer projected by the text is male, in my view, because of its masculine morphology. If the inclusive language argument is to have any validity, it must apply to ancient texts as well as to modern ones. In other words, the so-called impersonal masculine inflections merely establish the universality of the male perspective. It is overdetermined in our text by the assumed heterosexuality of the desire for the flower/fig. Feminizing the object of desire polarizes its subject as masculine.

48. The fig is a pervasive genital symbol.

unlike, yet compared with, the torrent and conquest. There is thus a threefold transfer, from the other to ourselves as the subject of desire; from ourselves to the fig as its object; and from destructiveness to enjoyment. Reversing the transfer, we are the figs, feminized and violated; our desire is in fact our death.

The other distinctive quality of the simile is temporal displacement. The fig is both in the hand and swallowed up; it is no sooner seen than picked; it is not quite summer, yet the sign of summer is already consumed. The humour arises from the surprised absurdity of the lagging consciousness, which acts before it is aware. The dialectic of presence and absence, the fruit tangible, visible, and vanished, pervades the passage, in which the illusory beauty of Ephraim, and, in vv. 3-4, its afterglow, merges with the advent of the destroyer.

Into this space, this gap, enters a new voice. *Bayyôm hahû'*, 'On that day', like *hôy*, 'Woe', and *hinnēh*, 'Lo', is the inscription of a textual threshold, that annuls the world created and destroyed in vv. 1-4. Like *hinnēh*, 'Lo', it is deictic, a sign of revelation; it announces, however, not the mode of destruction, but a polity which is not death. On the other side of the jouissance of violence is something else. *Bayyôm hahû'*, 'on that day', signifies both simultaneity and discontinuity.[49] The temporal caesura is the fundamental rupture in the poem. The poet is one who has a capacity to cross the caesura, to imagine what it is like on the other side.

'That day' offers stability, certainty, the order of truth instead of illusion. Its metaphors, however, represent a conjunction of opposites. The crown is now YHWH, invisible and beyond images. The judgment seat and the one who sits upon it are directed by justice and the spirit that animates it; the fourfold regress replaces the symbols of authority and the system of law with something intangible and uncontrollable. Likewise, the power that turns back war at the gate is presumably also immaterial.[50] Thus we move from the structured to the structureless,

49. Wildberger (1982: 1050) represents critical opinion in not taking 'on that day' literally. But that does not mean that it should not be taken seriously, as a poetic metaphor.

50. The intertextual link with Isa. 11.2 would suggest mere ellipsis of *ruah*, 'spirit'.

to the wind/spirit of YHWH as the motive force that counters violence, the immense storm that blew itself out in v. 2.

Conclusion

I began this essay by speaking of poetry as a covenant with death, as playing with death, in the transitional space between mother and child, and as going beyond death. The poet, like all of us, creates and speaks for a culture. But the poet also speaks, as Celan says, in the interest of an Other, for a radical alterity. Isaiah is a movement between that other voice and his own, inscribing the trace of the *rûaḥ*, the 'wind/spirit' on the other side of disaster. I would have liked to have discussed the rest of this extraordinary chapter: the parable in vv. 23-29, whose apparent clarity and reassurance is a guise for actual incomprehensibility, for God's marvellous but impenetrable wisdom; the nonsense syllables of vv. 10 and 13 that trap the people, and their relation to the child audience of v. 9; the covenant with death composed by the *mōšᵉlîm*, the 'rulers/proverb-makers' of Jerusalem, as a reflex of aphoristic wisdom; and the long and ceremonial description of Zion in v. 16 as a counterpart to that of Ephraim in v. 1. That will have to have to wait for another opportunity.

I will conclude with two reflections. The first is that there are two paradigms in the chapter. One is composed of the chain: drunkenness–excrement–nonsense–death, encompassed by the beauty of Ephraim and elegantly contrived speech. The other paradigm is that of the new age and its new language, of which the primary symbol in Isaiah is children. God's child language, in v. 9, replaces mother's milk as the nourishment of children. It is paronomastically linked to the vomit and excrement of v. 8, and is identical to the syllables that entrap the people in v. 13. The speech that gives knowledge to the children is that which appears to be nonsense to the adults. This suggests that the two paradigms are in fact identical, and that the oppositions, such as that between God and death, on which the chapter is based, are insidiously subverted. The second reflection is really a question. I have looked at *jouissance* in the poem purely from the point of view of God and the prophet, in other words from that of the author's experience, and I have assumed the reader's identification with them. But supposing the reader's pleasure, interest in reading, and psychosomatic processes are quite other than those of God and prophet? Can one imagine a reader who is entirely impersonal, unconcerned with the fate of Samaria, Judah, or the human race, who is simply fascinated by the creation and destruction of imaginary worlds? If one were to adopt a radical reader-response perspective, what difference would it make to the contract between author and reader of which the covenants with death and with God in our chapter are representations? The *mōš^{e}lîm*, rulers and aphorists, tell parables to a death invoked as a reader, a treaty-maker, simply because it cannot read; God and prophet speak to a people who cannot listen. In v. 12, God tells of a speech he once made: 'This is the resting place; leave it to the weary; this is the repose'. The speech suggests a narrative, such as the divine story, a location, and a way of life. Beyond that, however, it is the language, with its clarity and comfort, that creates the place, that deictically situates the people. If they refuse to listen, as the text says they did, all they can hear from outside the story are jumbled fragments: '*ṣaw l^{e}ṣāw, ṣaw l^{e}ṣāw, qaw l^{e}qāw, qaw l^{e}qāw;* a little here, a little there'.[51] And perhaps that is our situation also.

BIBLIOGRAPHY

Alonso Schökel, L.
1987 'Isaiah', in *The Literary Guide to the Bible* (ed. Robert Alter and Frank Kermode; Cambridge, MA: Harvard University Press; London: Collins): 165-83.

Barthes, Roland
1973 *Le plaisir du texte* (Paris: Editions du Seuil).

Blanchot, Maurice
1982 *L'écriture du désastre* (Paris: Gallimard).

Celan, Paul
Poems (selected and trans. Michael Hamburger; Manchester: Carcanet).

51. Interpretations of this enigmatic sequence abound. It is frequently assigned to the drunken prophets/priests; there is, however, no reason for this, as Halpern remarks (1986: 111-12). For a general discussion, see Wildberger 1982: 1053-54 and Watts 1985: 361.

Clements, R.E.
1980 *Isaiah 1–39* (NCB; Grand Rapids: Eerdmans; London: Marshall, Morgan & Scott).
Derrida, Jacques
1978 *Writing and Difference* (trans. Alan Bass; London: Routledge & Kegan Paul).
1982 'Of an Apocalyptic Tone Recently Adopted in Philosophy', *Semeia* 23: 63-97.
1986 'Shibboleth', in *Midrash and Literature* (ed. Geoffrey Hartman and Sanford Budick; New Haven: Yale University Press): 307-47.
Eco, Umberto
1979 *A Theory of Semiotics* (Bloomington: Indiana University Press).
Exum, J. Cheryl
1981 'Of Broken Pots, Fluttering Pots and Visions in the Night: Extended Simile and Poetic Technique in Isaiah', *CBQ* 43: 331-52.
1982 '"Whom Will He Teach Knowledge?" A Literary Approach to Isaiah 28', in *Art and Meaning: Rhetoric in Biblical Literature* (ed. David J.A. Clines, David M. Gunn and Alan J. Hauser; Sheffield: JSOT Press): 108-39.
Freud, Sigmund
1979 *Case Histories II* (The Penguin Freud Library, 9; trans. James Strachey; London: Penguin).
García-Treto, Francisco O.
1990 'The Fall of the House: A Carnivalesque Reading of 2 Kings 9 and 10', *JSOT* 46: 47-65.
Halpern, Baruch
1986 '"The Excremental Vision": The Doomed Priests of Doom in Isaiah 28', *HAR* 10: 109-21.
Heidegger, Martin
1975 *Poetry, Language and Thought* (trans. Albert Hofstadter; New York: Harper & Row).
Irwin, W.H.
1977 *Isaiah 28–33: Translation with Philological Notes* (Rome: Pontifical Biblical Institute Press).
Kristeva, Julia
1982 *Powers of Horror: An Essay in Abjection* (trans. Leon S. Roudiez; New York: Columbia University Press).
1984 *Revolution in Poetic Language* (trans. Margaret Waller; New York: Columbia University Press).
Lacan, Jacques
1966 *Ecrits* (Paris: Editions du Seuil).
1977 *Four Fundamental Concepts of Psychoanalysis* (trans. Alan Sheridan; London: Penguin).
Landy, Francis
1983 *Paradoxes of Paradise: Identity and Difference in the Song of Songs* (Sheffield: Almond Press).
1991 '*Jouissance* and Poetics', *Union Seminary Quarterly Review* 45: 51-64.

Mekilta de Rabbi Ishmael (trans. Jacob Lauterbach; Philadelphia: Jewish Publication Society of America, 1976).

Merleau-Ponty, Maurice
1968 *The Visible and the Invisible* (trans. Alphonso Lingis; Evanston: Indiana University Press).

Owen, Stephen
1989 *Mi-lou: Poetry and the Labyrinth of Desire* (Cambridge, MA: Harvard University Press).

Petersen, David L.
1979 'Isaiah 28, A Redactional Critical Study', *SBL 1979 Seminar Papers*, II, 101-22.

Taylor, Mark
1987 *Altarity* (Chicago: University of Chicago Press).
1992 'The Politics of Theory', *JAAR* 69: 1-37.

Turner, Victor
1977 *The Ritual Process: Structure and Anti-Structure* (Ithaca, NY: Cornell University Press).

Toorn, K. van der
1988 'Echoes of Judean Necromancy in Isaiah 28.7-22', *ZAW* 100: 199-217.

Vermeylen, J.
1977 *Du prophète Isaïe à l'apocalyptique: Isaïe I–XXXV, miroir d'un demi-millénaire d'expérience religieuse en Israël* (2 vols.; Paris: Gabalda).

Watts, John D.W.
1985 *Isaiah 1–33* (WBC, 24; Waco, TX: Word Books).

Wildberger, Hans
1982 *Jesaja 28–39* (BKAT, X/3; Neukirchen–Vluyn: Neukirchener Verlag).

Winnicott, Donald W.
1972 *Playing and Reality* (London: Tavistock).

Wyschogrod, Edith
1985 *Spirit in Ashes: Hegel, Heidegger, and Mass Death* (New Haven: Yale University Press).

MANASSEH AS VILLAIN AND SCAPEGOAT

Stuart Lasine

The most striking feature of the Deuteronomists'[1] portrait of Manasseh (2 Kgs 21) is that it is not the portrait of an individual at all. What this chapter portrays is the 'limiting case'[2] of an evil king. By taking the most sinful actions of his evil predecessors and magnifying them, the narrator produces a composite drawing of an evil king who is not only like the worst monarchs but the exact opposite of the best. As one reads from v. 3 to v. 9 one learns that Manasseh is the opposite of his father Hezekiah (v. 3) and like Ahab (v. 3; cf. v. 13). He is like Ahaz

1. I say 'Deuteronomists' because it is possible that more than one deuteronomistic author/redactor may have contributed to 2 Kgs 21 in its present form. Although I will discuss the views of scholars who espouse one or another of the various multiple redaction theories as well as those who assume one 'Deuteronomist', my goal is not to affirm any specific redactional scenario. However, my discussion of the rhetorical functions of 2 Kgs 21 does assume that the present text is addressed to an exilic audience.

2. Manasseh is the 'limiting case' of a villainous king in the sense that mathematicians call a circle the limiting case of a series of regular polygons with constantly increasing numbers of sides. While the circle is the limiting case of a polygon, in the strict sense it is not a polygon itself. In fact, 'to speak of a circle as being a regular polygon...is a convenient linguistic fiction' (M. Black, *A Companion to Wittgenstein's 'Tractatus'* [Ithaca: Cornell University Press, 1964], p. 229). Similarly, while 2 Kgs 21.2-16 describes the limit of royal villainy from the perspective of the Deuteronomistic History (= DH), the constantly increasing number of sins attributed to Manasseh by the narrator is so extreme that readers may no longer view the finished portrait as the depiction of a real villain. As I will discuss below, readers who view the Manasseh of 2 Kgs 21 as a 'convenient fiction' have reason to regard this pseudo-villain as a scapegoat whose role is to serve as the fictional cause of Judah's real demise.

(v. 6; cf. 2 Kgs 16.3) and the opposite of the Saul who banished spirit mediums in accordance with ritual laws (v. 6; cf. 1 Sam. 28.3, 9; Lev. 19.31; 20.6; Deut. 18.11). He is the opposite of David and Solomon when it comes to temple policy (vv. 7-8; cf. vv. 4-5). Finally, he is like Jeroboam in seducing the people and causing them to sin with 'his' idols (v. 9; cf. v. 11).

The narrator stresses the enormity of Manasseh's sins and their disastrous repercussions by listing each species of sin committed by the king and then expanding on the nature or extent of that sin or its consequences. For example, v. 2 likens Manasseh to the nations—v. 9 says he and the people did worse than the nations. Verse 2 reports that Manasseh did evil in Yahweh's eyes—v. 6 says he did much evil in Yahweh's eyes to provoke him. Verse 3 reports that he made altars for Baal—vv. 4-5 report that he put various altars in different areas of the temple. Verse 3 reports that he made an *'ᵃšērâ* like Ahab—v. 7 adds that he placed the *pesel* of the *'ᵃšērâ* he had made in the temple. Verse 4 states that Manasseh built altars in Yahweh's house where Yahweh had said his name would be put—v. 7 adds that Yahweh had said this to David and Solomon and reports what Yahweh had told them. Verses 10-13 then expand on vv. 2, 7 and 9 by detailing the dire consequences of Manasseh's many sins. When the narrator concludes his indictment in v. 16, he not only reiterates that Manasseh caused Israel to sin by doing what was evil in Yahweh's eyes (cf. vv. 2, 6, 9, 11), but expands on the king's evil deeds, charging that he shed so much innocent blood that he filled Jerusalem with it from one end to the other.

While the narrator of 2 Kings 21 leaves no doubt about the extent and variety of Manasseh's idolatrous actions, his portrait of the king does not include the person who performed these actions. Moreover, this faceless portrait is set against a blank background. The chapter includes no quoted speeches of the king, let alone descriptions of his emotions similar to those reported of his fellow-apostate Ahab. Nor does the narrator describe any interaction between Manasseh and the 'people', opposition parties, specific prophets, or rival leaders, as he did for Jeroboam and Ahab. While 2 Kgs 21.10 and 16 mention prophets and innocent blood spilled by the king, these allusions are so vague that they actually reinforce the schematic nature of the portrait rather than increase the verisimilitude of the story. Finally, in contrast to the Jeroboam and Ahab stories (as well as those of Hezekiah and Josiah), the narrator reports no interaction between Manasseh and any foreign nation. In light of the fact that scholars typically interpret 2 Kings 21 almost entirely in terms of Manasseh's assumed submission to Assyria, this omission is particularly striking.

One might argue that the narrator's explicit comparison between Manasseh and Ahab invites the audience to fill in the blanks left in Manasseh's portrait by going back to the presentation of Ahab's character in 1 Kings 16–22; after all, Ahab is the only sinful king to whom Manasseh is explicitly likened. Readers who accept this invitation may be surprised to find that the affinities between Ahab and Manasseh are rather limited, while those between Manasseh and Jeroboam are rather extensive and profound. In the first section of this paper I will isolate the defining traits of the biblical Jeroboam and Ahab, discuss their relationship to Manasseh within DH, and ask why Ahab is singled out for mention. I will also analyze postbiblical descriptions of Manasseh and Herodotus's portrayals of tyrants in order to gauge the significance and implications of DH's 'abstract' portrait of Manasseh.

If any synchronic analysis of the Manasseh narrative is to help one solve the complex historical and redactional problems surrounding 2 Kings 21 and related texts, it must address a number of issues concerning the 'authorial audience', that is, the audience for whom the texts were rhetorically designed.[3] For

3. The term 'authorial audience' was coined by Rabinowitz; see P.J. Rabinowitz, *Before Reading: Narrative Conventions and the Politics of Interpretation* (Ithaca, NY: Cornell University Press, 1987), pp. 21-30. Rabinowitz points out that readers of narrative play several audience roles at any given time. As members of the 'narrative audience', readers believe (or pretend to believe) what the narrator says (pp. 93-96). The narrative audience of 2 Kgs 21 would accept the narrator's portrait of Manasseh as believable and historically reliable. As members of the authorial audience, however, readers are also expected to possess a specific degree of knowledge and literary competence (e.g. knowledge of literary conventions, awareness of rhetoric devices, knowledge about historical persons and events alluded to in the narrative). Readers who view DH's Manasseh as

2 Kings 21 the most difficult problem is to determine whether the audience is expected to view the overwhelming inventory of Manasseh's sins as a case of tendentious 'overkill'. Almost all commentators[4] describe Manasseh as a 'foil' for Josiah, if not Hezekiah. But are ordinary readers of the chapter *also* expected to recognize the artificiality of this portrait of an arch-villain? Could the ancient audience have compelling reasons for accepting this depiction of the king as believable and historically accurate? In the second section of the paper I will attempt to answer these questions. I will consider the possibility that the account of Manasseh's reign given in Kings (as well as the report in Chronicles and later depictions of Manasseh as a persecutor of prophets) amounts to a posthumous scapegoating of that king designed for an audience that was struggling with the disastrous fall of Judah. These texts allow the audience to identify with their innocent ancestors whose blood was shed by Manasseh, and hence to view themselves as secondary victims of the evil king. For an audience coping with catastrophe and exile, a royal scapegoat-villain provides a more comforting explanation for

tendentious and unhistorical do so as members of the authorial audience. For an analysis of the difference between authorial and narrative audience reactions to the Jeroboam narrative, see S. Lasine, 'Reading Jeroboam's Intentions: Intertextuality, Rhetoric and History in 1 Kings 12', in *Reading between Texts: Intertextuality and the Hebrew Bible* (ed. D.N. Fewell; Louisville: Westminster/John Knox Press, 1992), pp. 138-39 and *passim*.

4. For example, H.-D. Hoffmann, *Reform und Reformen: Untersuchungen zu einem Grundthema der deuteronomistischen Geschichtsschreibung* (Zürich: Theologischer Verlag, 1980), p. 166; T.R. Hobbs, *2 Kings* (WBC, 13; Waco, TX: Word Books, 1985), p. 309. H. Spieckermann describes Manasseh as an 'antitype' to glorious Josiah (*Juda unter Assur in der Sargonidenzeit* [Göttingen: Vandenhoeck & Ruprecht, 1982], pp. 161, 196). B. Halpern makes a passing reference to Manasseh as 'the scapegoat of the books of Kings', but does not develop this idea or investigate the social functions of scapegoating ('Jerusalem and the Lineages in the Seventh Century BCE: Kinship and the Rise of Individual Moral Liability', in *Law and Ideology in Monarchic Israel* [ed. B. Halpern and D.W. Hobson; JSOTSup, 124; Sheffield: JSOT Press, 1991], p. 65). Neither does G.W. Ahlström, who states that the *Chronicler* makes Manasseh 'the scapegoat for the disaster of the country' (*Royal Administration and National Religion in Ancient Palestine* [Leiden: Brill, 1982], p. 79).

their plight than one based on the assumption that they and their ancestors are fundamentally corrupt.

Ahab, Jeroboam and Manasseh as Villains

The account of Manasseh's career in 2 Kings 21 includes two explicit comparisons between Manasseh and Ahab (vv. 3, 13). Yet the only *specific* feature that suggests that it is Ahab, and not Jeroboam (or Ahaz),[5] who most resembles Manasseh is that both Ahab and Manasseh are associated with Baal and Asherah. Moreover, none of the features that specifically characterize the Ahab of 1 Kings 16–22 is shared by the Manasseh of 2 Kings 21. While the author consistently stresses Ahab's weak will and infantile emotions, he displays no interest in reporting anything about Manasseh's feelings or internal thoughts in the way he had reported those of Ahab and Jeroboam. Furthermore, only Ahab[6] has a powerful foreign spouse who 'incites' him to worship Baal and who spills innocent blood.

One might argue that Ahab and Manasseh share one trait that is so important that any differences between them become negligible: both are emblematic of the evil king *sans pareil*. According to Ishida,[7] Ahab's house apparently 'became the symbolic name of Israel's most evil dynasty soon after its destruction'. If so, Ahab may have become the emblem of the evil king before the Deuteronomists made Jeroboam into the antitype of David and

5. Like Manasseh, Ahaz passes his son through the fire, emulates northern kings and repeats the abominations of the nations. For his part, the Chronicler increases the similarity between Ahaz and Jeroboam by describing Ahaz as the maker of molten images for the Baalim (2 Chron. 28.2; cf. 1 Kgs 14.9; 2 Kgs 17.16) and accusing him of having 'broken loose' (*pr'*, 28.19). While Jeroboam himself does not 'break loose', he follows in the footsteps of the calf-maker Aaron, who had caused the people to 'break loose' (Exod. 32.25). See Lasine, 'Reading Jeroboam's Intentions', pp. 144-45.

6. There are no grounds for McKay's 'suspicion' that 'Manasseh's Asherah cult, like Ahab's, was introduced as a consequence of a diplomatic marriage' (*Religion in Judah under the Assyrians* [SBT, 2/26; Naperville, IL: Allenson, 1973], p. 23). McKay goes so far as to suggest that 'Manasseh's wife was in all probability Arabian' (p. 24).

7. T. Ishida, 'The House of Ahab', *IEJ* 25 (1975), pp. 135-37 (136).

the cause of the fall of the Northern Kingdom. Of course, Ahab could have attained this status in spite of Jeroboam's role as antitype to David or his role as *Unheilsherrscher*.[8] In either case, the Manasseh of 2 Kings 21 may have been compared to Ahab not because the historical Manasseh was particularly akin to the Ahab of 1 Kings 16–22, but because Ahab had gained the reputation of being the epitome of the evil king.

At the same time, it is not at all clear that DH's Ahab *is* the symbol of the '*most* sinful dynasty' in the north.[9] Admittedly, the sins of Ahab that most recall the sins of Manasseh are introduced as though they were committed in addition to the sins of Jeroboam (1 Kgs 16.31). To this extent, Ahab's sins necessarily exceed those of Jeroboam, even if the bloody end of the two houses is identical (1 Kgs 14.10-11; 21.21-22, 24). Nevertheless, the fact remains that DH assigns the blame for the fall of the Northern Kingdom as a whole to Jeroboam, not to Ahab.[10] Two factors in particular suggest that the parallel between Manasseh and Jeroboam is more profound than that between Manasseh and Ahab. The first has just been discussed: it is the idolatrous king's role in causing the downfall of his kingdom. The second is the role played by the idolatrous king in causing his people to sin. Scholars often attempt to date portions of DH on the basis of whether they hold the king or the people responsible for popular idolatry. Texts that blame the people are typically assumed to be exilic in origin. Friedman goes so far as to claim

8. On Jeroboam as *Unheilsherrscher*, see C.D. Evans, 'Naram-Sin and Jeroboam: The Archytypal *Unheilsherrscher* in Mesopotamian and Biblical Historiography', in *Scripture in Context*. II. *More Essays on the Comparative Method* (ed. W.W. Hallo, J.C. Moyer and L.G. Perdue; Winona Lake, IN: Eisenbrauns, 1983), pp. 114-24.

9. Ishida, 'House', p. 136; emphasis added.

10. Considering the fact that Manasseh's key role in DH is to explain the fall of Judah, one would think that the parallel between Jeroboam and Manasseh would take precedence over the Ahab–Manasseh parallel. Some scholars actually discuss 2 Kgs 21 as though this were DH's strategy. For example, R.D. Nelson virtually ignores the explicit parallels with Ahab, preferring to characterize Manasseh as 'Judah's Jeroboam' (*First and Second Kings* [Interpretation: A Bible Commentary for Teaching and Preaching; Atlanta: John Knox, 1987], p. 247; cf. p. 249). Cf. Hoffmann, *Reform*, p. 158 n. 52.

that 'all responsibility is placed upon the people in every Exilic passage'.[11] On this basis he concludes that the prediction of the fall of Judah in 2 Kgs 21.8-15 is exilic because it 'places responsibility completely on the people, so that Manasseh is blamed primarily as a catalyst'.

Leaving aside for the moment that Manasseh is more than merely a 'catalyst' in 2 Kgs 21.8-15, one simply cannot date texts in DH solely on the basis of whether the king or the people is blamed. For one thing, DH, like Exodus–Numbers, implies that mass idolatry is the result of a process in which both the leader and the people play essential roles.[12] For another, this dating-clue is based on the assumption that an exilic text will necessarily address an audience that will be comforted by viewing the fall of their kingdom as just punishment for their collective sins, hoping in the efficacy of contrition and repentance rather than in the restoration of the monarchy. Yet it seems just as likely that exilic audiences would find coping strategies like scapegoating more palatable than an 'anthropodicy'[13] that blames their plight on their own abysmal guilt and corruption.

The fact that leaders and followers are typically assumed to be co-responsible does not mean that specific texts do not make crucial points by focusing almost exclusively on the role played by the leader or the people. In Jeroboam's case, for example, it is the king who is repeatedly said to have caused all Israel to sin and to have drawn them into idolatry. In contrast, Ahab is only once said to have caused Israel to sin (1 Kgs 21.22).[14] In the case of Manasseh the narrator manages to combine the people's and Manasseh's responsibility for doom in a way that spans many generations. In 2 Kgs 21.7-8 the narrator quotes what Yahweh

11. R.E. Friedman, *The Exile and Biblical Narrative* (HSM, 22; Chico, CA: Scholars Press, 1981), p. 33; cf. pp. 10-11.

12. See e.g. Lasine, 'Reading Jeroboam's Intentions', pp. 143-49.

13. On the use of the term 'anthropodicy' in this sense, see e.g. J.L. Crenshaw, 'Introduction: The Shift from Theodicy to Anthropodicy', in *Theodicy in the Old Testament* (ed. J.L. Crenshaw; Philadelphia: Fortress Press, 1983), pp. 1-16.

14. The other references to Ahab's apostasy focus only on his idolatrous actions (1 Kgs 16.31-33; 18.18; 21.26). These actions provoked Yahweh (*hik'îs*; 16.33; 21.22) but they are not explicitly said to have led Israel to follow its king in worshiping Baal and Asherah.

said to David and Solomon, that he would not make Israel 'wander from the ground I gave *their* fathers, if only *they*...' Verse 9 then begins with '*they* hearkened not'. Who are 'they'? The generation of 'Israel' to whom Yahweh referred when speaking to David and Solomon? The next words read: 'Manasseh caused *them* to go astray to do...' Clearly, the narrator is telescoping the 'Israel' to whom Yahweh had referred at the beginning of the monarchical period with the 'they' who still were not listening when Manasseh caused them to wander astray. Taken together, these verses imply that if 'they' had hearkened to what Yahweh told the kings who governed their ancestors, Manasseh's cultic reforms would not have been able to mislead them.

In spite of the profound similarities between Manasseh and Jeroboam and the limited affinities between Manasseh and Ahab, most scholars take their cue from 2 Kgs 21.3, 13 and conclude that Manasseh is 'Judah's Ahab'[15] or 'the Jezebel of the south',[16] not Judah's Jeroboam. Some ask whether the historical Ahab might have resembled the biblical and/or historical Manasseh in other ways as well.[17] Commentators often focus on the fact that both are said to have shed the blood of innocent people (1 Kgs 21.19; 2 Kgs 9.7, 26 // 2 Kgs 21.16; 24.4). Noting that 2 Kgs 21.16 comes soon after the reference to 'the prophets' castigations' in v. 10, Rofé concludes that Manasseh is associated with the persecution of prophets, as was Ahab.[18] He believes that 'an analogy between Manasseh and Ahab was clear to the people at the time of Menasseh [*sic*] and had found its way into the sources of the Books of Kings'. Rofé argues that the 'otherwise incomprehensible references to the murder of prophets' in the Ahab story are due to the fact that the Elijah

15. Hobbs, *2 Kings*, p. 311; cf. B.O. Long, *2 Kings* (FOTL, 10; Grand Rapids: Eerdmans, 1991), p. 250.

16. Y. Kaufmann, *The Religion of Israel* (trans. M. Greenberg; New York: Schocken Books, 1972), p. 141.

17. For example, Jones asserts that 'Manasseh's sins corresponded *in many respects* to those of Ahab' (*1 and 2 Kings*, II (NCB: Grand Rapids: Eerdmans, 1984], p. 596; emphasis added), although he does concede that even their cultic sins are not totally identical. Cf. Long, *2 Kings*, p. 248.

18. A. Rofé, *The Prophetical Stories* (Jerusalem: Magnes, 1988), p. 200; cf. pp. 189-90, 192.

'epic' was written during Manasseh's reign.[19] Unfortunately, 2 Kgs 21.10, 16 hardly constitute sufficient evidence to prove that prophets were persecuted during Manasseh's reign, let alone that the persecution was so severe that it influenced the way Ahab's reign was depicted. Nor does the apparent dearth of prophetic activity during Manasseh's reign allow one to conclude that the king had used violence to silence them. The phrase 'shed innocent blood' used in v. 16 refers to a much wider range of violent injustice than prophet persecution alone (see below). Finally, the vague reference to prophetic activity in v. 10 is never developed in Kings and is removed by Chronicles (2 Chron. 33.10). In fact, the narrator of 2 Kings 21 does not even say whether Manasseh himself heard the divine judgment conveyed by the prophets in vv. 11-15.

If 2 Kgs 21.3, 13 and 16 have prompted some commentators (and the authors of post-biblical legends)[20] to view Manasseh as an Ahab-like persecutor of prophets, these verses can also lead one to compare Ahab and Manasseh to the tyrants whose *pleonexia* and *koros* are so often described by ancient Greek authors. These monarchs indulge their insatiable desire for what rightfully belongs to another by committing acts of social injustice. From this perspective the innocent blood shed by Manasseh recalls the blood of the innocent Naboth and his sons (1 Kgs 21.19; 2 Kgs 9.26). At least one postbiblical work reinforces the image of Manasseh as a typical tyrant. In *2 Baruch* 64

19. Rofé, *Prophetical Stories*, pp. 189, 190.

20. See L. Ginzberg, *The Legends of the Jews* (Philadelphia: Jewish Publication Society, 1958), IV, pp. 278-79; VI, pp. 374-75; B.H. Amaru, 'The Killing of the Prophets: Unraveling a Midrash', *HUCA* 54 (1983), pp. 170-73. According to Josephus (*Ant.* 10.38), Manasseh slaughtered prophets daily [!]. In the *Martyrdom of Isaiah*, the satanic Belkira prompts Manasseh to saw Isaiah in half (5.1). Belkira is a descendant of the false prophet Zedekiah ben Chenaanah, the 'teacher...of the four hundred prophets of Baal' in the days of Ahab (2.12). The author interrupts the story of Isaiah's fate to detail Ahab's abuse of Micaiah and his son Ahaziah's killing the prophets of the Lord, including Micaiah (2.13-16). Clearly, in this work Manasseh the prophet-persecutor is Ahab *redivivus*. For the 'Martyrdom of Isaiah', see M.A. Knibb (trans.), 'Martyrdom and Ascension of Isaiah', in *The Old Testament Pseudepigrapha*, II (ed. J.H. Charlesworth; Garden City, NY: Doubleday, 1985), pp. 156-64.

Manasseh is described in terms that recall Otanes' portrait of the tyrant in Herodotus's famous 'Constitutional Debate' (3.80). According to Herodotus's Otanes, the insatiable and irresponsible tyrant will meddle with or remove ancestral customs and observances, force women, and kill men indiscriminately without trial. In 1 Kings 21, DH describes a king whose desire for the vineyard belonging to another is satisfied by executing the rightful owner of the property and his heirs. Far from giving Naboth a fair trial, Jezebel perverts justice in Ahab's name by convening a kangaroo court. The house of Ahab continues to be associated with acts of social injustice even in Mic. 6.9-16. For his part, the Manasseh of 2 *Bar.* 64.2 far exceeds Ahab in tyranny. He 'killed the righteous, and perverted judgment, and shed innocent blood, and violently polluted married women, and overturned altars, and abolished their offerings...'[21] While the author of 2 *Baruch* follows Chronicles in having Manasseh pray to the Most High, his impiety and *hubris* are apparently too great to allow him to avoid the punishment awaiting him at the end (64.8-10).

While the Chronicler deletes DH's tantalizing references to Ahab, in one respect his Manasseh has more in common with Ahab than does DH's unrepentant villain: both Ahab and the Manasseh of Chronicles 'humble themselves' (*kn'*) and repent of their sins (1 Kgs 21.27-29; 2 Chron. 33.11-13). Of course, one can dismiss these acts of penitence by arguing that the authors applied a formulaic 'schema of reprieve'[22] to these monarchs to explain why neither suffered for his egregious sins during his long reign. However, this common feature also links Ahab and Manasseh to Josiah, for whom Manasseh serves as foil in DH. Manasseh and Josiah are two of only four kings whom the Chronicler describes with *kn'*, his characteristic verb for humbling (cf. 2 Chron. 12.6, 7, 12; 32.26; 34.27), while Ahab and Josiah (2 Kgs 22.19) are the only two kings whose actions are described with this verb in DH. Oddly enough, Ahab resembles not only the repentant Josiah of 2 Chron. 34.27, but the Josiah

21. 2 *Bar.* 64.2; trans. A.F.J. Klijn, in *The Old Testament Pseudepigrapha*, I (ed. J.H. Charlesworth; Garden City, NY: Doubleday, 1985), p. 643.

22. Long, 2 *Kings*, pp. 227-28, 260-61; cf. Nelson, *First and Second Kings*, p. 143.

who disregards a divine warning, goes into battle in disguise, and is mortally wounded (1 Kgs 22.19-23, 30-37; 2 Chron. 35.22-24). If the exceedingly long reign of the sinner Manasseh presents a challenge to theodicy, the fact that he is the only one of these three kings to enjoy a peaceful death would seem to require that he be vilified and punished in other ways. In the next section I will argue that the portrayal of Manasseh as an arch-villain in 2 Kings 21 may have more to do with meeting the needs of the exilic audience than with justifying Yahweh's granting Manasseh such a long and successful career.

Manasseh as Scapegoat

Considering the stark contrast between DH's nuanced and extensive descriptions of Jeroboam and Ahab and his one-dimensional portrait of Manasseh, one might well ask whether the authorial audience could accept this report as believable and reliable. Could the author have *expected* his audience to view his 'Manasseh' as a tendentious fabrication? Scholars often evade such questions by regarding Manasseh as merely one of the several 'characterless, cardboard villains' created by an exilic editor who describes most of his villains in 'wooden phrases'.[23] Long[24] challenges this assumption, noting that advocates of the double redaction theory of DH typically identify the exilic editor's writing on the basis of their 'modern literary tastes', judging it to be simplified, imitative, wooden, vague or terse. Yet, in light of the fact that many sections of DH seem designed for an audience with considerable literary sophistication,[25] one must ask whether the ancient audience would find 2 Kings 21 to be any less cardboard and unbelievable than do their modern counterparts.

At the same time, one cannot simply assume that the narrative is *so* reductive that it must have been viewed as unbelievable. In her study of 'authoritarian fictions', Suleiman remarks that a

23. R.D. Nelson, *The Double Redaction of the Deuteronomistic History* (JSOTSup, 18; Sheffield: JSOT Press, 1981), p. 126; cf. p. 37.

24. B.O. Long, *1 Kings, with an Introduction to Historical Literature* (FOTL, 9; Grand Rapids: Eerdmans, 1984), pp. 17-18.

25. See Lasine, 'Reading Jeroboam's Intention', pp. 135-39, 145-46.

roman à thèse cannot accomplish its ideological goals if its characters are totally lacking in contradiction and totally predictable, and if its message is conveyed with excessive redundancy. Total closure undermines the verisimilitude—and therefore the believability—of the narrative. In fact, 'the more a *roman à thèse* is faithful to its didactic calling, the less it succeeds in making itself believed, that is, accepted as a reliable, truth-telling witness'.[26] While DH's Manasseh is certainly predictable and lacking in contradiction, it may nevertheless have been accepted as 'a reliable, truth-telling witness' by the ancient audience, if reader response to modern formulaic fiction is any indication. The villains in formulaic genres like the detective story are often accepted as believable by audiences who find the social chaos depicted in the work to be unsettling.[27] In general, the detective story conveys the comforting message that a single unambiguous villain is responsible for crime and disorder, not the prevailing social structure of which the reader is a part.[28] Here the fictional villain functions as a scapegoat for the real audience. In fact, Rabinowitz has demonstrated that when readers encounter ambiguous characters in sophisticated and 'disturbing' novels like Chandler's *The Big Sleep* they often reduce those characters to scapegoat-villains. He shows that Chandler's Carmen is more victim than villain, but that many readers employ a 'strategy that allows them to increase her monstrosity so that they can put enough blame on her to make her punishment cathartic'.[29] This procedure 'involves an act of scapegoating: in order to create a sense of resolution in a morally chaotic situation, someone must be seen as the wrongdoer and appropriately punished'.[30]

It is highly probable that 2 Kings 21 is also designed for an

26. S.R. Suleiman, *Authoritarian Fictions: The Ideological Novel as a Literary Genre* (New York: Columbia University Press, 1983), p. 189; cf. pp. 172, 194.

27. See S. Lasine, 'Solomon, Daniel and the Detective Story: The Social Functions of a Literary Genre', *HAR* 11 (1987), pp. 247-66.

28. J.G. Cawelti, *Adventure, Mystery, and Romance: Formula Stories as Art and Popular Culture* (Chicago: University of Chicago Press, 1976), pp. 104-105.

29. Rabinowitz, *Before Reading*, p. 205.

30. Rabinowitz, *Before Reading*, p. 203.

audience that was seeking to escape a 'morally chaotic situation'. If modern readers are quite willing to transform a complex victim like Chandler's into a villain, would not the exilic audience of Kings be all too happy to accept as believable a reductive portrait of an arch-villain like Manasseh? The fact that the reductiveness is so extreme could testify to the extent of the audience's anxiety concerning their situation. While the authorial audience of DH may have possessed considerable literary sophistication, it is entirely possible that this critical faculty would be laid aside if their need for a one-dimensional villain were strong enough. The temptation would even be stronger if the story also allowed the audience to identify with the victims whose 'innocent blood' was shed by that culprit. It is also possible that the reductiveness serves different functions for different readers, in the manner of an ambiguous figure drawing—except that this time the ambiguity would be the result not of indeterminacy in the narrative, but of its *hyper*determinacy. If this is the case, the excessively detailed depiction of Manasseh's sins would be believable to readers who need such a monstrous villain, while those who feel no such need would recognize that the emperor has no clothes—that is, no depth or reality—exposing the fact that the villain is actually a scapegoat.

If one is to determine whether the author or final redactor of 2 Kings 21 expected his target audience to accept his depiction of the villain Manasseh at face value, one must first ask what functions this portrait might have been designed to serve. The criteria employed by scholars to date texts like 2 Kings 21 involve the presumed social functions of those texts for their intended audiences as well as stylistic and thematic features. According to Cross,[31] the purpose of the exilic edition of DH was to transform the Josianic edition into 'a sermon on history' which 'overwrote' and contradicted the original theme of hope for a new golden age under a Davidic king, replacing it with 'a muted hope of repentance...and possible return'. The revised account of Manasseh's reign plays a particularly important role in this 'sermon', by 'conforming Judah's fate to that of Samaria and Manasseh's

31. F.M. Cross, *Canaanite Myth and Hebrew Epic* (Cambridge, MA: Harvard University Press, 1973), pp. 287-88.

role to that of Jeroboam'. In Nelson's formulation,[32] the theological lesson of the second edition is that salvation rests in 'an acceptance of the justice of Yahweh's punishment and in repentance'.

But can one assume that an exilic edition would necessarily be designed to function as an anthropodicy for an audience that would rather blame the loss of their world on themselves and their fathers than to question Yahweh's justice? While it might serve the interests of the literate elite who produced this text for an exilic audience to emphasize the role of the people as opposed to the role of leaders such as themselves, their addressees might well seek to cast off the shroud of responsibility spread over them by those in power. They might do so by projecting all responsibility onto the leaders. However, *all* parties in the exilic community could avoid the pall of guilt if they could agree that a specific individual from a distant generation was to blame for all their troubles. As Girard and others have shown, communities often regain stability and unity during religious and political crises by unanimously choosing a scapegoat whom all can affirm as the locus of guilt.[33] Considering the social function of scapegoating, the passages that heap all responsibility for the fall on King Manasseh are just as likely to have been composed as a means of coping with the exile as the verses that blame the people, in spite of the common assumption that passages that blame the king must be pre-exilic.

A review of the way Manasseh is described in DH, in Jer. 15.4, in Chronicles, and in postbiblical literature will indicate whether the biblical Manasseh is not only a 'foil' and 'antitype' to Josiah, but the audience's scapegoat. In *Violence and the Sacred*, Girard quotes Vernant's analysis of rituals in which a surrogate victim is sacrificed instead of the king. Here the community chooses an 'antisovereign' upon whom the king unloads all his negative attributes, creating a carnivalesque 'inverted image of himself'.[34] This double is expelled from the community

32. Nelson, *Double Redaction*, p. 123.

33. For a critical review of Girard's theories and their application to biblical texts, see S. Lasine, review of *Job: The Victim of his People*, by R. Girard, *HS* 32 (1991), pp. 92-104.

34. R. Girard, *Violence and the Sacred* (Baltimore: Johns Hopkins University Press, 1977), p. 109.

or put to death when the carnival is over, ending all the disorder symbolized by his topsy-turvy identity. I would suggest that the Manasseh of 2 Kings 21 is DH's 'antisovereign', the inverted image of a glorified Josiah. The fact that he is so extraordinarily and unequivocally evil indicates that his function is to represent the limiting case of an anti-king. This would explain why DH's Manasseh is about as believable as a figure in the topsy-turvy world of carnival. Only an unreal construct that 'embodies' the worst qualities of the Israelite tyrant could serve as the absent cause of the fall of Jerusalem and the exile.

Like Manasseh, the quintessential tyrant envisioned by Herodotus's Socles can overturn the order of the universe itself:

> Surely the heaven will soon be below, and the earth above, and men will henceforth live in the sea, and fish take their place upon the dry land, since you...propose to put down free governments...and to set up tyrannies in their stead. There is nothing in the world so unjust, nothing so bloody, as a tyranny (5.92).[35]

Manasseh's function as the emblem of a world-upside-down is signaled by the unique simile employed by Yahweh in condemning the Jerusalem debased by Manasseh: 'I will wipe Jerusalem as a man wipes a dish, wiping it and turning it upside down' (2 Kgs 21.13). Yahweh uses *hāpak*, the key verb for the world-upside-down *topos* in the Hebrew Bible, to describe how he will invert the city whose symbolic identity was perverted when Manasseh placed the *'ašērâ* image in the house and in the city in which Yahweh had placed his name (v. 7). If Job 'sought to turn the dish upside down' by accusing God of unjustly allowing the earth to be given into the hand of the wicked (*b. B. Bat.* 16a; Job 9.24), the hand of the wicked and unjust Manasseh has succeeded in turning the dish of Jerusalem upside down. When God turns it again he puts things right only in the sense of providing the right punishment for Manasseh's crimes.

According to Girard,[36] for a monarch himself to be

35. Translation by A. Ferrill, in 'Herodotus on Tyranny', *Historia* 27 (1978), p. 395. On the world-upside-down *topos* in the Hebrew Bible, see S. Lasine, 'The Ups and Downs of Monarchical Justice: Solomon and Jehoram in an Intertextual World', forthcoming in *JSOT*.

36. R. Girard, *Job: The Victim of his People* (Stanford: Stanford University Press, 1987), p. 88.

transformed into a scapegoat he must 'carry out the major social function of a "wicked man"'. That is, he must commit the 'imaginary "crimes" of the scapegoat'. Besides oppressing the people, he is 'expected to confess officially to a certain number of oedipal crimes' such as parricide or 'some well-concocted incestuous relationship with a mother or sister'. Is it merely a coincidence that postbiblical descriptions of Manasseh not only include vastly enhanced accounts of his victimization of the innocent (see below), but the charge that Manasseh committed incest with his sister (*b. Sanh.* 103b) like a biblical Cambyses, and parricide, by killing his grandfather Isaiah (*j. Sanh.* 10, 28c; *b. Sanh.* 103b)?[37] Those to whom such crimes are attributed are often viewed as having turned the world upside down like the biblical Manasseh. For example, the Christians persecuted by the Romans were often charged with committing *flagitia* such as incest and cannibalism, while witches in traditional societies are often accused of similar 'inverted world' behavior such as incest, cannibalism and infanticide—all the sorts of behavior that threaten the basic cultural categories that sustain social order.[38]

Ab. 5.9 links such crimes with exile in a manner which evokes the typical scapegoat pattern as well as Manasseh in particular: 'Exile ensues in the world on account of idolatry, because of incest, for spilling of blood, and on account of the release of the land'. The biblical Manasseh who causes the exile commits two of these four acts, while the postbiblical Manasseh commits three. In fact, the Bible itself connects all three of those sins with exile. Exile, bloodshed and idolatry are linked in Ezek. 36.18, and exile and incest are connected in Lev. 18.24-28. In addition, Manasseh's shedding of innocent blood (21.16)—a sin that can pollute not only a city (Deut. 21.8-9; Jer. 26.15) but also the nation (Deut. 19.10-13) and the land (Num. 35.33-34)—is

37. See further in Amaru, 'Killing', pp. 172-73.

38. On the *flagitia*, see e.g. G.E.M. de Ste Croix, 'Why were the Early Christians Persecuted?', in *Studies in Ancient Society* (ed. M.I. Finley; London: Routledge & Kegan Paul, 1974), pp. 233-34 and *passim*. On witches, see e.g. R.R. Wilson, *Prophecy and Society in Ancient Israel* (Philadelphia: Fortress Press, 1980), p. 74; S. Lasine, 'Jehoram and the Cannibal Mothers (2 Kings 6.24-33): Solomon's Judgment in an Inverted World', *JSOT* 50 (1991), p. 35 n. 1.

also associated with exile in one instance.[39]

Obviously, postbiblical descriptions of Manasseh in terms appropriate to the ritual monarch-scapegoat can only tell us how some postbiblical readers understood the Manasseh of 2 Kings. Yet Manasseh is also made to complete the typical career of the royal scapegoat within the pages of the Hebrew Bible, when his story is retold in 2 Chronicles 33. Like Oedipus and other typical scapegoats, this Manasseh is expelled from the community and confesses to his crimes (33.10-13). Like Oedipus, he brings blessings to a community before he dies. After Yahweh brings him back from exile Manasseh not only engages in building projects but purifies the very cult he had criminally defiled, commanding Judah to serve Yahweh (vv. 14-16). Perhaps the greatest 'blessing' he bestows is that he is no longer the cause for the exile. If the monstrous Manasseh of 2 Kings causes the exile of his people without going into exile himself, the exile and repentance of the Chronicler's Manasseh ensures that the people will *not* be exiled on his account.

The reference to Manasseh in Jer. 15.4 can also be construed as evidence that Manasseh is DH's scapegoat. The prophet quotes Yahweh as declaring that he will make the people a horror to all the kingdoms of the earth because of what Manasseh did in Jerusalem. Almost all commentators find this verse alien to Jeremiah's message precisely because it blames the catastrophe on Manasseh, whereas Jeremiah typically spreads the blame among the people as a whole and their leaders. As Clements puts it, in the book of Jeremiah 'no scapegoats are singled out as guilty'.[40] Carroll[41] explains the singling out of Manasseh by suggesting that the verse 'allows the Deuteronomists to settle an old score and round off the composition'. While 'the nation's

39. On the connection between shedding innocent blood and exile in Deut. 19.10, see A.D.H. Mayes, *Deuteronomy* (NCB: Grand Rapids: Eerdmans, 1979), p. 287. On the other cited passages, see J. Milgrom, *Numbers* (JPS Torah Commentary; Philadelphia: Jewish Publication Society, 1990), p. 509.

40. R.E. Clements, *Jeremiah* (Interpretation: A Bible Commentary for Teaching and Preaching; Atlanta: John Knox, 1988), p. 95.

41. R.P. Carroll, *Jeremiah: A Commentary* (OTL; Philadelphia: Westminster Press, 1986), p. 321.

destruction may seem excessively cruel...it is justified because King Manasseh was such a vicious and corrupt ruler'. Carroll finds it ironic that Manasseh was blamed for filling Jerusalem with innocent blood, 'when in reality Yahweh, under the guise of the Babylonians, did precisely that!'. In other words, Yahweh's shedding of innocent blood is justified not because the inhabitants of the city were in reality far from being innocent, but because they were doomed by their villainous king.

One can get a clearer idea of how the biblical Manasseh might have functioned as a scapegoat for readers of Kings and Jeremiah by exploring his alleged connections with prophet-persecution. On the basis of 2 Kgs 21.16 and 24.4, historians often conjecture that the historical king Manasseh did not merely shed the innocent blood of the oppressed, but instituted a 'reign of terror'[42] against persons who opposed his cult reforms. Others, like Rofé, argue that 2 Kgs 21.10, 16 constitute evidence that Manasseh persecuted prophets. As discussed earlier, Rofé conjectures that the Ahab narrative highlights prophet-persecution to a unique degree because it was composed during Manasseh's reign, when prophets had become 'persecutable' because they were no longer perceived as having access to divine power that rendered them inviolable. While the vagueness of 2 Kgs 21.10, 16 does not allow one to conclude that the narrator is alluding to prophet-persecution, these verses, together with the references to Ahab, probably did inspire the postbiblical portrayals of Manasseh as an Ahab-like slaughterer of prophets found in the *Martyrdom of Isaiah*, Josephus, and a number of rabbinic legends (see above). For example, the author of the *Martyrdom* attributes to Manasseh most of the sins described in 2 Kgs 21.1-10, 16, including 'persecution of the righteous' (2.5). To this list he adds an elaborate account of the king sawing in half the prophet Isaiah at the instigation of the satanic Belkira (5.1).

Perhaps Rofé could not connect the advent of prophet-persecution with Manasseh's period because the explicit allegations of prophet-persecution were generated by a later crisis, one triggered by the end of the monarchy, the loss of the temple, the exile, and the disruption of the sacrificial system. That the

42. Kaufmann, *Religion*, p. 435; cf. p. 141.

idea of prophet-persecution began to assume importance only during this crisis is supported by the portrayal of Jeremiah as near-scapegoat in the deuteronomistically edited book which bears his name. Jeremiah is also the only book composed before Nehemiah and Chronicles that alleges that a prophet was murdered.[43] At various times, all the authority figures as well as the people threaten or attack Jeremiah, although his provocative and seemingly seditious behavior does not lead the community to murder him. In contrast, one postbiblical tradition has the Jews stoning him to death in Egypt, while another has him inadvertently and involuntarily committing the scapegoat crime of incest (or, more precisely, impregnating his daughter *in absentia*).[44]

By stressing that the persecution of prophets becomes a major theme in texts composed or edited in response to the fall of Judah I am not suggesting that prophets were *actually* scapegoated at this time (or in postexilic times, as could be argued on the basis of the references to prophet-murder in Nehemiah and Chronicles). Rather, I am suggesting that the exilic (and possibly postexilic) communities shifted blame for the fall of Judah to former generations by claiming that earlier prophet-persecutions were a primary cause for the disaster. Exilic texts that harp on the people's spurning of Yahweh's 'servants the prophets'[45]

43. See Jer. 2.30 (a difficult text) and 26.20-23 (where the victim is Uriah son of Shemaiah); Neh. 9.26; 2 Chron. 24.20-22; cf. 2 Chron. 16.10. Compare *Pes. R.* 26, 129ab, according to which Jeremiah refuses Yahweh's call, asking, 'When lived there a prophet whom Israel did not desire to kill?' Considering that the Bible itself presents the prophet Jeremiah as a near-victim, it hardly seems a coincidence that this is the only prophetic book in which the villain Manasseh is specifically singled out as the cause of the exile, and the only book that links Manasseh's signature sins and phrases from 2 Kgs 21 with the scapegoat crime of cannibalism, which here becomes a punishment (Jer. 19.3-5 // 2 Kgs 21.3, 6, 12, 16; Jer. 19.9). Nor does it seem coincidental that the only biblical personage who explicitly applies the spilling of 'innocent blood' to the blood-polluting sacrifice of a prophet by the community at large is the potential victim Jeremiah (26.15).

44. On the former tradition, see e.g. Ginzberg, *Legends*, VI, pp. 399-400; on the latter see the *Alphabet of Ben Sira* 16-20 (in M.J. bin Gorion, *Mimekor Yisrael: Classical Jewish Folktales*, I [Bloomington: Indiana University Press, 1976], pp. 193-96).

45. See Nelson, *Double Redaction*, pp. 58-59.

imply that the exilic generation were also victims of their evil ancestors' persecution of prophets, in the sense that the warnings of those prophets, if heeded, could have averted the disaster that befell the exiles.

2 Kings 21 also affirms the audience's sense of innocence by representing it in the text in the form of innocent ancestors victimized by the king, as opposed to the guilty ancestors whose idolatry and violence are said to have caused the exile. The process of identification is facilitated by the vagueness of v. 16, which provides room for 'innocent' readers to find their true ancestors among Manasseh's victims. This understanding of the function of 21.16 is similar to Carroll's view of Jeremiah's soliloquies. According to Carroll, these speeches use language typical of individual and communal laments to represent the exiled communities who are pleading their innocence, even though the oracles in the same book continue to condemn the people as evil and deserving of destruction.[46]

This way of evading responsibility for the exile serves to verify both the general indictments of the people at large (the fathers did nothing but evil from the start [e.g. 2 Kgs 21.15; Deut. 9.7, 24; 1 Sam. 8.8; Jer. 32.30-31]) and the specific indictment of Manasseh as the direct cause of the catastrophe (2 Kgs 21.11-13; 23.26-27; 24.2-4). Admittedly, this coping strategy must concede that both the king *and* the people victimized the prophets whose mission was to implore the people to repent before it was too late. However, the good news is that the viability of repentance and the possibility of restoration remain

46. R.P. Carroll, *From Chaos to Covenant: Prophecy in the Book of Jeremiah* (New York: Crossroad, 1981), pp. 260-61; cf. pp. 28, 129-30. G. Garbini (*History and Ideology in Ancient Israel* [New York: Crossroad, 1988], pp. 114-19) also analyzes the late appearance of prophet-persecution stories and a 'victimistic' ideology in terms of the audience's need to identify with those whose innocent blood was shed. In his view, however, this is mostly due to the special interests of second-century 'Pharisees' whose hypocrisy on the subject of prophet persecution was exposed by Jesus in Mt. 23. Garbini does not consider the possibility that the reductive and one-dimensional nature of such portraits of Jewish villains might indicate that they are actually being made into scapegoats like DH's Manasseh. For an analysis of Stephen's speech as an attempt to transform his ancestors into scapegoat-villains, see Lasine, review of *Job*, by R. Girard, pp. 101-103.

intact for the victimized children of those who persecuted the prophets as well as for those whose fathers were slaughtered by Manasseh. By focusing on Manasseh's and the fathers' persecution of scapegoat-victims with whom the exilic audience can identify, the biblical historian has provided that audience with a more comforting explanation of their present plight than one based on the evil effects of their traditional—but illicit—cult practices or one predicated on the incorrigible perversity of their own nature. This is the ultimate blessing that the vilified Manasseh of Kings bestowed on the original readers and hearers of DH. This blessing would be denied only to those readers who were unable to accept the overblown portrait of the villainous Manasseh as true-to-life and to those who were reluctant to victimize their forebears by transforming them into villains in order to appropriate the role of victim for themselves. What remains unclear is whether or not the final redactor designed his description of Manasseh so that readers could view it as the portrait of a scapegoat as well as the portrait of a villain.

MOSES AND DAVID: MYTH AND MONARCHY

Peter D. Miscall

Pentateuchal narratives about Moses and the creation of the people Israel have a mythic quality because they are stories of origins in which God takes the initiative and because they have a narrative style marked by clarity and resolution. Moses, the hero, is introduced at the very start of the narrative. Narratives about the origins of kingship, including the book of Judges and 1 Samuel, do not have this mythic quality. They are marked by ambiguity and doubling, and it is impossible to speak of 'the very start of the narrative' of kingship. David, the hero, is not introduced until 1 Samuel 16. This contrast in narrative style reveals the author's inherent negative judgment on kingship. The author treats kingship at such length because it lasted 400 years and was an integral part of Israel's pre-exilic history.

In this essay, I compare and contrast narratives in Genesis–2 Kings, which I regard as a work written in the post-exilic period and not as an editorial compilation of already existing material.[1] Genesis–2 Kings is a particular interpretation of Israel's past, whether legendary or historical; it is not just the traditional view, i.e. 'Israel's perception of its history'. Many current works on this corpus (or on Deuteronomy–2 Kings, the Deuteronomic History) implicitly or explicitly point to its anti-monarchical stance, especially in (Judges) Samuel–Kings; kings and kingship are the problem and the main symptom, if not the cause, of Israel's turning from the Lord. Kings lead to the international treaties, intrigue and wars that eventually result in the fall of both Israel and Judah. As a counterpart to this particular interpretation, I assume that a post-exilic monarchist party would have presented a radically different story of David and the Davidic dynasty.

1. I do not take a stand on the question of specific authorship and date.

In this article I buttress this anti-monarchical reading of Samuel–Kings by contrasting aspects of the narratives about Moses in the Pentateuch with those about Samuel, Saul and David in 1–2 Samuel. (For the purposes of this study, I focus on the material in 1 Sam. 1–17.) My emphasis is on narrative style and mode of presentation, not on characterization and thematic content. In short, I maintain that the author expresses his anti-monarchical sentiments in how he tells his story and not just in the story he tells. The repetitive and disjointed style of 1 Samuel, when compared with the style of the story of Moses, is an integral part of the portrait of kingship. At the same time, I relate the length of the presentation of the kings in Samuel–Kings to the anti-monarchical author's acceptance of the fact that monarchy endured for centuries and gave Israel and Judah their identity and existence in the pre-exilic era.

In other cultures, kingship was connected with, if not equated with, the divine realm and with the creation of the world and its order, both cosmology and cosmogony. The stability of dynastic kingship mirrored the stability of the kingship of the creator god and his secure foundation of the universe. King, myth and ritual went hand-in-hand in the ancient empires. I cite the Babylonian example of Marduk in the *Enuma Elish* although examples from Egypt, Assyria and the Hittites could be included. Biblical parallels to this mix of myth and kingship occur mainly in the Psalms and the prophetic literature.[2]

In Genesis and Exodus, myth and ritual mix. Kings and kingship come later in the narrative and not with the creation of the world or of Israel. Parallels between Genesis and 1–2 Samuel, for example between David and both Abraham and Joseph, only emphasize the absence of kings in the former.

I use the term mythic to refer to narratives that deal with the origins of the world and of significant human institutions. These institutions are grounded in the created order and are due to the divine initiative; the creating deity acts and speaks and humans respond. In this sense much of Genesis and Exodus has a mythic aspect to it, especially the stories of the creation of the

2. See Heidel 1951 for the *Enuma Elish* and its biblical parallels. For contemporary discussions of biblical material, consult Ollenburger 1987 and Weinfeld 1983.

world and of Israel and its institutions. On the other hand, I do not want to over-stress this quality by presenting this material as purely mythic or as a myth. I am reading Genesis and Exodus as narrative with a mythic quality to it and not as a narrative myth. Genesis gives background to Exodus; the latter is not an absolute origin, for in it God 'remembers' his covenant with the forefathers (Exod. 2.24). The call of Israel in Egypt is not a *creatio ex nihilo* any more than the creation of the world in Genesis 1.

With these qualifications in mind, I focus on the narrative style and design of the story of Moses and compare them with those of the stories of the rise of kingship and of David in 1 Samuel 1–17.[3] This reading will reveal a contrast between the mythic quality and relative clarity of the Moses story and the non-mythic aspect and frequent obscurity of the narratives in 1 Samuel 1–17.

Exodus opens with Moses' birth and wondrous rescue. He is a man of destiny. Outside the Hebrew Bible, Sargon of Akkad and Cyrus of Persia, both kings, have similar birth stories.[4] God appears to Moses in the burning bush and calls him to lead Israel out of Egypt to take possession of the land of Canaan.[5] The call and commission of Moses are at God's initiative. The divine initiative contributes to the mythic aura of the narrative, and the aura continues in the plagues that present a mythic combat between the Lord and Pharaoh, the symbol of chaos.[6]

3. I am indebted to one of my students, Marilyn Thorssen, for drawing my attention to the absence of any mythic quality in the David story, particularly in his introduction in 1 Sam. 16–17 and in his death scene in 1 Kgs 1–2. Moses dies on a mountain top at 120 years of age 'with his sight unimpaired and his vigor unabated', whereas David dies in his bedroom at 70 years of age (2 Sam. 5.4) in bed with a young woman whom he knows not (1 Kgs 1.2).

4. For the Sargon legend, consult Pritchard 1969: 119; for the Cyrus legend, consult Herodotus, *The Histories* 1.11.108ff.

5. In view of Moses' being denied entrance into the promised land (Num. 20.2-14; 27.12-14; Deut. 32.48-52), I note that his commission in Exod. 3 is to confront Pharaoh and to bring the Israelites out of Egypt; the Lord will bring them into the land that he is giving them.

6. See Cross 1973: 112-44 and Levenson 1988: 3-50, for particulars on the myths of creation as combat and on their biblical parallels.

The event at the Sea, especially when coupled with the crossing of the Jordan on dry land (Josh. 3–4), is a Hebrew version of the Canaanite myth of Baal's defeat of the two-named god Sea–River.[7] The crossing of both sea and river are done at God's command. The revelation at Sinai in Exodus 19–Numbers 10, at God's initiative, sets in place significant social and cultic institutions. Again, divine initiative in origins and beginnings signals a mythic quality.

I emphasize the various aspects of the mythic quality of the material in Exodus–Numbers with the reminder that this is a mythic quality and not myth. The mythic occurs in the relative clarity and scope of the divine commission and revelation; these are stories of beginnings and foundings in which God takes the initiative. I say 'relative' because these are not absolute qualities. Josipovici (1988: 83-85, 193-200) speaks of the rhythm of life that begins in Genesis, but it is a rhythm that is threatened and upset by the narrative pattern of fairytale starts followed by the intrusion of harsh reality. Clines (1990: 93-98) notes the same pattern and refers to it as fair beginnings and foul endings. The wondrous Exodus and events at Sinai are both interrupted by and followed by stories of rebellion and death in Exodus, Leviticus and Numbers.

One final aspect of the mythic quality lies in the issue of leadership and succession. Moses is the divinely chosen leader of the people; he is succeeded by Joshua, who is likewise divinely appointed and installed in a formal ceremony (Num. 27.12-23; Deut. 31.1-8). Aaron is appointed priest and, at the time of his death, formally succeeded by his son Eleazar in the poignant scene in Num. 20.22-29.

Judges: The Centre Cannot Hold

The relative clarity and scope end with the book of Joshua. Joshua and Eleazar die without successors. Some elders live beyond Joshua but are not presented as successors (Josh. 24.31); Eleazar is buried at Gibeah, a town that belongs to his son Phinehas who, however, is not said to succeed his father

7. See Coogan 1978: 75-89, for comments on the Baal myth and a translation of its relevant part.

(Josh. 24.33; see Judg. 20.27-28). Judges opens with the Israelites, as a group of tribes without a leader, taking the initiative and inquiring of the Lord. Throughout Judges Israel splits into individual tribes or groups of tribes; 'all Israel' appears only when they do what is evil in the eyes of the Lord (e.g. 2.11; 3.7, 12; 8.22-28). The mythic aura is gone.

In her article 'The Centre Cannot Hold: Thematic and Textual Instabilities in Judges' (1990), J. Cheryl Exum captures the disintegration of the book in a mode of reading that directly relates the textual difficulties and obscurities of Judges to its content; the instabilities are both thematic and textual. I extend this mode of reading to the issue of kingship; the questioning, dissenting view of monarchy is revealed just as much in the narrative style of 1 Samuel as in its thematic content. Judges offers a way to overcome the disorder and violence. Several times the narrator notes that everyone did what they wanted because 'in those days there was no king in Israel' (17.6; 18.1; 19.1; 21.25). Perhaps a king can provide the missing leadership and unity.

At the close of Judges, Israel has two major problems. First are the Philistines. Samson only began the process of saving Israel from them (13.5)—if what he did can even be called a beginning. Second are the Israelites themselves. In Judges 17–21, the Israelites almost destroy themselves in civil war, and they save the Benjaminites by allowing them to raid other Israelite areas. Perhaps a king can save them from the Philistines and from themselves.

The previous stories of Gideon and his son Abimelech, however, already raise questions about the wisdom and effectiveness of kingship. Gideon refuses to rule over Israel because the Lord rules over them (8.22-23). After his death, his son Abimelech, whose name means 'My father is king', becomes king of Shechem in a bloodbath. A lone survivor, Jotham, issues a parable of the trees denouncing Abimelech. The king is like the bramble from which fire comes 'and devours the cedars of Lebanon' (9.15). The fire consumes Abimelech and the Shechemites and, in the long term, the dynasties and peoples of Israel and Judah. Judges ends with a divided opinion on a king. He can bring leadership and order but he can also bring trouble and destruction on the people he rules.

Judges 21 closes and 1 Samuel 1 opens at Shiloh, but we are given no indication of the time that separates the two chapters (Miscall 1986: 8). This uncertain gap indicates the uncertain relation between the two books. This contrasts with the break between Genesis and Exodus where there is a jump in time and a dramatic change from prosperity to oppression. Given the promises in Genesis, we expect divine help and it comes immediately in the first chapters of Exodus. This is not the case with 1 Samuel and the rise of monarchy. Judges leaves us uncertain as to what to expect and even, as Exum notes (1990: 431), uncertain about whether God is still with his people. The narrative in 01 Samuel does not begin immediately with the birth and call of the first king as Exodus begins with Moses. The 1 Samuel narrative is about origins but without any mythic quality; there is seldom clarity in the matters of the divine role and of the calls of the leaders. And if there is clarity, there is little scope or extent to it.

Doubles and Ambiguity

I focus the reading of my selected material in 1 Samuel (mainly chs. 1–17) around the issues of narrative clarity and scope, especially when the narrative is obscured by ambiguity and doubling. By doubling, I mean the situation when we are faced with the possibility of reading a single story or evaluating a character in two different ways, or when the text presents us with at least two different views or versions of a given event or character.[8] For example, the Israelites are threatened externally and internally, and a king may or may not be the answer to their problems.

1 Samuel 1 is a birth story, which begins with the barrenness of Hannah and the reactions of her husband Elkanah and her

8. I am working with a more limited understanding of narrative ambiguity than I did in either *Workings* or *1 Samuel*. In those two books, particularly the latter, I pushed ambiguity and its deconstructive twin, undecidability, to their limit and attempted to locate them in almost every part of the text; also I understood them to be necessary aspects of the text. Here I locate ambiguity in particular aspects of the text and relate it to some of the larger goals of the narrative.

co-wife Peninnah. There are two major parallels to the story. First is the birth of Moses. Second is the birth of Samson, who is anything but a Moses. 1 Samuel 1 has close ties with the latter story in the shared Nazirite themes. However, 1 Samuel 1 presents us with the misery of Hannah, not the misery of Israel in a situation of oppression, whether from Egyptians or Philistines. This is limited scope. Even if this is the birth of a child of destiny, what is he destined for?

The ambiguity continues in the naming process. Hannah 'called his name Samuel for "I have asked him (*šᵉ'iltîv*) of the Lord"' (1.20). The relevant verb *šā'al*, 'to ask' or 'dedicate', occurs nine times in reference to Samuel (1.17-28; 2.20). It takes interpretive gymnastics to regard *šā'al* as an etymology for Samuel (*šᵉmû'ēl*). There is an obvious relation to Saul (*šā'ûl*) and this exact form of the verb occurs in 1.28: 'he is *given* or *lent* to the Lord'. This is a birth story of two men, Samuel and Saul, or two birth stories folded into one.

Exodus moves quickly from oppression to the birth and call of Moses. 1 Samuel opens with the dual external and internal threat and matches it with the doubling of Samuel and Saul. The Lord's activity is muted. He remembers (*zākar*) Hannah and she conceives (1.19-20);[9] he remembered (*zākar*) his covenant and called Moses to save his people (Exod. 2.24). The former comment refers to Samuel's conception and birth and not to a divine plan or purpose for the child. There is one area in which the Lord's role is clearly delineated but with limited scope. He is against the house of Eli because of the corruption of Hophni and Phinehas (1 Sam. 2.12-17, 22-36); all three, father and sons, die on the same day (4.12-18). But this is the fate of the house of Eli, not the house of Israel,[10] and we are definitely not dealing with divinely appointed successors like Joshua and

9. The repeated statement in vv. 5-6 that the Lord had closed Hannah's womb may be the narrator's statement or the indirect discourse of Elkanah and Peninnah, i.e. the comment on divine action is ascribed to their judgment.

10. Absalom's fate is analogous. He is doomed 'because the Lord had ordained to cancel the wise counsel of Ahithophel so that the Lord could bring ruin on Absalom' (2 Sam. 17.14). Significantly, there is no mention of how this divine action relates to David and his future.

Eleazar. Ichabod, 'Where is the glory?', survives in the house of Eli (4.19-22).

Moses is 80 years old (Exod. 7.7) when called by the Lord. He questions who this God is and why he, Moses, should be accepted by the Israelites as their divinely called leader. Provided with answers, he still has the temerity and strength to object to and refuse the call; the Lord overrules the objections and sends Moses on his mission. Samuel, a mere youth, does not even realize that he is being called and must ironically rely on Eli, the object of the divine judgment, to direct him. The message to Samuel is a double of that delivered by the man of God (1 Sam. 2.27-36); it concerns the fate of Eli's house, not that of Israel, and Samuel only delivers it at the urging of Eli. This narrative on origins bogs down in doubling and ambiguity.

Samuel is the Lord's trustworthy prophet (3.19-21) and Saul, the Lord's king and anointed (2.10), lingers in the background. The narrative does not move to the establishment of monarch and prophet; instead it shunts both aside in favor of stories of Israel's defeat by the Philistines and the Philistines' defeat by the Ark. In large part I find Polzin's parabolic reading of this material compelling (1989: 55-79). The collapse and death of Eli prefigure the collapse and death of monarchy, and the powerful Ark on its own symbolizes Israel without kings and, I would add, without prophets. Samuel is far from the Mosaic successor predicted in Deut. 18.15-22 (also see 34.10-12).

Such a parabolic reading is fitting for a narrative of the rise of monarchy (and prophecy) that is unable to get to the point. Kings and kingship there will be even though defeat and death are their ultimate fate, yet the narrative provides this overview obliquely since we the readers provide the connection between the houses of Eli and David. The Lord retains power and control, yet this is symbolized by the story of the Ark wreaking havoc among the Philistines and then, surprisingly, among the Israelites (6.19; I am reading the number 50,070 that is in the Hebrew text).

In ch. 7, Samuel returns to the scene as effective leader to whom both Israel and the Lord listen. Israel puts away the Baals and serves the Lord; the Lord answers Samuel's cry and the Philistines are crushed. Samuel saves the Israelites from themselves and their enemies. He is a hero. The fairytale view is upset in the first four verses of ch. 8. Samuel has no divinely appointed successor and, like Eli, he is disgraced by his sons. The Lord does not step in to remedy the situation, and the elders take the initiative, demanding that Samuel 'set up for us a king to govern us, like all the nations' (8.5).

Kingship without Myth

Kingship and king are introduced in anything but a mythic mode. The people take the initiative and the Lord goes along. Chapter 8 is dominated by the negative themes of Israel's rejection of God (vv. 7-8), the harsh rule of the king (vv. 11-17), God's rejection of the people (v. 18), and the people's refusal to listen to their prophet (v. 19; see Deut. 18.15-22). Three times the Lord commands Samuel to listen to the people and to give them a king (vv. 7, 9, 22). Samuel does not obey. Unlike Moses who can object and question the Lord, Samuel simply does not act. He sends all the people home, and God has to send the future king to Samuel (Polzin 1989: 83-88).

A king, Saul, arrives on the scene in the folktale of the tall, handsome lad who goes in search of his father's asses and finds a kingdom. Samuel anoints him king and although the anointing is followed by signs, the signs are far from the signs and wonders of Exodus. Samuel recalls the Exodus in 10.17-19, and he again associates the king with the people's rejection of God. Saul, indeed, may be the only one affected by the signs; people around him muse, 'Is Saul also among the prophets?' (10.9-13).

The tale does not have the mythic qualities of the call of Moses. It is a fairytale or folktale, not a myth. I do not use these terms in a technical sense; I want to indicate the reduced scope of divine involvement in 1 Samuel, particularly involvement in the form of explicit and far-reaching statements and miraculous acts and wonders. Fairytale or folktale also captures the sense of a story in which things go particularly, and frequently surprisingly, well.

Saul comes late on the scene even though his 'birth story' is in ch. 1. The Lord speaks to Samuel and not to Saul. His statement echoes the assertion, in Exod. 2.23-25, that the Lord heard the

Israelites' groaning. Saul will save Israel from the Philistines; no further role is given.

> About this time tomorrow I will send you a man from the land of Benjamin, and you will anoint him leader [*nāgîd*][11] over my people Israel. He will save my people from the hand of the Philistines for I have taken note of my people because their cry has come to me (01 Sam. 9.16).

The folktale continues and doubles itself in 10.20-24 with the public selection as king of the tall lad who hides himself in the baggage. Negative statements on king and kingship in vv. 17-19 and 25-27 frame the tale. Narrative ambiguity arises in the doublets (and even triplets). God sends Saul; Samuel anoints him; finally, the people proclaim him king. King and kingship are both God's will and rejection of God.

Saul's fair beginning reaches its peak with his defeat of the Ammonites; this is his finest hour, yet it is tarnished by the fact that it is Ammonites, not Philistines. It is also marred by the parallel with Jephthah who defeats the Ammonites and who, like Saul in 1 Samuel 14, makes a rash vow that endangers his child. There is a parallel with Israel at the entrance to the land since at the end of Numbers they are near or even in Ammonite land; Samuel alludes to the exodus and the land in 1 Sam. 12.8. However, Saul is not an example of the divinely initiated transference of power witnessed in Eleazar's ordination (Num. 20.22-29) and Joshua's appointment (27.12-23).

Finally, there is the clouded and debated passage in 1 Sam. 11.12-13:

> The people said to Samuel, 'Who is saying, "Saul will reign over us"? Give us the men that we may put them to death'. And Saul said, 'Not a man will be put to death this day for today the Lord has wrought deliverance in Israel'.

However we interpret the passage,[12] it links king with death

11. I am not going to treat the issue of the distinction between *nāgîd*, leader, and *melek*, king. It is part of the non-mythic quality of this tale that *melek* is not used by the Lord in his first reference to Saul. See Miscall 1986: 53-59, 85-87, for discussion.

12. Polzin argues that Saul's victory over the Ammonites has established Saul as a judge-type leader and not as king; this scene is the people's last

and not with the death of a predecessor. The chapter closes with a repetition; the people, not Samuel or the Lord, make Saul king even though this is what the Lord commanded Samuel to do (8.22) and what, in the very next verse, Samuel claims to have done (12.1).

Samuel's address in ch. 12 compares with Moses' in Deuteronomy in its retrospective and anticipatory aspects, except that Samuel does not die immediately afterwards. We do not have a transfer of power as in Numbers and Deuteronomy; rather we have overlapping leaders and leaders in conflict. 1 Samuel 13–15 are the sad story of Saul's failure, both in collision with Samuel (chs. 13 and 15) and on his own (ch. 14). The Lord rejects Saul. He regrets that he has made Saul king; he changes his mind (15.10, 35; see Samuel's comment in v. 29). In other words, we are dealing with a divine mistake, a false start (Polzin 1989: 28-29), and not a long-range plan. Both the Lord and Samuel pronounce Saul's rejection as king, yet Saul remains as king until the end of 1 Samuel; the remainder of 1 Samuel is occupied by the conflict between Saul and David.[13]

We have leaders and stories in conflict. Even though I have spoken of Clines's pattern of fair beginnings and foul endings, there is no such clear succession in 1 Samuel 1–15. This narrative of the establishment of kingship (and of prophets) has a hard time beginning; it does not even get to the king until ch. 8. It starts and proceeds with doubles: Samuel and Saul; Samuel and Eli; king and prophet; and the Lord's anointed king (2.10) and the king as rejection of the Lord. It is better to speak of a conflict of fair beginning and foul beginning rather than a succession of beginning and ending.

David: Fairytale and Reality

Once Saul is rejected, David enters the picture. He is the youngest and the unexpected; unlike the others, he has no birth story. Fitting 1 Samuel, he has a double introduction: 16.1-13, Samuel

chance to turn away from the disaster of kingship (1989: 108-14).

13. The relationship between Saul and the Lord is one of conflict. Conflict was in place with Samuel and the house of Eli. The theme of conflict continues into 2 Kings, at least until the fall of Israel (2 Kgs 17).

anoints him king, and vv. 14-23, Saul makes him court musician and armor-bearer. This is the 'sweet psalmist of Israel' (2 Sam. 23.1 [RSV]; Gunn 1989: 133-34). The double introduction involves David with two families: Jesse's and Saul's (Jobling 1986 and Pleins 1992). David enters Saul's court and later Saul simply takes David 'and would not let him return to his father's house' (18.2).

With David, the Lord commands Samuel, 'Anoint him for this is he!' The spirit seizes David from that day on; an evil spirit comes upon Saul. Yet, to stress my point, this is not a transfer of power as in days of old. Rejection is not immediately followed by death or removal; anointing, by coronation. Can kingship save Israel from either themselves or their enemies, since there are two kings? In the long view the author of Genesis–2 Kings must answer 'no' to both parts of the question. Israel's and Judah's continuous rebellions against the Lord bring first Assyria and then Babylon down upon them. The Lord's false start and mistake with Saul are a parable for the false start and mistake of the institution of monarchy. On the other hand, the doubles and ambiguity of the narrative are a reflection of the fact that the doomed monarchy lasts for centuries. God is with his people either through or despite the monarchy.

In *The Workings of Old Testament Narrative* (Miscall 1983: 57-83), I deal at length with the ambiguity of the portrayal of David in 1 Samuel 17. Is he the pious young shepherd of tradition or a cunning, scheming contender for the throne? I conduct that reading with the focus on David. Here I place the tale of David and Goliath alongside the two stories of kingship.

In the fair beginning, the innocent and pious youth kills the fearsome and heavily armored enemy through his native skill and trust in God. This is analogous to Saul's first victory. From one view, the story contrasts the 'city' (Goliath) with the 'country' (David). David does not don the armor and sword of royalty (17.38-39); he is not a warrior like those of all the nations. His previous opponents have been lions and bears (vv. 31-37). David proclaims the powerful Lord, the God of the armies of Israel (vv. 26, 36, 45; see 2 Sam. 22). His victory is God's and knowledge of the Lord will be its result.

> This day the Lord will deliver you into my hand, and I will strike you down and cut off your head; I will give the bodies of the Philistine army today to the birds of the air and the wild beasts of the earth; that all the earth may know that there is a God in Israel, and that all this assembly may know that the Lord saves not with sword and spear; for the battle is the Lord's and he will give you into our hand (vv. 46-47).

The fairytale closes with the legendary killing of Goliath. The unarmed youth uses his sling to embed a stone in Goliath's forehead:

> David prevailed over the Philistine with a sling and with a stone; he struck the Philistine and he killed him. There was no sword in the hand of David (v. 50).

The victory belongs to God who saves without sword or spear.

The fairytale ends with v. 50, and the narrative doubles itself with a tale of a foul beginning, the reality of kings and kingship:

> David ran and stood over the Philistine; he took his sword, drew it out of its sheath, killed him and cut off his head with it. The Philistines saw that their hero was dead and they fled. The men of Israel and Judah, with a shout, immediately pursued the Philistines...The Israelites returned from the pursuit of the Philistines and plundered their camp. David took the Philistine's head and brought it to Jerusalem; his armor he put in his tent (vv. 51-54).

David kills with the sword and the victory is his; he claims the trophies and the spoils (Goliath's head is still with him when he later takes Jerusalem [2 Sam. 5.6-10]). The result of the death of Goliath is flight, death and plunder. No one, David, Philistine or Israelite, acknowledges that there is a God in Israel and that this battle is the Lord's.

The closing scene, 17.55-58, is terse and difficult. 'I do not know' contrasts with the repeated 'that they may know' of vv. 46-47. Chapter 18.1-5 follows immediately upon the scene. My concern is v. 4:

> Jonathan stripped off the tunic he wore and gave it to David with his armor including his sword, his bow and his belt.

David dons the 'city' symbols of royalty and war, and the foul and sad story of monarchy in Israel and in Judah follows. The sword never departs from David's house (2 Sam. 12.9-10). Seldom is there knowledge of the Lord, especially on the part of kings, but there is certainly knowledge of the sword. And the sword is frequently turned against themselves in the court intrigues, succession disputes, coups, etc. The fairytale of 1 Sam. 17.1-50 ends when David kills with the sword and girds himself with the royal armor and sword.

Indeed, the comment in 2 Sam. 21.19 shows it to be a fairytale. 'Elhanan the son of Jaareoregim, the Bethlehemite, slew Goliath the Gittite, the shaft of whose spear was like a weaver's beam'. David has probably taken credit for a warrior's accomplishment, as Saul did with Jonathan's victory (1 Sam. 13.2-4; Miscall 1986: 82-83). It is as though the narrator, at the close of 2 Samuel, says that after all the betrayal and violence involving David, he can no longer maintain the lie of David and Goliath even as a fairytale. Nevertheless David continues as king and his dynasty endures for almost 400 years.

The Fantastic

I have dealt with the Exodus material in my essay 'Biblical Narrative and Categories of the Fantastic' (forthcoming) and here focus on the relevance of one aspect of the fantastic for reading 1 Samuel. It derives from Tzvetan Todorov's *The Fantastic*. According to George Aichele,

> Todorov argues that literary fantasy arises as a moment of hesitation or uncertainty, in which one is unable to decide whether a given narrative phenomenon or set of phenomena belongs to the genre of the uncanny—bizarre occurrences, for which a natural explanation is nonetheless possible—or to the genre of the marvelous—for which only a supernatural explanation can be given (1991: 325).

The plagues and the revelation of Torah at Sinai are all marvelous; they are manifestly the work of the Lord. If, in the book of Exodus, we were to delete the speeches and events that are explicitly ascribed to the Lord, we would have little of the present book left. In 1 Samuel, on the other hand, if we were to conduct a similar deletion, we would have most of the present book left. With the exception of the Ark Narrative, the events and their sequence are not uncanny or bizarre occurrences.

The double story of kingship can be viewed from yet another perspective. First is the human story of a time of change when old institutions, priesthood and judgeship, die and new ones, prophecy and kingship, arise amidst controversy and conflict. The new institutions grow old and end in the disasters of the late eighth and early sixth centuries. This is the story of the academy told in scholarly histories of Israel. There is also the marvelous story of God's faithfulness to his people and his gracious direction of their history despite his periodic punishments of them for their rebellion. Samuel and David, like Moses, are heroes; Saul is the villain. Monarchy may end, but God remains with his people and brings them back from disaster and exile. This is the story of church and synagogue told in Bible histories, sermons and the countless retellings of the youth's defeat of the giant warrior.

The hesitation, the uncertainty, lies in the relation of the two stories. Kingship is both acceptable to God and rebellion against God. A king is God's choice (Deut. 17.14-20) and the people's choice (1 Sam. 8.18). Kingship endures, yet it is a mistake, a false start. The Lord remains with his people through or despite kingship. These views of kingship stand in tension, and that tension is captured in both the style and content of the narrative(s) of the rise of kingship. Contrasts with the Pentateuchal narrative of Moses and Israel highlight the ambiguity, the doubling, and the lack of resolution and scope of the Samuel narrative.

The tension is captured, not resolved. The differing views and stories of kingship stand in an uncertain relationship. This is a human story and a divine story. Supernatural explanations are given, but we are not always certain when God intervenes or why, to what extent and for what purpose. 2 Samuel ends with the story of the census-taking that encapsulates the doubling and uncertainty. The Lord is angry, punishing and relenting. It is David's sin (24.10, 17), although it is not at all clear in the Bible why census-taking is sinful. The Lord allows him to choose his punishment, but the plague falls on the people, not David. David is both means of punishment and death and means of atonement and relief. The relief, however, comes through more of a priestly than a royal function, and altar and sacrifices recall the sad story of Saul's failure in 1 Samuel 13–15.

Concluding Remarks

To summarize, the often noted ambiguity and doubling of the narratives in 1 Samuel 1–17 are not due to a secondary editing process that somewhat mechanically combined early and late materials that are, respectively, pro-monarchical and anti-monarchical. The narratives present kingship and its origins in a decidedly negative light through the doubling and through the narrative pace that delays the appearance of the first two kings, Saul and David. The style and pace stand out distinctly when we compare the Samuel material with the Pentateuchal story of Moses and its strong mythic quality. The same process of comparison and contrast reveals the negative judgment of kingship inherent in the style and pace of 1 Samuel 1–17; unlike the Exodus story, the story of the origins of kingship does not focus on the king, the one leader who can lead and save Israel from external and internal enemies. My title pairs Moses and David; the biblical narrative, however, parallels Moses with Samuel, Saul and David.

BIBLIOGRAPHY

Aichele, G., Jr
1991 'Literary Fantasy and Postmodern Theology', *JAAR* 59: 323-37.

Clines, D.J.A.
1990 'The Old Testament Histories: A Reader's Guide', in *What Does Eve Do To Help? And Other Readerly Questions to the Old Testament* (JSOTSup, 94; Sheffield: JSOT Press): 85-105.

Coogan, M.D. (trans and ed.)
1978 *Stories from Ancient Canaan* (Philadelphia: Westminster Press).

Cross, F.M.
1973 *Canaanite Myth and Hebrew Epic: Essays in the History of the Religion of Israel* (Cambridge: Harvard University Press).

Exum, J.C.
1990 'The Centre Cannot Hold: Thematic and Textual Instabilities in Judges', *CBQ* 52: 410-31.

Gunn, D.M.
1989 'In Security: The David of Biblical Narrative', in *Signs and Wonders:*

Biblical Texts in Literary Focus (ed. J. Cheryl Exum; Atlanta: Scholars Press): 133-51.

Heidel, A.
1951 *The Babylonian Genesis: The Story of Creation* (Chicago: University of Chicago Press, 2nd edn).

Jobling, D.
1978 'Jonathan: A Structural Study in 1 Samuel', in *The Sense of Biblical Narrative: Three Structural Analyses in the Old Testament* (JSOTSup, 7; Sheffield: JSOT Press): 4-25.
1986 'Deuteronomic Political Theory in Judges and 1 Samuel 1–12', in *The Sense of Biblical Narrative: Structural Analyses in the Hebrew Bible, II* (JSOTSup, 39; Sheffield: JSOT Press): 44-87.

Josipovici, G.
1988 *The Book of God: A Response to the Bible* (New Haven: Yale University Press).

Levenson, J.D.
1988 *Creation and the Persistence of Evil: The Jewish Drama of Divine Omnipotence* (San Francisco: Harper & Row).

Miscall, P.D.
1983 *The Workings of Old Testament Narrative* (Philadelphia: Fortress Press; Chico, CA: Scholars Press).
1986 *1 Samuel: A Literary Reading* (Bloomington: Indiana University Press).
1989 'For David's Sake: A Response to David M. Gunn', in *Signs and Wonders: Biblical Texts in Literary Focus* (ed. J. Cheryl Exum; Atlanta: Scholars Press): 153-63.
1992 'Biblical Narrative and Categories of the Fantastic', *Semeia* 60: 39-51.

Ollenburger, B.C.
1987 *Zion, the City of the Great King: A Theological Symbol of the Jerusalem Cult* (JSOTSup, 41; Sheffield: JSOT Press).

Pleins, J.D.
1992 'Son-Slayers and their Sons', *CBQ* 54: 29-38.

Polzin, R.
1989 *Samuel and the Deuteronomist: A Literary Study of the Deuteronomic History, Part Two: I Samuel* (San Francisco: Harper & Row).

Pritchard, J.B. (ed.)
1969 *Ancient Near Eastern Texts Relating to the Old Testament* (Princeton: Princeton University Press, 3rd edn).

Todorov, T.
1973 *The Fantastic: A Structural Approach to a Literary Genre* (trans. R. Howard; Cleveland: Case Western Reserve University Press).

Weinfeld, M.
1983 'Zion and Jerusalem as Religious and Political Capital: Ideology and Utopia', in *The Poet and the Historian: Essays in Literary and Historical Biblical Criticism* (ed. R.E. Friedman; HSM, 26; Chico, CA: Scholars Press): 75-115.

CURSES AND KINGS: A READING OF 2 SAMUEL 15–16

Robert Polzin

At Play in the Fields of the LORD: Paronomasia in 2 Samuel 15

2 Samuel 15–18 describes the rebellion of Absalom that ends with his death. Four years after David had allowed Absalom to return to Jerusalem, his son's revolt forces David himself to flee the city. That revolt and that flight comprise the main events of ch. 15, which wonderfully exhibits the interplay of esthetic brilliance and ideological complexity that characterizes the Deuteronomic History. Attention to a few apparently minor details in ch. 15 will help to introduce the more important issues driving the narrative along.

Verses 1–6 form a neat exposition of the four years preceding Absalom's revolt. In its frequent use of at least seven imperfective verb forms to indicate habitual, repeated or condensed action, this introductory section of the chapter recalls 1 Sam. 1.1-8, expository verses preparing the reader for the events surrounding the birth of Samuel.[1] Here in 2 Samuel 15, the narrator describes what Absalom *used to do* or *continued to do*, over the four years during which he 'stole the hearts of the men of Israel' (v. 6).[2] The importance of vv. 1-6 as exposition

1. See R. Polzin, *Samuel and the Deuteronomist* (Bloomington: Indiana University Press, 1993), pp. 19-20.

2. S.R. Driver remarks, 'Notice the pff. with *waw* conv., indicating what Absalom *used* to do. From 2b to 4, however, the narrator lapses into the tense of simple description, only again bringing the custom into prominence in v. 5, and 6a (*yb'w*)' (*Notes on the Hebrew Text and the Topography of the Books of Samuel, with an Introduction on Hebrew Paleography and the Ancient Versions and Facsimiles of Inscriptions and Maps* [Oxford: Clarendon Press, 2nd edn revised and enlarged, 1913], p. 310). We now believe that the narrator's 'lapse' into perfective verb forms in the MT of vv. 2-4 probably

introducing the revolt of Absalom lies in the efficiency with which these verses characterize Absalom and highlight major issues.

In the exposition, the narrator quotes Absalom, 'O that I were judge in the land! Then every man with a suit (*rîb*) or cause (*mišpaṭ*) might come to me, and I would justify him (*wᵉhiṣdaqtîw*)' (v. 4). The significance of this reported speech of Absalom in the expository material introducing his revolt can hardly be exaggerated.[3] By exposing the abiding motivation of Absalom as he played the politician before every Israelite who came to Jerusalem for a suit ('See, your claims are good and right' [v. 3]), vv. 1-6 move us to wonder on the one hand how Absalom's behavior corresponds to the Deuteronomic Lawcode concerning such matters, and, on the other, how such behavior compares to that of other characters in the story.

The relevant legislation in Deuteronomy is straightforward: 'When there is a dispute (*rîb*) between men and they come into court, then they shall judge them. They shall justify (*wᵉhiṣdîqû*) the righteous (*haṣṣaddîq*) and condemn the guilty' (Deut. 25.1). The contrast between this law and Absalom's behavior is clear and simple: Absalom ought to distinguish between the innocent and guilty, yet his practice, over many years, of declaring 'good and righteous' the claims of everyone coming to Jerusalem for judgment appears to contravene the Lawcode and to constitute a flagrant attempt to steal the hearts of his fellow Israelites in preparation for usurping the throne of his father.

With respect to the legislation of Deut. 25.1-3, Solomon's words in 1 Kings 8 contrast sharply with Absalom's. During the dedication of the temple, Solomon implores the LORD to judge those who come before the altar of the temple: 'Then hear thou

reflects more accurately a lapse of textual traditions instead. There is some evidence from LXX^{L}, 4QSam^{a} and 4QSam^{c} that even the perfective verbs used by the narrator in vv. 2-4 of the MT, like those of vv. 1, 5 and 6, were originally imperfectives denoting Absalom's habitual actions. See on this point P.K. McCarter, *II Samuel* (AB, 9; Garden City, NY: Doubleday, 1984), p. 354 and his reference there to Ulrich's work.

3. In *Samuel and the Deuteronomist*, pp. 60-63, I suggested that many details in 2 Sam. 15 look back to the ark account in 1 Sam. 4.1–7.2. Here I am concentrating on how ch. 15 looks forward to aspects of the story to come.

in heaven, and act, and judge thy servants, condemning the guilty by bringing his conduct upon his own head, and vindicating the righteous (*ûlhaṣdîq ṣaddîq*) by rewarding him according to his righteousness (*ṣidqātô*)' (1 Kgs 8.32).

What makes the language of justification in Moses's law and Solomon's prayer so important, as a context for Absalom's behavior here in 2 Samuel 15, is its distinctiveness within the History: Deuteronomy 25, 2 Samuel 15 and 1 Kings 8 are the only places in the History where the root *ṣdq* is used in a verbal form (whether *qal, nifal, hifil* or *hitpael*). Absalom's habit of declaring innocent *everyone* bringing a suit to Jerusalem appears, therefore, to focus upon and contravene the law of Moses, even as it stands in sharp contrast to the concept of divine justice voiced, if not practiced, by his half brother Solomon later on in the story. This aspect of Absalom's preparations for revolt succeeds in surrounding his *character zone* with a negative evaluation from the very beginning. Rather than being seen as trying to take over responsibilities somehow neglected by his father, Absalom is indirectly portrayed here as subverting Israelite law in order to curry favor with those whose support he will need if his revolt is to succeed. We see here how legal and literary context can transform an apparently innocent statement into an implicit condemnation of its speaker.

The chapter's tendency to play with language highlights a second aspect of Absalom's promiscuous justification of fellow Israelites during the period preceding his revolt. When David hears that 'the hearts of the men of Israel have gone after Absalom' (v. 13), the king quickly flees the city, and immediately meets three individuals, Ittai the Gittite (vv. 19-22), Zadok the priest (vv. 24-29) and Hushai the Archite (vv. 32-37). Wordplay immediately surrounds David's meeting with Ittai. The account of the meeting comprises only 4 verses, yet Ittai's name is connected to the circumstances surrounding David's flight in a number of interesting ways. The meeting begins with 'Then the king said to Ittai (*'ittay*) the Gittite, "Why will you also go with us (*'ittānû*)?"' (v. 19), and ends with 'So Ittai (*'ittay*) the Gittite passed on, with all his men and all the little ones who were with him (*'ittô*)' (v. 22). In between, Ittai's own oath indicates the thematic function that his name plays in the story itself, 'But Ittai answered the king, "As the LORD lives, and as my lord the king lives, wherever my lord the king shall be, whether for death or for life, there also will your servant be"' (v. 21). The narrative role of Ittai, therefore, whose very name suggests 'loyalty' or 'companion', is *to be with David wherever he goes.*[4]

Lest we assume that there is something haphazard about the appearance of Ittai in conjunction with the repeated usage of *'et* in vv. 19-22, a broader perspective suggests that paronomasia is a widespread feature of the story at this point. *'et*, as one of the two principal words meaning 'with' in Hebrew (*'im* is the other word), occurs 10 times in ch. 15 alone—more often than in any other chapter of 2 Samuel.[5] Moreover, if we look at the larger story of Absalom's revolt, we notice that the occurrences of *'et*, 'with', in these 5 chapters are much more frequent than anywhere else in the book.[6] It is safe to suggest, therefore, that wordplay involving the meeting of David and Ittai in 15.19-22 points to aspects of the narrative that transcend a merely esthetic connection of the name of Ittai to his abiding desire to be *with* David. A number of paronomastic details within the chapter not only structure its narrative events, but also indicate some authorial perspectives that shape the larger story of Absalom's revolt.

If we return to the expository material in vv. 1-6, we will see how wordplay can indicate something about the authorial motivation behind the three meetings recounted in this chapter. I have already remarked how rarely the History uses the language of justification employed by Absalom in v. 4 ('Then *every man* with a suit or cause might come to me, and *I would*

4. See M. Garsiel, *Biblical Names: A Literary Study of Midrashic Derivations and Puns* (Ramat Gan: Bar Ilan University Press, 1991), p. 219, who has independently noted the wordplay between Ittai and *'th/'tnw* in 2 Sam. 15.19. I am suggesting here that such wordplay extends beyond 15.19, and that its narrative functions are closely related to other textual features of chapters 15-20.

5. 15.3, 11, 12, 14, 19, 22, 24, 27, 30, 33. Twenty per cent of the occurrences of *'et* in 2 Samuel appear in 2 Sam. 15, a chapter that comprises only 5% of the book. *'et* occurs 53 times in 2 Samuel and 10 times in ch. 15 alone. (In terms of verses, ch. 15 constitutes about only 5% of 2 Samuel.)

6. In chs. 15–19, *'et*, 'with', occurs 29 times out of the 53 times it appears in 2 Samuel.

declare him righteous' [*wᵉhiṣdaqtîw*]). Besides providing the initial basis for a negative evaluation of Absalom in the story, his language also serves as esthetic play preceding further play on the name of the *second* person David will meet in ch. 15, Zadok the priest. Like Ittai, Zadok starts out with David, but unlike Ittai, whom David keeps with him, Zadok must return to Jerusalem. The one whose very name denotes 'companion' accompanies David as he flees Jerusalem. On the other hand, Zadok, whose name denotes 'justice' or 'righteousness', returns to the royal pretender, who for four years had unscrupulously justified or declared righteous every claimant he met at the gate of Jerusalem. There is something deliciously playful, yet intensely serious, about such verbal and narrative insinuations.

A third, albeit less direct, example of wordplay in ch. 15 involves the name of the last person David meets in the chapter, Hushai the Archite.[7] David's departure from Jerusalem is accomplished *in haste*: 'Then David said to all his servants who were with him at Jerusalem, "Arise, and let us flee; or else there will be no escape for us from Absalom; *go in haste* (*mahᵃrû*) lest he overtake us *quickly* (*yᵉmahēr*)"' (v. 14). It is more than accidental that Hushai, whose role in the History is confined to 2 Samuel 15–17, turns up in a story that emphasizes the dangerous haste surrounding David. *mhr* and *ḥûš* are synonyms,[8] so that there is an obvious correlation between the presence of Hushai and the increased usage of *mhr* in 2 Samuel 15–19.[9] The man whom we could call in English, 'Hasty the Archite', now comes upon the scene, because the semantic

7. Wordplay involving Hushai has already been pointed out by Garsiel (*Biblical Names*, p. 105). My discussion here widens the scope of his comments.

8. For example, in 1 Sam. 20.38: 'Jonathan called after the lad, "Hurry (*mᵉhērâ*), make haste (*ḥûšâ*), do not stay (*'al taᵃmōd*)"'.

9. *ḥûš* appears rarely in the History (Deut. 32.35; Judg 20.37 and 1 Sam. 20.38); the root *mhr*, 'to hurry', however, appears much more frequently—over 40 times in Deuteronomy–2 Kings. What is important for our purposes is, first, that 1 Sam. 20.38 uses the two words as synonyms and, secondly, that in the books of 2 Samuel–2 Kings, for example, *mhr* occurs only 10 times, yet 5 of these occurrences are here in 2 Sam. 15–19 (2 Sam. 15.14; 17.16, 18, 21; 19.17), precisely and only where we find Hushai, the hasty one, participating in the story (15.32, 37; 16.16, 17, 18; 17.5, 6, 7, 8, 14, 15). Hushai, as the father of Baaniah, is simply mentioned in 1 Kgs 4.16.

wordplay between *mhr* and *ḥûšay*, like that surrounding the names of Ittai and Zadok, indicates something more than simply the esthetic pleasure that comes from etymologizing. In such wordplay we encounter something close to a recurring signal about authorial motivations for shaping the story of Absalom's revolt in its present form. David keeps with him Ittai, his loyal companion from Gath, but returns the righteous one, Zadok, to Absalom, the one who had unrighteously declared all Israelites righteous. Then David returns Hushai, his hasty friend, to the royal pretender whose contemplated haste, David declared, would force the king to leave Jerusalem in haste (v. 14).

The names of those with whom David meets and speaks in 2 Samuel 15 tell us something about the ideological dimensions of the story: What does it mean to be *with* David—or *with* Absalom for that matter? Whose side is the side of *justice*, David's or Absalom's? And what evaluative accents surround the *hasty* comings and goings that constitute this part of David's story? Before I can attend to these important questions, further artful aspects within the story need discussing.

Between Mimesis and Artifice: Crossing Boundaries

So much has been written from the supposition that the author(s) of 2 Samuel 9–20 wrote from direct personal knowledge and with a wealth of realistic particulars, that I feel compelled to complement this picture by describing some signals of literary composition within it that highlight a central aspect of the story. The account of the succession to the throne of David, at this point at least, appears to possess two opposed stylistic characteristics: a narrative edifice exhibiting an elaborate façade of mimetic detail, yet in addition, a well crafted and highly stylized—even ritualized—account of Absalom's abortive attempt to succeed to the throne of David. In connection with the extensive wordplay that I am suggesting characterizes this part of David's story, the profoundly ritualized nature of David's hasty retreat from Jerusalem, and of his painful return, is an important signal of the extent of literary artifice that has gone into the final composition of this story. We can begin to cross over from mimesis to artifice by examining the many ways in which characters cross over (*'ābar*) boundaries—in this chapter and in those to come.[10]

We already know from the ritual procession in Josh. 3.1–5.1 that those in the History who cross the Jordan usually carry with them heavy ideological baggage.[11] To cross over (*'ābar*) into or out of the land is an especially appropriate action for Hebrews (*'ibrîm*), yet there are four sections within the History where occurrences of *'ābar*, 'to cross over', are particularly frequent: Deuteronomy 2–4; Joshua 3.1–5.1; Joshua 24; and here in 2 Samuel 15–20.[12] One ought not to be surprised, therefore, that the frequent use of *'ābar* in 2 Samuel 15–20 carries with it a number of important implications for our understanding of the story. If we concentrate on the choreography of 'crossing over' (*'ābar*) in the account of David's flight from Jerusalem in ch. 15, we can illustrate how the larger complex in chs. 15–20 uses highly stylized language to convey highly ritualized action.[13]

To begin with the *stylistic façade* of 2 Samuel 15, countless readers have commented on the general impression they have when reading 2 Samuel 9–20 that it is based upon an 'eye witness' account of the events unfolding within it. However, when such commentators try to give reasons for their impression, they rarely discuss those features which directly and obviously suggest that the narrator may be writing from direct personal

10. Ari Cartun has written an important article on topography as a literary template for 2 Sam. 15–99, thus highlighting an important aspect of this section's highly stylized language and composition: 'Topography as a Template for David's Fortunes during his Flight before Avshalom', *Journal of Reform Judaism* (Spring 1991), pp. 17-34. See also David Gunn, 'From Jerusalem to the Jordan and Back: Symmetry in 2 Samuel XV–XX', *VT* 30 (1980), pp. 109-13.

11. For an account of the ideological implications of crossing the Jordan in Josh. 3.1–5.1, see my remarks in *Moses and the Deuteronomist* (Bloomington, IN: Indiana University Press, 1993), pp. 91-110.

12. In Deut. 2–4, the root *'ābar* occurs 30 times in 115 verses; in Josh. 3.1–5.1, 25 times in only 42 verses; in Josh. 24, 8 times in 33 verses; and in 2 Sam. 15–19, 38 times in 165 verses. In 2 Sam. 15–19, a majority of occurrences (25) appear in the first and last chapters of the section.

13. This combination of stylization and ritualization is similar to the narrative style employed by the narrator to describe the events in 2 Sam. 2.

knowledge. Chapter 15's extensive use of imperfective verb forms in the reporting speech of the narrator is perhaps the most immediate, yet largely unrecognized, compositional reason for the 'eye witness' flavor of this portion of the story of Absalom's revolt. There may be no pericope within the books of Samuel that so abundantly employs verb forms whose function is to bring readers into the center of the action by presenting that action as if it were taking place before their eyes—in a manner similar to the temporal point of view of the characters themselves within the story world. David's flight from Jerusalem, from the conspiracy of Absalom mentioned in v. 12 to Absalom's subsequent taking over of the city in v. 37, is narrated from a predominantly *synchronic viewpoint* that succeeds in slowing the action down and giving readers the impression that they too are present as events unfold.

This obvious feature of the chapter's narrative style says absolutely nothing about anyone's actual knowledge of, or physical presence at, the events described therein. All we can really say is that the function of these imperfective verb forms is to convey an impression of such knowledge. Like the narrator's obvious omniscience, the text's synchronic perspective is simply a conventional literary feature that establishes for us, as Sternberg might say, the truth claim, but not the truth value, of the reliable, or eye witness, flavor of this chapter's happenings. And, as previous readers have so often remarked, what happens before our eyes in ch. 15 is as much a ritual performance as it is a strategic retreat.[14]

Besides, then, the imperfective verb forms which indicate habitual or condensed action in the exposition within vv. 1-6, an unusual number of other imperfective verb forms function in the chapter to draw the reader into the center of action and to represent the temporal perspective of characters rather than that of the narrator. The following narrative statements are synchronic rather than retrospective:

v. 12 And the people with Absalom continued to increase (*hôlēk wārāb*);

v. 18 And all his servants were passing by him (*'ōb^{e}rîm 'al*

14. See McCarter, *II Samuel*, pp. 375-76.

yādô) and all the Cherethites, and all the Pelethites, and all the six hundred Gittites were passing in front of the king (*'ōbᵉrîm 'al pᵉnê hammelek*);

v. 23 And all the land was weeping (*bôkîm*) with a loud voice and all the people were crossing (*'ōbᵉrîm*) and the king was crossing over (*'ōbēr*) the brook Kidron and all the people were crossing over (*'ōbᵉrîm*) in front of the road to (?) the wilderness;

v. 24 And lo (*wᵉhinnēh*) Zadok and the priests with him were carrying (*nōśᵉ'îm*) the ark.

v. 30 David was going up (*'ōleh*) the ascent, going up (*'ōleh*) and weeping (*bôkeh*), and his head is covered (*ḥāpûy*) and he was walking (*holēk*) barefoot, and they were going up (*wᵉ'ālû*) crying continually (*'ālōh ûbākōh*).

v. 32 Behold (*wᵉhinneh*) Hushai the Archite to meet him: his coat is torn (*qārûᵃ'*).

v. 37 And just when Hushai, David's friend, came to the city, Absalom was coming (*yābō'*) to Jerusalem.

Notice that these synchronic imperfectives are complemented by two occurrences of *hinnēh* ('behold', vv. 24, 32), by which the narrator further describes action from the various points of view of the characters themselves.

When one relates the distribution of these synchronic verbs to their specific content, it is clear that what is conveyed to readers as happening before their very eyes, as it were, is the series of events that begins with the *continuing increase* of Israelites who side with Absalom in v. 15 and ends with Absalom's *entering Jerusalem* in v. 37. In between these verses, the employment of at least 14 additional imperfective verb forms succeeds in making the action especially vivid and present to the reader. And yet, paradoxically, this recurring feature of the chapter is the clearest indication we have that such synchronicity is but a stylistic façade indicating the complex artifice that shapes the narrative at this point.

Listen to David after he hears that the hearts of the men of Israel have gone after Absalom: 'Arise, and let us flee; or else there will be no escape for us from Absalom; *go in haste*, lest he overtake us *quickly*' (v. 14). Yet, as the events of vv. 15-37 unfold, the synchronic, slow motion effect of all the imperfectives employed in these verses combines with the repetitive choreography of David's procession, and with his series of meetings, so that this combination turns the narrative into an account of a choreographed withdrawal rather than a hurried flight. What happens before our very eyes is a highly stylized account of the highly ritualized flight of David from Jerusalem.

Three features especially illustrate the stylization of language that embodies ch. 15 and foreshadows further details of the story to come. First, there are a couple of processional reversals using the verb *'ābar*, 'to cross over'. These reversals emphasize the ritualistic nature of the flight itself and prepare us for interesting and unexpected uses of *'ābar* later on in the story. Secondly, David's hasty journey out of Jerusalem in ch. 15 is interrupted by ideologically important conversations with Ittai, Zadok and Hushai, just as ch. 16 will narrate his meeting with Ziba and Shimei near the high point of his flight (16.1), and just as ch. 19 will describe David's meetings with Shimei, Mephibosheth and Barzillai prior to his crossing the Jordan back into the land in v. 39. This 3–2–3 configuration of meetings, as David goes up away from, and down toward, Jerusalem, intensifies the choreographed impression one has while reading this section. And, thirdly, there is a definite protocol involved in 'crossing over with the king' (*'ābar 'et* or *'im*) or 'making him cross over, escorting him over' (*'ābar* in the *hiphil*). Definite but complicated rewards or penalties await those who manage, or fail, to accompany the king on his processional journey.

Each of these three aspects of David's ritual procession out of and back into Jerusalem offer important ideological indications of what the text is saying at this point in the story.

First, the initial reversal in ch. 15 takes place in vv. 17-18, where, at first, the king is in front of the people, but then the procession halts and 'all his servants were passing by him (*'ōbᵉrîm 'al yādô*) and all the Cherethites, and all the Pelethites, and all the six hundred Gittites who had followed him from Gath were passing on before the king (*'ōbᵉrîm 'al pᵉnê hammelek*)'. A second processional switch occurs in v. 24, 'And behold, there were Zadok and all the Levites with him, who were carrying the ark of the covenant of God, and Abiathar also; they set down the ark of God until the people had all passed out (*'ad tōm laʿᵃbôr*) from the city'. The king and the ark precede, but at a crucial point in the journey, that is, at the last house in v. 17 and at the outskirts of the city in v. 24, the leaders stop and allow those behind them to cross over before them, as if following a rubric according to which king and ark are to precede *before* the crucial crossing, but then follow *after* it takes place.

Such ritual moves remind us immediately of the account of Israel crossing the Jordan into the land in Josh. 3.1–5.1. One has only to compare the following sets of verses to see that something similar is going on—whatever its ritual and ideological significance:

> And while all Israel were passing over (*'ōbᵉrîm*) on dry ground, the priests who bore the ark of the covenant of the LORD stood (*wayya'ᵃmᵉdû*) (Josh. 3.17).
> And they stood (*wayya'ᵃmᵉdu*) at the last house and all his servants were passing by (*'ōbᵉrîm*) and all the Cherethites and Pelethites and Gittites were passing over (*'ōbᵉrîm*) in front of the king (2 Sam. 15.17-18).
>
> ...until all the nation finished passing over (*'ad 'ᵃšer tammû la'ᵃbōr*) (Josh. 3.17).
> ...until all the people had passed out of (*'ad tōm la'ᵃbôr*) the city (2 Sam. 15.24).

This comparison of Joshua 3–5 and 2 Samuel 15–20 is especially relevant because both complexes center around crossing the Jordan. What 2 Samuel 15–20 adds, however, is a more ritualized and detailed procession and protocol, a stylized series of *crossover points* during the revolt of Absalom. In Joshua 3–5, the procession of Israelites crosses over the Jordan in only one direction (into the land), whereas here in 2 Samuel 15–20 David's crossing the Jordan is in both directions, out of the land in 17.22 and back into it in 19.40. Nevertheless, the journey in Joshua 3–5 is still described from two spatial points of view.[15] More importantly, the two features of crossing the Jordan in procession and of doing so according to a definite protocol concerning who leads or follows, are both repeated within 2 Samuel 15–20 through a number of wonderful narrative variations.

Secondly, ch. 15's account of David's meetings with Ittai,

15. See my remarks in *Moses and the Deuteronomist*, pp. 99-101.

Zadok and Hushai consists of a series of dialogues before the procession crosses over the Jordan, just as ch. 19's account of the king's meetings with Shimei, Mephibosheth and Barzillai interrupts his return journey just before the procession crosses over back into the land in 19.40 (English versification). And in between these triple meetings in chs. 15 and 19, David's meetings with Ziba and Shimei follow his 'crossing over the summit' in 16.1. The 3–2–3 configuration of meetings that I have already indicated, so obviously determined by the procession's location in reference to various crossover points throughout the journey, lends a stylized and ritualized cast to the account—a slant that is in tension with any 'realistic' features one may point to in the text.

As I suggested above, David's dialogue with Ittai, whose name suggests 'with me', or 'my companion', involves deciding whether he is to accompany David or return to Jerusalem; the conversation with Zadok, whose name suggests 'righteous' or 'just', leads to his and the ark's return to the royal pretender who habitually declared righteous every Israelite whom he met at the gate; and David's meeting with Hushai, whose name suggests 'hurry' or 'haste', is surrounded by numerous references to the dangerous haste caused by the revolt of Absalom. In short, these three meetings involve a paronomastic staging that makes this section of 2 Samuel a delight to read and a challenge to interpret.

Thirdly, the emphasis in chs. 15–19 on who is *with* or *not with* the king during his flight, on who escorts the king and who does not, and on the protocol that obtains within the procession itself—such emphasis is best understood when one considers that the local wordplay between Ittai's name and David's statement in 15.19 ('Then the king said to *Ittai* (*'ty*), "Why will you go, you *with us* [*'th 'tnw*]?"') is actually indicative of a much wider stylistic phenomenon within chs. 15–20. Concerning the two Hebrew words meaning 'with', *'et* and *'im*, each occurs with much greater frequency in chs. 15–19 than elsewhere in 2 Samuel.[16] This increased usage of 'with' obtains partly because

16. Simply put, the preposition *'et* occurs only 53 times in 2 Samuel, and 29 of these 53 occurrences are in chs. 15–19. Similarly, *'im* occurs about 70 times in 2 Samuel, and yet 16 of these occurrences are found in chs. 15–19.

the material itself is so much concerned with who is with David or not with him, with Absalom or not. Nevertheless, the lexical profile of chs. 15–19, insofar as wordplay involving *'et* or *'im* is concerned, corresponds to its thematic profile in ways that transcend the normal union of form and content found in everyday speech.

In short, the story that ch. 15 introduces is highly contrived and stylized, whatever its historiographic profile may be.

David's Flight: Ideological Directions

I have already described the emphasis, in chs. 15–19, on the theme of *being with the king* or not, a theme signalled by the marked increase and frequent wordplay of terms indicating 'accompaniment' in this section. The story of Ittai, Zadok and Hushai is one of being *with David* even though only Ittai physically and etymologically accompanies him across the Jordan into temporary exile; Zadok and Hushai remain with David despite faithfully and hastily returning to Jerusalem. There was also the definite protocol in the chapter concerning *how* one accompanies the king: in the procession out of the city and across the Jordan, those who are with the king are at times in front of him, at times behind him. To pass over with the king (*'ābar 'et* or *'im*) requires *following him* up to the boundary, but then *preceding him* across it. Finally, there are important implications in David's proposed return to Jerusalem.

The narrative, even before Absalom rebels, signals its coming preoccupation with matters of return by quoting Absalom's vow in v. 8. Absalom tells David, 'For your servant vowed a vow while I dwelt in Geshur in Aram, saying, "If the LORD will indeed bring me back to Jerusalem (*yāšîb yᵉšîbēnî*), then I will offer worship to the LORD"'. Absalom's vow introduces us to ideological issues of return (*šûb*) that will occupy the narrative until David's actual return to Jerusalem in ch. 19. Almost half of the occurrences of *šûb* in 2 Samuel appear in chs. 15–19, and the importance of 'returning to Jerusalem' is signalled by its intro-

Taken together, *'im* and *'et* occur in chs. 15–19 a total of 45 out of 123 times in the book. And this high proportion does not take into account the 8 occurrences of the name Ittai in chs. 15–19.

duction into the story even before the revolt begins.[17] Absalom is concerned about returning from exile to his own city, Jerusalem, and this is the first and most important function of the use of *šûb* in 2 Samuel 15–19.[18] After Absalom's revolt and David's flight have begun, David counsels Ittai the exile (*gōleh*), '*go back* [to Jerusalem], *go back* and take your brothers with you' (15.19-20). David then tells Zadok, '*Take the ark back* to the city; the LORD *may bring me back* to see the ark's habitation; *go back* to the city' (vv. 25, 27). The narrator reports that Zadok and Abiathar *brought the ark back* to Jerusalem (v. 29), and David finally tells Hushai, '*Return* to the city' (v. 34).

A second issue introduced by *šûb* is the restoration of the king(dom). If the eventual loser in the revolt is shown first returning to Jerusalem even before David—the eventual victor—is forced to start planning his own return in 15.19ff., it is Mephibosheth—already a loser—who first voices this second aspect of return: 'Today the house of Israel will *restore* to me the kingdom of my father' (16.3).[19] Will David, the supplanter of Saul, be returned to his throne after his exile across the Jordan?

A third function of *šûb* in this section concerns the question of divine recompense for David's actions. Again it is a loser in the story, here Shimei, who introduces us to this aspect of return. Shimei's curse to David states, 'May the LORD *return upon you* all the blood of the house of Saul' (16.8). David's response indicates the alternative that he, understandably, prefers: 'It may be that the LORD will look upon my affliction, and that the LORD *will return good to me* in place of this cursing of me today' (16.12).

These three facets of *šûb* in the story—returning to Jerusalem, restoring the king(dom) and repaying the king for his actions—help us to see something of the ideological point of view that permeates the story of Absalom's revolt. The various emphases

17. Some form of *šûb* occurs 57 times in 2 Samuel, and 25 of these occur in chs. 15–19: 15.8 (twice), 19, 20 (twice), 25 (twice), 27, 29, 34; 16.3, 8, 12; 17.3 (twice), 20; 18.16; 19.11, 12, 13, 15, 16, 38, 40, 44.

18. Returning to one's city, especially Jerusalem, is the focus of *šûb* in 15.8 (twice), 19, 20 (twice), 25 (twice), 27, 29, 34; 17.20; 19.38, 40.

19. This slant on 'return' is found in 16.3; 17.3 (twice); 19.11, 12, 13, 15, 16, 44.

on return in chs. 15–19 seem not so much required by David's flight from Jerusalem, as David's flight appears necessary in order to focus the story on some central issues: exilic return to Jerusalem; return with or without the king; following him across boundaries in deadly pursuit (17.22-24) or preceding him across as his loyal servants; exilic restoration of the kingship; and finally, divine retribution concerning the house of David. Here is a story wherein the hasty flight of David is slowed down in a highly stylized manner so that the central topic of his eventual return to Jerusalem may be addressed in terms that mirror the complex situation of discourse between a Deuteronomic voice and its contemporary audience.

For example, perhaps the ark is not allowed to cross the Jordan with David because it is no longer with Israel in Babylon. David may be stating to Zadok what many Israelites in Babylon hoped in their hearts, 'If I find favor in the eyes of the LORD, he will bring me back and let me see both [the ark] and his habitation' (15.25). As we earlier saw indications of the exilic situation of discourse lying behind references to exile or captivity in 1 Sam. 4.21-22, so also the use of *gālâ* here in 2 Sam. 15.19—where Ittai the exile (*gōleh*) is allowed to cross the Jordan with David into a kind of double exile—is the only other instance in the books of Samuel of *gālâ* denoting 'exile'.[20]

We find a final hint of the ideological dimensions of ch. 15 in David's command to his servants, 'Arise, and let us flee; or else there will be no escape (*pᵉlēṭâ*) for us from Absalom' (15.14). The root, *plṭ*, is found infrequently in the History, yet many of its occurrences concern issues of fratricide, whether familial, tribal or national in nature.[21]

The authorial perspective on survival during Absalom's revolt looks backward to the judicial period in Israel's history when, in the Jephthah story of Judges 12, the Gileadites smote their tribal brothers, the Ephraimites, and where both are called 'survivors of Ephraim (*pᵉlîṭê 'eprāyîm*)' (Judg. 12.4, 5). At the end of Judges, warfare between tribal brothers becomes so severe that the tribe of Benjamin nears extinction: 'And they said, "There

20. See my remarks in *Samuel and the Deuteronomist*, pp. 66, 237.

21. The root *plṭ* occurs in Josh. 8.22; Judg 12.4, 5; 21.17; 2 Sam. 15.14; 22.2, 44; 2 Kgs 9.15; 19.30, 31.

must be an inheritance for the survivors (*pᵉlēṭâ*) of Benjamin, that a tribe be not blotted out from Israel"' (Judg. 21.17). Here in 2 Samuel 15, during the monarchic period, 'survivors' refer to individuals within the house of David who are threatened by another member of the same house. And David himself will thank the LORD in 2 Sam. 22.44, 'Thou didst deliver me (*wattᵉpallᵉṭēnî*) from strife with my people'.

Ahead in the story, the conspiracy of Jehu against Joram also recalls Absalom's conspiracy against David. There, as here, the issue of escaping from the city is central to the plot, 'So Jehu said, "If this is your mind, then let no one escape (*pālîṭ*) from the city to go and tell the news in Jezreel"' (2 Kgs 9.15). The Deuteronomic issue of tribal and national survival will take one final turn during Assyria's assault on Israel. Isaiah will prophesy, 'And the surviving remnant (*pᵉlēṭat*) of the house of Judah shall again take root downward; for *out of Jerusalem* shall go forth a remnant, and out of Mount Zion a band of survivors (*šᵉʾērît ûpᵉlêṭâ*)' (2 Kgs 19.30-31). It is almost as if David's flight from Jerusalem in 2 Samuel 15 is a precursor of Isaiah's prophecy in 2 Kings 19, a ritual procession that looks forward to Israel's exile from the land even as it reverses, with similar choreography, Israel's original crossing of the Jordan *into* the land in Josh. 3.1–5.1.

2 Samuel 16: Considerations of Context

We begin with the ways in which the events in this chapter are structured. Verses 1-14 recount what happens after David crossed beyond the summit of the Mount of Olives, and vv. 15-23 report what transpires after Absalom came to Jerusalem. The parallels between these two halves are striking. First David meets Ziba and Shimei, then Absalom meets Hushai and Ahithophel. The chapter begins 'just beyond the summit', at the spatial highpoint in David's procession where David receives a couple of asses laden with food and drink—gifts from the servant of Mephibosheth 'for the king's house'. The chapter ends on the roof of David's palace, the spatial highpoint of Absalom's revolt, where Absalom went in to the concubines whom David had left 'to keep the house'. What happens on high to David in the chapter—beyond the summit and upon the roof—are ironic lowpoints in his career: he is cursed by Shimei and dishonored by Absalom. Finally and perhaps most importantly, the chapter characterizes Shimei's cursing of David and Absalom's taking of David's concubines in much the same way: Shimei curses David because, David believes, the LORD said he should (vv. 10, 11); and Absalom humiliates David in the sight of all Israel because Ahithophel said he should. Since David and Absalom consider the counsel of Ahithophel equivalent to the word of God (v. 23), not just the cursing of David in the first half, but even his humiliation in the second, are believed to be happening at the LORD's behest.

We should not forget that the narrator earlier has David saying, 'O LORD, I pray thee, turn the counsel of Ahithophel into foolishness' (15.31). Yet now we hear that David considers this counsel to be 'as if one consulted the word of God' (16.23). Are we to understand that David hopes to turn the *word of God* into foolishness? At any rate, we know that David hopes, 'Perhaps the LORD will look upon my iniquity and repay me with good for this cursing of me today' (v. 12). If the structural juxtaposing of these two events in ch. 16 has any obvious authorial point, it may be that David's hope in the first part is supposed to be dashed by Absalom's act in the second. It is not just David and Absalom who equate Ahithophel's counsel to the word of God. The narrative also does this here in as obvious a manner as one could expect. We know that David's sin in 2 Samuel 11 provoked God to prophesy in 2 Samuel 12: 'Behold I will take your women before your eyes, and give them to your neighbor, and he shall lie with your women *in the sight of the sun*' (v. 11). As countless readers have understood, Absalom's going in to his father's concubines *in the sight of all Israel* is an indication that Absalom's following of at least *this* counsel of Ahithophel is presented by the Deuteronomist as an obvious fulfilment of the word of God.

Besides this looking backward to the divine prophecy of 2 Samuel 12, the events in 2 Samuel 16 also refer to matters raised in 2 Samuel 9. There, David restored all of Saul's land to Mephibosheth, but here he returns it to Ziba. Further on, in 2 Samuel 19, David will again vary his position by halving the property between Ziba and his master. Taken together, these three occasions help to characterize David in a less than favorable light. The kindness (*ḥesed*) David shows Mephibosheth for Jonathan's sake in 2 Samuel 9 he now retracts in 2 Samuel 16—only to backtrack once more in 2 Samuel 19. Whether the LORD always shows steadfast love to his anointed, as David sings in 22.51, God's anointed is clearly inconstant in demonstrating *his ḥesed* to friend and foe alike.

Like 2 Samuel 15, 2 Samuel 16's continuing emphasis on synchrony within the first half of the chapter (through the use of imperfectives and *hinnēh*) may also help to formulate present aspects of the Deuteronomist's ideological perspectives. 2 Sam. 16.1-14 is filled with action that is presented by the narrator as if it were happening before our eyes, and represented by characters as if it were happening before theirs:

v. 1 Behold (*w^{e}hinnēh*) Ziba to meet him (*liqrā'tô*) with a pair of asses saddled (*ḥ^{a}bušîm*).
v. 3 'Behold (*hinnēh*) [Mephibosheth] is residing (*yôšēb*) in Jerusalem' (Ziba to David).
v. 5 Behold (*w^{e}hinnēh*) a man coming out (*yôṣē'*), and as he was coming out (*yōṣē' yāṣô'*) he was cursing (*ûm^{e}qallēl*).
v. 8 'Look at you (*w^{e}hinnēh*) in your ruin' (Shimei to David).
v. 11 'Behold (*hinnēh*) my son is seeking (*m^{e}baqqēš*) my life' (David to Abishai).
v. 13 And Shimei was going (*hōlēk*) along on the hillside opposite him, cursing and throwing stones (*hālôk way^{e}qallēl way^{e}saqqēl*) and flinging dust (*w^{e}'ippar*).[22]

When the narrative leaves David's 'ongoing' procession and returns to events in Jerusalem in the second half of ch. 16, it reverts to the usual *retrospective* presentation of events: in vv. 15-23 the narrator no longer employs imperfective verb forms and *hinnēh*. It is as if we are meant to see what happens in David's procession in the wilderness as somehow *still going on before us*, whereas Absalom's machinations in royal Jerusalem

22. On the frequentative or imperfective aspect of these verbs in v. 13, see Driver, *Notes on the Hebrew Text and the Topography of the Books of Samuel*, p. 319, and McCarter, *II Samuel*, p. 369.

have happened, and represent a stable past, one that is in contrast to the highly mobile, ongoing or durative aspects of David's stylized and ritualized procession in the wilderness. The contrast—between Israel wandering in the wilderness, while David's kingship is being threatened (vv. 1-14), and Israel's residing in Jerusalem (vv. 15-23), where Hushai ambiguously says to Absalom, '"Long live the king! Long live the king!' (16.16)—may represent the Deuteronomist's own perspective on the *synchronic*, ongoing dimension of Israel in exile (with their king in question and their geographic instability emphasized) and the *retrospective* aspect of royal Jerusalem (with its spatial and temporal permanence over with and done for).

One final feature of 2 Samuel 16 continues a concern of 2 Samuel 15: a heavy emphasis on who-is-with-whom during the constitutional crisis inaugurated by Absalom's revolt.[23] The paronomastic implications of *'et* continue in this section, and one wonders whether the increased usage of *'et* within 2 Samuel 15–19 has hermeneutic significance as well.

Cursing and Counselling Kings

The heart of 2 Samuel 16 lies in the complex interaction of two related themes concerning the house of David. First, is the house of David really cursed, and if so, in what ways? And second, what role does the king's counsel play in the cursing of the king? Chapter 16's response to these two matters suggests that there is an intimate connection between the LORD's cursing of the king and the king's reliance on human counselors. So

23. *'et* meaning 'with' occurs proportionately more often in these two chapters (2 Sam. 15–16) taken individually than anywhere else in the book. *'et* occurs 10 times in the 37 verses of 2 Sam. 15 and 7 times in the 23 verses of 2 Sam. 16. On average in 2 Samuel, *'et*, 'with', occurs about only once every 13 verses in the book as a whole, but increases to once every 5.7 verses in 2 Sam. 15–19, and once every 3.5 verses in 2 Sam. 15–16. There are people *with* David in the wilderness (v. 14); Ahithophel is *with* Absalom in Jerusalem (v. 15). Absalom questions Hushai's loyalty *with* David by asking, 'Why did you not go *with* your friend?' (v. 17). Hushai ambiguously promises to remain *with* the chosen of God (is this David or Absalom?) (v. 18), and Ahithophel counsels Absalom to make himself odious *with* his father so that the hands of those who are *with* Absalom may be strengthened (v. 21).

Shimei's cursing of David in the first half of ch. 16 appropriately precedes Ahithophel's counselling of Absalom in the second half.

The significance of Shimei's cursing of David rests upon the affinity that curses and kings have within the larger narrative. The object of curses, whether uttered or actualized in the History, can be nations, tribes or individuals.[24] When the accursed is a nation, that nation is exclusively Israel—most often the entire nation, but in 2 Sam. 19.44 the northern tribes alone. However, when the accursed are individuals, these unfortunates are almost always *royal figures*—and David turns out to be the History's favorite king to curse.

The narrative scope of Shimei's cursing of David, and its significance within the larger account of Absalom's revolt, is best seen against the backdrop of the History, which establishes an intimate connection between the cursing of individuals and the curse of kingship. Whether the individual instances combining kings and curses are explicit or not, in most cases accursed individuals are narrative stand-ins for the cursing of a nation.

Look first at the History's practice. The Deuteronomist's favorite objects of curses, *even before the onset of Israel's royal revolt in 1 Samuel 8*, are kings or those individuals who support them. Given the exquisite care with which the History has been fashioned, it is highly significant that the first person actually cursed through the use of a form of the root *qll* (to curse) is Abimelech, Israel's upstart king in the book of Judges: 'And [the men of Shechem] went out into the field, and gathered the grapes from their vineyards, and trod them, and held festival, and went to the house of their god, and ate and drank and cursed (*wayqal^{e}lû*) Abimelech' (Judg. 9.27). Later in this chapter, the narrator has God fulfilling the curse of Jotham upon the men of Shechem themselves because of their original support of Abimelech as king: 'And God also made all the wickedness of the men of Shechem fall back upon their heads, and upon them came the curse (*qil^{a}lat*) of Jotham, the son of Jerubbaal' (Judg. 9.57).

If kings are easy to curse, the History shows that David, of all

24. For nations, see Deut. 11.26, 28, 29; 23.5, 6; 27.13; 28.15, 45; 29.26; 30.1, 19; Josh. 8.34; 24.9; 2 Kgs 22.19. For tribes, see 2 Sam. 19.44. And for individuals, see Deut. 21.23; Judg. 9.27, 57; 1 Sam. 2.30; 3.13; 7.43; 2 Sam. 6.22; 16.5, 7, 9, 10, 11, 12, 13; 19.22; 1 Kgs 2.8; 2 Kgs 2.24.

Israel's kings, is the easiest one of all. No sooner is David anointed king in 1 Samuel 16, than Goliath curses him, 'And the Philistine cursed David by his gods' (1 Sam. 17.43). When David brings the ark of the LORD to Jerusalem in triumph, Michal's reproach provokes him to respond, 'I shall make myself even more accursed than this (*ûn*ᵉ*qallōtî 'ôd mizzō't*)' (2 Sam. 6.22). From then on in 2 Samuel, attention to cursing focuses exclusively upon Shimei's cursing of David (2 Sam. 16.5, 7, 9, 10, 11, 12, 13; 19.21). In his confrontation with Shimei, as with Michal in 2 Samuel 6, David gives what appears to be something like an authorial interpretation of such cursing. David rebukes Abishai, 'Let [Shimei] alone, and let him curse; for the LORD has bidden him' (2 Sam. 16.11).[25] Clearly, David is no longer the man after God's own heart.[26]

The History's practice of presenting kings as special objects of God's curse, therefore, suggests that the story of Shimei is significant in terms of the authorial perspectives refracted within it. David's statements to Abishai ('If he is cursing because the LORD has said to him, "Curse David", who then shall say, "Why have you done so?" Let him alone, and let him curse; for the LORD has commanded him') and David's hopes for the curse's reversal ('It may be that the LORD will look upon my iniquity, and will repay me with good for this cursing of me today' [16.10-12]) are striking. If the accursed himself admits that such cursing is from God, and if the substance of the curse is that 'the LORD has given the kingdom into the hand of your son Absalom' (16.8), then the understandable failure of Absalom to maintain his throne while he hangs from an oak would appear to corroborate David's prescience on both counts: David indeed

25. The boys who jeer Elisha, and are roundly cursed by him in 2 Kgs 2.24, are the only non-royal figures cursed in the History with the language of *qll*. The only other accursed characters in the History whom I have not yet mentioned are Eli's sons, whom God curses in 1 Sam. 2.30, and whom the narrator accuses of cursing God in 1 Sam. 3.13. See *Samuel and the Deuteronomist*, pp. 44-54 for my discussion of the royal dimensions of the cursing found in 1 Sam. 2 and 3.

26. In the execution of Absalom in ch. 18, the question of royal curses, especially as they are *indirectly* indicated in the story, once more becomes central to the story.

is cursed by God, but that does not mean that Absalom will not get cursed in turn.

We see that God's curse upon David differs from that imposed upon Saul. The cursing of David is like the cursing of his house: both involve the *continued existence* of the accursed. Yet within the house of David, the fates of father and son differ: the curse of Absalom involves hanging from a tree; God's particular curse for David, however, requires returning him to the throne.

Since the king, precisely as *royal head* of his people (1 Sam. 15.17; 2 Sam. 22.44), is a special carrier of the divine curse, it is ironically appropriate that the *character zone* of kings within the History is so often filled with heady violence and bloody heads. In fact, the characters within the History whose heads are bloodied or violently handled form something like an exclusive company of the royally damned—those who are unfortunate enough to get too close to the *character zone* of kings. Here is a listing of everyone in the History who literally suffers some kind of capital misfortune.

To be mentioned first are all those whose heads are somehow bloodied (*rō'š* plus *dām*) in the History:

1. The Amalekite who claimed to have slain Saul is executed by David with the words, 'Your blood be upon your head' (2 Sam. 1.16).
2. Joab is executed by Benaiah following Solomon's words to the executioner, 'The LORD will bring back [Joab's] bloody deeds upon his own head. So shall their blood come upon the head of Joab and upon the head of his descendants forever' (1 Kgs 2.32-33).
3. Shimei is executed by Solomon following these words, 'For on the day you go forth, and cross the brook Kidron, know for certain that you shall die; your blood shall be upon your own head' (1 Kgs 2.37).[27]

Next comes the procession of those in the History who suffer terminal violence to the head, whether by seizing, crushing, piercing, hanging, strangling or beheading.

27. Blood is called down upon the heads of God's enemies in Deut. 32.42, and upon those who go against the promise of Joshua's spies in Josh. 2.19, but these threats receive no narrative fulfilment in the story.

1. Sisera, the general of Jabin, king of Canaan (Judg. 4.2), dies at the hand of Jael: 'She struck Sisera a blow, she crushed his head, she shattered and pierced his temple' (Judg. 5.26).
2. Abimelech, the upstart king in Israel, dies after 'a certain woman threw an upper millstone upon Abimelech's head, and crushed his skull' (Judg. 9.53).
3. Goliath is beheaded by the newly anointed David (1 Sam. 17.51).
4. Saul is beheaded by the Philistines (1 Sam. 31.9).
5. The 24 Israelites at the pool of Gibeon act out the conflict between the royal houses of Saul and David by seizing one another's heads and killing each other (2 Sam. 2.16).
6. Ishbosheth is beheaded by the sons of Rimmon (2 Sam. 4.7).
7. The sons of Rimmon are executed by David, and hung beside the pool at Hebron (2 Sam. 4.12).
8. Ahithophel commits suicide by hanging himself, because Absalom did not follow the royal counsel (2 Sam. 17.23).
9. Absalom is executed while hanging by his head from a tree (2 Sam. 18.9, 10, 15).
10. Sheba is beheaded by the townspeople of Abel, with his head thrown over the wall to Joab (2 Sam. 20.22).
11. The seventy sons of Ahab have their heads cut off, put in baskets, and sent to Jehu at Jezreel.

It takes little imagination to see that there is a notable affinity, in the History, between the character zone of *royal* heads of nations, on the one hand, and the graphic language of doing bloody or terminal violence to anyone having the misfortune to come too close to these heads, on the other. Whatever the particular mix of unconscious mindset and esthetic motivation lying behind these dangerous linkages of physical and royal heads may be, the widespread tendency in the History to write of bloody heads and capital violence within an almost exclusively royal context argues for a good deal of conscious literary deliberation, else we would find such violences occurring more often in stories that do not have a royal cast to them. It is almost as if Moses' principle about the punishment fitting the crime—'If any harm follows, then you shall give life for life, eye for eye, tooth for tooth, hand for hand, foot for foot, burn for burn, wound for wound, stripe for stripe' (Exod. 21.23-24)—is taken over and given a literary application and a narrative form: Israel's capital crime is to have chosen a king to be their head (1 Sam. 15.17); their punishment now is *head for head*. There is a special connection between curses and kings, insofar as heady violence pertains particularly to the *character zone* of royal heads of nations.

This tendency in the History for royal overkill helps to explain why the story of Absalom's revolt is not a simplistic account of the eventual victory of those in the right over those in the wrong, but rather a complex and nuanced story of the doomed struggles of those whose lives are touched by the cursed sphere of kings. Almost all the characters in ch. 16 follow this rule about the dangers of getting too close to kings. Mephibosheth is sadly mistaken if he indeed said, 'Today the house of Israel will return the kingdom to me' (16.3). Ziba may receive all of his master's land, but he soon will lose half of it through David's caprice in ch. 19. Shimei's reward for cursing David, apparently at the LORD's behest, will be execution at Solomon's command. And finally, Ahithophel and Absalom, in chs. 17 and 18, will suffer the LORD's special curse by hanging.

The second half of ch. 16 (vv. 15-23) concerns the counsel of Ahithophel, and adds a second reason why the *character zone* of kings is so fraught with danger. If cursing the king eventually brings death to Shimei, counselling the king will do the same for Ahithophel. What is there about the counselling (*yā'aṣ*) of kings that connects it to the cursing of kings? Chapter 16 provides us with the beginning of an answer. Simply put, when royalty equates such counsel to consulting the oracle of God, as David and Absalom do in 16.23, then the counsel of kings, like kingship itself, is a threat to the rule of Yahweh.[28]

28. 'To counsel' (*yā'aṣ*) appears at least 34 times in the History (either in nominal or verbal form), and fully 31 of these 34 occurrences explicitly concern the giving of advice to or about kings. Most of this royal counsel (27 of 31 occurrences) is found in 2 Sam. 15–17, where Absalom does not follow Ahithophel's counsel and loses the throne he seized from David, and in 1 Kgs 12, where Rehoboam refuses to follow his elders' advice and loses the 10 northern tribes.

We learn from the two instances in the History where kings seek advice from royal counselors—Absalom in 2 Samuel 15–17, and Rehoboam in 1 Kings 12—that it is dangerous or risky for *kings to seek counsel*. In the story of Absalom's revolt, 'the LORD had ordained to defeat the good counsel of Ahithophel, so that the LORD might bring evil upon Absalom' (2 Sam. 17.14). When forced to decide between the counsel of Hushai and that of Ahithophel, Absalom unfortunately chooses Hushai's, and loses the kingdom. Similarly, when faced with the conflicting counsel of his elders and young men, Rehoboam 'forsook the counsel which the old men gave him, and took counsel with the young men' (1 Kgs 12.8)—and lost the northern kingdom.

Moreover, the History makes it clear that the danger of seeking counsel in a royal context rests upon the practice's opposition to more theocratic means of seeking advice. If prophetic inquiry constitutes a divinely ordained check upon unrestrained royal rule, then the introduction of 'the king's counselor' would appear to be a royal attempt to restrict the power of the prophet within the court.[29] When the narrator informs us that 'in those days the counsel which Ahithophel gave was as if one consulted the oracle of God; so was all the counsel of Ahithophel esteemed, both by David and by Absalom' (2 Sam. 16.23), then reveals to us that 'the LORD had ordained to defeat the *good counsel* of Ahithophel' (2 Sam. 17.14), and finally shows us this rejected counselor going home to hang himself (2 Sam. 17.23), the lesson is abundantly clear:

There are two passages that variously compare royal counsel to other kinds of royal consultation. In 2 Sam. 16.23, the narrator has David and Absalom equate the counsel of Ahithophel 'in those days' to the oracle of God, and during Hezekiah's reign, the story has the King of Assyria talking about Hezekiah's supposedly disastrous *counsel* in 2 Kgs 18.20, but the sequel contrasts this characterization with Isaiah's conveying of the (correct) *word of God* to Hezekiah in 2 Kgs 19. Clearly, *giving counsel* in the History is predominantly a *royal affair* (as in 2 Sam. 15–17; 1 Kgs 1.12; 12; 2 Kgs 6.8; 18.20), but not always (see Deut. 32.28; Judg. 19.30; 20.7).

29. See my discussion in *Samuel and the Deuteronomist*, pp. 99-100. Another attempt to limit the restrictions of the prophet would be to set up the royal prophets one against the other, as recounted in 1 Kgs 22.

whether good or bad, wise or foolish, merely human advice lacks the providential status and epistemological guarantees that result from seeking out or inquiring of the LORD.[30]

30. We also see in the History that inquiring of the LORD/God typically involves seeking out a prophet ('Formerly in Israel, when a man went to inquire of God, he said, "Come, let us go to the seer"; for he who is now called a prophet was formerly called a seer' [1 Sam. 9.9]) or priest (1 Sam. 14.36-37), whether in the presence of the LORD's ark (Judg. 20.27) or ephod (1 Sam. 30.7-8). If both David and his son consider their counsellor's advice like an inquiry of God, then it may be helpful to see how these two royal means of consultation, the one human the other divine, are distributed throughout 1–2 Samuel during the reigns of Saul's and David's houses.

SURVIVING WRITING: THE ANXIETY OF HISTORIOGRAPHY IN THE FORMER PROPHETS

Hugh S. Pyper

> In my opinion, the reader of a mystery is the only real survivor of the mystery he is reading, unles it is as the one real survivor that every reader reads every story (Saramago 1992: 12).

Why is it that in the great corpus of writings that we know as the Former Prophets there is so little reference to the process of writing itself? Given the multiplicity of references to messengers and messages, it is surprising to say the least that there are only five incidents in all the books from Judges to 2 Kings which involve a character in the act of writing. The fact that writing is mentioned at all makes it clear that it is available as a motif to the writers of these texts.

So why the reticence? One might intuitively expect that those who committed Israel's traditions to writing themselves would have a bias in favour of the process, and would, if anything, be inclined to overestimate its importance and its use in the wider population, and in earlier periods.

Something of the sort indeed has been suggested as the explanation for the case in Judg. 8.14, where a young man is captured and forced to *write down* the names of the leading men of Succoth. This is the only occurrence of the verb *ktb* 'to write' in the book of Judges, and it has occasioned more heat than light in the attempt to discover the level of literacy in ancient Israel.[1]

1. Does it indicate a widespread ability to write in the population at the time of Gideon, or is it a later retrojection of writing by the literate compilers of these traditions? Such an enquiry makes assumptions about the historicity of the text that are far from our concerns here, but which render the whole argument rather suspect. By an equally questionable inversion, this form of argument has been used to claim that the passage cannot be

The very singularity of this incident, however, makes the point. Why should the act of writing be mentioned explicitly just here? Whatever the historical situation that this passage reflects, if indeed it reflects any, what impels the reference to the act of writing?

The thesis which this study will seek to explore is that this repression of writing is tied to a wider anxiety, what we might call an anxiety of utterance. It centres on the paradox of survival. In the face of inevitable death, the individual can survive only through an act of utterance, either through the production of a text, or through begetting a child. Yet the corollary of this is that any such act of utterance carries with it the odour of death. The child or the text that bears our survival is also the sign of our death; our successor is our rival and our supplanter. Yet we in our turn by the fact of our existence testify to the death of those to whom we owe the debt of life, those who preceded and begot us. This leads to an ambivalence and to a repressed violence which we shall call on Freudian insights to illuminate. Such ambivalence has profound implications for the relation between the text and the reader, a relationship that in the case of the Former Prophets has been of exceptional scope, endurance and impact, spanning millennia and a vast range of interpretative communities. Our starting point, however, will be the small manifest symptom we have identified, the paucity of explicit mention of writing in these texts.

read as a scene of writing. See the note added by S.A. Hopkins to G.R. Driver's mention of this incident in his *Semitic Writing*: 'That any ordinary boy, met by chance on a journey, can have been able to write at such an early date is improbable; that he can have had the necessary tools with him and that Gideon can have had time to wait while he slowly painted on a sherd or engraved on a piece of stone, already prepared for the purpose, a list of 77 names is equally improbable. Why will he have needed to write and not recite them (Jud 8.14)?...Surely כתב "wrote" must in such contexts have had its primitive sense of "pricked, scratched, i.e. ticked off" e.g. the numbers on a piece of wood or stone to check them as he counted them' (Driver 1976: 242).

Writing in the Former Prophets

We will look in vain for a theoretical discussion of the nature of writing in these texts. What we can do is examine the other episodes where characters write, where there is a *mise en abyme* of the process of the production of texts. Such episodes allow the invisible process of writing which is the *sine qua non* of the existence of the texts to break surface, as it were. This may lead us to look again at the implications of the writtenness of the texts in which these embedded scenes of writing occur. What do these texts reveal of the conscious and unconscious reactions to the process of writing among those in Israel who took the momentous step of producing this unique written monument to their nation's past?

Such an examination reveals that the episode in Judges exemplifies a consistent pattern in the Former Prophets in which writing is linked with violence and death. The boy writes, if that is what he does, under duress, and Gideon uses the information to take the elders of Succoth and, as the RSV puts it, 'teach' them with thorns of the wilderness and briers (Judg. 8.16). Both the act of writing and its consequences involve violence.

A brief review of the few other incidents where writing is explicitly involved is enough to demonstrate this consistent association.[2] In 2 Sam. 11.14-15, we have the repeated insistence that David himself writes the letter that he sends with Uriah, who carries it as his own death sentence to Joab. The only other occasion when David's name is connected with writing is his instruction in 2 Sam. 1.17 that his elegy for Saul and Jonathan should be recorded—and the connection with death there need hardly be stressed.

1 Kgs 21.8-9 records Jezebel's letters to the elders of Jezreel, which instruct them to arrange for the fatal denunciation of Naboth. And in 2 Kings 10, Jehu sends two letters to the elders

2. This link is made by Nielsen (1954: 45), who in the course of examining the references to writing throughout the Old Testament turns to the matter of private letter writing and remarks that 'the known instances are all of a distinctly macabre nature'. Why this should be he does not discuss.

of Samaria who are harbouring Ahab's sons, the second of which is a direct command to bring the heads of these sons to him at Jezreel. The elders duly comply.

Every episode where a character explicitly writes in the Former Prophets involves the threat of death. This is borne out by the way in which the recipients of even ostensibly innocuous letters respond to them. Characters as readers react to writing as the harbinger of death. The most telling example of this is 2 Kgs 5.6-7. The king of Syria sends an innocuous letter to the king of Israel which reads: 'When this letter reaches you, know that I have sent you Naaman my servant that you might cure him of leprosy'. What is the king's reaction? 'And when the king of Israel read the letter, he rent his clothes and said, "Am I God, to kill and make alive, that this man sends word to me to cure a man of his leprosy? Only consider and see how he is seeking a quarrel with me."' Death does in fact overshadow the occasion of this writing, which embodies the attempt to save Naaman from the lingering decay of leprosy. The recipient, however, reads it as a threat to his own existence.

In 2 Kings 10, Jehu's letter is also on the face of it quite innocuous. He writes ostensibly to invite the elders of Samaria to appoint one of Ahab's sons as king. Their reaction contains an element of paradox: 'We are your servants, and we will do all that you bid us. We will not make any one king; do whatever is good in your eyes'(2 Kgs 10.5). The one thing Jehu has instructed them to do, however, is precisely to make someone king. They are obeying him by disobeying him. It is the threatening power of the written word that induces their anxiety and their realization of the ironic import of his request that is borne out in the brutal directness of his second letter.

Such anxiety is also seen in the crucial episode of the rediscovery of the book of the law in 2 Kings 22. This leads to utter consternation and a great display of mourning on the part of king and people. By far the largest category of references to writing, however, is the oddity of the repeated appeal in the books of Kings to other written sources, almost always couched in the form of a question: 'Is this not written in...?' It is almost as if the claim of these texts is being evoked only to be repressed. They exist in the half-world of the interrogative. The reader, especially the modern reader, to whom these books are lost, is left with a sense of deprivation. All we know of these books is their absence, the fact of their loss of which we only become conscious through this reminder.[3]

Paradoxically, the very physicality that gives a text its possibility of endurance is also the point at which it can be assailed. An oral account cannot be obliterated in the same way, simply because it has no 'body' to be destroyed. Books may be burnt, but oral narrative is much more difficult to suppress. Only a written text can be lost. The possibility of its survival serves as a reminder that the threat of destruction is ever present.

Writing and Death

The association between writing and death is discussed by Walter Ong (1977). Writing, so he argues, entails death in two aspects. It makes the continued life of its author irrelevant, so he or she is 'as good as dead' (1977: 235), but its own ability to outlast its author comes at the price of a fixity, a 'death' of the language that it contains. This death, however, is also freedom, as Paul Ricoeur would have it. By becoming detached from its context and from its author, writing becomes open to reinterpretation,

3. In her recent article, 'Joseph's Bones and the Resurrection of the Text: Remembering in the Bible', Regina Schwartz discusses the relationship between remembering and forgetting in the biblical text, and especially in its 'scenes of writing'. Loss and destruction leading to rewriting attend these incidents. The book of the law is lost and rediscovered (2 Kgs 22–23); Jeremiah's scroll is burnt and rewritten (Jer. 36); the tablets of the law are smashed and reinscribed. See also the recent article by Conrad (1992), who urges the *rhetorical* importance of the mention of 'books' in the Old Testament, and the danger of anachronistic assumptions engendered by the translation of *sefer* as 'book' rather than 'letter' or 'document'. He concludes his study as follows: 'In summary, then, books in OT are for the *ear*, not for the eye of the silent reader; unlike the proverbial child, they should be heard and not seen. Furthermore, when the OT mentions 'books that are lost for the reader, it is not referring to 'books' 'out there' in the world external to the text, but to books that the implied audience is encouraged to remember and recreate by the only means available—the 'book' they are hearing in the present whose narrator (represented by the one reading it aloud) gains authority as the one who has known more than the audience can ever know' (1992: 59).

to survival in new contexts. In his essay, 'Phenomenology and the Theory of Literature', Ricoeur writes:

> Primarily, through writing, the discourse slips away from the speaker, since writing has the power to preserve the discourse after the destruction and disappearance of the speaker. So there is an autonomy of text in relationship to the occurrence of the discourse, which is at its origin and enables the text to have a destiny distinct from that of its author. The writer dies, but the text pursues its career, continues and produces its effects from time to time, which is vaster than the time of a human life. Therefore, if we go back to this triangular relationship, in regard to the author of the discourse, the text frees itself from the boundary of the history of its production and survives the occurrence of speech (1991: 442).

But the very fact that writers need to write to ensure their survival, and know that the text may survive them, itself acts as a reminder of the inevitability of their death.

Quite apart from the effect on the author and on the language of the text, Ong also refers to the very real sense of threat that the encounter with writing may bring to its potential readers. Writing, seen as the key to culture by the literate, can also be seen as the death of oral culture, of memory and of immediate communication.[4] On the one hand, there are those such as Eric Havelock who see the introduction of alphabetic writing in ancient Greece as the necessary condition of the revolution in human consciousness that underlay that culture's unparallelled philosophical and scientific sophistication (1986: 98-116).[5] On the

4. 'Those reared in a highly literate culture, where literate habits of thought are acquired shortly after infancy, commonly have little if any memory of entry into writing as a cutting loose from oral thought processes, as a kind of death. For those dominated through adolescence by the functional orality of subcultures in our American cities or some of our rural districts, the situation is quite different. They feel writing as a threat, a destruction of their psychic world, however desirable writing may be' (Ong 1977: 237). The classic expression of this distrust of writing is Plato's *Phaedrus*.

5. This view is not uncontroversial. Havelock is following a line that derives from the classic article by Goody and Watt (1968). For another view, which sees literacy as an aspect, but not a cause, of this intellectual revolution, see Street 1984; a summary of the issues of this debate is to be

other hand, a writer such as Claude Lévi-Strauss records with sorrow the introduction of all the malaise of Western civilization into the lives of a primitive people when they are first exposed to the technology of writing (1961).

In *Tristes tropiques*, he recounts an incident where the leader of the Nambikwara people grasps the significance of writing as an act. He asks for a note pad and traces wavy lines on the page in imitation of Lévi-Strauss's note-taking. He then proceeds to read out to Lévi-Strauss this 'list' of objects to be distributed to his people. As Lévi-Strauss sees it, he has understood the social function of writing, which is to mark out relations of power, to mark a difference between those who have the ability or authority to write and those who do not. Writing becomes an élite means of communication which confers power on those who possess the secret, a power that becomes the legitimation of violence and oppression. Lévi-Strauss offers this hypothesis: 'the primary function of writing, as a means of communication, is to facilitate the enslavement of other human beings' (1961: 292).

Lévi-Strauss's position is subjected to a detailed criticism by Jacques Derrida in a section of his *Of Grammatology* entitled, significantly, 'The Violence of the Letter' (Derrida 1976). Derrida has gained notoriety for his championship of writing in the face of the 'logocentric' bias of Western culture. In this case, he argues that the swiftness with which the Nambikwara chief adopted the technology of writing suggests that what he calls 'writing in the narrow sense', i.e. inscribing visual signs on a substrate, is a manifestation of something that was already at work in Nambikwara culture, long before Lévi-Strauss and his notebook arrived. This he calls 'arche-writing'. Lévi-Strauss's sense of guilt as a bringer of corruption depends on the Rousseauesque fiction of the innocent savage, of a culture without violence built on the immediacy of oral communication, of the possibility of presence. On the contrary, at the heart of all human culture and language in Derrida's world is what he calls *différance*, the artificial compound of 'difference' and 'deferral' that sidesteps any claim to certainty and presence. The

found in Ong 1988: 78-116. In ancient Israel, the situation is complicated by the debate over whether the Semitic alphabet is a true alphabet or a syllabary (see Havelock 1986: 100; Ong 1988: 89-91).

technology of 'writing in the narrow sense' merely expresses this tension, this absence, this ceaseless slippage.

Neither Havelock nor Lévi-Strauss gets to the heart of the matter in their attempt to find some causal connection between writing and culture as Derrida would have it. 'Writing in the narrow sense' expresses rather than creates the conditions of human interaction. This leaves us with the question of what leads to the manifestation of this technology; why does 'arche-writing' display itself as 'writing in the narrow sense'? In the concrete terms of our discussion, what leads Israel to embody its own history and origins in the unique series of texts which now comprise the Former Prophets?

Derrida himself connects the birth of 'writing in the narrow sense' to a condition of anxiety: *genealogical* anxiety, the anxiety about one's place in the social and familial world. 'The genealogical relation and social classification are the stitched seam of arche-writing, condition of the (so-called oral) language, and of writing in the colloquial sense' (1976: 125).

The Anxiety of Genealogy

Genealogical anxiety has two aspects, depending on whether we see it from the point of view of the parent or the child. For the parent, the anxiety reflects the knowledge that there can be no 'conscious begetting', as Joyce's Stephen Daedalus puts it.[6] The biblical text is full of fathers who cannot engender the heirs they, or their wives, seek because of the exigencies of the reproductive process. Their position can be summed up in the words of Jacob in response to Rachel's plea for sons: 'Am I in the place of God, who has withheld from you the fruit of the

6. 'Fatherhood in the sense of conscious begetting, is unknown to man. It is a mystical state, an apostolic succession from only begetter to only begotten. On that mystery and not on the madonna which the cunning Italian intellect flung to the mob of Europe the church is founded and founded irremovably because founded, like the world, macro- and microcosm, upon the void. Upon uncertitude, upon unlikelihood. *Amor matris*, subjective and objective genitive, may be the only true thing in life. Paternity may be a legal fiction. Who is the father of any son that any son should love him or he any son?' (Joyce 1961: 207).

womb?' (Gen 30.1-2). No man can ensure that he fathers a child merely by so willing. David's unlooked-for child by Bathsheba represents the other side of the same coin; fatherhood may be unwilled.

The reciprocal anxiety of the child finds its expression in that oddest of biblical questions, Saul's enquiry as to whose son the young man who has killed Goliath may be (1 Sam. 17.55-58). To know the name of one's father[7] is to know one's place in the social order of Israel. The importance of the father as against the mother in this reflects the old legal tag, *pater semper incertus est, mater certissima* (paternity is always uncertain, maternity is most certain).[8] There is an uncertainty about the relationship

7. The concept of the 'Name-of-the-Father' is also a central reference point in Jacques Lacan's theory of symbolization, not something susceptible to easy summary. Lacan himself acknowledges its origins in the return of the dead father in *Totem and Taboo* (1977: 199). The conjunction of fatherhood and death in Lacan's thought gives rise to the insight that both of these states can only be known through the agency of the signifier.

Though Lacan does not specifically indicate this, the common factor for both states is *absence*. Both the father and the dead are lost to the present, the father through the gap between copulation and the emergence of the faculty of recognition in the child. Just as the child can have no knowledge of those who die before its birth without the facility of symbolization and language, the ability to make the imaginative leap to assimilate what can only be told to him or her by others, to acknowledge the father as progenitor demands the same faculty.

The Name-of-the-Father becomes the essential anchor point for language, and for the speaking subject itself. The child is located socially by relation to the father and the bearing of his name, notably so in the Hebrew Bible. Not only that, however, the child is located as a speaking subject by the access to the metaphoric process that the recognition of the father brings. Recognizing one's father is the paradigm case for recognizing one's place in the biological, social and linguistic networks of exchange that constitute network of relations designating the self. The father is seen as the origin of law, of authority, with, in Lacan's account, the same possibilities for destruction and creation that we have seen any form of repression can engender (see Lacan 1977: 179-225).

8. Quoted in Freud 1909: 223. Freud in this essay refers to the fantasies that children have of being the unacknowledged offspring of the rich and famous, perhaps something to be set against the fantasies that parents have of their children succeeding in becoming what the parents failed to be (see the discussion of the work of Leclaire and Gunn below).

between father and child that does not exist in the case of the mother, an uncertainty intrinsic in the delay between copulation and child-birth, and the hiddenness of the process of development. The male can only have direct knowledge of the moment of ejaculation and of birth, not of the moment of conception. The consequences of this sense of the uncertainty inherent in fatherhood are profound.

The parallels between this and the uncertainties in the process of communication are the key to our discussion here. Fatherhood in the Hebrew Bible is initiated by an irrevocable act of utterance, of outpouring of seed, which is then consigned to the body of the woman, and whose subsequent fate it is not in the power of the father to determine. Writing too is an act of utterance, and irrevocable commitment of particular signs to the physical form of the document, whose subsequent fate and interpretative history are no longer in the power of its author to determine, for good or ill. The two aspects come together in the story of 2 Samuel 11, in which David writes the fatal letter to Joab that arranges for its bearer's death.

David's secret moment of lust will be literally brought to light in the embodiment of Bathsheba's child. A night of passion can be forgotten; a child must be lost or destroyed. A child is an exposed secret, which can be read and interpreted by those around. The child is the sign of sexual congress, the clue that leads to whispering about what went before. Like a text, the child is the visible, tangible evidence of a secret moment of creation and of utterance.[9]

In the text of 2 Samuel 11, David's secret letter is exposed to the reader's gaze. It is at once secret and vulnerable. The reader of the biblical text is put in the position of reading his secret document 'for Joab's eyes only'. How do we come to see it? Is what we are reading a transcript of a document that Joab preserved and somehow fell into the hands of those who wrote

9. Ricoeur remarks, 'A text opens up an audience, which is unlimited, while the relationship of dialogue is a closed relationship. The text is open to whoever knows how to read, and whose potential reader is everyone' (1991: 442). Writing is at once secret and open: secret in that it is incomprehensible to those who cannot read, open in that those who can read need no other permission to read it.

whatever version of the story eventually found its way into our present text? Is it a reconstruction? Is it in fact a literary device, originating not with David but with the teller of David's story? Is it merely a coincidence that the only reference to writing in the *Iliad* is also to a letter which commands the death of its bearer?[10]

The text provides no answers to these questions. What is clear is that the one piece of writing that is unequivocally attributed to David in the text that inscribes the rise and final death of his dynasty's rule over Israel is the shamefully exposed secret, intended 'for Joab's eyes only', of his attempt to repress the knowledge of his paternity. The consequence of this escapade is the birth and subsequent death of the child born to David. The prophet Nathan makes it clear that the child dies in some sense in David's stead (2 Sam. 12.13-14).

David's reaction to the child's illness is sufficiently unusual to be the subject of comment by his servants. While the child is dying he mourns and fasts. On receiving the news of his death, he gets up and dresses himself and eats. When he is challenged, his response is enigmatic: 'While the child was still alive, I fasted and wept; for I said, "Who knows whether the Lord will be gracious to me, that the child may live? But now he is dead; why should I fast? Can I bring him back again? I shall go to him, but he will not return to me"' (2 Sam. 12.22-23). This speech has provoked contradictory reactions in commentators. Is this the highest degree of spiritual resignation and wisdom, or is it the cynical response of a man who has got off with his own life?

10. *Iliad* 6.167ff. Bellerophontes is sent off by king Proitos to the king of Lycea with a tablet that instructs his death. This is done at the instigation of Proitos's wife Anteia, who had failed in an attempt to seduce Bellerophon. In his comment on Rousseau's dismissal of this episode as a later interpolation, Derrida draws attention to the fact that 'the only piece of writing in Homer was a letter of death' (Derrida 1976: 349). Not only that, but there is a sexual motivation as well. In the light of our subsequent discussion, this observation may be of some significance. Why should the two great bodies of ancient literature that the eastern Mediterranean has bequeathed to us both contain such a story as one of the few references to writing? The fact that the motif of the messenger who bears his own death sentence is known in other folk-literatures could indicate that it reflects some deep-rooted anxiety (see Gunkel 1984: 15).

Again the text gives no answer, and indeed goes out of its way to emphasize the ambivalence of the situation. The reactions of commentators may reveal more about their own anxieties than about the text. Yet there is an underlying ambivalence here in terms of the survival of the father at the expense of the son. Whose death does David truly desire—his own or his son's?

Son as Heir/Son as Rival

The complex dynamics of father–son relations in a biblical text have recently been explored by Devorah Steinmetz (1991). She begins with an overview of the attempts to relate psychoanalytic theory to the social anthropological study of kinship relations in various societies. She summarizes the ambivalence of the relation of father to son as follows:

> Fathers live on through their sons, passing down together with physical substance, possessions, ideals and customs. Whatever the father has accomplished will die with him if he has no son to take over. It is here that the ambivalence lies. As an extension of the father, the son ensures his immortality, yet as successor, the son usurps his place—he can continue for his father only on his father's death.
>
> To the father then, the son represents both the ultimate promise and the ultimate threat, immortality and death, and the father responds both by claiming the son and by rejecting him, in being torn between nurturing and killing him (1991: 21).

For the son, too, the father represents both promise and threat. The father has engendered the position that the son will inherit, but is also the obstacle to his achieving that position. How much more is this true in the situation of a hereditary monarchy, where questions of political power are overlaid on this already fraught structure of compromise between father and son.

Every episode of writing that we have looked at in the former prophets displays exactly this anxiety. On each occasion it is an issue over inheritance that prompts both the violence and the writing. This is obvious in the case of David and Uriah, where the paternity of Bathsheba's child is the problem, but the same holds true of the other incidents we have explored. Jehu is concerned to cut off the threat to his rule posed by the legitimate heirs of the king against whom he has rebelled. His purpose is to eradicate all the sons of Ahab who might revenge themselves upon him for the murder of Joram and who might use the argument of heredity to regain the throne. His letters result in the deaths of Ahab's seventy sons and all his former retainers.

It is a problem caused by the implications of inheritance that, in turn, gives Jezebel the motive to write. The reason Naboth gives for refusing to sell Ahab his land is that it is 'the inheritance of my fathers' (1 Kgs 21.3).[11] Jezebel seeks to override that customary right in this case, but, by doing so, could be seen as undermining a fundamental pillar of Israel's social organization, coherence and continuity. Even the episode in Judges 8 is part of the story of Gideon's vengeance on the murderers of his brothers. Writing, death and genealogy seem inextricably linked in these books of the Hebrew Bible, linked by the ambivalence of the need to risk the vulnerability of utterance, sexual or linguistic, in order to ensure survival. This ambivalence is manifested in the attitude of parents to children, and of writers to texts, of children to parents, and of readers to authors. It is also manifested in the attitude of a community to its history, the history that gave rise to it, and the history it propounds.

Fathers and Sons/Readers and Texts

The violence implicated in the ambiguous attitude of parents to children is explored by Daniel Gunn (1988). He draws on the writings of Serge Leclaire (1975), in which he describes the necessity for each of us 'to kill a child'. In each of us, there is a fight to the death between the real child and the 'marvellous child', the fantasy child who is compounded out of the long-suppressed narcissism of the parents. If this child is not killed, then the real child is doomed to a life of unreality. Yet the parents seek to resurrect in their child the child that died in their own early lives. In doing so, they evince an unconscious hostility to the

11. We might note here that Jehu represents himself as the avenger of the death of Naboth when he orders that Joram's body be cast on Naboth's plot in accordance with a divine oracle (2 Kgs 9.25-26). He thus respects the rights of inheritance in the middle of his revolt against the hereditary monarch.

concrete limitations and potentials of their real child. What is needed is the conscious choice to repudiate the fantasy in order to ensure the survival of the child, at the price of affirming the death of that fantasic child in the parents. Yet Leclaire urges the need for this child to be mourned repeatedly. 'Whoever does not mourn and mourn repeatedly the marvellous child he would have been, remains in limbo, in a milky light shed by a hopeless state of waiting that casts no shadow' (1975: 12).

Gunn himself relates the necessity of this choice to the anxiety of writing. In order to achieve any writing, the 'marvellous book' must be killed, the book that will end all books, that will have the final word.

> Only in the acceptance of the partiality and awful finality of words, and from the ashes of the narcissistic fantasy of universality, can writing emerge which will leave a place not for all readers but for any single reader and this individual reader's desire (1988: 46).

Writing involves the choice of one set of words among infinite possibilities, one narrative structure among infinite structures. Writing history demands the choice of one past among many possible pasts. In Gunn's words,

> Poets utter, writers write and as a consequence readers read, as it were backwards...The present tense of literary production is constantly a return to a worrying beginning which is itself inferred from the traces of death, completion and fulfilment. Ending seems to be written into narratives by the very ambivalence of the fact of telling, when telling is founded in the need to tell, and the suppression of the infinite possibilities which any particular telling necessitates (1988: 123-24).

Yet it must also involve a mourning for the past that has been killed, the ideal past that has perished in the compromises and failures of the present and whose monument is the written history—a monument both in the sense of a marker of death, but also as a reminder, which becomes the possible dream of the future.

The fascination of the Former Prophets is that they display a dizzying synthesis of content, form and expression in regard to this fundamental anxiety. Their subject matter is the growth and decline of the hereditary monarchy, the familial metonym for the community. Here political and sexual tensions are not just related, but become fused. Absalom, for instance, rebels against both his father and his king. Yet the text that records this itself arises out of the genealogical anxieties of the community. It reaches its final form in the anxieties of the exile, among a community that seeks both to establish its continuity with its past and yet to reject it, to remember and forget, to mourn the fact that the ideal was never realized and to use that ideal to ensure the survival of the real community.[12]

Regina Schwartz (1990: 47-49) argues that this dialogue between memory and forgetting resembles the relationship of repression and interpretation as expounded by Freud in his *The Interpretation of Dreams*. Her point is that these two apparently contradictory movements of destruction and restoration are bound up with one another. Repression is not only bound to destruction and forgetting but also to remembering and interpretation. Put negatively, repression makes interpretation necessary in the attempt to recover what is missing; positively, it creates the space in which interpretation as re-creation becomes possible. The act of remembering is also an act of repetition, which itself entails both continuity and discontinuity:

> discontinuity, because there must be a break to enable something to be repeated, just as something must be lost to be recovered, forgotten to be remembered; and continuity, because the fact of repetition, recovery, memory ensures a living on (1990: 55).

In *The Interpretation of Dreams*, Freud makes intriguing references to the importance of writing in his account of dream

12. See on this Brooks (1984: 227-28), who writes the following in his discussion of the transference relationship as applied to the interaction between the text and the reader: '[T]here is in the dynamics of the transference at once the drive to make the story of the past present—to actualize past desire—and the countervailing pressure to make the history of this past definitely past: to make an end to its reproductive insistence in the present, to lead the analysand to the understanding that the past is indeed past, and then to incorporate this past, as past, within his present, so that the life's story can again progress'. In this sense, the historiographer could be seen as attempting though writing the cure of a society traumatized by the loss of its monarchy, as offering a way that the story may yet continue.

interpretation. He offers an account of one of his dreams which contains a baffling phrase. In a footnote, he remarks,

> This description was unintelligible even to myself; but I have followed the fundamental rule of reporting a dream in the words which occurred to me as I was writing it down. The wording chosen is itself part of what is represented by the dream (1900: 598 n. 2).

What actually matters in the end is the wording. It is in the act of writing that the dream-text which is the object of analysis comes into being. In writing, the conscious mind makes choices, and represses the unconscious, which nevertheless makes its presence felt through oddities and aporiae. It is the questioning of the text in its immutable display of repression and displacement at work in its language that leads to a reading of the dreaming subject, and access to the unconscious, the repressed, the forgotten.

Freud, Historiography and the Death of the Father

The link between this and the topic under consideration is made by Freud himself, who ventured an analysis of biblical historiography in his *Moses and Monotheism* (Freud 1939).[13] In this work he defended the thesis that the core of Judaism was an act of repression of the murder of Moses. He presents Moses as an Egyptian nobleman who had enforced an ethical monotheism derived from that of the heretic Pharaoh Akhenaton on a recalcitrant band of slaves. In an upsurge of resentment at his repressive demands, the Israelites turned on him and killed him. Their collective guilt at this deed perpetrated against one whom they both loved and feared as a father-figure led to repression of this memory. The God Yahweh displaced Moses as the origin of their ethical and cultic practices. The re-emergence of ethical monotheism now associated with Yahweh in later Judaism is an

13. See on this Yerushalmi (1991: 29): 'What readers of *Moses and Monotheism* have generally failed to recognize—perhaps because they have been too preoccupied with its more sensational aspects of Moses the Egyptian and his murder by the Jews—is that the true axis of the book, especially of the all-important part III, is the problem of tradition, not merely of its origins but above all its transmission'.

instance of the return of the repressed, to be compared with the same phenomenon in the aetiology of neurosis.

Though Freud's historical reconstruction is far-fetched to say the least, what is interesting from our point of view is his analysis of the role of writing in this development. The historical books of the Hebrew Bible that offered a very different account of Moses' career from that offered by Freud. How is this to be explained? According to Freud, the biblical text as we now read it is the product of two opposed forces: (a) an impulse to revise and mutilate the historical record in the interests of repressing the knowledge of the primal murder; and (b) a solicitous piety that preserves the resultant text despite its manifest contradictions. 'In its implications', he writes, 'the distortion of a text resembles a murder; the difficulty is not in perpetrating the deed, but in getting rid of the traces' (1939: 283).

This adds another murder to the story. The 'murder' of the text is the attempt to suppress the evidence of the original murder of Moses. This, however, is itself the re-enactment of an even earlier murder: the murder of the father of the primal horde which Freud discusses in his *Totem and Taboo* (Freud 1913).[14]

Yet any act of writing, not simply Israel's historiography, bears witness to an act of murder, as Derrida points out in his extended reaction to Plato's familial metaphors in the discussion of writing in the *Phaedrus*. Socrates there dismisses the written text as the errant son:

> ...once a thing is committed to writing it circulates equally among those who understand it and those who have no business with it; a writing cannot distinguish between suitable and unsuitable

14. In *Totem and Taboo*, Freud argues that an original despotic patriarchy was overthrown by the banding together of the subjugated sons to murder their father and gain access to the women he controlled. This fraternity, however, was still haunted by the figure of the absent father who was loved and admired as well as hated, and so set up in his place a particular animal as a totem, and renounced in a process of 'deferred obedience' their rights to the women they had gained. Over time, the repressed knowledge of this murder resurfaced in the form of a gradual evolution toward the worship of a single all-powerful generative God, the idealized replacement for the father of the primal horde.

> readers. And if it is ill-treated or unfairly abused it always needs its parent to come to its rescue; it is quite incapable of defending or helping itself (1973: 97).

As Derrida reads Plato's account, writing is in need of its father, but also exists without him. It is at once a pitiable orphan and the despicable wastrel, one who could stand accused of parricide. It carries with it the taint of murder. Writing acknowledges its 'father' only by calling attention to his absence, and by calling down upon itself the reproaches due to the murderer of the father. The figure of the parricide is the only witness to the existence of the dead father.[15]

This means that the relation of generation is reversed. Although biologically the father generates the son, the existence of the son is the datum that follows the inference of the existence of the father. In the same way, the reader of the text must engage in an act of generation of its author, through the body of the text. The 'author' of a text is a fiction that the reader engenders, a 'father', a point of origin, that the reader fathers, with all the complexities of emotional interaction that that might give rise to. In biblical studies, historical criticism has engendered a plethora of such authorial figures, all of whom have been engaged in the rewriting, the suppression of their predecessors' work, just as critics have built upon and thereby assailed the work of their precursors.

This process is exposed by Harold Bloom in *The Book of J*, an attempt to resurrect the writing of the female author J, whom he regards as responsible for the strongest biblical writing in the canon (Rosenberg and Bloom 1991: 19).[16] Bloom has great fun in

15. Derrida explains, 'Writing can thus be attacked, bombarded with unjust reproaches (*ouk en dikei loidoretheis*) that only the father could dissipate—thus assisting his son—if the son had not, precisely, killed him. In effect, the father's death opens the reign of violence—and that is what it's all about from the beginning—and in violence against the father, the son—or patricidal writing—cannot fail to expose himself, too. All this is done in order to ensure that the dead father, first victim and ultimate resource, not be there. Being-there is always a property of paternal speech' (1981: 146).

16. Bloom's work on the anxiety of influence in which he sees all creative work as marked by a strong misreading of the author's precursors and a wresting of the tradition from those who went before is here belatedly acknowledged as a significant influence on the present text. In

describing this delightful woman, who carries the whole erotic weight of his response to these astonishing texts. She is, of course, a fiction, and he makes no bones about this:

> This J is my fiction, most biblical scholars will insist, but then each of us carries about a Shakespeare or a Tolstoy or a Freud who is our fiction also. As we read any literary work, we necessarily create a fiction or metaphor of its author. That author is perhaps our myth, but the experience of literature partly depends on that myth. For J we have a choice of myths, and I boisterously prefer mine to that of the biblical scholars (1991: 19).

J's text has not ensured her survival; she survives as a fiction, the mother fathered by Bloom on the text. She can only be resuscitated at the expense of the murder of the text as we have it, the dissection out of its layers, and the killing of the figure whom Bloom refers to as R, the Redactor, the equally fictional author of the final form of the texts. R himself has suppressed the existence of those authors whose work he draws upon, and yet paradoxically it is the fact that his text survives in the canon that enables his precursors to be raised to a ghostly life.

Text and Reader: The Fight for Survival

As such, R embodies, as must any author, the guilt of the survivor.[17] Tellers of stories display an incident from the past.

particular, Bloom's reading of Freud and his use of the category of repression is of great interest. See on this Bloom 1982 and, specifically on *Moses and Monotheism*, Bloom 1991.

17. The term 'survival guilt', now widely used in the examination of the trauma that affects those who live through major disasters, and with a special application in the experience of those who were liberated from the death camps of Nazi Germany, is generally traced back to Freud himself, who identified it as a component of his own reaction to the death of his father. See *The Complete Letters of Sigmund Freud to Wilhelm Fliess 1887–1904* (ed. J.M. Masson; Cambridge, MA: Harvard University Press, 1985): 202 (Freud to Fliess, 2 November 1896), where Freud, in the course of describing his reaction to his father's death, talks of the 'self-reproach that appears regularly among the survivors'. He alludes to this more fully in his open letter to Romain Rolland ('A Disturbance of Memory on the Acropolis' [1936] *Standard Edition* 22, 239) where he traces the strange feeling of dissociation he experienced on his first visit to the Acropolis to a

When they talk of danger, they talk of danger survived, or else there would be no story and no teller. When they speak of death, they speak of the death of another. They display the life of the dead in order to affirm their own life. The narratorial voice has survived the events of the narrative in such a way as to be able to bring those events into narrative form.[18] That guilt finds expression in the writer's abuse of writing, the unreliable bearer of the writer's own survival. For the historiographer, this resentment is compounded by the knowledge that survival has been bought at the price of murder and repression. The historiographer turns against the process of writing that ensures his own survival. It is this process that leads to the suppression of writing through its association with death and violence that prompted the question with which this study began.

Yet that tale of survival can only be resurrected by the reader, who comes after, who has survived the author in that he or she is addressed by the text which bears witness to the supplanting of the author. The voice of the author which the reader fathers upon the text becomes a voice that speaks from and for the dead who lay claim to life through the reader.

So R, the fictional figure of the compiler of the historical books, embodies the anxiety of those who have survived the exile and of the demise of the Davidic kingdom. The very act of writing of that kingdom as past confirms that demise. The monarchy dies when its passing is committed to writing. The terror of that act of pious killing imbues the act of writing itself with a sense of doom, just as the movement towards a hereditary kingship in Israel presaged its own doom through its activation of the tensions between father and son. The suppression of the act of writing and its association with death and genealogy within the

feeling of guilt in having surpassed his father.

18. On the topic of the relation between the survivor and the story, see Felman and Laub 1992 for their fascinating exploration of the testimony of the survivors of the Holocaust. They summarize the relationship between these elements as follows: 'The story of survival is, in fact, the incredible narration of the survival of the story, at the crossroads of life and death' (1992: 44), and later: 'The survivor survives in order to tell the story, but also must tell the story in order to survive, in order to have a sense of who they are—we *are* our testimony' (1992: 78).

text reflect the complex of anxieties that culminated in the act of generating the text, an act that itself involves both murder and procreation.

The final compiler of these texts commits murder against the texts of his predecessors, a murder that he suppresses, but that becomes the only means of survival of these precursors. In his turn, his name is lost to history. But that death also becomes the possibility for reinterpretation after the exile. The writer is effaced by his own text, and has in turn effaced other writers. The community that his text now enables to survive is unrecognizably different from the one he knew, which in turn is radically different from the one he created as he wrote. His text is silent testimony to the murders on which it is based, and yet it is a testimony to the possibility of survival. As such, it places on its readers a choice. It claims us as its progeny, as members of the community that will ensure its survival. It promises the survival of the community in order to induce the community to ensure the survival of the text.

Ultimately, the power of these texts for the reader is that that they activate the reader's anxiety of genealogy. The text claims its status as father, as the key to the genealogy of the community and the reader, with all the implied aggression that that claim embodies. Is the reader to assent to the claims of the text, to accept the possibilities it offers and so to enter the community whose survival is symbiotic with that of the text, to father the next generation of readers? Or will the reader repudiate the text as the harbinger of death, the father that seeks to kill the possibilities of the reader's existence, in its frustrated attempt to create an ideal child?

BIBLIOGRAPHY

Bloom, H.

1982 'Freud's Concepts of Defense and the Poetic Will', in *Agon: Towards a Theory of Revisionism* (New York: Oxford University Press): 119-44.

1991 'Freud and Beyond', in *Ruin the Sacred Truths: Poetry and Belief from the Bible to the Present* (Cambridge, MA: Harvard University Press): 143-204.

Brooks, P.
1992 *Reading for the Plot: Design and Intention in Narrative* (Cambridge, MA: Harvard University Press).

Conrad, E.W.
1992 'Heard but not Seen: The Representation of "Books" in the Old Testament', *JSOT* 54: 45-59.

Derrida, J.
1976 'The Violence of the Letter: From Lévi-Strauss to Rousseau', in *Of Grammatology* (trans G.C. Spivak; Baltimore: Johns Hopkins University Press): 101-40.
1981 'Plato's Pharmacy', in his *Dissemination* (trans. B. Johnson; Chicago: University of Chicago Press): 63-171.

Driver, G.R.
1976 *Semitic Writing: From Pictograph to Alphabet* (rev. S.A. Hopkins; Oxford: Oxford University Press [1944]).

Felman, S. and D. Laub
1982 *Testimony: Crises of Witnessing in Literature, Psychoanalysis, History* (New York: Routledge).

Freud, S.
1900 *The Interpretation of Dreams* (The Pelican Freud Library, 4; Harmondsworth: Penguin, 1976).
1909 'Family Romances', in *On Sexuality* (The Pelican Freud Library, 7; Harmondsworth: Penguin, 1977): 217-26.
1913 'Totem and Taboo', in *The Origin of Religion* (The Pelican Freud Library, 13; Harmondsworth: Penguin, 1985): 43-224.
1936 'A Disturbance of Memory on the Acropolis', in *On Metapsychology: The Theory of Psychoanalysis* (The Pelican Freud Library, 11; Harmondsworth: Penguin, 1984).
1939 'Moses and Monotheism', in *The Origin of Religion* (The Pelican Freud Library, 13; Harmondsworth: Penguin, 1985): 237-386.

Goody, J.R. and I. Watt
1968 'The Consequences of Literacy', in *Literacy in Traditional Societies* (ed. J.R. Goody; Cambridge: Cambridge University Press): 27-84.

Gunkel, H.
1984 *The Folktale in the Old Testament* (trans. M.D. Rutter; Historic Texts and Interpreters, 5; Sheffield: JSOT Press).

Gunn, D.
1988 *Psychoanalysis and Fiction: An Exploration of Literary and Psychoanalytic Borders* (Cambridge: Cambridge University Press).

Havelock, E.A.
1986 *The Muse Learns to Write: Reflections on Orality and Literacy from Antiquity to the Present* (New Haven: Yale University Press).

Joyce, J.
1961 *Ulysses* (New York: Modern Library).

Lacan, J.
1977 'On a Question Preliminary to Any Possible Treatment of Psychosis', in *Ecrits: A Selection* (ed. and trans A. Sheridan; London: Tavistock; NewYork: Norton).

Leclaire, S.
1975 *On tue un enfant: un essai sur le narcissisme primaire et la pulsion de mort* (Paris: Seuil).

Lévi-Strauss, C.
1961 *Tristes tropiques* (trans J. Russell; London: Hutchinson).

Masson, J.M. (ed.)
1985 *The Complete Letters of Sigmund Freud to Wilhelm Fliess 1887–1904* (Cambridge, MA: Harvard University Press).

Nielsen, E.
1954 *Oral Tradition: A Modern Problem in Old Testament Introduction* (London: SCM Press).

Ong, W.J.
1977 'Maranatha: Death and Life in the Text of the Book', in *Interfaces of the Word: Studies in the Evolution of Consciousness and Culture* (Ithaca, NY: Cornell University Press): 230-71.
1988 *Orality and Literacy: The Technologizing of the Word* (London: Routledge): 78-116.

Plato
1973 *Phaedrus and Letters VII and VIII* (trans. W. Hamilton; Harmondsworth: Penguin).

Ricoeur, P.
1991 'Phenomenology and Theory of Literature', in *A Ricoeur Reader: Reflection and Imagination* (ed. M. Valdés, New York: Harvester Wheatsheaf): 441-47.

Rosenberg, D. and H. Bloom
1991 *The Book of J* (London: Faber & Faber).

Saramago, J.
1992 *The Year of the Death of Ricardo Reis* (trans. G. Pontiero; London: Harvill).

Schneidau, H.N.
1976 *Sacred Discontent: The Bible and Western Tradition* (Baton Rouge: Louisiana University Press).

Schwartz, R.M.
1990 'Joseph's Bones and the Resurrection of the Text: Remembering in the Bible', in *The Book and the Text: The Bible and Literary Theory* (ed. R.M. Schwartz; Cambridge, MA: Blackwell): 40-59.

Steinmetz, D.
1991 *From Father to Son: Kinship, Conflict and Continuity in Genesis* (Louisville: Westminster/John Knox).

Street, B.
1984 *Literacy in Theory and Practice* (Cambridge: Cambridge University Press).

Yerushalmi, Y.H.
1991 *Freud's Moses: Judaism Terminable and Interminable* (New Haven: Yale University Press).

DAUGHTERS AND FATHERS IN GENESIS... OR, WHAT IS WRONG WITH THIS PICTURE?

Ilona N. Rashkow

While it is not surprising that biblical narratives depict a definable family structure, what *is* surprising is that conspicuously absent is a figure lurking beneath the text, a figure repeatedly subjected to erasure, exclusion, and transformation. Genesis lacks daughters. Narrative after narrative describes the desire for male children, the lengths to which women would go to have sons,[1] the great joy surrounding the birth of a boy, and father–son relationships.[2] The birth of a daughter, on the other hand, by no means creates such attention. As Archer notes, biblical genealogical tables 'indicate a startling disparity in the ratio of male:female births, a disparity which can in no way reflect a demographic reality' (Archer 1990: 18). The tables do, however, reflect the attitude towards daughters. Inscribed within Genesis

1. So important were sons that barren women sometimes resorted to having children by their handmaids (Gen. 16.2; 30.3). For the use of concubines and handmaidens in this early period and the legitimacy of offspring from such unions, see Archer 1987: 4.

2. This is most obvious in the covenant between the deity and Abraham (Gen. 17.9-10), the implicit symbolism of circumcision powerful in its patriarchal reverberations. A son was regarded as a special blessing, more often than not the direct result of divine intervention in a couple's life. Eve, for example, the first to give birth (significantly, to a boy) triumphantly declares: 'I have gotten a man *with* [*the help of*] *the Lord*' (Gen. 4.1); Abraham, convinced of Sarah's sterility, is informed by God: 'And I will bless her, and moreover *I* will give you a son of her' (Gen. 17.15); similarly, in Gen. 30.21-24, 'God remembered Rachel and God hearkened to her and *opened her womb*. And she conceived and bore a son.' The passage ends with Rachel's plea for more sons: 'And she called his name Joseph, saying "The *Lord add to me* another son"'.

is something more than a general disregard of women: the *daughter* is specifically absent. Since the daughter's presence is normal and necessary to the biological realities of family, her narrative absence is significant and calls attention to itself. My conclusion is that beneath the surface father–son narration lies a suppressed daughter–father relationship.

Perhaps because I cannot help thinking that Genesis is more ambivalent than a narration of disinterested fathers, I read daughters in a more paradoxical way. Instead of measuring what the daughter may or may not materially contribute to the family, I consider what she threatens to subtract from it. The most obvious answer, of course, is that while yet within her father's house the daughter is the only member of the family who does not participate in extending the patronymic line. But that answer is too superficial. By aligning feminist analyses of Freud's rejection of the seduction theory with the suppressed daughter–father biblical construct, a subtext is uncovered: what makes the nearly absent daughter so central in this otherwise emphatically masculine epic is her potential to determine and expose a threat to the father's power and patriarchal rule.

Many biblical narratives describe a daughter's transgression against and departure from the closure of her father's house.[3] The text in effect becomes a code for what is subliminally the father's story of the sins of the daughter. Decoded, the accusations might read: *because* of the daughter's sin against the father, sons must henceforth leave their father's control ('This is why a man leaves his father' [Gen. 2.24]);[4] *because of* the daughter's disobedience, daughters likewise leave the protective enclosure and become maternal figures. Daughters are subsumed as mothers,[5] and the text 'reads itself through a chain-male linkage'

3. See, for example, Jephthah's daughter, whose departure from her father's house is viewed by Jephthah as a transgression *against him* (Judg. 11.35). Dinah 'goes out', is raped (Gen. 34.1-2), and is then narratively banished from the text (Rashkow 1990: 98-100).

4. Northrop Frye's comment on this verse is that 'the chief point made about the creation of Eve is that henceforth man is to leave his parents and become united with his wife. That parent is the primary image...that...has to give way to the image of the sexual union of bride-groom and bride' (Frye 1982: 107).

5. When her identity as daughter is exchanged for wife, she is still the

(Boose 1989: 22). These repeated biblical narratives of a daughter's 'transgression' seem to be prototypical of Freud's narrative of the 'catastrophe' that leaves 'the path to the development of femininity...open to the girl' (Freud 1925: 241). Significantly, the 'catastrophe' Freud describes is father–daughter incest. If read from that perspective, the daughter can be seen as locked into a conflicted text of desire and sanction.

Lévi-Strauss's well-known analysis argues that the most significant rule governing any family structure is the ubiquitous existence of the incest taboo which imposes the social aim of exogamy and alliance upon the biological events of sex and procreation. Genesis nearly constitutes a meditation on the questions from where wives for the patriarchy should come, how closely they should be related to 'us', and how 'other' they should be (Pitt-Rivers 1977: 128, 165). Within the patriarchal sagas, Abraham twice acknowledges his wife to be his sister,[6] and his son, Isaac, marries his father's-brother's-daughter. Isaac's son, Jacob, acquires two wives, sisters who constitute a lineal double of each other. That is, Jacob marries two of his father's-father's-brother's-son's-son's-daughters, who are simultaneously his mother's-brother's-daughters and thus again connected back to Abraham. In the next generation, Reuben sleeps with his father's second wife's maid, symbolically violating family purity laws, and Judah sleeps with his daughter-in-law. Is there a pattern here? *Contra* Lévi-Strauss, familial and sexual integrity across Genesis seems to be observed more in the breach than in the maintenance.[7] Why?

While many elements of the conventional vocabulary of moral deliberation (such as 'ethical', 'virtuous', 'righteous', and their

alien until she has once again changed her sign to 'mother of new members of the lineage', which by implication means mother to a son.

6. Not all scholars view intercourse between siblings as incestuous (see Fokkelman's discussion of the story of Amnon and Tamar, 1981: 103). As Landy points out, however, this might be another example of a royal family that 'feels itself too good for the world' (Landy 1983: 307 n. 63).

7. Within Genesis, Adam–Eve, Noah–Ham, Lot–his daughters, Reuben–Bilhah, and Jacob–Tamar are examples of parent–child incestuous congress or exposure; Adam–Eve and Abraham–Sarah are brother–sister unions (as is Amnon–Tamar in 2 Sam.); Isaac–Rebekah and Jacob–Leah–Rachel are cousin marriages.

opposites) are largely alien to the psychoanalytic lexicon, the concepts of 'guilt' and 'shame' do appear, albeit in technical (and essentially non-moral) contexts (Smith 1986: 52). 'Guilt' and 'shame' are described as different emotional responses, stemming from different stimuli, reflecting different patterns of behavior, and functioning in different social constructions, although the two are often related. Their primary distinction lies in the internalized norm that is violated and the expected consequences.

Guilt relates to internalized societal and parental *prohibitions*, the transgression of which creates feelings of wrong-doing and the fear of punishment (Piers and Singer 1953). Shame relates to the anxiety caused by 'inadequacy' or 'failure' to live up to internalized societal and parental *goals and ideals* (as opposed to internalized prohibitions), expectations of what a person 'should' do, be, know, or feel. These feelings of failure often lead to a fear of psychological or physical rejection, abandonment, expulsion (separation anxiety) or loss of social position (Alexander 1948: 43). The person shamed often feels the need to take revenge for his or her humiliation, to 'save face'. By shaming the shamer, the situation is reversed, and the shamed person feels triumphant (Horney 1950: 103).

The difference between guilt and shame is subtle but important in the context of this paper. Within the biblical text, 'shame' is a powerful and prevalent emotion and sanction indicated by the number of Hebrew words used to convey the violation of goals and ideals,[8] although in translation the differences in

8. 'Shame' is expressed in Hebrew by the verb בוש, 'to shame' (and the nouns בושה, בשנה, בשת, 'shame'); the verb כלם, 'to humiliate/shame' (and the nouns כלמה and כלמות, 'humiliation/shame'); the verb קלה (niphal; perhaps a form of קלל, 'to be light', 'to be lightly esteemed or dishonored/shamed') (and the noun קלו, 'dishonor/shame'); the verb חפר, 'to be ashamed, blush'; the verb שפל, 'to be low, abased, be humiliated' (and the noun שפלה, 'lowliness, humiliation'); the verb מכך, 'to be low, humiliated'; and the nouns נבלות, 'shamelessness', and נבלה, 'disgrace'.

'Shame' words are often accompanied by phrases that express 'shame on the face' (blushing), shame expressed in body position (hanging the head), or a reduction in social position in one's own eyes and in the eyes of others (e.g. Jer. 48.39; 2 Sam. 10.5; Isa. 16.14; Jer. 50.12). The verb חרף, 'to reproach/verbally shame' (and the noun חרפה, 'reproach/verbal shame'); the verb קלס, 'to mock/shame' (and the noun קלס, 'derision/shame'); the

meaning among these words are often hard to discern. Strikingly, the vocabulary for 'guilt' is far less extensive than that of 'shame'.[9] It would appear that the text is less concerned with the violation of societal prohibitions—in this case, incest—than with the failure to achieve internalized goals, that is, the idealization and perpetuation of patriarchy and family prestige.

It is within this framework that the father–daughter relationship becomes problematic, complex in ways that even the mother–son dynamic is not, despite the same asymmetries of age, authority and gender-privilege that work to separate mother and son. On the one hand, daughters are property belonging exclusively to the father;[10] like Laban's daughters, Leah and Rachel, they are bartered for economic profit. And as the Genesis narrative of Jacob's daughter Dinah makes clear, rape is not considered a violation of the daughter so much as a theft of property from her father that necessitates compensation to him. On the other hand, although the daughter is clearly regarded as legal property inside the family, she is not a commodity to be bartered in the same way as an ox or an ass. She is explicitly *sexual* property acquired from the father's sexual expenditure and his own family bloodline, not by economic transaction. Her presence as daughter resexualizes the family

verb לעג, 'to mock/shame' (and the noun לַעַג 'derision/shame'); the verb לִץ, 'to scorn/shame' (and the noun לצון, 'scorning/shaming') denote verbal shaming, taunting, mocking or scorning with insulting words. The main oppositional term to shame is the root כבד (signifying 'honor' or 'heaviness'); that is, honor increases 'heavy' esteem, while shame decreases it, causing 'light' esteem (see קלה above) (Bechtel 1991: 54).

9. 'Guilt' is expressed by the verb אשם, 'to offend/be guilty/commit iniquity' (and the nouns אשם, 'offense/guilt/iniquity', אשמה 'wrong-doing/guiltiness', and the adjective אשם, 'guilty'); the verb רשע, 'to be wicked/condemn as guilty' (and the adjective רשע, 'wicked/guilty'); and the noun עון 'iniquity/guilt/punishment' (Bechtel 1991: 55).

10. Since it is the father who controls the exchange of women, the woman most practically available to be exchanged is not the mother, who sexually belongs to the father, nor the sister, who comes under the bestowal rights of her own father, but the daughter. Other anthropological models do exist, however. Among the Nuer, for example, 'fatherhood' belongs to the person in whose name cattle bridewealth is given for the mother (Rubin 1975: 169).

configuration and necessitates a detailed taboo, codified in Leviticus 18, which ostensibly defines illicit congress. Virtually every family female (mother, sister,[11] aunt,[12] cousin, sister-in-law, niece, daughter-in-law, granddaughter, and so on[13]) is off-limits. Conspicuously, the only one not included is the daughter. As Judith Herman points out,

> the wording of the law makes it clear that...what is prohibited is the sexual use of those women who, in one manner or another, already belong to other relatives. Every man is thus expressly forbidden to take the daughters of his kinsmen, but only by implication is he forbidden to take his own daughters (Herman and Hirschman 1981: 61).

Of all possible forms of incest, that between father and daughter is overlooked. The daughter's presence within the father's house retains a figuratively, if not literally, incestuous option that implicitly threatens the family structure.[14]

Since the text lacks this specific taboo, the father–daughter relationship has no internalized prohibitions (hence no 'guilt'). But because the purity of a wife is the law of first priority upon which patrilineage depends, it is at the juncture of the daughter's

11. Lev. 18.9, 11; 20.17. Paternal half-sister prohibition was of special concern to Ezekiel (22.11), and his concern shows that the practice continued.

12. Lev. 18.12-13; 20.19. We might note that Moses and Aaron were both born of such a union (Exod. 6.20; Num. 26.59).

13. The other incestuous relations itemized in Lev. 18 and 20 belong to the category of incestuous adultery (that is, group-wife prohibitions) or pertain to polygamy and are therefore not our concern here. For a full analysis of the incest laws in Leviticus, their function and origin, see Bigger 1979 and Fox 1967. For a situating of these laws in the wider context of historical shifts in Jewish social structure and the changing position of women, see Archer 1990 and 1983.

14. In fact, it can even be argued that the relations of biblical daughters and fathers resemble in some important ways the model developed by Judith Herman and Lisa Hirschman to describe the family situations of incest victims: a dominating authoritarian father; an absent, ill or complicitous mother; and a daughter who, prohibited from speaking about the abuse, is unable to reconcile her contradictory feelings of love for her father and terror of him, her desire to end the abuse and fear that if she speaks out she will destroy the family structure that is her only security (Herman and Hirschman 1981: esp. chs. 1, 4–7).

marriage and transfer of proprietary rights from father to son-in-law that father–daughter incest would point a finger directly at the character whom the text privileges, the one status role that the narrative repeatedly goes out of its way to exempt from blame of any sort, the father. The biblical daughter becomes dangerous to the father's authority, and her existence within the 'safety' of her family ambivalent. It is in this context that the elaborately detailed punishment for the accused bride in Deut. 22.13-21 makes sense. All of the numerous proscriptions codified in Deuteronomy are essentially purification laws to 'banish evil from Israel'. This one, however, is unique in thrusting the father to the very center of the drama, making him a special actor, *protected* by a formulaic dialogue yet placed in the role of *defendant* against the son-in-law's charges of the daughter's impurity.[15] Implicitly, the husband has accused the father, the man who gave him this woman, of having taken the husband's property (her virginity) in advance. If evidence of virginity exists, the groom is flogged and must pay the father one hundred shekels 'for publicly defaming a virgin of Israel'. But the payment is made to the father, so perhaps we should read 'for publicly defaming a virgin's *father*'. If the bride's virginity cannot be substantiated, 'they shall take her to the door of her father's house and her fellow citizens shall stone her to death for having committed an infamy in Israel by disgracing her father's house'. This crime is not merely 'an evil' to be 'banished from the midst'; it is 'an infamy in Israel' that disgraces the father's house (the place from which the punishment implies it emanated) by tacitly accusing him of incest. It then masks the accusation by transposing cause and effect: because no hymeneal blood was shed

15. In the three sex laws that follow this one in Deut. 22, the father is either not mentioned or minimally important. If, for instance, a man forcibly seizes an unbetrothed virgin and 'they are found', he must pay her father fifty shekels and marry her (22.28-29). If a man lies with a betrothed virgin inside the city, the two offenders are to be taken outside the gate of the town and stoned to death, she for not 'crying out' and he for 'violating the wife of his neighbor'. The father is not involved here, but the male violator (as well as the female property that is now 'soiled') must die since the future rights of another man have been stolen. See Mary Douglas's chapter 'Internal Lines' (1966) for an examination of the connections between social pollution and cultural ideas of 'dirt'.

in her husband's house, the daughter's blood is to be shed on her father's door. A threat to the father's reputation (and hence his power) is averted by deflecting blame for sexual misconduct, real or imagined, from the privileged patriarch onto the powerless daughter. The shamed thus shames the victim.

A parallel construct exists in Freud's abandonment of his seduction theory. When Freud first began working with hysterical patients, in every case he found an account of childhood sexual abuse by a member of the patient's own family, and it was almost always the father.[16] On this evidence, Freud developed his 'seduction theory', that hysterical symptoms have their origin in sexual abuse suffered in childhood, which is repressed and eventually assimilated to later sexual experience. Within a year, however, Freud wrote that he 'no longer believe[d] in neurotica' (quoted in Froula 1989: 118). At this point, Freud founded psychoanalytic theory upon the Oedipus complex.

This change was crucial. As several feminist critics have argued,[17] Freud turned away from the seduction theory because he was unable to come to terms with his discovery: the abuse of paternal power.[18] The issue for Freud was credit versus

16. The editors of the *Standard Edition* trace (without critique) the vicissitudes of Freud's acknowledgement of sexual abuse on the part of fathers in a note to 'Femininity' (Freud 1933: 120 n.).

17. See, for example, Alice Miller (1984), as well as Herman and Hirschman (1981) who present clinical evidence; Marie Balmary (1979) for a psychoanalytic reading of the 'text' of Freud's life and work; and Florence Rush (1980) for a historical perspective. See particularly David Willbern's examination (1989) of the chronological complexities and fluctuations in Freud's theorizing about fathers and daughters, including a discussion of Freud's discounting of the seduction theory and his strangely unprofessional alteration of several case testimonies in which the father had been identified as the incestuous seducer of his daughter.

18. The cases of Anna O., Lucy R., Katharina, Elisabeth von R., and Rosalia H., described in *Studies on Hysteria* (Freud and Breuer 1893–95), all connect symptoms with fathers or, in Lucy's case, with a father substitute. In two cases, however, Freud represents the father as an uncle, a misrepresentation that he corrects in 1924. His reluctance to implicate the father appears in a supplemental narrative of an unnamed patient whose physician-father accompanied her during sessions with Freud. When Freud challenged her to acknowledge that 'something else had happened which she had not mentioned', she 'gave way to the extent of letting fall a single

authority—whose story to believe, the father's or the daughter's.

While many analysts have simply followed Freud in rejecting the seduction theory for the Oedipal theory, others have tried to explain and resolve the apparently contradictory ideas of 'seduction-as-fact' and 'seduction-as-fantasy' by means of Freud's concepts of '*psychic* reality' and '*primal* fantasy'.[19] That is, seduction can be a *representation* of the father's repressed and deflected sexual desires, or even a metaphor for power ('primal fantasy'). *Actual* incest ('reality') need not enter the picture, thus bridging the gap between the actual and the imaginary, the very structure of fantasy.

Conveniently, this brings us to the creation narrative in Genesis 1–3. While almost all interpretations of this text acknowledge its sexual nature, traditional exegesis has concentrated on 'Adam's Fall'. But the familiar story masks two interwoven subtexts: Freud's sexualized father–daughter narrative in which the Adam material appears merely as a re-narration, and a feminist narrative of an unacknowledged daughter's rebellion by means of her appropriating the forbidden fruit that stands 'erected' at the center of the enclosed garden. Read from this perspective, the father has planted an invitation to transgress (a metaphoric seduction) accompanied by a prohibition against doing so. The ambivalence of the father's part in the 'Fall', the focus of considerable theological commentary, perhaps can be seen as Freud's 'catastrophe', with its dangerous potential inherent in the daughter's 'transition to the father object' (Freud 1925: 241). The father desires yet forbids desiring; he simultaneously wants but does not want the transgression he has provoked, a transgression he will deny and punish. This ambivalence is textually revealed by its most psychologically accurate defense. Just as Freud, by abandoning the seduction

significant phrase; but she had hardly said a word before she stopped, and her old father, who was sitting behind her, began to sob bitterly'. Freud concludes: 'Naturally I pressed my investigation no further; but I never saw the patient again' (Freud and Breuer 1893–95: 100-101 n.).

19. See, for example, Laplanch and Pontalis, for whom the daughter's seduction story is a fantasy, its reality 'to be sought in an ever more remote and hypothetical past (of the individual or the species)' (1968: 17).

theory, deflects guilt from the father to (variously) the nurse, the mother, and, by way of the Oedipus complex, the child herself (Gallop 1982: 144-45), so the father projects his seduction onto others and thus denies paternal complicity. The seduction is displaced first onto the (phallic) serpent,[20] and then onto the daughter herself in her seduction of Adam. Thus, the chain of deflections to protect the father begins. It was not the father but the serpent who seduced the daughter and, by the end of this narrative, it is the daughter who seduced her father! Once again, the shamed shames the shamer. To effect this, however, the narrative subjects itself to a labyrinth of self-exposing transformations.

Some feminist biblical scholars see Genesis 1 as a mitigating authorization for women's equality.[21] I disagree. Every authorization of equality in Genesis 1 is subsequently repressed and erased by chs. 2–3. In fact, the juxtaposition of the two accounts of creation exposes the shadowed family construct and highlights the subtext of deflected paternal desire. The syntax in Gen. 1.26-27, which implies that man and woman are created simultaneously and equally, constructs Adam and Eve as son and daughter. Typical of the defense mechanism associated with projection and denial, this narration is an attempt to reconstitute the family into a desired model. However, this makes the deity overtly a father who authorized his children's implicitly incestuous union, and therefore necessitates a re-narration which repeatedly shows the marks of backward erasure and exclusion. When ch. 2 recreates man and woman, it erases the parallelism of the ch. 1 account and dissociates the deity entirely from the parentage of the woman, further distancing the original father–

20. The indicator I see for the serpent's phallic symbolism is based less on Freud's association than on two other factors. First, since Hebrew has no neuter gender, nouns must be either masculine or feminine, and the word for serpent is, indeed, masculine. Second, the narrative function of the serpent, and his description as 'the most wily of the beasts of the field which the Lord God had made' anticipates, and seems embedded in, Augustine's famous use of the fall to explain the frustrating unruliness of the male sexual organ.

21. See, for example, Phyllis Trible, who points out that in Gen. 1 the masculine exists no more than does the feminine (1983).

daughter relationship.[22] Adam's paternal parentage remains, and even his maternal parent is implicitly present in the earth from which he is shaped and from which his name is derived, but Eve, who is born from Adam's body, has a lineage lost in ambiguities. No matter how her creation is read, what does seem clear is that the text has tried to detach her genealogy from the father and place it with Adam. Ironically, however, in an attempt to mask the threat of deflected desire that is posed, the text inadvertently reconstitutes it. Because of the emphasis placed on Eve's derivation from Adam's side, and therefore Adam's implied paternity, the narrative re-enforces the paradigm of a tacitly condoned but overtly disclaimed act between father and daughter. The original father–daughter story that has been so problematic is repressed but remains visible in Adam. Adam, the acknowledged son, becomes the father, making father and son analogous.

At the same time, the text also contains the subtext of Eve's appropriation of the forbidden fruit, a mythology of the daughter's rebellion into sexual maturity, a 'seizing' of her fruitfulness.

In replacing his seduction theory with the Oedipus complex, Freud explains that a daughter's attachment to the father parallels a son's attachment to his mother; but for the girl, attachment to the father is 'positive', following an earlier 'negative' phase in which she learns that her mother has not 'given' her a penis. She turns in despair to the father, who may be able to give her some of its power (Freud 1925 and 1931, *passim*).[23] If read from Freud's perspective, the 'seed-bearing fruit' on the father's tree might signify the father's self, the 'father's Phallus', in both its Lacanian meaning as a symbol of paternal authority and its

22. Simultaneously, it erases the incestuous implications of the son–daughter union by eliminating the Gen. 1 license for the human children to be fruitful and eat unrestrictedly of all the 'trees with seed bearing fruit'.

23. René Girard's theory of language and culture explains the marginal situation of biblical daughters in a way that also challenges Freud's theory of the Oedipus complex. Girard argues that violence has its roots in 'mimetic desire', an approach/avoidance concept that describes the drive to imitate a respected and feared model. While the desire is to imitate, there is the recognition that a complete reproduction would result in an implicit rivalry, the extreme form of which would be displacement and, ultimately, elimination. On the other hand, if this rivalry is rejected and repressed, the subject then stands in a slave relationship with the master (Girard 1986).

Freudian significance as the physical sign of 'presence' and biological superiority. The taboo on plucking/eating this knowledge of good and evil forbids the daughter from obtaining the father's potency and privilege.

This symbolism becomes clearer if we follow the time-honored exegetical practice of reading the Bible intertextually. Just before the children of Israel are to enter the Promised Land, a recapitulation of 'the Father's original garden' (Frye 1982: 72), the fruit taboo resurfaces, and, with it, its phallic significance: 'When you enter the land and plant all [manner of] trees for food, you will regard its fruits as *uncircumcised*. For three years it will be to you a thing *uncircumcised*, and it *will not be eaten*' (Lev. 19.23). Placed into this context, Genesis 3 seems to narrate the daughter's desire to acquire the father's knowledge and power through the (phallic) sign that has been denied her, and to dramatize the threat to patriarchy that daughters represent. By asserting her desire for the sign that confers exclusive rights to the male, the daughter symbolically challenges the privilege of the gender system that the phallus signifies.

Since the text is confronted with a daughter's desires that have no legitimate place in its patriarchal order, it mutes them by denial and displacement. By reasserting the primacy of the father–son relationship, the story represses the more threatening material of its father–daughter text. Thus, Eve gives the 'seed-bearing fruit' to Adam and becomes the medium through which this symbol of potency and privilege (the Phallus in both Freudian and Lacanian meanings) is passed from father to son. Once Eve has transferred the fruit to Adam's possession, she transfers also her narrative centrality. Eve as *daughter* disappears into the margins of the story. Eve as *mother* effectively banishes the female transgressor of the father's garden. Her denied desires are perpetuated into a frustrated 'yearning'—what Freud would have called 'penis envy', or the daughter's 'recognition of absence'. But it is also a recognition of what Freud's feminist interpreters have defined as another kind of knowledge, the knowledge of the way that 'cultural stereotypes have been mapped onto the genitals' (Rubin 1975: 195). If in the 'phallic phase', as Freud asserts, 'only one kind of genital organ comes to account—the male' (1923: 142), then Eve's act of aggression is a representation of her desire to get beyond the prohibitiveness of the Phallus, its rule as standard, what Irigaray calls 'the reign of the One, of Unicity' (1977: 43): the Father.

Eve's choice to give fruit, the conventional symbol of female sexuality, to another male may represent the daughter's ultimate dispossession of her father, and reveal this family member as *the* dangerous threat to paternal power, the reason for narrative absence. The daughter's act is a violation cursed by the father and resulting in a permanent barrier of separation. At the daughter's instigation, the son has cast aside perpetual security, in an outright rejection of the father and his authority. It is at this junction that the two interwoven subtexts merge.

The original commandment to be fruitful and multiply is transformed into the structures of taboo, transgression and punishment. Adam is now a laborer, and Eve is ordered into the *creation* of family, her presence as daughter permanently eliminated. Significantly, from now on (with the exception of the anomalous story of Ibzan [Judg. 12.9]), biblical fathers assiduously avoid ever giving daughters away. In fact, the Hebrew Bible avoids daughters almost altogether. Indeed, a father and daughter do not re-enter Genesis until the incestuous tale of Lot.[24] By then, however, the text has rationalized deflected desire: Lot is 'blamelessly' seduced by his daughters, just as Adam was unwittingly seduced by the woman he fathered.

BIBLIOGRAPHY

Alexander, Franz
1948 *Fundamentals of Psychoanalysis* (New York: W.W. Norton).

Archer, Léonie J.
1983 'The Role of Jewish Women in the Religion, Ritual and Cult of Graeco-Roman Palestine', in *Images of Women in Antiquity* (ed. A. Cameron and A. Kuhrt; London: Croom Helm): 273-87.

24. Another mention of 'daughter' precedes that of Lot, but it is the 'collective catastrophe' of Gen. 6.4 brought about by the '[generic] daughters of men'. In this odd (and obscure) fragment, the 'sons of God' are seduced by desire for the 'daughters of men', and their corrupt but heroic offspring provide the motive for God's decision to destroy humanity by the flood. Here, as elsewhere, the problem revolves around a woman positionally coded as 'daughter'.

1987 'The Virgin and the Harlot in the Writings of Formative Judaism', *History Workshop: A Journal of Socialist and Feminist Historians* 24: 1-16.

1990 *Her Price is beyond Rubies: The Jewish Woman in Graeco-Roman Palestine* (JSOTSup, 60; Sheffield: JSOT Press).

Balmary, Marie

1979 *Psychoanalyzing Psychoanalysis: Freud and the Hidden Fault of the Father* (trans. Ned Lukacher; Baltimore: Johns Hopkins University Press).

Bechtel, Lyn M.

1991 'Shame as a Sanction of Social Control in Biblical Israel: Judicial, Political, and Social Shaming', *JSOT* 49: 47-76.

Bigger, Stephen F.

1979 'The Family Laws of Leviticus 18 in their Setting', *JBL* 98: 187-293.

Boose, Lynda E.

1989 'The Father's House and the Daughter in It: The Structures of Western Culture's Daughter–Father Relationship', in Boose and Flowers 1989: 19-74.

Boose, Lynda E., and Betty S. Flowers (eds.)

1989 *Daughters and Fathers* (Baltimore: Johns Hopkins University Press).

Douglas, Mary

1966 *Purity and Danger: An Analysis of the Concepts of Pollution and Taboo* (New York: Praeger).

Fokkelman, J.P.

1981 *Narrative Art and Poetry in the Books of Samuel.* I. *King David (II Sam 9–20 & I Kings 1–2)* (Studia Semitica Neerlandica, 20; Assen: Van Gorcum).

Fox, Robin

1967 *Kinship and Marriage* (Harmondsworth: Penguin).

Freud, Sigmund

1923 *The Standard Edition of the Complete Psychological Works of Sigmund Freud.* XIX. *The Infantile Genital Organization* (trans. and ed. James Strachey *et al.*; London: Hogarth and The Institute for Psycho-Analysis, 1965).

1925 'Some Psychical Consequences of the Anatomical Distinction between the Sexes', in Freud 1923: 241-60.

1931 *The Standard Edition of the Complete Psychological Works of Sigmund Freud.* XXI. *Female Sexuality* (trans. and ed. James Strachey *et al.*; London: Hogarth and The Institute for Psycho-Analysis, 1965).

1933 *The Standard Edition of the Complete Psychological Works of Sigmund Freud.* XXII. *Femininity* (trans. and ed. James Strachey *et al.*; London: Hogarth and The Institute for Psycho-Analysis, 1965).

Freud, Sigmund, and Joseph Breuer

1893–95 *The Standard Edition of the Complete Psychological Works of Sigmund Freud.* II. *Studies in Hysteria* (trans. and ed. James Strachey *et al.*; London: Hogarth and The Institute for Psycho-Analysis, 1965).

Froula, Christine
1989 'The Daughter's Seduction: Sexual Violence and Literary History', in Boose and Flowers 1989: 111-35.

Frye, Northrop
1982 *The Great Code: The Bible and Literature* (New York: Harcourt Brace Jovanovich).

Gallop, Jane
1982 *The Daughter's Seduction: Feminism and Psychoanalysis* (Ithaca, NY: Cornell University Press).

Girard, René
1986 *The Scapegoat* (Baltimore: Johns Hopkins University Press).

Herman, Judith Lew, and Lisa Hirschman
1981 *Father–Daughter Incest* (Cambridge, MA: Harvard University Press).

Horney, Karen
1950 *Neurosis and Human Growth* (New York: W.W. Norton).

Irigaray, Luce
1977 *Ce sexe qui n'en est pas un* (Paris: Minuit; trans. Catherine Porter and Carolyn Burke as *This Sex Which Is Not One* [Ithaca, NY: Cornell University Press]).

Landy, Francis
1983 *Paradoxes of Paradise: Identity and Difference in the Song of Songs* (Sheffield: Almond Press).

Laplanch, Jean, and J.B. Pontalis
1968 'Fantasy and the Origins of Sexuality', *International Journal of Psychoanalysis* 49: 1-18.

Miller, Alice
1984 *Thou Shalt Not Be Aware: Society's Betrayal of the Child* (trans. Hildegarde Hannum and Hunter Hannum; New York: Farrer, Straus & Giroux).

Piers, G., and M. Singer
1953 *Shame and Guilt* (New York: W.W. Norton).

Pitt-Rivers, Julian
1977 *The Fate of Shechem, Or, The Politics of Sex* (Essays in the Anthropology of the Mediterranean; Cambridge Studies in Social Anthropology Series; Cambridge: Cambridge University Press).

Rashkow, Ilona N.
1990 *Upon the Dark Places: Anti-Semitism and Sexism in English Renaissance Biblical Translation* (Sheffield: Almond Press).

Rubin, Gayle
1975 'The Traffic in Women: Notes on the "Political Economy" of Sex', in *Toward an Anthropology of Women* (ed. Rayna Reiter; New York: Monthly Review): 157-210.

Rush, Florence
1980 *The Best-Kept Secret: Sexual Abuse of Children* (Englewood Cliffs, NJ: Prentice–Hall).

Smith, Joseph H.
1986 'Primitive Guilt', in Smith and Kerrigan 1986: 52-78.

Smith, Joseph H., and William Kerrigan (eds.)
1986 *Pragmatism's Freud: The Moral Disposition of Psychoanalysis* (Baltimore: Johns Hopkins University Press).

Trible, Phyllis
1983 *God and the Rhetoric of Sexuality* (Overtures to Biblical Theology, 20; Philadelphia: Fortress Press).

Willbern, David
1989 '*Filia Oedipi*: Father and Daughter in Freudian Theory', in Boose and Flowers 1989: 75-96.

INDEXES

INDEX OF REFERENCES

OLD TESTAMENT

INDEX OF AUTHORS

JOURNAL FOR THE STUDY OF THE OLD TESTAMENT

Supplement Series

95 God Saves:
Lessons from the Elisha Stories
Rick Dale Moore
96 Announcements of Plot in Genesis
Laurence A. Turner
97 The Unity of the Twelve
Paul R. House
98 Ancient Conquest Accounts:
A Study in Ancient Near Eastern and Biblical History Writing
K. Lawson Younger, Jr
99 Wealth and Poverty in the Book of Proverbs
R.N. Whybray
100 A Tribute to Geza Vermes:
Essays on Jewish and Christian
Literature and History
Edited by Philip R. Davies & Richard T. White
101 The Chronicler in his Age
Peter R. Ackroyd
102 The Prayers of David (Psalms 51–72):
Studies in the Psalter, II
Michael Goulder
103 The Sociology of Pottery in Ancient Palestine:
The Ceramic Industry and the Diffusion of Ceramic Style
in the Bronze and Iron Ages
Bryant G. Wood
104 Psalm Structures:
A Study of Psalms with Refrains
Paul R. Raabe
105 Establishing Justice
Pietro Bovati
106 Graded Holiness:
A Key to the Priestly Conception of the World
Philip Jenson
107 The Alien in the Pentateuch
Christiana van Houten
108 The Forging of Israel:
Iron Technology, Symbolism and Tradition in Ancient Society
Paula M. McNutt
109 Scribes and Schools in Monarchic Judah:
A Socio-Archaeological Approach
David Jamieson-Drake
110 The Canaanites and their Land:
The Tradition of the Canaanites
Niels Peter Lemche

111 YAHWEH AND THE SUN:
THE BIBLICAL AND ARCHAEOLOGICAL EVIDENCE
J. Glen Taylor

112 WISDOM IN REVOLT:
METAPHORICAL THEOLOGY IN THE BOOK OF JOB
Leo G. Perdue

113 PROPERTY AND THE FAMILY IN BIBLICAL LAW
Raymond Westbrook

114 A TRADITIONAL QUEST:
ESSAYS IN HONOUR OF LOUIS JACOBS
Edited by Dan Cohn-Sherbok

115 I HAVE BUILT YOU AN EXALTED HOUSE:
TEMPLE BUILDING IN THE BIBLE IN THE LIGHT OF MESOPOTAMIAN AND NORTH-WEST SEMITIC WRITINGS
Victor Hurowitz

116 NARRATIVE AND NOVELLA IN SAMUEL:
STUDIES BY HUGO GRESSMANN AND OTHER SCHOLARS 1906–1923
Translated by David E. Orton
Edited by David M. Gunn

117 SECOND TEMPLE STUDIES:
1. PERSIAN PERIOD
Edited by Philip R. Davies

118 SEEING AND HEARING GOD WITH THE PSALMS:
THE PROPHETIC LITURGY FROM THE SECOND TEMPLE IN JERUSALEM
Raymond Jacques Tournay
Translated by J. Edward Crowley

119 TELLING QUEEN MICHAL'S STORY:
AN EXPERIMENT IN COMPARATIVE INTERPRETATION
Edited by David J.A. Clines & Tamara C. Eskenazi

120 THE REFORMING KINGS:
CULT AND SOCIETY IN FIRST TEMPLE JUDAH
Richard H. Lowery

121 KING SAUL IN THE HISTORIOGRAPHY OF JUDAH
Diana Vikander Edelman

122 IMAGES OF EMPIRE
Edited by Loveday Alexander

123 JUDAHITE BURIAL PRACTICES AND BELIEFS ABOUT THE DEAD
Elizabeth Bloch-Smith

124 LAW AND IDEOLOGY IN MONARCHIC ISRAEL
Edited by Baruch Halpern and Deborah W. Hobson

125 PRIESTHOOD AND CULT IN ANCIENT ISRAEL
Edited by Gary A. Anderson and Saul M. Olyan

126 W.M.L.DE WETTE, FOUNDER OF MODERN BIBLICAL CRITICISM:
AN INTELLECTUAL BIOGRAPHY
John W. Rogerson

127 THE FABRIC OF HISTORY:
TEXT, ARTIFACT AND ISRAEL'S PAST
Edited by Diana Vikander Edelman

128 BIBLICAL SOUND AND SENSE:
POETIC SOUND PATTERNS IN PROVERBS 10–29
Thomas P. McCreesh

129 THE ARAMAIC OF DANIEL IN THE LIGHT OF OLD ARAMAIC
Zdravko Stefanovic

130 STRUCTURE AND THE BOOK OF ZECHARIAH
Michael Butterworth

131 FORMS OF DEFORMITY:
A MOTIF-INDEX OF ABNORMALITIES, DEFORMITIES AND DISABILITIES IN TRADITIONAL JEWISH LITERATURE
Lynn Holden

132 CONTEXTS FOR AMOS:
PROPHETIC POETICS IN LATIN AMERICAN PERSPECTIVE
Mark Daniel Carroll R.

133 THE FORSAKEN FIRSTBORN:
A STUDY OF A RECURRENT MOTIF IN THE PATRIARCHAL NARRATIVES
R. Syrén

135 ISRAEL IN EGYPT:
A READING OF EXODUS 1–2
G.F. Davies

136 A WALK THROUGH THE GARDEN:
BIBLICAL, ICONOGRAPHICAL AND LITERARY IMAGES OF EDEN
Edited by P. Morris and D. Sawyer

137 JUSTICE AND RIGHTEOUSNESS:
BIBLICAL THEMES AND THEIR INFLUENCE
Edited by H. Graf Reventlow & Y. Hoffman

138 TEXT AS PRETEXT:
ESSAYS IN HONOUR OF ROBERT DAVIDSON
Edited by R.P. Carroll

139 PSALM AND STORY:
INSET HYMNS IN HEBREW NARRATIVE
J.W. Watts

140 PURITY AND MONOTHEISM:
CLEAN AND UNCLEAN ANIMALS IN BIBLICAL LAW
Walter Houston

141 DEBT SLAVERY IN ISRAEL AND THE ANCIENT NEAR EAST
Gregory C. Chirichigno

142 DIVINATION IN ANCIENT ISRAEL AND ITS NEAR EASTERN ENVIRONMENT:
A SOCIO-HISTORICAL INVESTIGATION
Frederick H. Cryer

143 THE NEW LITERARY CRITICISM AND THE HEBREW BIBLE
Edited by J. Cheryl Exum and D.J.A. Clines